T0264972

War
and
Democracy

The editors and publisher of this volume gratefully acknowledge the generous support of the Korea Foundation. The opinions, findings, and conclusions or recommendations expressed in this publication are those of the authors and do not necessarily reflect the views of the Korea Foundation.

War
and
Democracy

A
Comparative Study
of the
Korean War
and the
Peloponnesian War

David McCann and Barry S. Strauss
editors

AN EAST GATE BOOK

M.E. Sharpe

Armonk, New York
London, England

An East Gate Book

Copyright © 2001 by M. E. Sharpe, Inc.

All rights reserved. No part of this book may be reproduced in any form
without written permission from the publisher, M. E. Sharpe, Inc.,
80 Business Park Drive, Armonk, New York 10504.

Library of Congress Cataloging-in-Publication Data

War and democracy : a comparative study of the Korean War and the Peloponnesian War
/ David R. McCann and Barry S. Strauss, eds.
    p. cm.
  Includes bibliographical references and index.
  ISBN 0-7656-0694-1 (alk. paper)—ISBN 0-7656-0695-X (pbk. : alk. paper)
    1. Korean War, 1950-1953. 2. Greece—History—Peloponnesian War, 431-404 B.C. 3.
Democracy. I. McCann, David R. (David Richard), 1944- II. Strauss, Barry S.

DS918 .W37 2000
938′.05—dc21   00-059588

Printed in the United States of America

The paper used in this publication meets the minimum requirements of
American National Standard for Information Sciences
Permanence of Paper for Printed Library Materials,
ANSI Z 39.48-1984.

BM (c) 10   9   8   7   6   5   4   3   2   1
BM (p) 10   9   8   7   6   5   4   3   2   1

To Sang-duk Yu, former vice president of the Korean Teachers' Union, for his
dedication to the humane human spirit
—D.R.M.

To Donald Kagan, teacher and friend
—B.S.S.

Contents

Introduction

David R. McCann and Barry S. Strauss

At first glance it seems odd to compare the Peloponnesian War and the Korean War. One conflict was ancient, one modern; one took place among city-states and empires; the other, among modern nation-states; one was fought with pre-industrial technology, the other included both human waves of soldiers and the most modern weapons of war; one was fought in the Mediterranean, the other the Pacific. The Peloponnesian War lasted twenty-seven years, from 431 to 404 B.C.; the Korean War lasted only three years, from 1950 to 1953.[1] The Peloponnesian War is part of Western civilization, while the Korean War is both a war in Asia and also a chapter in the history of East-West encounters. The Peloponnesian War is a famous conflict, perhaps one of the most carefully studied wars in military history. The Korean War is a historical orphan: Americans call it "the forgotten war"; Koreans remember it but until recently have hedged it with taboos and censorship; China still refers to its troops as "volunteers"; finally, although the archives have begun to open since 1989, scholars remain hampered by the large number of documents that still remain secret.

Opposites attract; such contrasts themselves can stimulate new approaches to each of the case studies. It was, however, the similarities between the two cases that first led us to the project.

The geographical similarities first, if superficial, are striking. Both Greece and Korea are peninsulas, rugged, mountainous, and surrounded by thousands of islands and islets. Korea (North and South) is about 85,000 square miles in size, Greece (the modern state) about 50,000 square miles. Each sits at roughly the same latitude, each bisected by the thirty-eighth parallel, which runs not only just north of Seoul but also through Athens. Each provides challenging terrain for agriculture, but in each country at the time of the war in question, most of the population eked out a living from the soil.

Other similarities are less superficial. Both wars were phases in a longer conflict whose main feature was an ideological divide between an open society—a democracy and its allies—and a closed society—Sparta's garrison state in the one case, and communism in the other. For the United States, the Korean War was a brief but violent episode in the Forty Years' War—the Cold War that stretched from the Truman Doctrine of 1947 to the fall of the Berlin Wall in 1989. For Korea the war of 1950–1953 was a similarly brief, violent conflict within a sequence of struggles that had begun in 1947, the roots of which can be traced back to the Tonghak uprisings of the nineteenth century. The Peloponnesian War was also an episode in the longer Atheno-Spartan conflict from 460 to 371.[2] Both wars were turning points. The Peloponnesian War ended Athens' empire, brought Sparta to the peak of power in the Aegean, and allowed Persia to re-enter Greek interstate affairs, after having been evicted by a united Greece three generations previously. Unlike the Peloponnesian War, the "Korean conflict," as many called it, was a limited war—a mere "police action," in Truman's phrase. Yet few actions in the history of the Cold War proved more decisive. As a result of the Korean War the United States rearmed to fight the USSR by tripling its military budget and doubling its troops in NATO. A rift between the United States and China was in place for the next twenty years. Meanwhile, the Chinese-Soviet alliance was strengthened, and the United States committed itself to the defense of Taiwan.[3]

Both wars are case studies in unpredictability. In 431 B.C. Athens' leading statesman, Pericles, seems not to have expected a long war, let alone one that would witness a massive epidemic, the reintroduction of Persian power into the Aegean, and the transformation of Sparta into a naval power that matched and ultimately beat the greatest sea power in Greek history, Athens. It would take Thucydides himself to do justice to the twists and turns of policy during the Korean War. The American side of the story alone—from the apparent public abandonment of South Korea in January 1950 to the attempt to conquer even North Korea just ten months later—is a tale of miscalculations, reversals of fortune, cliff-hangers, tragedies, heroism, massacres. No wonder that General Omar Bradley famously warned that America's involvement in Korea ran the risk of turning into "the wrong war, in the wrong place, at the wrong time, and with the wrong enemy."[4]

Both conflicts proved highly destructive to life and property. In the Korean War, among soldiers, an estimated 900,000 Chinese and 520,000 North Koreans were killed or wounded, while the United Nations suffered 400,000 casualties, almost two-thirds South Korean. The United States had 37,000 dead and 103,000 wounded. Including civilians, an estimated 3 million people, about one-tenth of the population of the Korean peninsula, were killed, wounded, or missing in the war. Because of the epidemic of the 420s B.C., the Athenian casualty rate may have been even higher, while Sparta suffered considerable battle casualties. The Korean War left most of Korea's cities in ruins, its countryside ravaged. Ancient technology could not match the modern technique of destroying property: although Athenian farmhouses had been stripped clean by raiders, Athenian grain, olive, and wine production suffered limited damage, and while 20,000 Athenian slaves ran away, the domestic economy was largely intact.[5]

Yet the central similarity between the two wars may lie in what many observers thought they were fought for. The Korean War proved to be a test of strength and power, in which one state nearly annihilated the other in order to try to accomplish reunification. Yet in the West, the Korean War no less than the Peloponnesian War was conceived as a test of democracy. The Peloponnesian War achieved this status famously in the work of Thucydides; the Korean War, less obviously but no less trenchantly, in contemporary media and polls and in historiography since. As Thucydides conceived it, the Peloponnesian War put Athenian democracy on trial. The charges are several. The first is aggression. With its fleet and its activist citizen ethos, Athens represented not only the spirit that had saved Greece from Persian invasion at Salamis in 480, but also the bottomless ambition that destabilized the traditional Greek order and the oligarchic virtues it engendered. Thucydides was at one with most of the Greek elite in bemoaning the decline of "the ancient simplicity" (Thuc. 3.83.1) in the new world of democracy.

The second charge against democracy is inefficiency. "A democracy is incapable of empire," the Athenian politician Cleon says despairingly (Thuc. 3.37.1) and Thucydides would seem to agree. He argues that Athens was at its best at the start of the Peloponnesian War under Pericles, when the regime was, in his opinion, "a democracy in name only but in fact, rule by the first man" (Thuc. 2.65.9). Afterwards, without a strong man at the helm, democracy degenerated into mob rule and Athens all but defeated itself: so Thucydides argues.

The third charge against democracy is atrocity. Chronicler of realpolitik though he was, Thucydides is in the final analysis a moralist, charting the crushing of the old Hellenic virtues in the crucible of war. He shows Athens descending from the moral high ground of Pericles' Funeral Oration to the ethical chaos of the so-called Plague (in fact, a still-unclassified epidemic disease) to the close call of making—and then undoing—a decision to massacre all adult male Mytilenians as punishment for rebellion to deciding finally to carry out just such a punishment on the adult males of Melos. In his *Trojan Women*, Euripides too seems to question the morality of an Athens that would engage in such behavior.[6]

The Korean War, in turn, put American democracy on trial. For many at the time, the central question was whether a liberal society could compete with the global threat of communism. How could the United States protect its still-prostrate allies in Europe and a devastated Japan without bankrupting itself and adopting a standing army? How could the country defend its traditional freedom of speech when communist spies, who were bound by no such niceties at home, had penetrated its government and were stealing its military secrets? Indeed, how could the United States keep such secrets and remain an open society? And how could the United States catch communist spies without going after innocent, patriotic people and without engendering a witch hunt?

The Korean War raised large questions about leadership in American democracy. The Truman administration faced a huge political task in justifying to the American people a war in a country that few could place on the map, and which the administration itself had seemed publicly to write off just six months before the war began. While its decisions to avoid a congressional declaration of war, involve the United Nations, and downplay the seriousness of the "police action" may have worked in the short run, they could not prepare the nation for three years of combat, and they arguably cost Truman a chance to run for re-election in 1952. In the longer run, they raised constitutional questions about the war powers of the presidency. Truman himself raised issues about character and leadership. He had the broad vision, combativeness, and decisiveness to commit America immediately to war in the wake of the North's attack on South Korea; without such a quick decision the war would have been lost at the outset. Yet his own ignorance of East Asia and his inflexibility led him to follow MacArthur into a disastrous adventure north of the thirty-eighth parallel. Nor, for all the drama of his eventual sacking of MacArthur, did he move quickly enough to defend his constitutional prerogatives against an insubordinate general; the result was a political crisis that could have been avoided.

Finally, although it was perhaps less evident at the time than in retrospect, the Korean War raised big questions about the relationship of American democracy to another culture. How could the United States fight to defend democracy in a country—South Korea—that seemed to have virtually no indigenous democratic tradition, and whose very legitimacy, as only half of what had been the kingdom of Chosŏn (the pre-modern Korean state, 1392–1910), was contested by many of its inhabitants? Similar questions would arise a decade later when America intervened with massive force in Vietnam.

There are, therefore, major differences but also considerable ideological, rhetorical, moral, political, and international-systemic similarities between the two far-off conflicts of the Peloponnesian and Korean wars. Comparisons do not lead to a direct transfer of insights, but they can sharpen our awareness and let us think in ways in which we would not have thought before. What, then, can a student of the Peloponnesian War learn from a comparison with the Korean War?

The comparison takes us back first to the perennial question for the classicist: Was ancient Greece similar to the modern West or radically different? In particular, how similar or different were ancient democracy (*dêmokratia*) and modern democracy? We cannot hope to do justice here to a subject on which many books have been written.[7] Still, the contrast is well known and can be sketched briefly. For better or worse—make that *and* worse, because it was an atypical polis— Athens must stand for the democratic Greek polis generally, because it is the only case for which sufficient documentation survives. Not that documentation for other poleis is completely absent; Aristotle's *Politics* offers numerous data, for example. These are by and large aperçus, however: only Athens presents a broader picture.

Ancient democracy was direct democracy, modern democracy is representative; the ancients believed that self-government required that citizens take part in the legislature and alternate in holding public office, modern democracy delegates these jobs to deputies; ancient democracy chose its officers through lottery, modern democracy chooses them through election; ancient democracy considered citizenship a precious commodity, like ownership of stock, and restricted it to a male birth elite, while modern democracy extends the citizenship to both sexes and often to immigrants too; ancient democracy was acephalous, lacking either president or prime minister, while modern democracy always has a chief; ancient democracy had so little bureaucracy that some prefer not to think of it as a state at all, modern democracy is bureaucratic and concentrates power in the state. While not necessarily engaging in class warfare, ancient democracy insisted that the interests of the poor and ordinary people be served; modern democracy is consistent with a "trickle-down" theory of the public good. Although ancient democracy went hand-in-hand with a market economy, it was a pre-industrial economy with relatively simple associations and rooted in slave labor; modern democracy is capitalist and the state co-exists with large corporations. The culture of ancient democracy often betrayed its elite, aristocratic roots, whereas modern democracy has a liberal culture with massive leveling tendencies. An ancient democrat transplanted to a modern democracy might judge it to be an oligarchy because the people did not participate directly in the government; a modern democrat might judge *dêmokratia* to be an oligarchy because it excluded women and immigrants and tolerated slavery. In short, the differences are so great that one may wonder what the two regimes have in common, besides etymology.

But that, one might reply, is a counsel of despair. Freedom and equality are the hallmarks of both ancient and modern democratic regimes, and both particularly prize freedom of speech. *Dêmokratia* and democracy each aim at subordinating the rule of men to the rule of law. Among the competing regimes of its own day, each form of democracy offers the common man the best hope of prosperity. Each glorifies the ideal of the active, public-spirited citizen. Both ancient and modern democracy make fearsome armies when aroused to war, as Hanson argues, and they are often so aroused. Pericles declared that Athens' enterprising spirit "had forced an entry into every sea and into every land" (Thuc. 2.41.4); Americans and their allies might have felt no less adventurous as they fought in far-off Korea, for most of them a place located halfway around the world. Ancient democracy had no truck with pacifism; modern democracy is often anti-war, but only after experiencing the terrible destructiveness of modern warfare, like France and England after 1918, and only when its security is assured, like the European democracies under NATO, and only when it is not provoked by a need to avenge past defeat, like Germany after 1918. And modern democracies are far more peaceful toward each other than to competing regimes. American democracy was neither pacifist during the Cold War nor has it become pacifist since 1989, not with the rising challenge of international anarchy, not with state after state seeking nuclear, biological, or chemical weapons. South Korea has democratized in the last decade but, with North Korea not merely continuing to threaten armed conquest but to add nuclear (and some would say, biological and chemical) weapons to its arsenal, it can hardly disarm, despite significant recent progress toward reconciliation.

Yet if they are powerful warriors democracies enjoy no simple bloodlust. Although ancient democracy lacked modern democracy's universalism or its belief in human rights, it expressed moral hesitation about the extreme acts it committed in war. Americans killed many civilians during the Second World War and expressed moral doubts because of universalist beliefs about human equality. Athenians had no such beliefs, but they understood that indifference to mistreatment of non-Athenians was risky. The Athenians, for example, resorted to mass executions during the Peloponnesian War (at Scione and Melos) but not without wavering or dissent: witness Diodotus' speech in 427 on the proposed punishment of Mytilene (Thuc. 3.36–50) or Thucydides' Melian Dialogue or Euripides' tragedy of 415, *The Trojan Women* (the latter perhaps a protest of Athenian atrocities against civilians). Although they professed hierarchical political values, that is, that freeborn adult Athenian males had no equals, Athenian society shaded the differences between categories: in Athens, metics (i.e., resident aliens), young males, and, to a lesser degree, even women and slaves enjoyed certain freedoms. One reason is the cosmopolitan freedom of a sea power, but another is the nature of ancient democracy, whose egalitarian logic tended to level the very hierarchies it erected. Melians were not Athenians but they were Greeks and free; might not the very logic that allowed them to be killed en masse be turned someday on poor Athenians by rich ones? In short, by its very nature, democracy, whether parochial or universalist, tends to problematize killing.[8]

Yet a major difference in the way that ancient and modern democracies make war emerges. In Athenian democracy the people or demos made war immediately, fettered by few if any entangling institutions. The assembly debated policy openly, without benefit of watchdogs from the Pentagon or intelligence agencies. Soldiers and sailors openly complained to their generals on campaign and expected them to give speeches justifying their strategy. In modern democracy, by contrast, military mutiny is punishable by death and political discourse is often scaled back "for the duration," either by consensus or explicit action, such as the Defence of the Realm Act in Britain or the suspension of the right of habeas corpus by Abraham Lincoln during the U.S. Civil War.

Modern democracy offers a clear and hierarchical chain of command during wartime. The state designates a civilian war leader, be it a president, prime minister, or minister of defense. If the war leader stumbles and lets go of the reins of power, as Truman did in tolerating MacArthur's subordination, they are there for him to pick up again. Truman finally did fire MacArthur. The general struck back in the court of public opinion, but the "man on horseback" rode into town and the people escorted him back out again. There was in the last analysis nothing that he could do against the established power of the presidency. By contrast, leadership was much more fluid and up for grabs in democratic Athens.

Pericles, for example, was the most influential leader in Athens at the time of the outbreak of the Peloponnesian War, and the chief war hawk. The people followed him into rejecting the Peloponnesian ultimatum, but when the war went badly, Pericles had no presidency to fall back on. He was, true enough, an elected general, but he was only one of ten generals and served only for an annual term. In 430 B.C., he was either fined or ejected from office (the sources disagree; Thuc. 2.65.3, Plut. *Per.* 35.4) when the insult of Spartan raiders harming Athenian property with impunity and the devastation of an epidemic behind Athens' city walls made a hash of his policy of restraint. A generation later, not only was Pericles' nephew Alcibiades forced to share command of the Athenian expedition against Syracuse, which he had championed, with the very man who had stormed against it, Nicias, but he was unable to prevent his political rivals from sabotaging his position. No sooner had the expedition sailed when, because of the scandalous affairs of the Herms and Mysteries, they managed to have him indicted and recalled to stand trial in a capital case.

The difference between Truman, on the one hand, and Pericles or Alcibiades on the other, underlines the peculiarly communal emphasis of the polis. In its political structure and particularly when it comes to arrangements regarding war, modern democracy follows the pattern of a republic: a hierarchy of elected officials to whom great powers are delegated. By contrast, the polis stayed much closer to its roots in tribal egalitarianism. Athens willingly sacrificed efficiency for democracy, just as it sacrificed efficiency for protecting lives. The assembly would declare war and the warriors, of whom a core were the very citizens who sat in that assembly, would insist that their commanders respect their lives. For example, after the naval battle of Arginusae in 406, Athenian generals pursued the enemy fleet without ensuring the safety of the thousands of Athenian seamen who had survived the first engagement. A storm suddenly blew up and the men were drowned. Although Athens had won that engagement, the generals faced trial, were convicted, and executed. The Truman–MacArthur controversy could never have happened in democratic Athens because Truman would not have lasted in office long enough. For going to war without consulting the people, Truman would have been impeached, tried, convicted, and executed all in one day.[9]

Yet, surprisingly, the polis succumbed to a greater cult of personality during wartime than did the modern republic. Alcibiades came closer than MacArthur, much less Truman, to sweeping his country by storm; so did Lysander. Both Alcibiades and Lysander were driven from power by rivals who feared their ambition: Alcibiades in 415, Lysander in 403. Nonetheless each man managed to come back to power. Lysander died commanding a Spartan army in battle in 395; Alcibiades, after losing power a second time, died in exile ca. 403, but his behavior before the battle of Aegospotami in 405 shows that he was no old soldier content just to fade away. The comic poet Aristophanes commented in 406 that the Athenian people "loves him, hates him, and wants to have him" (*Frogs* 1425). MacArthur, by contrast, was all but forgotten by the American public a year after his forced return from the field in 1951.

What made *dêmokratia* more vulnerable to the appeal of the charismatic leader than was American democracy in the 1950s? For one thing, Americans traditionally have prized sobriety and distrusted glamor. For another thing, the citizenry of the polis was closer to the battlefield than the American electorate. Few Americans could imagine MacArthur as their commander in the field; many Athenians and Spartans not only could imagine following Alcibiades or Lysander into battle but actually had done so. One is more likely to believe in, even to idolize, a man who holds one's life in his hands than one who does not. Finally, neither Athens nor Sparta was ever as suspicious of individual idiosyncracy as America is, or at least was in the fifties. For the Greeks, such suspicion was the hallmark of a tyrant who, like Thrasybulus of Miletus, looked to "lop off the tallest ears of grain" (Herodotus 5.92f.1–92g.1). For an American, it was the outward sign of inner devotion to the public good. Perhaps the infamous conformism of the fifties was the price Americans had to pay for the failure of something like the MacArthur balloon to float.

For all their superficial similarity, the two "witch hunts" of the wars—the Affair of the Herms and Mysteries in Athens in 415 and the McCarthy phenomenon of the 1950s—seem to betray much the same difference. Both responded to real, if often vague-seeming threats, be they pro-Spartan oligarchs and sneering aristocrats on the one hand, or communist subversion on the other. Both were initiated by elites and struck a chord with ordinary people. Yet it was the *object* of investigation whom Athenians remembered—that is, Alcibiades—while it is the *author* of the investigation whom Americans remember, that is, Joseph McCarthy. Conservative critics have taken liberals to task for later downplaying the reality of communist subversion and exaggerating McCarthy's faults. Perhaps so, but the ease with which liberals made their case suggests how little McCarthy's flamboyant individualism was appreciated by the public. Americans opposed communism but they had little taste for a crusader; what they wanted was a steady chief like Eisenhower. Athenians opposed oligarchy and tyranny but they preferred Alcibiades, with his imperial flair, to the good grey men who brought him down.

And they made no bones about their taste for empire. Athenians had none of the reticence of Americans for ruling an empire. They believed, of course, that they deserved the empire, and said so (e.g., Thuc. 1.76.2). Yet, the naked reality of power was not something they merely conceded when forced; rather, in the harsh realism of their political discourse, they seemed almost to enjoy such concessions. Don't expect people to give up power entirely—that would be superhuman; ask only that they pay some attention to justice while exercising power (Thuc. 1.76.2). Athenian rule over its allies is like a tyranny; it may have been wrong to take this power but it would be dangerous to let it go (2.63.2). In foreign affairs, justice is less important than self-interest (3.44.1–2). So spoke the Athenian politicians.

Americans were, and are, different. Most deny that their global hegemony is an empire and those who do profess a desire to give it up. But are such denials hypocrisy? Or do they merely demonstrate that, however imperfect still, democracy has indeed learned to deal with outsiders more justly than in the days of its ancient forebears?

And what of the objects of Athenian imperialism and of American imperialism or hegemonialism? In this matter, the ancient historian is at the mercy of his sources, and they are not generous. For all their eloquence, the ancient sources leave vast silences. Inscriptions, for example, can record Athenian tax exactions in the empire; orators speak of Athens' landgrab there. Thucydides and perhaps Euripides can register the protest of Athenian intellectuals against the brutality of the Athenian people in deciding, in 416 B.C., to punish the rebels of Melos by enslaving the women and children and executing the men. They cannot speak for the Melians, however. To take another example, Aristophanes can put on stage and make fun of the plight of the Megarians caught between the hammer of Athens and the anvil of the Peloponnesians early in the Peloponnesian War (*Acharnians* 729–835, 425 B.C.), but his characters' voices are not the Megarians'. Much less are the Koreans' voices Megarian or Melian, but they are tremendously useful for the ancient historian nonetheless.

For most of its history, Korea has been a Pacific Megara, caught between two or more great powers. In the sixteenth and nineteenth centuries, China and Japan fought in or over control of the Korean peninsula. Over the years, other powers to Korea's north have also intervened in the peninsula, including the Mongols, the Manchus, and the Russians. The Korean War itself arose in the power vacuum created by the collapse of Japanese colonialism in 1945 and the subsequent intervention by the traditionally interested powers, China and Russia, as well as by a more recent player on the peninsular board, the United States. Koreans trenchantly characterize their nation's geostrategic vulnerability by citing the proverb, "When whales fight, the shrimp's back breaks." The whales respond by pointing out, in effect, that a land bridge extending from China and Russia to a mere hundred miles from the Japanese coast is hardly a shrimp. Japanese have referred to Korea as "a dagger pointed at Japan," while Mao Zedong called Korea "the lips to China's teeth."[10]

* * *

How might Korea engage the foregoing issues and questions arising from a consideration of the subject, democracies at war? There are several ways to reframe the matter from a Korean perspective. In the history of American relations with Korea, did or does the United States seem to operate on a moral basis, or on the basis of realpolitik? In terms of Korean history and literature, how might it be possible to register the relative significance of local rather than internationalized historical narratives? In that connection, does starting from a Korean perspective on events necessarily lead to an interpretation of the Korean War as a civil war? Can local history, or a Korean historical perspective, balance the narrative of the war as a chapter in U.S.-Asia relations? Or can Korean voices speak in any way for Megara and Melos? Do any of the concerns and motives that seem operative at the level of U.S. or Russian or Chinese foreign policy seem operative at the local level in Korea as well? And just what is local in Korean history? Can either of the combatant states be described, finally, as a democracy at war?

In Western historiography, there is a fundamental difference between scholars who work with Korean materials and those who do not. In the present volume, for example, Kathryn Weathersby has used Russian archival material to argue that the war was a major-power, hegemonic struggle, less the result of inter-Korean rivalries between regimes than the unleashing of the potential combatants, armed to the teeth, by their American and Russian patrons. Bruce Cumings's two-volume study, *The Origins of the Korean War*,[11] published in 1981 and 1990, was pioneering in two senses: it was the first ever by a non-Korean historian to make extensive use of Korean documents, and it confronted a prevailing, powerfully ideological reading of the war as nothing but an act of communist aggression.

Among variations in Korean perspectives on the war, four can be noted here. North Korea has consistently denounced the American involvement as nothing more than imperialist aggression, and South Korea as nothing more than America's puppet. In South Korea, during the first two decades after the war the most apparent attitude was a sense of gratitude and appreciation for the American commitment and sacrifice in defending the South; this, and a mirror image description of the war as an act of outside, North Korean, communist aggression. This attitude found critics in the 1970s among those who came to conclude that the American commitment might have been based not so much (or only) upon a concern to defend the (South) Korean people as upon a readiness to make Korea the place to hold the line against what once was called the specter of international communism. Coupled with a growing attention to American support for the repressive Rhee and Park regimes, such questioning of American motives became transformed into quite focused anti-Americanism during the decade following the 1980 Kwangju incident, when South Korean troops were deployed to the city of Kwangju where they killed between two hundred and one thousand or more citizens.[12]

American involvement in Korea has been complicated and contradictory for much of its century-or-more history: on the one hand, positive, and dramatically generous, while on the other, cruelly indifferent; at times informed by awareness of local conditions, at other times constrained by entirely different, international relations priorities. Prompted in part by continuing questions concerning the level of American awareness or involvement in Kwangju in 1980, the contradictory nature of that history has become one focal point for current-day Korean research on American involvement and responsibility in Korean affairs prior to the outbreak of the Korean War in June of 1950. A case in point, the series of massacres on Cheju Island during the course of the campaign waged by South Korean army troops and paramilitary units against a so-called leftist or communist insurgency: by the time it was over, somewhere between ten thousand and sixty thousand men, women, and children had been killed. The numbers of those killed are difficult to ascertain because the dead were declared to have all been communists, and under the provisions of the 1948 National Security Law, the bedrock foundation of the South Korean state, to make inquiries, or even to seek to bury such dead, was an indictable offense.

The killings on Cheju Island began in March of 1948, when police in Cheju

City fired on demonstrators entering the city after a rally to commemorate the March 1, 1919, Independence Movement and to protest the United Nations–sponsored elections scheduled for October. When the Cheju City police action prompted a series of protests and counterattacks, someone made the decision that Cheju Island was full of communists, and that the best course of action would be to eradicate them. The records of the massacres that followed remain incomplete, as yet. Whatever the results of current efforts to fill in the narrative of events, it is the case that the U.S. Military Government in Korea held overall responsibility for governance south of the thirty-eighth parallel until August of 1948. The two histories, American and Korean, are intertwined but difficult to understand. One may wonder whether the United States was trying to help the Syngman Rhee regime demonstrate its strength on Cheju Island or whether it was already in 1948 taking a stand against international communism. The question, in other words, was whether American concerns were local, international, or both.

At the very least, the history of Korean affairs prior to the actual outbreak of the war makes undeniably clear that the military conflict between north and south had been preceded by intense, violent struggle within both North and South. The local history of this period has, until recently, been literally buried with the dead, but emerging documentation from the South complements the narratives in Korean literary works, such as those described in Dong-Wook Shin's chapter in the present volume, which repeatedly articulate a sense of the role of ideology as a causal factor in the war, and of local cultural identity and cohesiveness as its only anodyne.[13]

*   *   *

For all the differences between them, the two case studies ask similar, broad questions. Do democracies make efficient fighting states or do they stand at a disadvantage against more tightly controlled regimes? In order to fight oligarchies, tyrannies, or totalitarian regimes, must democracies cease (at least temporarily) being democratic? Can democracies meet the exigencies of wartime without trampling on freedom at home and engaging in atrocities abroad? Can small states caught between great powers maintain a degree of freedom? Can they afford the luxury of democracy? Can a democracy lead an alliance and exercise enormous power without becoming an empire? Is any attempt by a democratic hegemon to respect its allies' autonomy doomed to fail, either because it leaves the alliance lax and disorganized or because it runs afoul of the human lust for power? How does war affect democratic culture? Because democracies value free and unfettered knowledge, should they accept realpolitik as the accurate, if cynical, analysis of international relations, or should they reject it as superficial and simplistic?

The analysis of this volume moves from interstate relations to domestic structures.

The first section, "Democracy: Bellicose, Imperial, or Idealistic?" takes up the theme of democracy as a wartime regime. While Victor Hanson argues for the general efficacy of democratic regimes as warriors, Ronald Steel and Robert Kagan argue a more specific question: Was America during the Cold War an imperial power or a benevolent hegemon?

In his "Democratic Warfare, Ancient and Modern" (chapter 1), Victor Hanson cautions about assigning any scientific status to the oft-made comparison between the Peloponnesian War and the Cold War. Yet he concludes that analogies between ancient and modern democracy do indeed focus the mind. For all the differences, the two regimes share an emphasis on popular power, a willingness to accept and even embrace change, and a tendency to encourage economic expansion. Both ancient and modern democracy proved dynamic, pragmatic, and populist, which in turn rendered them both warlike. In his judgment, ancient democracy was the most bellicose, revolutionary, and militarily successful of all ancient regimes. In the case of Athens, democracy, because of its dynamism, led to dramatic military innovations, in particular, the move to naval power, the incorporation of fortification and siegecraft, the erosion of census standards, the diminution of agriculture, and the prominence of capital. Hanson suggests that, like its ancient predecessor, modern democracy tends to use armed force more flexibly and successfully than an authoritarian state, which lacks the economic dynamism and popular support of democracy. He concludes that the more that South Korea creates a free society and open economy the greater its comparative military advantage over North Korea will grow.

Can a democracy be an empire and still remain a democracy? Athens wrestled with this question, and Ronald Steel maintains that America has as well. In chapter 2, "The American Imperium," Steel argues that the United States today controls a hegemonic empire, which it governs indirectly through economic links and military and political subordination rather than directly through annexing territory and administering colonies. Begun before 1945, the empire only reached global proportions during the Cold War: the Korean War proved the decisive step in America's imperial leap. Korea turned communism from a remote force into a tangible threat and turned anti-communism from an ideology into a course of action, now funded by massive American expenditure and pursued by force not merely in Europe but around the world. In order to win public and bureaucratic support for what amounted to an expensive global crusade, the Truman administration exaggerated the extent of the threat and mistook what was largely a civil war on the Korean peninsula for a proxy attack on the United States by the Soviet Union. As in the Peloponnesian War, conflict among small states brought the great powers to blows, although it was China rather than the Soviet Union that played Sparta to America's Athens, while the Soviets tended to fight America indirectly, rather the way Persia fought Athens in the earlier conflict. An analogy between America's alliances and the Delian League also comes to mind, since both were formed voluntarily but neither tolerated later withdrawal. As hegemonic powers, both Athens and the United States preferred (and sometimes forced) their allies to adopt democratic regimes similar to their own. Yet if America is indeed an empire as was Athens, there is a significant ideological difference. While Americans tend to believe that their global role is not merely powerful but moral, Athenians rather openly spoke the language of realpolitik.

Was America, however, indeed an empire? In chapter 3, "The American Empire: A Case of Mistaken Identity," Robert Kagan argues the contrary. As a great power or even a superpower,the United States has since 1945 resembled an empire without being one: its relationship with its allies is utterly different from that between an imperial master and its subjects. Kagan believes America and its allies resemble the early Delian League but not the later Athenian *arkhê* ("empire," as it is sometimes translated, but literally "rule"): that is, not an empire but a voluntary league of autonomous states led by the strongest of them aimed at a specific purpose. That purpose, Kagan states, was the containment of communism. With the collapse of communism the allies might have taken their own course, and yet the attraction of America remains strong. Kagan contrasts Athens, whose imperial rule many Greeks wanted to escape, with America, whose hegemony, he says, most states have embraced. The United States has been an idealistic and lenient hegemon, particularly when compared to Athens.

We move next, in the second section, "Categorizing Wars: Civil or Hegemonic, Decisive or Cyclical?" to a discussion of just how to characterize the two wars at issue. Bruce Cumings argues that by its intervention in Korea, America mistook what was a civil war for a hegemonic conflict with its Soviet rival. Kathryn Weathersby maintains, by contrast, that in the last analysis the Korean War was less a civil war than a hegemonic conflict. Paul Cartledge tends implicitly toward a similar conclusion, if only by his argument that both the Korean War and the Peloponnesian War need to be understood as part of the larger cycle of great-power wars.

In his "When Sparta Is Sparta but Athens Isn't Athens: Democracy and the Korean War" (chapter 4), Bruce Cumings takes the argument that the Korean War was a civil war to a next logical position. If North Korea can be likened to Sparta, but South Korea was not an Athenian-style—or any style—democracy, as seems to be the case also according to Oh and Suh, then the war was not fought to save a democracy. Rather, American intervention helped to support a police state, another Sparta, because that state happened to be anti-communist. Signs of democratization may be glimpsed in Korea in the periodic reform measures that Kongdan Oh refers to, or in the land reform efforts that Cumings describes. How, though, were American democratic institutions affected by the war? Historians are just now beginning to return to the era of Joseph McCarthy, to a period when for perhaps good reason, as some argue now upon examination of Russian archives and their lists of spies, or for not sufficient reason at all, when weighed against the personal misfortunes created by the national effort to control communism, the political left was for the most part eradicated as a viable force in the United States. Cumings takes note of this period, but in addition notes the virtual palace coup in which Dean Rusk and Dean Acheson engineered the commitment of American forces to the defense of South Korea. From such a beginning, he suggests, it was not many more steps to the "imperial presidency."

In her "Stalin and the Decision for War in Korea" (chapter 5), Kathryn

Weathersby considers whether we can explain the outbreak of the Korean War by using Thucydides' model of hegemonic warfare. As Weathersby understands him, Thucydides grounds the outbreak of the Peloponnesian War in the so-called security dilemma, that is, the fear by one state (in this case, Sparta) that the power of its rival state (Athens) was growing so strong that only war could stop it, however little the first state wanted war. Basing her argument largely on new documents from the Soviet archives, she argues that indeed the security dilemma was also responsible for the outbreak of the Korean War. The war came about because one hegemon (the Soviet Union) concluded, though erroneously, that it could gain strategic advantage by allowing its client state (North Korea) to conquer its rival's client state (South Korea) before that rival (the United States) could intervene. She further argues that the Soviet Union came to this conclusion not out of any objective aggressive threat by the United States but rather because of its own mistaken ideology. Unlike, however, the Peloponnesian War—at least on a likely, if not universal, reading of Thucydides—the Korean War was not inevitable. Had the United States made clear its willingness to intervene militarily in the Korean peninsula, then the ever-cautious Stalin would not have approved North Korea's attack. In addition to these arguments, Weathersby also weighs in on the debate as to whether the Korean War was essentially a civil war. Although the impetus for war came from the two Koreas, each of which was eager to attack the other and unite the country, the final decision came from the great powers, without whose approval neither Korean state would have dared fight. Without Stalin's decision early in 1950 to permit the attack, North Korea would not have moved on the south in June 1950. Weathersby concludes, therefore, that in the last analysis the Korean War was less a civil than hegemonic war.

In chapter 6, "The Effects of the Peloponnesian (Athenian) War on Athenian and Spartan Societies," Paul Cartledge makes a strong revisionist case against the widespread thesis that the Peloponnesian War was a watershed in ancient Greek history, leading to dramatic and general change. He questions the historicity of a single war in the period 431–404 B.C.: it was Thucydides' brilliant but revisionist notion. More important, Cartledge argues that just as the Cold War was, in many ways, a continuation of World War II by other means, so too the conflicts of post–Peloponnesian-War Greece continued those of before the Peloponnesian War; the war aggravated and accelerated these conflicts but did not change them. Unlike World War II or the Korean War, the Peloponnesian War was not a total war, and so did correspondingly less damage to Athenian or Spartan society, culture, or politics than did World War II to the defeated. Much less did the Peloponnesian War cause a general crisis in the Greek world.

Continuing a structural analysis of the conflicts in question, the third section, "Third Forces, or Shrimps Between Whales," focuses on the predicament of small states caught between great powers. Gregory Crane focuses on the ancient case of Plataea, while Kongdan Oh and Dae-Sook Suh examine respectively South and North Korea.

In chapter 7, "The Case of Plataea: Small States and the (Re-)Inventions of Political Realism," Gregory Crane offers a sensitive analysis of the position of a small state, ancient or modern, caught between great powers. Plataea's geostrategic position between Athens and Thebes (the latter a close ally of Sparta) resembles Korea's location between China and Japan, with each situation made more complicated by the intervention of a third power: Athens in the Peloponnesian War, the United States in the Korean War. To preserve its independence from Thebes, Plataea became an Athenian client, but to preserve its self-respect it blurred the status. Plataeans expressed pride not only in their contribution to Athens' victory at Marathon in 490, but also in their history of friendship with Sparta and above all in the location in their territory of the great battle against Persia of 480, a Panhellenic victory that freed Greece. Unfortunately, what Crane calls the spreadsheet logic of great powers allows no such ambiguity. Because the Plataeans proved to be the enemies of their friends, the Spartans agreed to Thebes' demand that Plataea be destroyed and its men massacred or driven into exile. Caught between a similarly relentless political realism, small states in the Korean War—that is, North and South Korea—might have learned from Plataea just how little room to maneuver they had.

Dae Sook Suh, in his "Korean War and North Korean Politics" (chapter 8), analyzes the direct connections between internal, domestic political struggles in the north, especially between Kim Il Sung and Pak Hŏn-yŏng, and the planning for a war of reunification. As leader of the partisan group, Kim looked to full-scale military action as the road to recapture the south and reunify the peninsula under his authority. Pak, leader of the so-called domestic group, was less enthusiastic about unilateral military action, hoping instead that some form of intervention would encourage popular uprisings in the south that would topple the Rhee regime. With the resultant new state having its capital in Seoul rather than Pyongyang, Pak, originally from the south, imagined that his own chances for leadership would be improved. The rivalry between the two groups led ultimately to an unsuccessful coup attempt by the domestic group in early 1953. Kim's successful weathering of the domestic challenge, and the shift in his international alignments from Russia, which had promised much but delivered little, to China, which had provided personnel and matériel in abundance, both during and after the war, are two of the significant legacies of the war in the north.

Both the two Koreas and the various Greek city-states had political systems forged in the crucible of a terrible, hegemonic war, be it in the 1950s or fifth century B.C. There, however, the similarity ends, as Kongdan Oh argues in "The Korean War and South Korean Politics" (Chapter 9). The leading Greek city-states enjoyed relative independence from foreign influence, relative freedom of debate and, in poleis such as Athens, democracy (at least for a citizen elite). At the time of the war, the two Koreas had no experience of even limited democracy and little freedom of action from their Cold War sponsors. Oh describes the development of authoritarian regimes in the south under Syngman Rhee and Park Chung

Hee as a historic irony: the colonial Japanese policies of control and repression were continued by the political regimes that followed the 1945 liberation. Oh describes an intensely divisive struggle among the various political parties between 1945 and 1948, when the two separate regimes were in effect ratified by the October election. She makes clear that the South Korean state was no democracy. The war's legacy of bitterness and suspicion between the two Koreas remains as a serious impediment to reunification efforts as well as to historical research. Only in the late 1980s did democratization come to South Korea. Its growth proved an unintended consequence of the regime's dependence on the United States, because America's liberal culture proved infectious. Constantly exposed to American culture in the media, South Koreans reasoned that they enjoyed the same political freedoms as the United States that was protecting them. No doubt Thucydides, who witnessed many an ally reach for the freedom enjoyed by its hegemon, would have enjoyed the irony.

The final two sections of the volume focus on the domestic level of analysis. Section IV, "Demagogues? or Domestic Politics in Democracies at War," examines the use and abuse of populism by leaders in wartime democracies. Ellen Schrecker and Stephen Whitfield look at McCarthyism and its aftermath, while Jennifer Roberts compares ancient and modern would-be men on horseback.

Can a liberal democracy maintain its commitment to individual liberty under the pressure of war? For most American historians, McCarthyism—as the campaign against the communist-linked left is often if misleadingly known—provides reason for pessimism. McCarthyism was the product of the Cold War. As Ellen Schrecker demonstrates in chapter 10, "McCarthyism and the Korean War," however, it intensified dramatically after the war's outbreak in summer 1950. Schrecker argues that the Korean War transformed the American political climate by vindicating anti-communist scenarios. Newly hyper-patriotic, the country accepted an increase in attacks on individuals and organizations associated with the Communist Party. Perhaps such attacks would have come about eventually even without the Korean War, Schrecker argues, but not with the rapidity that the war fever permitted—and perhaps not with the same success. In the event, McCarthyism did nothing less than destroy the institutional structure of the American left, Schrecker argues, from labor unions to popular front groups. The Korean War marked a watershed in that process.

In his "Korea, the Cold War, and American Democracy" (chapter 11), Stephen J. Whitfield analyzes what he sees as a problem in the history of modern American democracy, that is, the rapid alternation between repression and tolerance. On the one hand, America during the Korean War (and throughout the 1950s) appears to be a case study in the excesses of democratic populism or the tyranny of the majority. On the other hand, within ten years, America witnessed the explosion of individual liberty that was the sixties. During the Korean War, demagogues—most famously, Joseph McCarthy—threatened individual freedom in the name of public safety. At times the excesses of American anti-communism recalled, at least to some observers, the excesses of totalitarianism itself. It is no accident, Whitfield suggests, that it was in 1958 at the height of the Cold War that Sir Isaiah Berlin published his famous distinction between negative and positive liberty: the latter maximized the pursuit of virtue, the former protected individual freedom. Given the dangers of official orthodoxy, Berlin preferred negative liberty. Yet the world looked different a decade later, when young Americans were reveling in an ethos of "do your own thing." To Whitfield, the explanation of the change lies in the underlying continuity between the 1950s and 1960s: affluence. American and Western abundance ultimately doomed the Soviet Union to lose the Cold War. At the time of the Korean War, prosperity bred American complacency and a retreat from public into private life; a decade later, it underlay the triumph of private interests over public issues. Athenian democracy, with its positive liberty—to use Berlin's terms—was capable of McCarthy-style intolerance (witness the Affair of the Herms) but the "dropout" mentality of the American 1960s never caught fire there (except for a small, oligarchic coterie).[14]

In her "Warfare, Democracy, and the Cult of Personality," Jennifer Roberts examines in chapter 12 what she calls the paradox of democratic leadership—that larger-than-life heroes play so prominent a role in the most egalitarian of political regimes. Arguing that both the similarities and differences between the two cases can be instructive, she focuses on one example from America during the Korean War and one from Athens during the Peloponnesian War. Douglas MacArthur and Alcibiades differed in many ways, but both men stood out as military leaders to whom a mystique attached itself. Each man was distinguished by talent, good looks, communication skills, a distinguished family pedigree, and a long sojourn abroad that increased his appeal at home. Each man had a charisma that was only increased by what Roberts sees as the inherent erotic tension of the warrior. Each spoke the language of patriotism but was, arguably, out for himself, with little regard to principle. Each ultimately fell afoul of his country's institutions of accountability and each was driven from office. Yet Alcibiades' is a case of Athenians' propensity to indict and convict public officials—he was driven from office twice—and MacArthur's of Americans' hesitation to impeach, let alone convict. The difference between direct and representative democracy lies at the root of the distinction: a people that makes and unmakes leaders itself feels less restraint than elected representatives or magistrates who must sit in judgment on the people's choices (if chosen only indirectly through public opinion polls). If modern democracy is more institutionalized and less participatory than its ancient ancestor, it is less likely to be swayed by a "man on horseback."

The fifth and final section, "Realism, Militarism, and the Culture of Democracies at War," examines the culture of democracies at war. Josiah Ober offers a comparative case study in the worldview of policymakers, ancient and modern. Kurt Raaflaub sketches the militarized culture of fifth-century Athens. Dong-Wook Shin examines the response to the traumatic Korean War of Korean literary works.

In his "Thucydides Theoretikos/Thucydides Histor: Realist Theory and the

Challenge of History" (chapter 13), Josiah Ober argues that there are two identifiable voices in Thucydides' text, a historical as well as a theoretical voice. As a theoretician, particularly in the beginning of his account, Thucydides is a realist: that is, Thucydides argues for a world of unitary, self-confident state actors, each attempting to maximize power. Athens had an advantage in this pursuit because it enjoyed a triad of material factors that tended to maximize national strength: walls, ships, and treasure. Yet the complex reality of the Peloponnesian War gives the lie to this tidy theory—as Thucydides himself shows. Using Thucydides' account of the Mytilenian Debate as a case study, Ober demonstrates a world of competing loyalties, errors of communication, and the intertwining of domestic and foreign politics. Realist theory turns out to be inadequate to explain state behavior. A similar analysis may be applied to the American experience in Korea and the Cold War more generally, as Ober suggests tentatively. Had American policymakers been more flexible then they might have perceived in advance such things as the danger of Chinese intervention north of the Yalu in 1950, the non-monolithic nature of the communist bloc, and the long-term weakness of the Soviet Union. Perhaps Thucydides' most basic lesson is the necessity of intellectual humility in a world of complexity and idiosyncracy.

In chapter 14, "Father of All, Destroyer of All: War in Late Fifth-Century Athenian Discourse," Kurt A. Raaflaub raises troubling questions about the limits of public discussion of war in democratic Athens. In Athens, unlike most modern states, those who engaged in politics, that is, the citizens, had usually fought in war as well. In other words, war was part of the school for citizenship; we might equally say that much of Athenian culture was a school for war. A walk through the city of Athens and its port, Piraeus, which Raaflaub takes the reader on as a framing story, offered—through buildings and monuments, inscriptions and images, rituals and drama—constant reminders of Athens' empire and the military victories upon which it was based. From youth on, the Athenian was all but conditioned to consider war inevitable and even desirable. Although playwrights, particularly Euripides and Aristophanes, often challenged official ideology with questions of Athenian wisdom and morality, their advice was honored mainly in the breach. Theater served as a kind of safety valve, balancing public discourse without changing it. Although Athens was a democracy and prided itself on freedom of speech, Athenian culture stacked the deck in favor of war. To many, the range of opinion will appear disappointingly limited.

In his "Characters and Characteristics of Korean War Novels" (chapter 15), Dong-Wook Shin describes the effort in Korean literary works to recover a sense of meaningful cultural identity through which to overcome the psychic destruction caused by the war. One might suggest that many of the works in fact register a sense of the loss of such an identity. The works described in Shin's chapter locate the causes of the war in the externally imposed, ideological division between communist and anti-communist, and invoke a sense of "stoic reserve," which the author suggests may be linked to the Korean experience of colonial occupation by Japan, to overcome it.

* * *

This volume grows out of a conference held at the Woodrow Wilson International Center for Scholars in 1995. All of the authors participated in the conference as paper-givers or respondents, but none of the chapters herein appears as it did then: each has been either thoroughly revised or, in some cases, written expressly for this volume. Other participants in the conference were Chung-Moo Ch'oi, Peter Euben, Charles Hedrick, Ross King, Ian Morris, John K.C. Oh, Hunter W. Rawlings, and Philip West. At an earlier planning session for the conference, valuable comments came from Laurel Kendall, Josiah Ober, Kurt Raaflaub, Robert Kagan, Joel Silbey, and Arthur Waldron in addition to other participants who later took part in the conference.

We would like to thank James Morris and Susan Nugent, current or former staff of the Woodrow Wilson Center, whose help and encouragement for both the volume and the conference have been extraordinary. Sandra Kisner of the Cornell Peace Studies Program has provided invaluable editorial assistance.

Notes

1. Unless stated otherwise, all three-digit dates in this book are B.C., all four-digit dates A.D.

2. On the longer conflict, see Barry S. Strauss, "The Problem of Periodization: The Case of the Peloponnesian War," in Mark Golden and Peter Toohey, eds., *Inventing Ancient Culture: Historicism, Periodization, and the Ancient World* (New York: Routledge: 1997), 165–75.

3. Don Oberdorfer, *The Two Koreas: A Contemporary History* (Reading: Addison-Wesley, 1997), 9.

4. On Pericles' strategy at the start of the Peloponnesian War see Donald Kagan, *The Archidamian War* (Ithaca: Cornell University Press, 1974), 24–42. "Wrong War": Omar Bradley, testimony to the Committee on Armed Services and Committee on Foreign Affairs, U.S. Senate, May 15, 1951.

5. Casualties in the Korean War: Oberdorfer, *The Two Koreas*, 9–10; *Time,* June 12, 2000. Casualties in the Peloponnesian War: Barry Strauss, *Athens After the Peloponnesian War: Class, Status, and Policy, 403–386* B.C. (London: Croom Helm and Ithaca: Cornell University Press, 1986), 179–82; Paul Cartledge, *Agesilaos and the Crisis of Sparta* (London: Duckworth, 1987), 37–43. Property devastation in the Peloponnesian War: Victor Davis Hanson, *Warfare and Agriculture in Classical Greece* (Pisa: Giardini, 1983). Psychological effect of invasions of Athens: Josiah Ober, *Fortress Attica: Defense of the Athenian Land Frontier, 404–322* B.C. (Leiden: Brill, 1985), 51–68.

6. On morality and realpolitik in Thucydides see W.R. Connor, *Thucydides* (Princeton: Princeton University Press, 1984). On chronological grounds it would have been difficult for Euripides to refer specifically to the atrocities on Melos in *Trojan Women,* though not impossible, *pace* A.M. van Erp Taalman Kid, "Euripides and Melos," *Mnemosyne* 15 (1987): 414–19. By the time the play was produced the audience would have known what had happened on Melos and may have had the events in mind. In any case, Athens' policy of massacring males in a rebel city had begun several years earlier, at Scione in 421 B.C. (Thuc. 5.32.1), and Euripides may have been alluding to this event in *Trojan Women.*

7. The reader might begin with M.I. Finley, *Democracy Ancient and Modern* (New

Brunswick: Rutgers University Press, 1985); M.H. Hansen, *Athenian Democracy in the Age of Demosthenes: Structure, Principles and Ideology* (Oxford: Blackwell, 1991); J. Ober and C. Hedrick, eds. *Dêmokratia: A Conversation on Democracies, Ancient and Modern* (Princeton: Princeton University Press, 1996); P.A. Rahe, *Republics Ancient and Modern* (Chapel Hill: University of North Carolina Press, 1992).

8. On the leveling tendencies of Athenian democracy, see J. Ober, "Quasi-rights: Political Boundaries and Social Diversity in Democratic Athens" (forthcoming); Strauss, "Genealogy, Ideology, and Society in Democratic Athens," in Ian Morris and Kurt Raaflaub, eds., *Democracy 2500? Questions and Challenges,* Archaeological Institute of America, Colloquia and Conference Papers, No. 2 (Dubuque: Kendall/Hunt, 1997), 141–54.

9. Kurt Raaflaub, at the 1995 conference, held at the Woodrow Wilson International Center for Scholars, from which this volume began.

10. "Whales and shrimp": The saying in Korean is *"korae ssaume saeu tŭng t'ŭjinda,"* literally, "in a whale fight, the shrimp back breaks." For a slightly different version, see William Stueck, *The Korean War: An International History* (Princeton: Princeton University Press, 1995), 13. "Lips to China's teeth": Michael Hunt, "Beijing and the Korean Crisis, June 1950–June 1951," *Political Science Quarterly* 107, no. 3 (1992): 464.

11. Bruce Cumings, *The Origins of the Korean War*, vol. I, *Liberation and the Emergence of Separate Regimes, 1945–1947,* and vol. II, *The Roaring of the Cataract, 1947–1950* (Princeton: Princeton University Press, 1981 and 1990).

12. On Korean attitudes toward the United States, see Kim Kyung-dong, "Korean Perceptions of America," in Donald N. Clark, ed., *Korea Briefing 1993* (Boulder: Westview Press, 1993), 163–84; and Donald N. Clark, "Bitter Friendship: Understanding Anti-Americanism in South Korea," in Donald N. Clark, ed., *Korea Briefing 1991* (Boulder: Westview Press, 1991).

13. Also see Uchang Kim, "The Agony of Cultural Construction," in Hagen Koo, ed., *State and Society in Contemporary Korea* (Ithaca: Cornell University Press, 1993), 185–92, on "Repossessing History"; and Marshall R. Pihl, "Contemporary Literature in a Divided Land," in Clark, *Korea Briefing 1993*; David R. McCann, "Our Forgotten War: The Korean War in Korean and American Popular Culture," in Philip West, Steven I. Levine, Jackie Hiltz, eds., *America's Wars in Asia: A Cultural Approach to History and Memory* (Armonk, NY: M.E. Sharpe, 1998).

14. On these oligarchs see L. Carter, *The Quiet Athenian* (New York: Oxford University Press, 1986).

Part I

Democracy: Bellicose, Imperial, or Idealistic?

1

Democratic Warfare, Ancient and Modern

Victor D. Hanson

Political thinkers from Machiavelli and Hobbes to George Marshall have written of pragmatic wisdom gleaned from their own Thucydides.[1] Yet the time-honored practice of applying the paradigm of his history of the Peloponnesian War to specific contemporary issues is a treacherous one, prone to false assumptions and rife with unfounded analogies. To be frank, I have no idea whether bipolarity or multipolarity is a more stable system of world alliance, or whether Thucydides himself knew the answer—or even anymore whether the historian was an objective realist dutifully outlining the bleak world of interstate relations or a sophisticated moralist whose examples are subjective, deliberately sifted to confirm his own preconceived notions of human behavior and ethics, or whether he was at times neither or both.[2]

Still, many historians and political scientists have believed that our fifty-year standoff with the Soviet Union is explicable through analogy with the rivalry between Athens and Sparta. In this view, following a successful alliance against a tyrannical would-be enslaver (Persia/Germany), the inevitable differences between Athens/America and Sparta/Soviet Union led the two former allies into armed opposing hegemonies. In these bipolar worlds, the former are democratic, free societies, based on market economies and strong fleets; the latter inward-looking, static cultures, which rely on well-disciplined, if not brutal, infantry regimens.[3]

In this well-worn model, small Third World states such as Melos/Korea might be seen as either caught in, or manipulated by, the imperial aspirations of morally neutral superpowers. Further similarities between ancient and modern hegemonies extend to the distortion of their own respective domestic societies during decades of growing polarization: Sparta and the USSR are forced to acquire navies and soon doomed to open their societies to compete on the world scene, while

Athens and America often betray the spirit of their own constitutions to match the realpolitik of their autocratic adversaries.

But many scholars have countered that the entire analogy is simplistic and upon close analysis fails. The presence of nuclear weapons as final arbiter of conflict makes such ancient and modern parallels simplistic: in antiquity miscalculation, desperation, accident, and misfortune did not lead to instantaneous annihilation of civilization itself. Ancient renegade regimes, with small populations and without allies, did not warrant deference due to the possession of a small nuclear device or two.

The modern Western notions of Christianity and Marxism further distort any facile correspondence and show that intent in America and Russia has been guided by ideologies, religious and otherwise, quite different from their supposed ancient counterparts' allegiance to either democracy or oligarchy. More paradoxically still, not all would accept the identification of Sparta with the Soviet Union and Athens with America. Sparta's more egalitarian Peloponnesian League is far closer to the collective security of NATO; the Athenian ironclad confederation surely more akin to a Warsaw Pact that likewise followed the lead of its imperial hegemon. And the more pragmatic critic will simply point out that, unlike the victorious authoritarian Spartans, the Soviet Union *lost* their cold war against an imperial democracy, proving that 2,500 years of history leave only the thinnest veneer of similarity and tell us nothing about the unchanging nature of the collision between repressive and liberal states.[4]

Thus I cannot explain how Korea follows or does not follow the Thucydidean paradigm, whether it was a modern Melos that was dragged into a bipolar struggle only to suffer horrendous losses in lives and natural treasure in the backwaters of a larger war, or whether, like the tiny village of Plataea, its slim chance for safety against foreign thuggery lay under the protection of a democratic and benevolent big brother.[5]

Instead, I wish to learn from the Greeks about the more general question of how democracies conduct themselves in war—and why they often win. There is a considerable body of modern research on the role between democracy and warfare of the last three centuries: in general the evidence suggests that democracies are *no* less likely to go to war than their autocratic counterparts. In the words of Quincy Wright: "Peace produces democracy rather than that democracy produces peace." The only encouraging conclusion of the social scientists is that belligerency *between like democratic states* is infrequent, giving hope to Kantian idealists that should the world become a pandemocratic association of nations, war itself might be rarer. In the end, however, such utopian theories remain unproven, and our view of warfare in democracies—which most effectively and most quickly voice the will of the citizen majority—apparently depends on how we view the nature of man himself.[6]

In any case, our ancient evidence is unequivocal: the first and greatest democracy in the West widened, amplified, and intensified the conduct of conflict in a manner unseen before in the history of Greek warfare. In fact, of all forms of ancient government—from monarchy to aristocracy to tyranny to oligarchy and moderate timocracy—for nearly two centuries democracy was the most warlike, revolutionary, and successful of all in its practice of arms.

For purposes of brevity in this ancient investigation, I concentrate on Athens, the first, largest, and best known of the Greek democracies. By "democracy," I mean rule by the majority of all resident, adult male citizens whose vote in the assembly alone marked the declaration of war. In practical terms, in the Greek world, extreme or radical democracy meant an absence of a property qualification, the empowerment of the poor, the use of the lot, and near full citizen public participation in courts, magistracies, and the assembly regardless of property, wealth, class, or birth (Arist. *Pol.* 3.1279b-80a; 6.1317a16–b40; b2–b10; cf. Hdt. 3.80.6).[7]

First however, we must review quite briefly the status of Greek warfare prior to the rise of fifth-century Athenian democracy. That two-century history of hoplite battle in the seventh and sixth centuries B.C. reveals just how radical a departure was the democratic way of war.

Pre-Democratic Warfare Among the Greek City-States

Greek warfare of the city-state emerged somewhere around 700 B.C. as an agrarian enterprise, one made up of farmer-citizens fighting on farmland over disputed farmland (Arist. *Pol.* 4.1291a31–33; cf. Xen. *Oec.* 5.12–18; Xen. *Oec.* 5.4–5; 6.6.7. 9–10; [Arist.] *Oec.* 12. 1343b2–6; Pl. *Rep.* 2.374C). Indeed, often farmland itself was assessed in terms of how great a number of hoplite infantry that region might theoretically support (Arist. *Pol.* 1270a17–32; Plut. *Mor.* 413F–414A; cf. Dem. 23.199; *FGrH* 115 225). One-day battles were not designed for aggression or sustained expeditions (Thuc. 1.141.5; Dem. 1.27; Ar. *Pax* 1183). From an exclusively agricultural point of view hoplite warfare was an extremely *frugal* enterprise (Dem. 9.48; Polyb. 13.3.2–4), where property and income taxes were not raised from the rural populace to pay for lengthy campaigning. Few, if any, significant wars on land took place. Most smaller disputes were usually between neighboring city-states, whose hoplites killed each other over small tracts of border ground (Thuc. 1.15.2; 5.41.2). Those lands, ironically, were not always valuable in a purely agricultural sense, but were important to farmers as symbols of agricultural prestige, promoting the growing agrarian chauvinism of the community at large, and reflective of individual agrarians' own endless haggling with neighboring farmers (cf. Hom. *Il.* 12.421–24).[8]

Once this system of hoplite warfare arose in the late eighth and seventh centuries B.C., the traditional social and economic expenditures of fighting, so common earlier in the palace economies of the Near East and in Egypt—property taxes and rents, technology, soldiers' salaries, extensive fatalities, lost agricultural productivity, lengthy training and preparation, sieges, permanent officers and planners, destruction of entire cultures, unity of political and military authorities—were minimal. Indeed, such military cargoes were often nonexistent throughout a vast area of the Greek-speaking world, a break between early Western and Eastern practice that cannot be strongly enough emphasized.[9]

Instead, simple durable weapons were largely standardized for nearly two centuries throughout the insular Greek city-states. The extensive and unmatched protective cover of such bronze armor, the deliberate exclusion of missile-weaponry, cavalry, skirmishers—all the usual sources of fatal motion and speed on the battlefield—and the accompanying rules that limited fighting in a concrete and moral sense, reduced drastically the number who were killed in any given battle. Like restrictions against land accumulation and political aggrandizement, land warfare of the Greeks between the eighth and fifth centuries strove to ensure the equilibrium of the agrarian patchwork of middling property owners. So unlike their more sophisticated successors, Greek hoplite landowners of the seventh and sixth centuries B.C. were fighters of a day; they may not have been democratic in the later Athenian sense, but their property census was not high, and anywhere from a third to a half of the native-born citizenry enjoyed a remarkable egalitarianism.

More specifically, there was neither a priestly class of unproductive military professionals nor an otherwise unemployed military intelligentsia. The formal study of tactics as either a command or academic enterprise did not even exist before the late fifth century B.C. Agricultural devastation was a trigger to start fighting rather than a means to ruin a community. Plunder and booty were incidental, not essential, to the mobilized army's existence, at least until the fourth century B.C. Formal hoplite battle left relatively little imprint on the local civilian environment. Since the key was to maintain capital and political authority among a tight-knit group of yeoman, warmaking did not evolve beyond the farmer's peculiar and blinkered notion of time and space.[10]

To the modern military mind, very little was left to chance in hoplite battle. Most often it was simply a case of winning the fray with your strong right wing—"horns," as the agrarian infantrymen called them—before suspect and less reliable allied militias on the left collapsed (Thuc. 5.73; Xen. *Hell.* 6.4.14)—sometimes, as in the case of the so-called tearless battle (Plut. *Ages.* 33.3; cf. Xen. *Hell.* 3.217; 4.3.12), before even meeting the enemy. Most rudimentary tactics that did emerge were simple, rather than complex, variations of head-on assaults in column. Given the absence of reserve troops, specialized units, the surprise attack, the night engagement, and the concealed ambush, there was no desire until the fourth century B.C. for elaborate pre-battle tactical planning or even real battlefield command. An Alexander, Napoleon, Patton, or Rommel might have sent cavalry to punch holes in the enemy phalanx, javelin-throwers and archers to volley at his wings, followed by light infantry feints, encircling columns, and surprise attacks in the rear—all preliminaries to the main assault held in reserve, awaiting the opportune moment of enemy weakness. But to the polis Greeks before the advent of Athenian imperial democracy, these generals would have been a distasteful bunch, tinkerers and manipulators unwilling, at first sight of the enemy mass, to grab the shield and run anonymously with their men to death.

Yet, I do not intend to minimize the brutality of early battle. One need only read the poet Tyrtaeus' "beating waves of assault," "the dead man in the dust," "toe-to-toe and shield against shield," and the hand "holding the bloody groin" (Tyrt. 10.21–25; 11. 31–34; 12.23), or glance at early Corinthian vase-painting to recoil from the mayhem and bloodletting among those in the front ranks. For those few hundreds and occasionally thousands of combatants, battle was an especially horrific experience (cf. Pindar fr. 120.5).

The Athenian Military Renaissance

The rise of radical Athenian democracy in the fifth century altered the entire nature of hoplite fighting and thus the Greek way of war. It mutated agrarian warfare to reflect new democratic ideology and values and new challenges beyond the Greek mainland, and so left in its wake a sophisticated and dynamic military practice that had a great deal to do with the destruction of the Greek city-state itself. True, infantry battle between phalanxes still continued throughout the slow decline of the free city-state well into the middle and later fourth century B.C., as the famous murderous encounters at Coronea (394 B.C.), Nemea (394 B.C.), Leuctra (371 B.C.), Mantineia (362 B.C.), and Chaeronea (338 B.C.) attest. Those battles, fought magnificently with hoplite infantry and seemingly oblivious to the military revolutions of the times, were in themselves not much different from phalanx fighting of the seventh century B.C.—even though such collisions were no longer any part of a rigorous agrarianism. And the arrangement of spearmen in columnar formation was commonplace even during the subsequent Macedonian period in Greece, as the engagements between heavy pike-bearing phalangites at several later clashes confirm (e.g., Sellasia [222 B.C.], Cynoscephelae [197 B.C.], Pydna [168 B.C.]).

So technically, the charge of a phalanx, whether armed with the traditional hoplite panoply or, in the third through second centuries B.C., modified tactically and equipped with the Macedonian *sarissa* (the fourteen to twenty-foot pike) and lighter body armor, remained an option for Greek military commanders well into Roman times; on its chosen ground no formation could withstand its onslaught. The legacy of phalanxes remains with us in the West today, seen in our preference to fight it out quickly, brutally, and at all costs decisively, a military spirit that arose originally in the Greek countryside.

But "warfare" is usually a much more inclusive term than "battle." It not only suggests the myriad rules, regulations, and practices that surround conflict, but, far more important, "hoplite warfare" denotes the supremacy and the exclusivity of infantry battle between small farmers as the *only* real means of resolving conflict between the Greek city-states of the early polis. For the insular world of the hoplite farmer to make any sense, for his polis to remain agrarian, his peculiar showdown between rows of armored agrarians could not be simply *a* theater of operations. It had to be *the* theater of most military conflict. Infantry battle, as in the seventh and sixth centuries B.C., had to remain the equivalent of war—and it could as long as the enemies of the Greeks were exclusively those who dared to tread over their farmland. Anything less would call into question the premise of the closed system of agrarian monopoly: land-based timocracy, agricultural self-sufficiency, egalitarianism in property-holding—a system that had characterized the majority of the Greek city-states for more than two centuries (700–490 B.C.). The rise of the Greek city-state itself is in large part explicable by these ingenious efforts to limit and direct military activity.

But soon after the Persian wars (490, 480–479 B.C.) the elements of agrarian and hoplite supremacy began to erode throughout Greece. The reason is easy to see: the unique conditions of relative historical isolation that had marked the peculiar birth of the city-state came abruptly to an end in the detritus of the Persians' withdrawal from the Aegean and in the ascendancy of democratic imperialism at Athens.[11]

After the Greek victories at Marathon (490 B.C.), Salamis (480 B.C.), Plataea (479 B.C.), and Mycale (479 B.C.), the Athenian record of military engagement on land and at sea in the next nine decades of the fifth century is truly remarkable. Athenian triremes ranged over the Mediterranean from Cyprus to Egypt to Sicily, as her land forces marched throughout Greece. The litany of her warmaking need not be reviewed again, but its frequency is not in doubt, and the now-worn cliché that Athens was at war two out of every three years in the fifth century is absolutely correct. Perhaps most emblematic of this assertion is the extant casualty list on stone of a single one of the ten tribes at Athens for the year 460/459 B.C.: 177 men from the tribe Erechtheis were killed fighting that season in Cyprus, Egypt, Phoenicia, Halieis, Aegina, and Megara—that is, on land and at sea over much of the Eastern Mediterranean (Meiggs-Lewis 33).[12]

Still, it is not entirely clear what caused this radical departure from past Greek military practice, whether preexisting domestic unrest led to military experimentation and adventurism in the Persian War, or more likely battle-hardened Athenian veterans of overseas naval campaigning returned home to demand representation commensurate with their new military clout. Did the opportunities after the Persian withdrawal in the Mediterranean require greater Athenian naval forces and flexibility in response, reverberating at home in increased liberalization and expanded power to the landless class who staffed the navy? Or, in contrast, was it a political rather than military impetus: did the diminution in the influence of the propertied class in Athens encourage the demos to find opportunity in the Aegean manning triremes and conducting sieges?[13]

The fifth-century political transformation at Athens to a more radical democratic society has been well chronicled; scholars generally cite most prominently the end of property qualifications, the rise of paid juries, the use of the lot, and a wide range of checks on the elite that encouraged popular control of the legislative and judicial aspects of government. But less has been written about the simultaneous Athenian alteration of Greek warfare, even though an array of ancient observers remarked—usually in a romantic context—on the simplicity, nobility, and economy of a lost Greek land warfare prior to the Athenian military revolution in the late fifth and fourth centuries (Polyb. 18.3.3; Aechin. 2.115; Plut. *Philop.* 13.6; Thuc. 4.40.2; 5.41; Isoc. 8.48–54; Pl. *Rep.* 5.469B–471B; *Menex* 242D).

Some scholars, who have recognized an Athenian break with traditional Greek military thought, have attributed such innovation largely to Pericles, and in particular to his reliance on sea power and the abandonment of the Athenian countryside during the initial years of the Peloponnesian War, as if these developments were not necessarily themselves the logical manifestation of democracy. My point, however, is broader: radical innovations in Athenian military practice were *not* so much imperial or Periclean as they were democratic: military adventurism reflected the nature of Athenian society in general and not the career or policy of any one man. Remember, that Pericles was simply expanding on a prior democrat Themistocles' strategy of abandoning the farms of Attica in face of the Persian onslaught. And much later in the fourth century, well after the heyday of the empire, the Athenian fleet grew to its greatest size, Attica continued to be abandoned and evacuated in times of national peril, fortifications grew rather than shrank, and the economy became more market oriented and still less dependent on agrarianism.

In general, I can identify seven general tenets of a new warfare instituted by the Athenian demos that were relatively unknown previously in Greece, but which were soon to doom the old agrarian manner of conflict and so change the course of Greek history as a whole. All such changes *antedated* Periclean strategy, were first envisioned in the first half of the fifth century, and continued to characterize Athens until the end of the democratic state.[14]

Naval Supremacy

The introduction of organized navies and the trireme itself were late phenomena in the history of the Greek city-state (Thuc. 1.13–14; Hdt. 7.144), and largely at odds with the agrarian genesis of the polis. While other states such as Corinth and Corcyra by the mid-fifth century manned extensive navies, none built, manned, and maintained a fleet the size of the fifth-century Athenian navy, which various sources tell us even in peacetime was kept at about 300 ships with crews numbering about 40,000 men (Thuc. 2.13.8; Aeschin. 2.175; [Xen.] *Ath. Pol.* 3.4; Ar. *Ach.* 545). In the fourth century—*after* the supposed decline of Athenian imperial power—the fleet grew to its largest extent in the history of the democracy, perhaps between 300 and 400 ships (cf. Dem. 14.13; Diod. 18.10.2; *IG* II³ 1611, 1627, 1629), and nearly 60,000 sailors.

Moreover, the fleet itself was but part of an entire naval bureaucracy at Athens that included overseas commerce, imperialism, ship construction, enormous dockyards (Pl. *Gorg.* 519A), the transference of capital from allies, and the ascendancy of an entire underclass of thetes who were to power the triremes: military and peacetime maritime service became absolutely inseparable ([Xen.] *Ath. Pol.* 1.19–20; Ar. *Ecc.* 197–8), creating the *nautikos ochlos* so criticized by unsympathetic moderate and conservative thinkers (e.g., Thuc. 8.72.2; Arist. *Pol.* 5.1304a22–27).

And perhaps unlike other navies, a majority of the crews themselves at least during peacetime were Athenian citizens; slaves, allies, mercenaries, and others who were without a stake in the policy of the city comprised a *minority* of rowers.

The results of this shift to the sea were an increase in importation of foodstuffs ([Xen.] *Ath. Pol.* 2.7, 12), maritime commerce, and a lightning-quick ability to transport men and material rapidly to about any point in the Aegean. Indeed, the navy was the concrete manifestation of the vote of the assembly, a mechanism to translate instantly the verbal megalomania and bellicosity of the demos into deed (Plut. *Them.* 19.4; Arist. *Ath. Pol.* 27.1; [Xen.] *Ath. Pol.* 1.2). No wonder Aristophanes criticized the demos' desire to extend their hegemony throughout the Mediterranean through the triremes of their navy (Ar. *Eq.* 236–38; 797–800; 1303–4; cf. Thuc. 6.15.2; 6.34.2).

There was also a widespread belief that naval service was qualitatively *different* from the amateurism of hoplite fighting, specifically that it involved training and skill (*technê*), not mere nerve and strength ([Xen.] *Ath. Pol.* 1.19–20; Thuc. 1.142.6–9; cf. 7.75.7; Plut. *Per.* 11.4; *Cim.* 11.2–3). Therefore, mastery of rowing, naval maneuvers, and maritime raiding were far more disruptive of traditional Greek military practice and the entire notion of amateurism associated with the citizenry of the polis (e.g., Pl. *Laws* 4.707A–D). The necessary expenditures for dockyards, ship construction and maintenance, training, and wages led to taxation, which led to a transference of resources away from landed capital. (Recall Gibbon's lament: "All taxes must, at last, fall upon agriculture.") Again, this metamorphosis was not felt to be the policy of any one Athenian, but rather the natural expression of democracy in general, which favored the more numerous with fewer material resources and less capital (Pl. *Laws* 707A; Arist. *Pol.* 6.1321a14–16; 7.1327b5–17; *Ath. Pol.* 27.1).

In Greek tradition, Athens shifted her defense from land-based hoplites to maritime power during the invasion of Xerxes and in the immediate aftermath of the Persian Wars (Thuc. 7.21.3; Arist. *Pol.* 5.1304a20–25). Thus, at the outbreak of the Peloponnesian War (431 B.C.) Pericles himself acknowledges a prior half-century of Athenian naval policy (Thuc. 1.142.7–9). Again, the relationship between democracy and sea-power characterized the entire history of Athens and was not the responsibility of any one man. Navies by design extended Athenian power abroad and cemented democracy at home: ships meant transfer of capital within the population, steady employment, the empowerment of an entire underclass, and unchecked military hegemony. The old triad of farmer, hoplite, and council member gave way to landless day laborer, rower, and democrat.[15]

Fortification

Wall building had a long tradition in Greece extending back through the Dark Ages to Mycenean times, and originally probably had little significance in political terms. Yet, the construction and use of fortifications were always controversial during the life of the polis (Thuc. 1.90.1). Later Greek philosophers and strategists debated the wisdom and morality of fortifying a city-state (Pl. *Laws* 6.778D–779A; Arist. *Pol.* 7.1330b33–1331a19): some worried that it bred complacency among the citizenry, created faction between landed and poor, and discouraged hoplite battles altogether. Others, like Aristotle, were concerned that the absence of walls left the citizen populace naked before surprise attacks. But no city-state until Athens in the fifth century built such vast fortifications aimed at incorporating its own port at the expense of the surrounding countryside. Even during the fourth century, fortification expanded out into the countryside of Attica, the idea being that mobile light-armed forces might man garrisons and better protect the borders than ritual collisions of hoplite infantry.

Thus fortification at all periods at Athens took on entirely new ramifications far beyond mere self-defense in times of hostility: in the fifth century, the city circuit and long walls suggested that the polis proper and its fleet, not the countryside and its yeomen hoplites, must be saved at all costs; in the fourth, rural infrastructure might be protected by forces other than heavy infantry. Yet even in the post-imperial fourth century, when an enemy entered Attica in strength, the population was once again—as it was a century and more earlier—brought back inside the city wall and the land abandoned, confirming as in the past that the urban population need not fight for the farmland of Attica (Diod. 17.4.6; Arr. *An.* 1.10.2).

So walls under democracy were designed no longer as a last resort to save the population in times of national defeat, but rather as the deliberate *first* option to ensure that hoplite battles would not be necessary. No wonder that contemporary Greeks felt that Athenian fortification was a dangerous precedent for Greece as a whole, understanding correctly that the abandonment of the countryside freed Athens from traditional agrarian checks on fighting (cf. Thuc. 1.69.1; 1.143.5), threatening to undermine the entire ideological basis of Greek warfare.

In short, long walls connecting port facilities with the city proper became synonymous with Athenian-style democracy, and their construction elsewhere in the Greek world was seen as tantamount to the triumph of the multitude (Thuc. 1.91–93; 5.52; 5.82–83; Plut. *Alc.* 15.2–3). In a military sense, like navies, fortification gave the polis new options concerning the occasion and scope of hostilities, for it did not predicate survival on the success or failure of its landholding infantry. Wars, not battles, now mattered, as defeat lay not in the collapse of infantry within an afternoon, but in the lengthy and expensive subjugation of an entire people within the walls. Like navies, the construction of fortifications required vast expenditures of public monies and enormous commitments of manpower, and thus became a logical democratic mechanism to redistribute income from landed capital to the government and the poor. Sparta, the archconservative, then naturally remained unwalled, as a reactionary symbol of the old infantry prowess.[16]

Siegecraft

Previously in Greek warfare, city-states customarily marched out to battle to the open challenge of their landed adversaries. Those too timid or too wise to muster a hoplite army, if they possessed stout walls, were essentially safe from attack by agrarian infantry, who had neither the desire, expertise, time, nor capital for such lengthy operations. Lengthy sieges were still rare, since most agrarian populations resented the notion that foreigners were ensconced on their own farms outside the walls. Most often they preferred to fight.

Athens however—and this was so recognized in antiquity—mastered the arts of storming palisaded camps and cities (Thuc. 1.102.2; Diod. 12.28; cf. Hdt. 9.70.1–2), and used that expertise as an element of state policy. That expertise too was a complete anathema to traditional Greek warfare, for it elevated specialization—circumvallation, logistics, and specialized technology—to important military sciences in their own right. It ensured that Athens would carry war well beyond the confines of the traditional agrarian battlefield and the territory of the Greek mainland itself, and could now successfully challenge islands and commercial centers less dependent upon agriculture. In the great majority of its sieges Athens was successful in reducing targeted city-states despite enormous expenditures in time, manpower, and money (e.g., Thuc. 1.65.3; 1.98.1–2; 4.69; 4.130; 4.133.4; 5.3.4; 5.32.1; 5.116.4). Thucydides recognized the Athenian innovation in promoting sieges against entire peoples when he remarked that the assault on and subsequent enslavement of Naxos in 466 was the *first* against an allied city and "contrary to established custom" (Thuc. 1.98.4). But by 415 B.C., a mere fifty years later, the Athenians could warn the Melians that "the Athenians have never in a single instance withdrawn from a siege due to the fear of any opponent" (Thuc. 5.111.1–2).

More importantly still, siegecraft, along with fortification, was an admission that civilians were now essential to both military offensive and defensive operations, ensuring that they would pay dearly in defeat through either execution or enslavement (cf. Pl. *Rep.* 5.469B–C; Xen. *Ages.* 7.6; *Hell.* 1.6.14; Aeschin. 2.115)—something not characteristic in the aftermath of hoplite pitched battles. Siegecraft, then, brought the citizenry of the Greek state into war both in a concrete sense as casualties and in the more abstract as exhausted taxpayers who met the tab for lengthy and expensive operations of a year and more. In some sense, siegecraft and its twin fortification were the logical and ultimate military manifestations of democracy, which sought to give the majority of its population a role in every aspect of political and military affairs. Typically the democratic assembly back at Athens had absolute control over the fate of the vanquished civilian populace (e.g., Diod. 13.2.6; 13.67.7; Thuc. 3.28.2). And when the Athenian navy was finally destroyed at Aegospotami (404 B.C.), Xenophon has a moving portrait of the ensuing hysteria at Athens, where the entire citizenry now expected a blockade and full retribution for the barbaric sieges its democratic assembly had sanctioned for others (Xen. *Hell.* 2.2.3–4).[17]

Decline of Agrarianism

Closely connected with naval supremacy, fortification, and siegecraft, then, was the loss of influence among the rural residents of Attica, those who traditionally formed the phalanxes of the city and raised the community's food in the country-side outside the city walls ([Xen.] *Ath. Pol.* 2.14–15). While I have argued else-where that Athens sought to incorporate her hoplite landowners into the democratic fabric of the city on a wide variety of economic, political, and military fronts, and that many Athenian agriculturalists prospered in the short term under democracy (Xen. *Oec.* 20.22; Dem. 42.21; *Hell. Oxy.* 12.4–5; cf. Thuc. 2.14–16; 2.65.2), it is also true that Greek agrarianism—as distinct from agriculture itself—as a whole never recovered from the Athenian democratic experiment: radical democracy was always considered at odds with agriculture and the idea of a timocratic agrarian state of landowning peers (Arist. *Pol.* 6.1318b10–35). Only at Athens might a permanent enemy garrison be ensconced relatively unchallenged in the farmland near Decelea while the city sent a massive armada overseas to Sicily (Isoc. 8.83–5; cf. Lys. 34.8–10). Hoplite battle as exclusive Greek warfare was now dead, as was all that was implied therein: equal-sized plots, timocratic government, immu-nity from taxation, and exclusive dependence on locally grown food.

But in purely military terms, this was a *liberating* experience for the Greeks (cf. Plut. *Per.* 9.1). The entire notion of hoplite battles of a day fought largely over farmland was a limiting and restraining influence on the conduct of war (Hdt. 7.9.1–2), and had no place in fighting on a Mediterranean scale, where capital, logistics, generalship, tactics, strategy, and the incorporation of light-armed infan-try were crucial on the new battlefield (Thuc. 1.141.–144; cf. Xen. *Oec.* 8.6)—an arena that might range from a city's harbor, to her hills, to her streets and houses themselves, involving slaves, women, and children. An agrarianism whose first concerns were egalitarianism, amateurism among the landowning citizenry, and restrictions on infantry battle were obstacles both to dynamic warfare in general and to Athens in particular. Quite simply, farmers do not want to leave home, and they do not wish to be taxed for others to do so.

Once agriculture was removed from the military equation, where, when, and how fighting was to take place was not predicated on the farmers' blinkered and limited notions of time and space, nor on their snobbish notion of who might be allowed to defend the city. And just as democratization increased specialization and bureaucratization, so too the amateur nature of hoplite fighting had no place in a complex governmental hierarchy, where more than 20,000 citizens were em-ployed outside of agriculture (e.g., Arist. *Ath. Pol.* 24.3; Ar. *Vesp.* 709).

Perhaps the most deleterious aspect of democratic military evolution on agri-culture was the need to raise capital from the countryside to pay for urban fortifi-cations, dockyards, and overseas sieges that eventually meant more capital and labor divorced from the countryside, with all the ensuing political ramifications. This trend only continued throughout the Hellenistic period; both archaeologists and historians have seen that increasing financial burden as a prime cause of rural depopulation and disintensification of agriculture practice in later Greece.

In short, ancient democracy was at odds with a property-qualification and a constitution of landowning peers, at odds then with the entire notion of the social and cultural limitations on Greek warfare. Almost every aspect of Athenian military policy was opposed to the agrarian state and seen as such by Athenians themselves (Thuc. 1.143.5; 2.62.2; Plut. *Arist.* 24.1). And because the influence of traditional agriculture was reduced under democracy, the effectiveness and frequency of war grew in an almost inverse proportion.[18]

Redeployment of Heavy Infantry

Despite the expansion of warfare, hoplites as warriors per se, and not as emblems of agrarian exclusivity, continued to have their uses. The Athenians recognized the intrinsic military value of the hoplite and attempted to remove him from the exclusive domain of agrarian warfare fought on and over plains. During the Peloponnesian War hoplites were transported by sea and then expected alone to occupy ground or seek out enemy light and heavy infantry. While ostensibly marines in the sense that they arrived in enemy territory on ships, rather than independently by land, it is better to describe these new hoplites as "expeditionary forces" ([Xen.] *Ath. Pol.* 2.4–5). Often, for example, their activity was far from the coast, independent, and not always supported by combined naval strategy (e.g., Thuc. 3.98.4; 4.44.6; 5.2.1; 6.43; 7.20.7; 7.87.6; Diod. 12.65.6).

Also, Athenian hoplites in the fifth century B.C. were used by the democracy in smaller numbers as amphibious marines or as more daring seaborne raiders (e.g., Pl. *Laws* 4.706B–C). During the Archidamian War, they skirted the coast of the Peloponnese, disembarked, ravaged, plundered (e.g., Thuc. 2.17.4; 2.23.2; 2.25.30; 2.26; 2.32; 2.56; 2.58.1), and then, in the wake of enemy reprisals, fled to the safety of the ships. Although these troops probably rarely formed up in the phalanx for decisive single battles or engaged in drawn-out sieges, there is little doubt they were employed as hoplites proper, *not* armed as light-armed skirmishers.

More controversial still is the actual status of the trireme's standard seaborne contingent of ten *epibatai*, the naval "marines" permanently attached to Athenian vessels quite independent of any accompanying hoplite forces that were being transported from theater to theater. Given their exalted status in the literary and epigraphical sources, the nature of their training and armament, and their apparent recruitment from the hoplite muster roll, it is logical—if these strict census differentiations still always applied in the late fifth and fourth centuries B.C.—that most *epibatai* should not have been landless thetes. Marines were often instead those who were eligible and liable to see service as hoplites proper (Arist. *Pol.* 7. 1327b10–12). From the record of their employment during many naval engagements, Athenian marines were not mere adornments. Often they became the decisive element in battle as both naval boarders and defensive troops who kept the enemy from gaining access to the decks of their own ships (e.g., Hdt. 8.90.2; Diod. 13.40; 45–46; 77–80; Thuc. 7.70; cf. especially Plut. *Cim.* 12.2).

There were domestic ramifications as well in the reinvention of hoplites. Aristotle remarked on the political advantages for democracies in mixing their heavy infantry in with lighter-armed contingents to prevent hoplite exclusivity leading to property dominance in politics (*Pol.* 6.1321a17–21). Democracy, in other words, had expropriated the cultural privileges of heavy infantry, ending its connection with agriculture, and in the bargain making it a more effective and flexible purely military resource. No wonder the Old Oligarch remarked that Athenian infantry, while deliberately made weak in traditional terms, was nevertheless adequate to fulfill Athenian strategic aims ([Xen.] *Ath. Pol.* 2.1–5)—aims that had nothing to do with the protection of agriculture.[19]

Military Capital and Finance

Of all the city-states, Athens best understood the possibilities of capital in the new warmaking that had evolved far beyond martial gallantry and muscular power. Her entire system of imperial tribute was part and parcel of her potential to wage war in the fifth century, followed by an even more elaborate array of domestic devices to raise capital from the wealthy in the ensuing fourth century (Dem. 1.19). Demosthenes confidently asserted that in the mid-fourth century Athens had as much privately held wealth as all the other city-states combined (Dem. 14.25). In contrast, before the war with Athens, Sparta and her agrarian Peloponnesian allies essentially had no available capital at all (Thuc. 1.19; 1.80.4; 1.141.3–5; 1.142.1; cf. Arist. *Pol.* 2.1271b10–17).

A sophisticated thinking arose at Athens that accumulated capital was essential to fighting (Thuc. 2.13.3–6), and hence taxation followed. Funds extracted from subject states (Thuc. 2.69.1; 3.19.1; 3.39.8; 3.46.3; 4.50; 4.75.1) were required to provide for an entire range of military assets—public ownership of arms, state pay for troops, hiring of mercenaries, permanent garrisons—in a way not seen before in Greek warfare, whose prior capital consisted solely of the physical presence of men and their private arms at a given time and place.

Power gradually shifted from the individual, who had previously provided his own panoply (*IG* I³ 1.7–12), to the democratic state, which built and outfitted ships, paid wages, and even kept a ready supply of infantry weapons in state armories (Arist. *Ath. Pol.* 42, 46; cf. *Pol.* 4.1297b9–12). Demosthenes remarked that the war tax fell particularly hard on the farmers (Dem. 22.65; 24.172), illustrating that agrarians' money rather than merely the bodies of farmers themselves were what the polis most wanted. The decision of when and where to fight no longer rested with the militia of landed citizenry but now lay in the hands of specialists who understood the new costs of war and predicated military expeditions on the ability of the state to raise sufficient funds (Xen. *Vect.* 5.11–13).

In short, Athenian democracy introduced the entire idea of the Greek state's military interest in taxing and spending and thus redistributing income between social and economic classes. That ability to expropriate private capital to use on state military expenditures made the entire amateur and seasonal nature of traditional Greek warfare absurd. But such flexibility was consistent with democratic policy that military practice should further radical egalitarianism and be an instrument of both national aggrandizement and capital accumulation.

Few, if any, other states could match Athens' ability at raising capital and so most would be at a decided disadvantage in any theater of conflict (Thuc. 1.80.3–4; 1.141.5; 1.142.1; 1.143.1–2; 3.19.1). Thus the Spartan king, Archidamos, is forced to admit on the eve of the Peloponnesian War that war with Athens would become "not so much a question of arms as of money, through which arms become serviceable" (Thuc. 1.83.2–3). Hermocrates the Sicilian said about the same thing of the Athenians: "They have acquired an enormous amount of gold and silver, by which war and other matters are facilitated" (Thuc. 6.33.2–3). Indeed, the only alternative to the democratic state's ability to raise capital was simply for agrarian states to loot Panhellenic sanctuaries (1.143.1). No wonder in his reactionary utopia Plato addressed the issue of capital's role in warfare and naively assumed that hard training and discipline might negate the need for money and therefore avoid the corrupting influence of wealth on the fighting force and society at large (Pl. *Rep.* 4.422A–E).[20]

Mixing of the Census Classes

One of the great tenets of Greek military practice had been the close association between military service and social status: the wealthy rode horses and skirmished before the main hoplite fight as elite cavalry; the larger middle strata of yeomanry formed the ranks of the phalanx and provided the chief defense arm of the state as landed infantry; the landless poor were mostly a rag-tag bunch who covered the wings and joined in the pursuit once the main fighting between infantry was over.

Often by the fifth century these rubrics had to be somewhat artificially maintained, as the manpower employed at each tier did not always reflect military realities. Cavalry were on occasion needed in numbers beyond what the wealthy could provide, or, alternatively, the numerical abundance of the poorer classes in interior states was not fully utilized through mere skirmishing and pursuit. Infantry battles became rarer, causing problems for city-states since the hoplite class had traditionally been the backbone of the state. Again, the city-states' strong tie between social status and military responsibility was an artificial one, with increasingly little relationship by the fifth century to tactics, strategy, or military need.

Once Athens blurred those distinctions (Cf. [Xen.] *Ath. Pol.* 1.2)—first in the fifth century B.C. and increasingly so in the fourth—she was free to direct manpower to where it was most needed regardless of the particular civilian status of the combatant. Thus we hear of hoplites used as cavalry (Arist. *Ath. Pol.* 49) and rowers (Thuc. 3.16.1; 3.18.3–4; 3.18.3–4; Arist. *Pol.* 7.1327b13–15; [Dem.] 50. 6, 16, 47). Wealthy knights sometimes fought as hoplite infantrymen (Lys. 16.13; Aeschin. 2.168–70; Plut. *Cim.* 5.2–3), skirmishers, and even rowers (Xen. *Hell.* 1.6.24–5). The poor could serve as heavy and marine infantry (Thuc. 6.43.1 Ar. fr. 232 Kock; Antiphon fr. B6; cf. Thuc. 6.72.4). Lysias, for example, implied that by 403 B.C. many hoplites and cavalrymen must have held no property at all (Lys. 34.4).

Because Athens' landless were employed as rowers, she fielded no specialized corps of light-armed skirmishers. Rather her hoplite arm was often followed by a mongrelized ad hoc conglomeration of noncitizens who were not uniformly equipped (Thuc. 4.94.1), but indicative of the city's policy that all residents had some role in her defense. Some idea of the enormous number of various contingents that Athens might field at one time is clear from her invasion of Megara. In 431 such a multifaceted army marched south—10,000 hoplites strong, accompanied by 3,000 resident alien heavy infantry, and "a not small body of lighter-armed"—at precisely the same time 3,000 Athenians were to the north besieging Potidaea, and 100 ships with 20,000 sailors were circling the Peloponnese (Thuc. 2.31.2–3). Clearly, propertied and propertyless, citizen and noncitizen, were serving simultaneously in a variety of different and similar capacities: the erosion of the military census resulted in a much greater and more flexible fighting force by "the greatest of all the Greek city-states" (Thuc. 4.60.1).

In the fourth century, all four census classes were called up to the Athenian *ephebia*, and military service finally had little if anything to do with social background. Aristotle thought it significant to note, when talking about hoplite losses in early fifth-century Attica, that at the time "the army was drawn from the muster roll" (*Pol.* 5.1303a10)—clearly, he felt the need to explain such an anachronistic requirement.

While the chief impetus for blurring class distinctions in war may have been political and social as much as military, there is no doubt that Athens' available military forces increased and her options in using that power multiplied. Nor should we speak of this fluidity in purely political terms; the inflation of the economy, colonies overseas, and cleruchies all tended to raise a substantial number of formerly landless and poor into what would have been formerly the *zeugite* class anyway (Arist. *Pol.* 1308a35–40; *Rhet.* 1387a21–26; *Ath. Pol.* 7. 4; [Dem.] 42.4; Pl. *Laws* 754D).

Athens may have seen the erosion of the military census as beneficial to the landless free and a logical development of democracy, but in reality the divorce of fighting and social status eventually doomed the entire parochial notion of the city-state in favor of sheer military dynamism. From elimination of census rubrics to the establishment of the mercenary system of the fourth century and later was but a small step. And while it was logical of democracy to end hierarchies among the military forces and so field a city-state in arms, that move tended to expand and diversify the military as never before and to widen and barbarize warmaking itself.[21]

Democracy at War: An Appraisal

In the context of the fifth century, these alterations in traditional Greek warfare of the past two centuries explain Athens' extraordinary success in creating and maintaining a maritime empire for much of the fifth century, waging war at one time or another against most of the Greek-speaking world for twenty-seven years, and obtaining a surprising military ascendency in the subsequent fourth century B.C. The military antithesis of a democratic state, then, would have a large hoplite army (Thuc. 1.121.3; 1.143.5), fight in pitched battles (Thuc. 5.41.2–3), see strategy as largely agricultural devastation (Thuc. 1.143.5; 2.10.3–5), and neglect both siegecraft and fortification in the expectation that opponents would do likewise (Thuc. 1.90.1–3). Its army would be organized along strict census rubrics, and be largely amateur (Thuc. 4.94.1–3); indeed there might be little if any capital on hand in the city's coffers (Thuc. 1.141.3), if coffers even existed (Arist. *Pol.* 2.1271b10–15). Navies would be small, if not nonexistent (Thuc. 1.142.6–9). In other words, military efficacy would be less important than the social and political exclusivity of hoplite landowners.

By the fifth century, states such as Thebes or those in the Peloponnese (Thuc. 1.141.2; 142.1) would share many of just these characteristics; but to fight in the Peloponnesian War successfully, Sparta and her allies would be forced either to follow the Athenian example—with dangerous domestic ramifications for their hoplite constituents in doing so—or lose the contest. That other city-states eventually adopted the example of Athenian military practice explains both why Greek warfare grew more savage and frequent—and why so many city-states unraveled in the fourth century.[22]

Yet do such Athenian military adaptations enlighten us about democratic warfare in general, or teach us anything of lasting significance beyond the confines of fifth-century Greece? The move to naval power, the incorporation of fortification and siegecraft, the erosion of census standards, the diminution of agriculture, and the prominence in capital may seem to us of antiquarian value alone. Yet, these changes did not take place in a vacuum. What lay behind the Athenian expansion of warfare was not merely social policy or utilitarianism brought on by the needs of a vast and unique imperial structure. Rather, the revolution in Greek warfare at Athens reflects the dynamic nature inherent in democracy itself, and that resulting flexibility in military enterprise is a logical result of ancient democratic practice— not necessarily of Athenian culture per se or any of her leaders in particular.[23]

Again, I am not so much interested in the morality of democratic conduct, as in the frequency of conflict and the degree of democracy's effectiveness in conducting military operations once they begin. I also grant that other nondemocratic societies very quickly followed suit and adapted many of the tenets of democratic warfare. My point, then, is that the nature of ancient democracy was best suited to transform war as never before seen in the Greek world. In general, at least three practices inherent in democracy allowed the Athenians to expand the nature and purpose of Greek warfare. Nearly all those traits, I think, are also characteristic of modern democratic custom and may perhaps explain why our own manifestation of democratic warfare has became both so dynamic and lethal.

The Voice of the People

We can start with deliberation and open discussion of military policy. Once defense policy is separated from an imperial elite, an aristocracy, or even a broad landholding oligarchy, fighting becomes the ideological domain of the entire populace and the private preserve of none. Ostensibly, what is militarily efficacious is undertaken by the civic body, not a particular tiny cadre (Thuc. 6.39.2). Whether or not this is always true in practice, psychologically it lends the veneer of cohesion to military enterprises and suggests that the state's campaigning reflects the will of the people in toto to go to war—a mob hard to mobilize but then difficult to stop (cf. Thuc. 6.24.3–4; 7.16.1). Thus foreign expeditions, sieges, naval expenditure, and fortification per se become logical resolutions of the populace, not moves to reward and punish strata of domestic society—although of course they do that as well (Thuc. 6.13.1–2). And while there are inevitably charges that certain policies benefit various members of the community (Thuc. 6.24.4), there is at least the implicit knowledge that the state as a whole has decided on military campaigning and will be responsible for its outcome.

Consequently, Athenian commanders on the eve of battles often appeal for each man to fight for the state and the commonwealth of Athens—in contrast to the Peloponnesian generals who direct their speeches to the self-interest of the individual who must be responsible singly for the reproach of cowardice or the rewards of heroism (Thuc. 2.87.9–10; 5.69.3; cf. 2.11.3, 9; 1.70.9). That civic spirit of democracy led Aristotle to state that of all forms of government, democracy was the least likely to suffer domestic insurrection after military setbacks abroad, ostensibly because the would-be wealthy insurrectionists are in smaller number and the government's responsibility for blame is shared among a wider strata of the people (*Pol.* 5.1303a11–12). That is exactly the Old Oligarch's complaint against democracy as well: in democracy too many people are involved in the decision-making to lay blame on any one person or faction ([Xen.] *Ath. Pol.* 2.17). It is a common strain in Athenian literature that democracy brought the poor into the civic body with positive results for all, through both greater social cohesion and the greater pool of talent and expertise (Dem. 21.67; 24.59; Eur. *Supp.* 350–53; 429–31; 438–41; *Phoen.* 535–47; Thuc. 2.37.1; 6.39.1; Hdt. 5.78). In his Funeral Oration, Pericles says of the war dead that even less reputable Athenians can at last find renown through their ultimate sacrifices to the state (Thuc. 2.43.1–6).

Even the disastrous Sicilian expedition does not bring down lasting revolution at Athens, since it was clear to the Athenians that they had no one to blame but themselves, after having voted in open debate and with a clear eye to plunder and personal advancement (6 31.3–5). "In short, in the immediate panic, the Athenians, *as is the case in a democracy*, were prepared to show discipline in every regard" (Thuc. 8.1.4; emphasis added). Thucydides later points out that after the Athenian disaster off Euboea and the trouble at Samos, Sparta lacked the dynamic spirit and imagination necessary to take advantage of Athens' reversals. This was in direct contrast to the democratic Syracusans, who, he adds, "were the most similar in character to the Athenians" and "made war against them the most successfully" (Thuc. 8.96.5), a people "with a democratic constitution like themselves [the Athenians]" (Thuc. 7.55.2). The implication is clear that the democratic approach to war was the most flexible in attack and resilient from setback.

Especially important was ancient democracy's ability to harness the capital of the rich to contribute to expeditions that were not in their immediate self-interest (Lys. 19.11; 27.1; 30.22). At Athens, the wealthy often bragged about their own contributions to the democracy's foreign expeditions, even during the harsh years of the fourth century when the absence of previous imperial bounty obligated the wealthy to pay for the bulk of Athenian military rearmament and expenditure (Dem. 21.153; 51.4–7; Isoc. 8.128; Is. 5.37–8). Thucydides, again, in highlighting the advantage of democracy over oligarchy during the Peloponnesian War, points out that even the wealthy and powerful serve more efficiently, since they attribute their political defeat to the understandable ignorance or envy of their inferiors, and so yield to the will of the people in a way impossible under oligarchy (Thuc. 8.89.3). In a democracy—more so than in an oligarchy or autocracy—military activity can tap the resources of the wealthy (Xen. *Oec.* 2.5–6), and so use military expenditure to redistribute income with clear advantages to the people (Thuc. 6.24; cf. [Xen.] *Ath. Pol.* 1.1–2; 2.) and the efficiency of the armed forces in general.

There is little support for the ancient charge (cf. Thuc. 3.37.1) that democracy was less willing than other forms of government to shoulder the burden of defense. More often the exact opposite is true: after calamitous reversals, the people make enormous sacrifices to re-equip the fleet and muster a new army (cf. Thuc. 2.65.8–13; 7.16.1–2; 8.1.3–4). Themistocles, remember, had little trouble in convincing the citizenry to forgo a public donative from the silver mines at Laurion in order to build and man a fleet of 200 ships to attack Aegina (Hdt. 7.144).

Among the other advantages enjoyed by democracy at war is the inherent presence of dissent. Such opposition is not merely the private grumbling of those of military age or a worried cabal of potential usurpers—although again it can be all that as well. Just as frequently attacks on state policy are made in deliberative bodies—the council, the assembly, even the courts—and are the stock of playwrights, orators, and poets, and often generally recognized as healthy to the formation of democratic policy (Dem. 20.106; Thuc. 2.37.2; 7.69.2). Again, in a strict military sense, this can bring greater dividends to the state as a whole: loss of secrecy is outweighed by the free participation of those with different views whose criticism can just as easily enhance and improve campaigning as prevent it (Thuc. 2.39.1–4).

True, ancient authors such as Thucydides and the Old Oligarch at times complain that the fickleness of the mob and the gratuitous criticism of orators robbed democratic expeditions of much needed support (Thuc. 2.65.8–11; 3. 37.1; [Xen.] *Ath. Pol.* 3.1–5), and encouraged private agendas and lack of adherence to majority decisions (cf. Pl. *Rep.* 8.557E; *Laws* 12.955B–C). Common complaints are democracy's abject reversal of policy and ignorance of the brutal world outside their assembly hall, when the mob is misled by clever demagogues (e.g., Arist. *Ath. Pol.* 34; Thuc. 2.65.8–11; cf. 3.36–49; Xen. *Hell.* 1.7.34–5), leading democracy's chief spokesmen such as Pericles, Cleon, or Alcibiades to dub majority rule a "folly" or "tyranny" (Thuc. 3.37.2; 2.63.2–3; 6.89.6).

But just as often ancient literary evidence proves the opposite: fighting is reassessed constantly by both private and public oversight, which tends to refine and focus in an effective way the application of military power, and to check and recall unwise expenditures of campaigns that are not militarily efficacious or serve poorly state interests in general. Self-interest even seems to outweigh ideology, and there is little evidence that Athens ever let idealistic democratic concerns influence her decision to go to war: she was quite willing to fight other democracies and to lend military aid to the most repressive of regimes when she felt it in her own national interest (Thuc. 1.102). Moreover, her assembly could be every bit as harsh on military failure as any autocrat (Thuc. 7.48.2–5; cf. Andoc. 3.33), and her legendary vindictiveness against hapless returning commanders drew the ire of an array of critics (Arist. *Ath. Pol.* 34.1; Diod. 13.102.5). It is hard to find any Athenian general who at one time in his career was not fined, exiled, had his property confisicated, or put on trial for his life.

The free exchange of ideas rarely provides an idealistic brake for the conduct of war (Thuc. 6.39–40). Just as often, popular voices demand risky military escalation (Thuc. 6.24.1). In the heated debate over Pylos, for example, Cleon's more radical plan to forgo peace and commit forces in the Peloponnese won the day, and eventually proved to be the most unlikely but greatest Athenian success of the entire war (Thuc. 4.28.2–5; 4.39–40). Herodotus saw the power of group bellicosity when he remarked that the Athenian assembly might be easier moved by Ionian requests to go to war than more narrow oligarchies since at Athens it "proved far easier to persuade 30,000 than one" (Hdt. 5.97.2). In general, the Athenian demos was much more likely to commit enormous resources for expeditions than it was to send too little; usually timocratic moderates like Nicias were in the minority in their reluctance to commit forces abroad (Thuc. 6.24.1–25.2; cf. 4.28.2–3; 5.16.1–2; 7.8). There is evidence that had Athens been successful in Sicily, her democratic leaders would have advocated further expansion beyond into Italy and perhaps south to Carthage as well (Thuc. 6.15.2, 34.2, 88.6, 90.2, Ar. *Eq.* 236–38, 797–800, 1303–4). Aristotle, again with Athens in mind, believed that imperial states remained hardened when they were fighting to expand and maintain their empire, but lost their edge, like an iron blade not in use, when they were at peace (*Pol.* 7.3334a8–10).

The charge of timidity and fear of committing sufficient resources usually is more often lodged against Sparta or agrarian states such as Thebes and those in the Peloponnese (Thuc. 1.70–72; 4.91; Hdt. 6.106.3; cf. Thuc. 3.15.1; 3.16.1–2; 3.89.1; 4.6.2). There is little evidence in the ancient world—for all the complaints of the critics of democracy—that the rule of the assembly proved to be a disadvantage to the conduct of war. Rather, the evidence suggests just the opposite: rule by majority brought a revolutionary and chauvinistic fervor to Athenian foreign policy that made the democracy more likely to intervene in the affairs of other states.[24]

Acceptance of Change

It was a well-worn complaint against the Athenians that they were fickle and changeable, the *mobile vulgus* who could order execution one day, reprieve the next.

True, there is much support for the charge in ancient literature, especially when we see that Athenian decisiveness, daring, and risk-taking lead directly to military reversals (Thuc. 1.70.4–7; 2. 65.10–12; 7.87.5–6). Yet, such unsteadiness was not without military value. The same restlessness inherent within the assembly also explains the adventurousness in Athenian military practice, and even the willingness to embrace new technologies and routines, ranging from extended sieges, the abandonment of its countryside, or the creation of seaborne forces. Since public decisions can be reversed without individual responsibility, the assembly feels free to change and adopt policy given the immediate needs of the present without the baggage of past custom and tradition. The repudiation of agrarianism—the historic tie with reactionary conservatism—was inherent in radical democracy, making the state less tied to the past, and more open to new trends in fighting outside the small world of hoplite battle.

It is no accident, then, that the most innovative of fifth- and fourth-century warmakers such as Themistocles, Pericles, Alcibiades, Demosthenes, Iphicrates, and Chabrias were all Athenian. The Spartan Brasidas is deemed successful by Thucydides precisely because he does not act like a Spartan (Thuc. 4.81.1, 3; 4.108.7). He and other unblinkered Spartans such as Gylippus and Lysander owe much of their daring to the demands of fighting against the Athenians. If democracy breeds recklessness and fickleness, those same characteristics also promote innovation and adaptability to new circumstances (Thuc. 1.102.3; Plut. *Per.* 27.3–9; *Cim.* 12.2), providing a far greater range of military options on the battlefield. The resourceful use of light-armed troops on Pylos (425 B.C.; Thuc. 4.34–5), the ability of skirmishers to destroy a Spartan regiment at Corinth (390 B.C.; Xen. *Hell.* 4.5.12–17), and the halt of an entire Spartan expedition by the well-trained peltasts of Chabrias (378 B.C.; Diod. 15.32.5–6; Polyaen. *Strat.* 2.1.2) were typically all Athenian enterprises that successfully pitted originality and ingenuity against blinkered Spartan conservatism.

In 432 B.C. the Corinthians complained at Sparta that the Athenians "are addicted to innovation" in "both conception and execution" (Thuc. 1.70.2–3). After reciting a litany of stereotyped national characteristics that promote change and instability, they conclude that Athens' adversary, Sparta, is "old fashioned," a lethal characteristic since "constant necessities of action must be accompanied by constant improvement of methods"—something that the Athenians excel at (1.71.2–3). Indeed, Alcibiades had used precisely that argument to sail to Sicily in the first place (6.18.7): even if Athens had wanted to, it could not adopt a reactionary policy at odds with its natural democratic and imperial character. Pericles earlier had bragged that the sheer liberality of Athenian life and education made them more, rather than less, ready to meet their opponents in the field. In his democratic way of thinking, courage is best inculcated by a voluntary rather than enforced patriotism, something again possible only in a democracy (2.39). In both comedy and tragedy the theme is repeated: the natural buoyant and lively charter of a free people proves invaluable in meeting the sudden changes of war (Eur. *Supp.* 321–

25). Because war requires constant improvements in weapons and tactics, and an ability to change plans and objectives, democracy is by nature the most flexible constitution for those challenges.

Economy and Demography

For nearly a century, the nature of the ancient economy has been a real source of debate between modernists and primitivists, between those who argue that the Greek city-state functioned in a rational economic sense roughly analogous to modern governments, and others who stress instead the moral and spiritual dimension in Greek life that ensured backwardness and unsophistication in economic development. These primitivists argue that profit and loss, supply and demand, interest, dividend, and the means of production in the city-state were not conscious activities, but rather "imbedded"—governed by more household forces other than the sheer desire to create capital and affluence, and so functioned outside the natural laws that govern exchange.[25]

That controversy need not be restated here, although I note that recent scholarship has to a large degree proven that the ancient economy—that of Athens particularly—was not so nearly primitive as was thought twenty years ago, and that citizens of the polis were aware of economic development and the notion of capital formation per se. In any case, no one argues much anymore that the Athenian economy was not different from its nondemocratic counterparts: greater trade, greater numbers divorced from agriculture; greater use of coinage; greater presence of noncitizen metics engaged in mining, small-scale manufacturing, and commerce.

The so-called Old Oligarch had deplored the bustling ruckus that he witnessed on the streets of Athens, but like other ancient observers recognized the diversity and dynamism of the city's economic life as unique in Greece ([Xen] *Ath. Pol.* 2.7–12), a view largely supported by Xenophon (*Vect.* 3.1–90), Plato (*Rep.* 8.557c), Aristotle (*Ath. Pol.* 24.1), and Thucydides (Thuc. 1.141–2; 2.38.2). Aristophanes ridiculed democratic urban man's tendency to view the world in terms of buying and selling, removing more traditional and aristocratic criteria from the judgment of human character (Ar. *Ach.* 35–40).

So the ability of Athens as a democracy and a liberal state to encourage economic activity had enormous military dividends in the raising of men and material. Contemporary critics of Athens, at any rate, felt that radical democracy had even extended its liberality to foreigners and slaves, some of whom were beneficiaries of the general dynamism of the Athenian economy. And, of course, imperialism was seen as an integral part of the democratic cargo, the charge repeatedly leveled that Athens could fight continuously and in a variety of theaters only due to the money expatriated from her subjects (Thuc. 1.81.4; 3.13.5–6; 3.31.1). Incidentally, later observers as varied as Adam Smith, Karl Marx, Tocqueville, and Thomas Jefferson recognized that democratic governments were able to foster larger and more vibrant economies, which brought clear military dividends during wartime.[26]

Closely related was the increase in population. Most scholars now agree that Athens grew steadily in the fifth century. This expansion was not merely the increase in citizen population brought on by affluence and urbanization, nor indicative of Greece as a whole during this period. Instead, the resident population of legal aliens and slaves alike also grew ([Xen.] *Ath. Pol.* 1.12; Pl. *Rep.* 8.563B), as the exploitation of the mines, the intensification of agriculture, and small workshops drew in servile workers. Similarly, foreign tradesmen, shippers, and bankers flocked to Athens, followed by an array of itinerant doctors, traders, and teachers who found economic activity not infringed by an absence of citizenship. Besides the increase in specialized skill that this influx added, there was also a new resource of manpower to be tapped for the army and navy (Thuc. 2.13.6; 2.31.2; 4.28.4; 5.8.2; 7 57.2; Xen. *Vect.* 2.1–6; 4.40–42).

We read in a variety of sources that metics and slaves took part in Athenian land and naval operations and helped to bring numerical superiority to the side of Athenian forces (Thuc. 2.13; 2.31.1; 4.90, 92; Xen. *Hell.* 1.6.24; Lycur. *Leocr.* 41). When the Athenian army and navy went forth together it was an impressive spectacle like none other in Greece (Thuc. 2.31.2–3; 6.31). In general, Athenian statesmen and supporters assume that all members without exception who hold citizenship fight in some capacity on the city's behalf (Thuc. 1.70.6; 2.42.4; 2.39.3). Aeschylus, for example, remarked in his *Suppliants* that every citizen of the polis takes an active part in the defense of their country (*Supp.* 365–69; 398–401, 483–85). Clearly, the absence of stark hierarchies gave the lower classes a greater sense of belonging and encouraged those outside the traditional boundaries of the polis to feel more at home in a democracy (cf. the complaints at democratic license and liberality in Pl. *Rep.* 557B–558C; [Xen.] *Ath. Pol.* 1.10–13).

It is no accident then that the antithesis of Athenian democracy, Spartan oligarchy, suffered from continuous depopulation, its city as quiet as Athens was congested (Thuc. 1.10.2). A common theme in Aristotle's *Politics* is the depopulation of nondemocratic regimes through falling birthrates and emigration (Arist. *Pol.* 2.1270a30–40; 1273b18–21). Aristotle also assumes that states with maritime economies and large navies had a growing population (*Pol.* 6.1327b7–9), and that landed oligarchies were much more vulnerable after significant wartime losses given their smaller citizen populations (*Pol.* 5.1303a9–12).

Conclusion

Some will object that Athens was stalemated in the so-called first Peloponnesian War (ca. 460–445 B.C.), lost the second, was reduced to second-rank military status in the fourth century, and then beaten or bullied into submission by Macedon. A discussion of Athens' success or failure in two centuries of war-making is beyond the scope of this small investigation. But we should recognize that rarely in the history of her democracy was Athens at war with a single city-state, at a single locale, and on a single occasion. The successful war against Persia was waged against Persians and

Greeks, at sea and on land, in Greece and in Asia. Her initial stalemate with Sparta in the 450s was a result of her wide-scale fighting on a variety of fronts ranging from Boeotia to the Peloponnese to Egypt and Cyprus. The mystery of the Peloponnesian War is not that Athens lost, but rather, as Thucydides saw (2.65.12–13), that it held out against the Peloponnesians, Boeotians, and Persians for as long as it did. The triumph of Philip and Alexander over the Greek city-state involves questions greater than the efficacy of democratic warfare; but let it be said that most Greek resistance to Macedon, material and spiritual, centered around Athens, and her demise was due not so much to her democracy as to the entire inappropriateness of a single city-state waging war on an international scale.

The ancient appraisal is nearly uniform in its assessment that the Athenians both fought extremely effectively under democracy and were a constant source of death and destruction among the Greek-speaking peoples. The Old Oligarch despised the nature of Athenian democracy, yet granted the logic of political and military cohesion and the dynamism of its naval supremacy ([Xen.] *Ath. Pol.* 1.1–3; 2.1–7). Xenophon argued that of all war-weary city-states Athens had the most to benefit by peace, though he worried that many there believed she could recover her ascendancy only through war (*Vect.* 5.5–6). Herodotus remarked that Athenians were no better than their neighbors in battle until they threw out the tyrants. But after the establishment of the democracy by Cleisthenes (507 B.C.) they became the "best," as each citizen eagerly strove for his own success (Hdt. 5.78). Aristotle lamented that during the mid-fifth century the Athenian army campaigned "continually" and lost between 2,000 and 3,000 hoplites per expedition (*Ath. Pol.* 26.1–2); he went even further and suggested these losses by land accounted for the further radicalization of democracy (*Pol.* 5.1303a9–12). Pericles bragged that Athens had the greatest reputation among the states of Greece because it had "expended more lives and energy than any other city and had won the greatest power" (Thuc. 2.64.3). Thucydides has Alcibiades say on the eve of the Sicilian expedition (to the apparent approval of the demos) that Athenians would be quickly ruined by peace, inasmuch as such unaccustomed inactivity would be lethal to such a belligerent and restless power (Thuc. 6.18.6–7). Both Isocrates and Plutarch record that the Athenian reputation for enterprising and constant military operations gave alarm to the rest of Greece throughout the fifth and fourth centuries (Plut. *Cim.* 17.2; 18.1; Isoc. 1.68–71).

My argument thus far has been relatively straightforward: through naval power, fortification, siegecraft, diminution of agrarianism, the innovative use of capital, and the divorce between military and social classes Athens waged war unmatched by any regimes in the area. This complete refutation of prior Greek military practice was made possible by the very nature of democracy: its view that military power should serve the needs of the greatest number of the population; its open, frank discussion about military enterprises which encouraged an array of diverse opinions; its natural character to welcome innovation both technological and social; its creation of an economic thinking where morality and tradition gave way to a more rational discussion of how funds could be created and expended for military needs. Those changes were not the result of any one individual or an accident of history but the logical fruition of democratic ideology. No wonder the other large democracy of antiquity, classical Syracuse, fought like the Athenians and so became their most difficult adversary.

Thus far, however, I have not answered more important questions that have underlined much of our discussion and worries over democracy: if Athens was more dynamic than its neighbors, was she by extension a more moral state either in her aims or conduct of war-making? Was her newfound power, then, a source of stability among her neighbors or a catalyst for greater death and mayhem among the Greek city-states? The question goes right to the heart of our investigations, for the ancient evidence suggests *not* that democracies tend to lose military confrontations and cannot match the resources of more authoritarian states, but rather that their superior ability to project military power, and the political and moral sanction that such projection of force receives from the people, tend to make democracies both more flexible and more successful in their use of armed force.

If the ancient world has any bearing on the modern, the study of early democratic states suggests that South Korea, to the degree that it creates a free society and an open economy, will wage war more effectively than its northern neighbor. To the modern post-enlightenment mind with its greater confidence in the innate nature of man, the idea that democracy can choose its own fights and then fight well for the people offers hope. But to the Greeks who had a far bleaker view of human nature it was a discomforting thought that the military aspirations of the man in the street could be realized almost instantaneously on the battlefield.

Notes

1. On the use of Thucydides by Machiavelli and Hobbes, see Ste. Croix, *Origins*, 24–30. George Marshall's address at Princeton on February 22, 1947, is discussed by Connor, *Thucydides*, 3. Daniel Bell, *Cultural Contradictions*, 200–201, saw Thucydides as instructive about the conduct of contemporary democracies during wartime, especially their inability to maintain unity and to refrain from the temptations of imperial aggrandizement. Most recently, Donald Kagan (*Origins*, 8) has again studied the origins of war, ancient and modern, and found the common catalyst in all conflicts similar to those originally outlined by Thucydides at the outbreak of the struggle between Athens and Sparta: "honor (*timê*), fear (*deos*), and self-interest (*ôphelia*)" (Thuc.1.76.2).

2. On the questions of bipolarity and Thucydides, see, for example, Connor, "Polarization," 54–60; Santoro, "Bipolarity and War," 76–81. On the nature of Thucydides' historical technique, use of evidence, and his attitude toward democracy, see the review of the general problems in Hornblower, *Thucydides*, 73–109; 155–190; Gomme, *Commentary*, 1–87.

3. For a summary of strengths and weaknesses of the Athens/America and Sparta/Soviet Union paradigm, see Gilpin, "Cold War," 37–49; Evangelista, "Democracies," 215–230.

4. On problems of facile comparisons between the fifth-century Greeks and the postwar period, see Sabin, "Athens," 236–239; Kagan, *Origins*, 76, n. 10; 571: "They [the

Greeks and Romans] knew nothing of ideas such as would be later spoken in the Sermon on the Mount, and they would have regarded them as absurd if they had."

5. For a synopsis of American motives and good(?) intentions gone bad, see Cumings, *Origins*, 440–444. His view of American-Korean relations reflects the now mostly orthodox criticism of American foreign policy during the Cold War: a well-meaning but unsophisticated and blinkered diplomacy assumed that all Third World peoples shared the same cultural aspirations as Americans, and so through either coercion or bribes could be made eventually to adopt the tenets of capitalist democracy—which was, we alone knew, far better than communism for their long-term interests. Such a censure of containment, to my mind, is too charitable to indigenous communist movements. That Korean or Vietnamese socialists were less bloodthirsty than Stalin or Mao is no guarantee against the gulag and relocation camps—not to mention boat people—which were the logical dividends of such systems. For those scholars who see a rather benign North Korea as victim of American imperialist aggression, it is obligatory upon them to advocate the immediate withdrawal of all American troops from Korea in expectation that such pernicious foreign influences might be eliminated and thus indigenous Korean socialism might at last run its natural and positive course.

6. See Wright, *War*, 841. There is an enormous body of research on the relationship between democracies and warfare—nearly all of it post-classical. In general, more recent studies have upheld Wright's view that democracies are as aggressive as other states. Even attempts to link bellicosity with capitalism—the premise that the economic system distorts the natural peaceful tendencies of democracies—have met with little success. In general, social scientists and historians agree only that states of opposite natures are more likely to go to war; the more akin their constitutions, the less likely two nations are to fight each other. For a concise summary of the relationship between constitutions and bellicosity, and a history of the question, see Cashman, *What Causes War*, 124–159, and cf. Sabin, "Athens," 238–239. In antiquity, supporters of democracy—the Attic orators mostly—optimistically viewed man's innate nature—in contrast to the philosophers like Plato and Aristotle, who held a bleak view of humanity and thus any constitution that discarded cultural constraints to free expression and commerce like birth, property, and wealth.

7. It is, of course, an abstraction to speak of "Athenian warfare" or "democratic warfare" when we refer to over 170 years of changing Athenian democratic practice between Cleisthenes and Chaireonia. Yet, before the fifth century there is little reason to believe the Athenian military under tyranny or oligarchy was very different from most other agrarian city-states, among whom overseas expeditions were rare and fighting consisted mostly of border squabbles between heavy infantry. See Frost, "Athenian Military," 292–294; Starr, *Sea-Power*, 27–30. But enormous changes in military practice occurred in the fifty years after Salamis, as Thucydides saw, and accelerated during the Peloponnesian War and its aftermath. For a brief synopsis of the evolution of Athenian warfare and its connection to democratic practice, see Hanson, *Other Greeks*, 329–352; 369–375; Meier, "Die Rolle," esp. 563–587. On the explosive changes in military innovation on the part of Athens during the Peloponnesian War and after, see Vidal-Naquet, "Tradition," 94–95; Garlan, *Recherches*, 19–86; Hanson, *Warfare and Agriculture*, 78–83; Ober, *Fortress*, 69–86; Munn, *Defense*, 32–33.

8. On the direct connection between farming and early hoplite warfare, see Hanson, *Other Greeks*, 239–244; *Western Way of War*, 27–39. Scholars often refer to hoplite warfare as ritualistic; such a description is correct only to the degree that it characterizes the limitations placed on land warfare aimed at preserving the agrarian basis of the Greek city-state.

9. The direction in classical scholarship most recently has been to identify a strong Near Eastern and Egyptian strain in Greek culture. In the case of military affairs, there is no doubt that the Greeks learned much about mass attack, body armor and weaponry, and the principles of fortification and siegecraft from the East (cf. Gabriel and Metz, *Sumer to*

Rome, 1–26; Ferrill *Origins*, 33–89; Snodgrass, *Early Greek Armour*, 194–197). Yet the Greek polis was an entirely unique social entity, founded on completely different economic and political assumptions about its citizenry from the palace economies of the Near East, Egypt, or earlier Greece. Consequently, true heavy infantry, the diminution of horses, secondary importance of missile troops, civilian oversight of military operations, separation of military leadership from religious authority, private ownership and possession of weapons, the absence of a permanent military aristocracy in control of political power, the lack of coercive military taxes, tithes, and rents, and organized opposition to war-making are just a few of the characteristics that separate the polis from its earlier Mediterranean predecessors. See Drews, *End*, 242–245; Hanson, *Other Greeks*, 400–402. And cf. most recently the antithesis between the Western and non-Western approach to warfare discussed in Keegan, *History of Warfare*, 386–392, and Seabury and Codevilla, *War*, 18–19.

10. On the relative economy of hoplite fighting and the percentage of native-born of the polis who may have participated in hoplite warfare as citizen-militiamen, see Hanson, *Other Greeks*, 291–323, 208–213.

11. In most Athenian literature Marathon was seen as the last gasp of pre-democratic warfare, the shining example of Athenian agrarian infantry that had repelled the invader in a fair fight between massed armies, without need of the fleet or the landless. See Vidal-Naquet, "Tradition," 91–92; Loraux, *Invention*, 161–164; 170–171. Cf. the fabricated nature of the hoplite contribution during the battle of Salamis: Fornara, "Hoplite Achievement," 51–54; cf. Wardman, "Tactics and Tradition," 59–60.

12. For the relative absence of extensive Athenian military campaiging before the fifth-century democracy, see Connor, "Land Warfare," 6–8; Frost, "Athenian Military," 292–294. The pattern of constant campaigning continued nearly unabated in the fourth century, see Stockton, *Athenian Democracy*, 104–105; JACT, *World of Athens*, 246: "From 497–338, Athens was at war for three out of four years." The Persian War, not the Peloponnesian War, saw the first manifestation of the new Athenian war-making, which reached its logical fruition during the latter conflict. For those who see military restlessness as inherent in democracy, see also W.A. Williams's (*Empire*, 102–110) study of the United States, in which American overseas military activity appears ceaseless between 1829 and 1898.

13. The ancient opinion of the sequence of events is clear: the naval victory at Salamis gave the landless unwarranted prestige and led to their undue domestic status and influence (Plut. *Arist.* 22.1–2). No wonder that Aristotle dubbed Salamis an "accident" (*sympôtma*), as if the decision to abandon Attica and face the Persians at sea changed the course of what had been theretofore a more moderate democracy along the Solonian model (Arist. *Pol.* 2. 1274a12; cf. his views on the domestic ramifications in general of military success and failure at *Pol.* 5.1304a22–35). For the original notion that naval success *first* caused democratic radicalization, not vice versa, see also Plato, *Laws* 4.707B–C, where the bizarre argument is made that Salamis, unlike Marathon and Plataea, made the Greeks worse as a people. See, too, Plut. *Arist.* 22.1 and Xen. *Ath. Pol.* 1.2, who assume the people at Athens have power because the navy is more important than the infantry. Themistocles' decision to wall off Athens was sometimes in antiquity equated with the desire to do away with hoplites as serious forces altogether, cf. Arist. *Ath. Pol.* 24.3; 27.1. Cf. in general, Gomme, *Commentary*, 1. 266–267.

14. For two previous and excellent treatments of Athenian changes in prior Greek military practice, see in general Ober, "Classical Greek Times," 20–26; Meier, "Die Rolle," 563–587. Ober provides a useful table of twelve generally accepted rules of hoplite warfare before the Athenian military revolution, and recognizes their demise was largely due to the rise of democracy. Yet, he sees Athenian flexibility and response under the new system of warfare encouraging, attesting to resilience and adaptability of democratic government. I agree, but interpret Athenian resurgence as yet further proof that ancient democracy was both more competent and more bellicose than other constitutions. For these rules of hoplite war, see also Lonis, *Guerre*, 25–40; Romilly, "Guerre et paix," 215–216.

15. On the social and political rise of the thetes and the connection of their ascendancy to Athenian sea-power, see Amit, *Athens*, 38–40; Jordan, *Athenian Navy*, 210–211; Strauss, "Athenian Trireme." The navy's crews were probably composed largely of poorer citizens, augmented as the fleet increased by resident aliens, foreign mercenaries, and slaves. The relative presence in the fleet of the four respective groups of rowers—citizen, metic, mercenary, and slave—is controversial and contingent often on the circumstances of any given year of the war. Cf. Amit, *Athens*, 28–50; Jordan, *Athenian Navy*, 240–268; Graham, "Thucydides," 257–270. On the differences in training and technique between naval and hoplite service and the influence of sea-power in the evolution of Greek warfare, cf. Vidal-Naquet, "Tradition," 92–94. Again, we must remember that in the sixth century B.C., the Athenians possessed few ships and the thetes were not important either politically or militarily; cf. Starr, *Sea-Power*, 27–30; Ste. Croix, *Origins*, 393–395. Kallet-Marx, *Money*, 21–69, outlines the rise of naval power in the Greek world and its relationship with capital accumulation.

16. For an ample collection of ancient sources on fortification and its ramifications for Greek social policy, see Garlan, *Recherches*, 19–86. Ober, *Fortress*, 68–86, and Munn, *Defense*, 3–33, also detail the increasing reliance at Athens on walls during the fifth and fourth centuries. Whether the policy was city defense or border defense, whether the enemy was Persian, Spartan, Theban, or Macedonian, the Athenians for nearly two centuries sacrificed Attica to ensure the safety of the citizenry inside the walls—a policy of diminution of agriculture that incensed the philosophers and orators (Arist. *Pol.* 7.1330a 16–25; Dem. 19.86; Aeschin. 2.139).

17. On siegecraft and the reputation of Athenian mastery in taking fortified positions, see Gomme, *Commentary*, 1.16–19; Garlan, *Recherches*, 20–44; 105–147. Grundy, *Thucydides*, 286–287 has a convenient table of Athenian sieges during the Peloponnesian War. Potidaea was reduced after two years (Thuc. 2.70.1; 3.17). Thucydides' description of that siege—whose authenticity has often been questioned—is presented in the general context of her financial exhaustion. Cf. Kallet-Marx, *Money*, 130–134.

18. See Hanson, *Other Greeks*, 357–403; *Western Way*, 27–39. By agrarianism, I mean the centuries-old idea in the polis that the surrounding countryside should be divided up into roughly equal plots and maintained within a family, such a parcel being the touchstone for service in the phalanx and voting privileges in a timocratic council.

19. On the new role of Athenian heavy infantry outside the domain of pitched battle, see Hanson, *Other Greeks*, 369–374. The diminution in the power of the polemarch and the institution of the strategia brought the Athenian phalanx under the control of the entire citizenry and democratized what must have been previously a largely conservative source of timocratic and oligarchic values; see Fornara, *Athenian Board*, 8–10. Aristotle, who equated military and political practice with the inculcation of virtue, was surely thinking of Athens' diverse forces when he complained in the *Politics* that "a polis that sends off to war a large number of base men, but few hoplites cannot be a great *polis*—for a great polis and one with a large population is not the same thing" (*Pol.* 7.1326a23–26; cf. Pl. *Laws* 4.706B–C). On the large numbers of combatants that Athens was able to field, cf. Vidal-Naquet, "Tradition," 92–93. On the versatility of infantry in Athenian counterattacks against the Peloponnese and their allies, see de Wet, "Defensive Policy," 105–117; Westlake, "Seaborne Raids" 84–85. See Rosivach, "Athenian Fleet," 54–55 for the idea that poorer farmers might row on a regular basis during slack periods of the agricultural years. The problem here, though, is that there is very little evidence, either archaeological or literary, for a large body of Attic "peasants" or small landholding residents below the hoplite census; see, for example, Jameson, "Class," 62–63.

20. On the growing practice of state armories and the supply of public weaponry to hired troops, see McKechnie, "Mercenary Troops," 301–305. On the effect of inflation, mining, and overseas settlement raising the number who met the *zeugite* census, see Jones,

Athenian Democracy, 166–177. Hodkinson, "Imperialist Democracy," 54–55, reviews the liturgies, the *eisphora*, and the general ability of the democracy to tap the rich to pay for much of the cost of Athenian campaigning, especially in the fourth century. Finance, of course, became more acute in the fourth century; see Ober, *Mass and Elite*, 97–100; Jones, *Athenian Democracy*, 54–57. Kallet-Marx, *Money*, 11–12, traces the rise of Athenian naval mastery and its ability to infuse money into the military sphere as marking the true revolution in Greek warfare.

21. For the increasing irrelevance of the census classes at Athens and the logical democratic intent behind eliminating strict hierarchies in military service, see Hanson, *Other Greeks*, 350–351; 385–386. For the composition of the *ephebia* in the fourth century and the absence of class distinction, see Hansen, *Athenian Democracy*, 108–109.

22. For the effect of the Peloponnesian War on Spartan society, see Cartledge, *Agesilaos*, 167–170; Hodkinson, *Crisis*, 49–54, 148–150. For the city-state's structural inability to react to the demands of a new, non-hoplitic warfare, see Runciman, "Doomed," 348–367.

23. We have little information as noted about other ancient democracies. But it is significant that the one large democracy other than Athens, Syracuse, was almost identical to Athens in her military practice. Indeed, Thucydides saw the two democracies as unique powers far more adept in war than the Peloponnesians under Sparta or any other city-states in Greece—sharing a nearly identical relationship between military and political power (Thuc. 7.55.2; 8.96.5; cf. 6.20.3). From what we know, the navy of Syracuse was extremely large (Hdt. 7.158; Thuc. 6.34.4; 7.12.3–4), her rowers and landless politically influential in her democracy (6. 17.2–5; 7.55.2), her walls, harbor fortifications, and dockyards substantial (6.75.1–2; 6.99.1–3), her navy excellent (7.55.2), her population large (Thuc. 6.37.1–2), her hoplite infantry poor (Thuc. 6.17.5–6; 6.72.3–4; 6.98.2–4), her census classes not critical to military service (Thuc. 6.72.4), the defense of her countryside less important (6.49.3–4), and her ability to raise capital significant (Thuc. 2.20.4)—in short the precise elements that led to Athenian military dynamism beyond the norms of the Greek city-state.

24. See Sabin, "Athens," 238–246, for a discussion of how Athens in fact did *not* suffer from the supposed timidity inherent in a democratic society. In general, she was more than willing to devote enormous amounts of men and matériel in consistent policies that were void of what we might now call liberal considerations. Vagts, *Militarism*, 76–77, showed how democratic cries for each citizen to do his own part for the state easily lead to conscription, and in fact a greater body of manpower available for warfare. See Evangelista, "Democracies," 226–228, for the advantages of open democratic societies such as Athens and America in meeting changing international situations and technological innovation.

25. On the shortcomings of M.I. Finley's insistence on the primitive nature of the Greek economy and his belief that Aristotle did not understand the nature of an economy, see Meikle, "Aristotle and the Political Economy," 67–73. Cartledge, "Trade and Politics," 13–15, traces the history of the debate. On the market-oriented nature and sophistication of the Athenian economy, see, in general, Cohen, *Athenian Economy*, 3–25, Hodkinson, "Imperialist," 7. Kallet-Marx, *Money*, 18–20, has a good discussion of how surplus wealth at Athens was integral to naval mastery and democratic imperialism and so completely at odds with the values of the traditional Greek city-state.

26. See Wright, *War*, 200, and cf. 842: "In wars of attrition their superior economies give them [democracies] an advantage."

Bibliography

Amit. M. *Athens and the Sea. A Study in Athenian Sea Power*. Brussels: Latomus, 1965.
Bell, Daniel. *The Cultural Contradictions of Capitalism*. New York: Basic Books, 1976.

Cartledge, Paul. *Agesilaos and the Crisis of Sparta*. Baltimore: Johns Hopkins University Press, 1987.

———. "Trade and Politics Revisited: Archaic Greece." In Peter Garnsey, Keith Hopkins, and C.R. Whittaker, eds., *Trade in the Ancient Economy*. Berkeley: University of California Press, 1983, 1–15.

Cashman, Greg. *What Causes War?* New York: Lexington Books, 1993.

Cohen, Edward E. *Athenian Economy and Society. A Banking Perspective*. Princeton: Princeton University Press, 1992.

Connor, W. Robert. *Thucydides*. Princeton: Princeton University Press, 1984.

———. "Early Greek Warfare as Symbolic Expression." *Past and Present* 119 (1988): 3–29.

———. "Polarization in Thucydides." In Richard Ned lebow and Barry S. Strauss, eds., *Hegemonic Rivalry: From Thucydides to the Nuclear Age*. Boulder: Westview Press, 1991, 53–69.

Cumings, Bruce. *The Origins of the Korean War*. Princeton: Princeton University Press, 1981.

Drews, Robert. *The End of the Bronze Age*. Princeton: Princeton University Press, 1993.

Evangelista, Matthew. "Democracies, Authoritarian States, and International Conflict." In Lebow, Richard Ned, and Barry S. Strauss, eds., *Hegemonic Rivalry: From Thucydides to the Nuclear Age*. Boulder: Westview, 1991, 213–234.

Ferrill, Arthur. *The Origins of War*. New York: Thames and Hudson, 1985.

Fornara, Charles W. *The Athenian Board of Generals from 501 to 404*. Wiesbaden: F. Steiner, 1971.

———. "The Hoplite Achievement at Pystalleia." *Journal of Hellenic Studies* 86 (1966): 51–54.

Frost, F. "The Athenian Military Before Cleisthenes," *History* 33 (1984): 283–294.

Gabriel, Richard, and Karen S. Metz. *From Sumer to Rome: The Military Capabilities of Ancient Armies*. New York: Greenwood Press, 1991.

Garlan, Yvon. *Recherches de poliorcétique grecque*. Athens: Ecole Française d'Athèns, 1974.

Gilpin, Robert. "Peloponnesian War and Cold War." In Richard Ned Lebow and Barry S. Strauss, eds., *Hegemonic Rivalry: From Thucydides to the Nuclear Age*. Boulder: Westview, 1991, 31–52.

Gomme, A.W. *Historical Commentary on Thucydides*, vols. I–V. Oxford: Clarendon, 1945–1970; vols. IV–V with Antony Andrewes and Kenneth J. Dover.

Graham, A.J. "Thucydides 7.13.2 and the Crews of Athenian Triremes." *TAPA* 122 (1992): 257–270.

Grundy, George B. *Thucydides and the History of His Age*. Oxford: Blackwell, 1948.

Hansen, Mogens H. *The Athenian Democracy in the Age of Demosthenes*. Oxford: Blackwell, 1991.

Hanson, Victor D. *The Other Greeks. The Family Farm and the Agrarian Roots of Western Civilization*. New York: Free Press, 1995.

———, ed. *Hoplites: The Classical Greek Battle Experience*. London: Routledge, 1991.

———. *The Western Way of War. Infantry Battle in Classical Greece*. New York: Knopf, 1989.

———. *Warfare and Agriculture in Classical Greece*. Pisa: Giardini, 1983.

Hodkinson, S. "Warfare, Wealth, and Crisis of Spartiate Society." In John Rich and Graham Shipley, eds., *War and Society in the Greek World*. London: Routledge, 1993, 146–176.

———. "Imperialist Democracy and Market-Orientated Pastoral Production." *Anthropolozoologica* 16 (1992): 53–61.

Hornblower, Simon. *Thucydides*. Baltimore: Johns Hopkins University Press, 1987.

Joint Association of Classical Teachers (JACT). *The World of Athens*. Cambridge: Cambridge University Press, 1984.

Jameson, M.H. "Class in the Ancient Greek Countryside." In Panagiotis N. Doukellis and Lina G. Mendoni, eds., *Structures rurales et sociétés antiques*. Paris: Belles Lettres, 1994.

Jones, A.H.M. *Athenian Democracy*. Oxford: Blackwell, 1957.

Jordan, Borimir. *The Athenian Navy in the Classical Period*. Berkeley: University of California Press, 1975.

Kagan, Donald. *On the Origins of War and the Preservation of Peace*. New York: Doubleday, 1995.

Kallet-Marx, Lisa. *Money, Expense and Naval Power in Thucydides' History 1–5.24*. Berkeley: University of California Press, 1993.

Keegan, John. *A History of Warfare*. New York: Knopf, 1993.

Krentz, Peter. "Casualties in Hoplite Battles," *Classical Antiquity* 4 (1985): 50–61.

Lebow, Richard Ned, and Barry S. Strauss. *Hegemonic Rivalry: From Thucydides to the Nuclear Age*. Boulder: Westview, 1991.

Lonis, Raoul. *Guerre et religion en Grèce à l'époque classique*. Paris: Belles Lettres, 1979.

Loraux, Nicole. *The Invention of Athens. The Funeral Oration in the Classical City*. Trans. A. Sheridan. Cambridge: Harvard University Press, 1986.

McKechnie, Paul. *Outsiders in the Greek Cities in the Fourth Century B.C.* London: Routledge, 1989.

_____. "Greek Mercenary Troops and their Equipment." *Historia* 43:3 (1994): 297–305.

Meier, C. "Die Rolle des Krieges im Klassischen Athen." *Historische Zeitshrift* 251 (1990): 555–605.

Meikle, Scott. "Aristotle and the Political Economy of the Polis." *Journal of Hellenic Studies* 99 (1979): 57–73.

Millet, P. "Warfare, Economy and Democracy in Classical Athens." In John Rich and Graham Shipley, eds., *War and Society in the Greek World*. London: Routledge, 1993, 177–196.

Munn, Mark H. *The Defense of Attica. The Dema Wall and Boiotian War of 378–375 B.C.* Berkeley: University of California Press, 1993.

Ober, Josiah. "Classical Greek Times." In Michael Howard, George Andreopoulos, and Mark R. Shulman, eds., *The Laws of War: Constraints on Warfare in the Western World*. New Haven: Yale University Press, 1994, 12–26.

_____. *Mass and Elite in Democratic Athens: Rhetoric, Ideology, and the Power of the People*. Princeton: Princeton University Press, 1989.

_____. *Fortress Attica: Defense of the Athenian Land Frontier*. Leiden: E.J. Brill, 1985.

_____. "Thucydides, Pericles, and the Strategy of Defense." In John W. Eadie and Josiah Ober, eds., *The Craft of the Ancient Historian: Essays in Honor of Chester G. Starr*. Lantham, Md: University Press of America, 1985, 171–188.

Pritchett, W. Kendrick. *The Greek State at War*, Parts I–V. Berkeley: University of California Press, 1971–1991.

Romilly, J. de. "Guerre et paix entre cités." In Jean-Pierre Vernant, *Probl`emes de la guerre en Grèce ancienne*. The Hague: Mouton, 1968, 207–220.

Rosivach, Vincent J. "Manning the Athenian Fleet, 433–426 B.C." *American Journal of Ancient History* 10, 1 (1992): 41–66.

Runciman, W.G. "Doomed to Extinction: The Polis as an Evolutionary Dead-end." In Oswyn Murray and Simon Price, eds., *The Greek City: From Homer to Alexander*. Oxford: Oxford University Press, 1990, 348–367.

Sabin, Philip A.G., "Athens, the United States, and Democratic 'Characteristics' in Foreign Policy." In Richard Ned Lebow and Barry S. Strauss, eds., *Hegemonic Rivalry: From Thucydides to the Nuclear Age*. Boulder: Westview, 1991, 235–250.

Ste. Croix, *G.E.M de. The Class Struggle in the Ancient Greek World*. Ithaca, N.Y.: Cornell University Press, 1981.

————. *The Origins of the Peloponnesian War*. Ithaca, N.Y.: Cornell University Press, 1972.

————, ed. *Studies Presented to Victor Ehrenberg*. Oxford 1966, 109–114.

Santoro, Carlo M. "Bipolarity and War: What Makes the Difference?" In Richard Ned Lebow and Barry S. Strauss, *Hegemonic Rivalry: From Thucydides to the Nuclear Age*. Boulder: Westview, 1991, 71–86.

Seabury, Paul, and Angelo Codevilla. *War: Ends and Means*. New York: Basic Books, 1989.

Snodgrass, Anthony M. *Early Greek Armour and Weapons*. Edinburgh: University Press, 1964.

Spence, I.G. "Perikles and the Defense of Attika During the Peloponnesian War." *Journal of Hellenic Studies* 110 (1990): 91–109.

Starr, Chester G. *The Influence of Sea-Power on Ancient History*. Oxford: Oxford University Press, 1989.

Stockton, David. *The Classical Athenian Democracy*. Oxford: Oxford University Press, 1990.

Strauss, Barry S. *Athens After the Peloponnesian War: Class, Faction, and Policy 403–386 B.C.* Ithaca, N.Y.: Cornell University Press, 1986.

————. "The Athenian Trireme, School of Democracy." In Josiah Ober and Charles Hedrick, eds., *Dêmokratia: A Conversation on Democracies, Ancient and Modern*. Princeton: Princeton University Press, 1996, 313–326.

Vagts, Alfred. *A History of Militarism: Civilian and Military*. London: Hollis and Carter, 1959.

Vidal-Naquet, Pierre. "The Tradition of the Athenian Hoplite." In Pierre Vidal-Naquet, *The Black Hunter*. Baltimore: Johns Hopkins University Press, 1986, 85–105.

Westlake, H.D. "Seaborne Raids in Periclean Strategy," *Classical Quarterly* 39 (1948): 75–84.

de Wet, B.X. "The So-Called Defensive Policy of Pericles." *Acta Classica* 12 (1969): 103–119.

Wardman, J. "Tactics and Tradition of the Persian Wars." *Historia* 8 (1959): 49–60.

Williams, William A. *Empire as a Way of Life*. Oxford: Oxford University Press, 1980.

Wright, Quincy. *A Study of War*. Chicago: University of Chicago Press, 1942.

2

The American Imperium

Ronald Steel

For more than half a century the United States has been not only the most powerful nation in the world, but also the one most able to forge its political and economic order. Since 1991, with the demise of the Soviet Union, it has been the undisputed global hegemon, with the power to exert unparalleled influence over the world's destiny.

Since 1940, and its emergence from the guarded neutrality of the interwar period, the nation has fought three major wars: against Japan and Germany from 1941 to 1945, against China in Korea from 1950 to 1953, and against Vietnamese communist-led nationalists from 1963 to 1975. It has intervened directly with military force in a number of nations both hostile and friendly, and it has used its clandestine organizations or proxy armies to install, prop up, or topple governments from Iran to Greece and Guatemala. It maintains controlling alliances with the world's major industrial powers, and a network of bases that encircles the earth.

The American imperium is not a traditional territorial empire in the style of the British, French, or even Soviet empires, for it eschews formal colonies and the direct administration of distant peoples. Rather, its control is indirect: through economic links and military and political dependency. It is a hegemonic imperium.

This imperium is, in its post-1945 extension outside the Western hemisphere, a product of the cold war. It is at once a military, economic, geopolitical, and mental construct that was forged in the conflict across the Korean peninsula. That conflict, which took place only months after the consolidation by the communists in China, and the successful testing of the atomic bomb by the Soviet Union, merits special attention. In it we can find the genesis of the imperial authority that fought, won, and has so far survived the cold war.

Unlike great imperial states of the past, the United States does not have formal colonies. Its preferred military relationship with friendly states is that of alliance. In some cases the allied country, like those in Western Europe, may be fully autonomous. In other cases, as in the island-states of the Caribbean, and the planta-tion-states of Central America, they may be dependencies, limited in the policies they pursue by the importance Washington officials place on their military, eco-nomic and political objectives. Between these two extremes—formal allies and informal colonies—lie a great number of developing (or nondeveloping) states where American influence is extremely powerful or even dominant. The policies they pursue are influenced, and often determined, by how the American govern-ment will react.

In the nineteenth century the exercise of such power would have been deemed to be imperial—and proudly so. In 1895, as the United States was in the process of seizing and annexing the Hawaiian Islands, Secretary of State Richard Olney de-clared, with regard to a border dispute with Great Britain over Venezuela, "Today the United States is practically sovereign on this continent, and its fiat is law upon the subjects to which it confines its interposition." Anti-imperialists considered such a statement to be blatantly jingoistic. Nonetheless, it expressed a simple real-ity. Americans took it for granted and believed it to be natural that the United States would exert preeminence (or what others would call dominance) in the Caribbean and Latin America.

After World War II—when the old global balance of power had been destroyed—the United States emerged not only as the world's greatest military and industrial power, but overwhelmingly so. Its previous rivals—Germany, Japan, Britain, and the Soviet Union—were devastated by war. The ensuing competition in Europe between Washington and Moscow ensured that the United States would not choose political withdrawal, as after World War I, but on the contrary would seek to orga-nize the world along lines congenial to its interests.

This is a policy that other great powers of the past have pursued, and inevitably this has led to accusations of "imperialism." The line between imperialism and great power dominance is not always easy to define. From outside American shores the United States is often viewed as marshalling a degree of influence in other nations that amounts to effective control over key elements of foreign and eco-nomic policy. This is not an accusation so much as the recognition of a condition. If a dominant state has de facto veto power over the policies of a weaker one, it enjoys an imperial relationship to that state.

The American sphere of influence is not regional, but global. It embraces the entire Western hemisphere, the world's oceans, and most of the Eurasian land mass. The United States has the means to destroy whole societies and rebuild them, to create as well as topple governments, to impede social change or to cause it, to defend its friends and devastate its enemies. To this day, although the cold war has long since ended, the nation spends more on arms than much of the rest of the world combined. Without a serious enemy, it maintains what its leaders deem to be important "security obligations," which entail the upholding of an interna-tional political-economic-military system.

There is considerable precedence for this. In the words of George Liska, "An empire, or imperial state . . . being a world power and a globally paramount state, becomes automatically a power primarily responsible for shaping and maintaining a necessary modicum of world order."[1] This indeed has been the goal of American policymakers since the end of World War II. The containment of the Soviet Union was only part of that policy, and the demise of that state has not changed the larger goal. One of the difficulties for American leaders in the post–cold war period is to justify to the American people a level of global engagement that had for so long been linked to communism and Soviet power.

Although the cold war was about power primarily, rather than ideology, its theology was ideological. The ideology of communism was challenged by the counterideology of anti-communism. This counterideology became the justification for a complex of alliances with dependent, but strategically located, states; for extensive programs of military and economic aid; for a global network of bases; for a host of military interventions, sometimes overt, sometimes clandestine; and for two divisive and costly land wars in Asia.

These activities were viewed, of course, not as imperialistic, but as a form of global philanthropy, inspired by altruism rather than by such base motives as power, profit, and influence. American policymakers saw themselves engaged, in the words of Undersecretary of State George Ball in 1965, shortly after the United States launched the bombing campaign against North Vietnam, in "something new and unique in world history—a role of world responsibility divorced from territorial or narrow national interests."[2]

While sincerely disavowing imperial ambitions, American leaders have used classical imperial methods. They have established military garrisons in remote and distant outposts, sustained client governments and their leaders, applied sanctions and military force against recalcitrant states, and sent abroad an army of administrators (in effect, a colonial service) working through various government agencies. What began as the military containment of the Soviet Union and its ideology became transmuted into the establishment of a powerful international system directed from Washington and geared toward the preservation and expansion of American power and influence.

This is not surprising, nor should it be a cause for regret or indignation. This is what great powers do. Like ancient Athens, the United States saw itself not as exerting power, but as promulgating a noble ideal. What America promised to the world—freedom, democracy, prosperity—was promoted with the energy and conviction of a religious faith. In fact, it was a kind of faith. John Foster Dulles recognized this when he noted that all empires had been "imbued with and radiated great faiths" such as "Manifest Destiny" and the "White Man's Burden." Arguing that religious faith was a way to sell America, the secretary of state insisted that Americans "need a faith that will make us strong, a faith so profound that we, too, will feel that we have a mission to spread it through the world."[3]

Such ambitions must be built not only on faith, but on necessity. The nation's outward thrust has, for the past century, rested on the belief that by expansion

America was both promulgating liberty and ensuring security. Thomas Jefferson's vision of an "empire of liberty" that would ultimately reach across the vast continent of an unmapped America has been echoed by every generation of political leaders. When John Kennedy in 1963 declared that "we in this country, in this generation, are—by destiny rather than choice—the watchmen on the walls of world freedom,"[4] he was expressing a powerful motivating force of American interventionism. By advancing the cause of freedom throughout the world, by seeking to "make men free," in Woodrow Wilson's phrase, the expansion of American values and American influence became moral imperatives.

While traditional imperialism has led to economic exploitation, it has also brought material rewards. For better or worse indigenous peoples are yanked into the modern world. This brings literacy, roads, hospitals, and education: the panoply of benefits that liberalism promises. The whole concept of progress over the past two centuries has been linked to imperial expansion. It is virtually an article of faith, even among anti-imperialists, that imperialism provides benefits not only for those who impose, but those who are imposed upon. It was, after all, Karl Marx who asked whether "it was such a misfortune that glorious California has been wrenched from the lazy Mexicans who did not know what to do with it."[5]

Yet it is on liberty, not progress, that the main justification for America's global thrust has rested. The nation intervenes abroad, officials ritualistically declare, for a noble idea: democracy. Indeed, the United States is surely alone among the world's powers in declaring the promulgation of democracy to be a major foreign policy objective. This crusade for democracy has three aspects. The first is an affirmation of the political ideology of America. The nation "conceived in liberty" affirms its qualities by extending them to others, as republican France did in the late eighteenth century by fostering revolutionary movements across Europe. Such ambition is a form of self-affirmation; it is the banner under which the nation legitimizes itself. This sets it apart from less idealistic states. It is in essence evangelical. Liberty is not merely a private blessing, but a grace to be offered, or thrust upon, others. Thus the export of democracy, in the free market version, is central to America's sense of mission.[6]

The second aspect of the crusade for democracy rests on the belief that democratic governments are less prone to aggression (or at least against other democratic states) than authoritarian ones. This assumption has become virtually an article of faith among professors and policymakers. Yet democracy is relatively recent in much of the world, and it may be too early to tell. But we do know that democracies go to war frequently against authoritarian regimes. From this one may conclude, as Winston Churchill said, that democracies are the worst form of government, except for all others, and while their virtues are many, nonbelligerence should not be put at the top of the list.

The third aspect of democracy as a crusading belief is its utility for actions that might otherwise be considered excessively meddlesome, self-aggrandizing, or belligerent. During the cold war, and since, the United States has intervened in the affairs of many nations—whether through invasion, coercion, manipulation, or persuasion—to defend its interests as it saw them. Invariably these interventions were said to be inspired by a concern for moral principle as well as security. Lyndon Johnson expressed this when in 1965, after ordering air attacks on North Vietnam, he declared: "What America has done, and what America is doing now around the world, draws from deep and flowing springs of moral duty, and let none underestimate the depth of flow of those wellsprings of American purpose."[7]

While Johnson's rhetoric may sound insincere, it should not be dismissed. The virtues of democracy in American diplomacy are sincerely believed, by the wider public, if not always by the professionals. Americans want to believe that their crusades are virtuous. Avowedly realpolitik diplomacy of the kind normally practiced by Europeans makes them uncomfortable. That is why they must use a German word even to convey such a concept.

An interventionist American foreign policy rests not only on the impulse to spread democracy, but also on the vague but compelling demands of "national security." Even in the nineteenth century, territorial expansion across the continent was often justified in terms of security, as were interventions in the Caribbean. By the turn of the century Manifest Destiny (which earlier inspired a war of conquest against Mexico), reinforced by the formulas of Alfred Thayer Mahan and other apostles of geopolitics, bolstered Christian uplift with hard considerations of security and defense. The frontier expanded westward in pursuit of "national interest" as well as of trade. The Puritan compulsion to purify American society through the export of religion and civilization, as expressed by such figures as Josiah Strong, was reinforced by Mahan's insistence that the nation build powerful navies to bolster prosperity and ensure invincibility.

While the theologians provided the uplift, the "active carriers" of the imperial thrust, as Arthur Schlesinger Jr. has written, were "politicians, diplomats and military leaders."[8] None was more energetic than Theodore Roosevelt. He viewed expansion as the key to world power, and justified it in terms of safety from foreign predators. "If we shrink," Roosevelt said following the acquisition of Hawaii and the Philippines, "then the bolder and stronger peoples will pass us by, and will win for themselves the domination of the world."[9]

The anti-imperialists responded in terms of morality. As the war to subdue Philippine resistance to the American occupation became ever more contentious and bloody, William James spoke for many in declaring that America was beginning to "puke up its ancient soul and the only thing that gave it eminence among the nations."[10] But such complaints did not quench the belief that the nation's quasi-imperial ventures were justified by higher moral purpose.

With the election of 1912 a moralist entered the White House, one who declared that he would teach the recalcitrant Mexicans "to elect good men," by force if necessary, and who believed that he could "do no other" than to take the nation into the European war. "The right is more precious than peace," declared Woodrow Wilson in justifying his decision to come to the aid of one group of imperial powers over another. The United States would fight in Europe "for democracy, for the right of those who submit to authority to have a voice in their own government, for the rights and liberties of small nations . . . and to make the world itself at last free."[11]

A generation later Franklin Roosevelt, a supreme realist who nonetheless realized the resonance of moral principles upon the electorate, insisted that World War II would be fought for a democratic purpose. In the Atlantic Charter of August 1941 he insisted upon, and a dubious Winston Churchill agreed to, a declaration that their two countries would respect the right of all peoples to choose their own forms of government, and would ultimately restore "sovereign rights and self-government . . . to those who have been forcibly deprived of them."

It was a sweepingly ambitious pledge (from which Britain and France later exempted themselves) that at the end of the war provided the moral justification for Washington to forge a politically and economically congenial global environment. American policymakers were intent on avoiding the detachment from power politics that had marked the interwar period. Officials were convinced that the United States must assume the major responsibility for a new world order. A key element of this order was the global economic structure. The trade restrictions of the 1920s and 1930s were believed to have been instrumental in causing the political crisis that led to war. A peaceful world meant one open to unrestricted trade and investment. This was of course particularly important to the United States as the world's major trading nation.

Policymakers took their cue from previous imperial powers. During the nineteenth century Britain had exported its capital and provided a stable environment for a global trading system from which it grew rich. American leaders were now ready to perform a similar task for the United States as Britain's heir to world power. American prosperity was linked to that of the entire capitalist system. This required an open economy incorporating all nations. The problem posed by the Soviet Union (apart from its political domination of Eastern Europe) was its refusal to participate in such a system. Instead of open markets it pursued protectionism; instead of the free flow of capital and investment it maintained planned economies. Thus the political conflict that emerged in Europe after 1945 had an economic as well as a political basis.

In pursuing their goal of an open international system, American policymakers understood that political nationalism posed a major threat. It impeded trade and investment. The system worked most smoothly when directed by a single great power, or hegemon. The hegemon would ensure security for all, thereby preventing the degeneration of national rivalries into war. In order to play this role the hegemon would have to maintain military superiority over its rivals, and political superiority over other states within the system.

American officials have supported this reasoning, albeit in somewhat different form. Dick Cheney, secretary of defense in 1992, argued that the United States "cannot afford to step down from our world leadership role" because the "worldwide market that we're part of cannot thrive where regional violence, instability, and aggression put it at peril. Our economic well-being and our security depend on a stable world in which the community of peaceful, democratic nations continues to grow."[12]

A confidential Pentagon planning document of the same year, later leaked to the press, made a similar, if starker, argument. The United States, the planners insisted, must "discourage the advanced industrial nations from challenging our leadership or even aspiring to a larger regional or global role." This was justified both on security grounds and by the argument that prosperity and trade require tranquility, and tranquility can best be assured by a benevolent hegemon keeping the peace. In the Pentagon's language, the United States must "retain the preeminent responsibility for addressing . . . those wrongs which threaten not only our interests, but those of allies or friends, or which could seriously unsettle international relations."[13]

The breakdown of the wartime alliance between Washington and Moscow, reinforced by fears that radical forces of the left might gain control of some governments of Western Europe, led American policymakers to intervene politically, economically, and ultimately militarily. The first arena was Greece, where communist-led factions threatened the conservative, British-supported government. In March 1947 president Harry Truman urged Congress to provide aid for the beleaguered regime. But he also phrased the request in far more expansive language, declaring that "it must be the policy of the United States to support free peoples who are resisting attempted subjugation by armed minorities or by outside pressures." In this sentence lay the doctrine that was to become the formula for intervention globally against radical and revolutionary forces.

Few at the time realized the full implications of the new doctrine. Columnist Walter Lippmann, however (although supporting the aid program to Greece and Turkey) warned that a "vague global policy, which sounds like the tocsin of an ideological crusade, has no limits."[14] Initially American postwar activism remained largely confined to Europe. That same year the administration announced the Marshall Plan for the economic reconstruction of Europe. Because the Soviet Union refused to participate or to allow its satellites in Eastern Europe to do so, this resulted in the economic and political partition of the continent. In 1949 Washington expanded this economic plan into the military realm by forging the North Atlantic Treaty Organization: an integrated military alliance by which the United States organized and guaranteed the defense of Western Europe.

By the end of the decade the major elements of American policy toward Europe came together. But that policy was not yet fully global. This occurred in 1950 with the Korean War. The sense of urgency generated by that conflict made it possible to implement the provisions of a planning document known unofficially as NSC-68. Its purpose was to forge a consensus within the bureaucracy regarding a more expansive and confrontational policy toward the Soviet Union. Its prognosis was grim, its language florid and rhetorical, for its purpose was to persuade key figures in the bureaucracy of the need for a more militant stance and a greatly expanded military force.

Soviet leaders, the document's State Department authors argued, were different from "previous aspirants to hegemony." Powerfully "animated by a new fanatic faith antithetical to our own," they sought the "complete subversion or forcible destruction of the machinery of government and structure of society in the countries of the non-Soviet world." To deal with this challenge the only American response should be the "rapid buildup of the political, economic and military strength" of the noncommunist world. Faced with a determined will and a superior military force, the Kremlin would be compelled to "accommodate itself, with or without the conscious abandonment of its design, to coexistence on tolerable terms with the non-Soviet world."[15]

To implement this plan the authors urged a fourfold increase in the military budget, then at $13 billion. The hyperbolic language of NSC-68 was designed to provide the incentive for such measures. It was, as stated by Secretary of State Dean Acheson, intended "so to bludgeon the mass mind of 'top government' that not only could the president make a decision but that the decision could be carried out." As Acheson later admitted in his memoirs, to "explain and gain support for a major policy" it is sometimes necessary to make arguments that are "clearer than truth."[16]

Initially NSC-68 was simply a bureaucratic proposal. Only a few months later, however, in the summer of 1950, it became a course of action. The stimulus was the outbreak of war between the two Koreas. The administration responded by direct military intervention in the conflict, and Congress provided the money that soon led to the quadrupling of the military budget that NSC-68's authors had desired. "Truman's prompt response to the attack," McGeorge Bundy wrote some years later, "helped make a reality of the American-led defense of the West that had been only a matter of political alliance and secret planning papers before June 25."[17] Or as Acheson himself later wrote, Korea "came along and saved us."

The Korean conflict transformed the cold war. Until then the fear of communism had been remote and based on ideology. But after American troops began fighting first North Korean, then Chinese communist armies, it became tangible and urgent. The threat of communist aggression no longer was merely rhetorical. Within hours after the fighting began between the two Koreas, Truman sent military aid to the South. Further, in a step heavy with repercussions, he dispatched the Seventh Fleet to defend Chiang Kai-shek's forces on Formosa, and extended aid to France for use against the communist-led national independence struggle in Indochina. Acheson called the United Nations Security Council into emergency session and obtained from it (in the absence of the Soviet delegate) a resolution branding North Korea as the aggressor. Within less than a week American ground troops had been dispatched to the battle—locked in combat against the extension of communist control over a terrain never previously considered to be central to America's own security interests.

Ultimately an estimated 3 million people died in the war, including 37,000 Americans. The conflict put the United States into a war with China, a nation on which Americans had long looked benevolently and protectively. The decision to protect Chiang's defeated army on Formosa involved the United States directly in China's ongoing civil war. And the despatch of aid to the French launched the long slide toward America's own later war against communism in Vietnam.

The Truman administration presented the war as a direct challenge to the United States by the Soviet Union operating through an obedient proxy. This explanation, however, ignored the local causes of what was above all else a civil war.[18] The conflict was not only between the North and the South, but within South Korea itself. As in Vietnam later, a civil war became an international war when outside forces intervened. This is not to lessen the importance of outside involvement and encouragement before the outbreak of hostilities. The recent opening of the Soviet archives reveals that the North Korean leader, Kim Il Sung, sought and received the guarded approval of Stalin before launching his attack against the South.[19] Soviet premier Nikita Khrushchev stated that the Russians had only reluctantly given the green light after having assured themselves—on the basis of statements made by Acheson himself—that the United States did not consider Korea within its "defense perimeter." Another view was offered by George Kennan, who later wrote that Soviet approval may have been governed by the American decision in 1950 to sign a separate peace treaty with Japan that locked out the Soviets and turned the islands into an American military base.[20]

Parallels could be drawn with the Peloponnesian Wars, viewing North Korea as Corinth and China as Sparta, with Japan (which had just signed a military alliance vital to the United States) as Corcyra. The Soviet Union, which had given the green light to the North Koreans only on the assumption that the Americans would not intervene and quickly ceded the role of protector to China, is more analogous to Persia than to Sparta. Since Washington viewed Moscow and Beijing as a double-headed Sparta, it responded to the challenge as Athens to preserve its position as hegemonic guarantor.

The American decision to enter the war and then, once the North Korean invasion was repulsed, to unify the country by force has long troubled students of the event. Acheson, in a January 1950 speech, had seemed to leave South Korea in a strategic limbo, leading Stalin and Kim Il Sung to believe that the Americans would not intervene. The reason for the first intervention seems to have been rooted in concerns about credibility. The administration feared that even though South Korea might not have been strategically vital, a failure to react would make both the Chinese and the Soviets more bold in other areas. Once the event occurred, the pressure to intervene was overwhelming, and any other course of action became improbable.

The second intervention—the American invasion of North Korea—rested on opportunity. The invading army was in flight, the Pyongyang regime in a state of disarray, the Russians eager to avoid any entanglement in an affair that had gone wildly out of control, and the Chinese too weak (or so Washington thought) to do anything but bluster. Furthermore, the administration, smarting under Republican attacks that it had been insufficiently militant against communism, wanted to demonstrate that the Democrats—not their rivals—were the ones who were able to "roll back" communism, rather than merely talk about it. The temptation to destroy the reeling communist regime in the north was all but irresistible—particularly because it promised to be so easy.

The Korean conflict was a turning point in the cold war. It changed the conflict with communism from an ideological abstraction to a costly and impassioned war. It cemented the anti-communist consensus that was to be the hallmark of policy for decades. It effectively silenced dissent from the liberal left. It destroyed any possibility of enlisting China as a balancing force against the Soviet Union—at least for the next twenty-two years. It led to a vast expansion of American geopolitical engagement and military power. And it shut off conservative critics of an expanded militarized republic, such as Senator Robert Taft. With the outbreak of the Korean War a galvanized and frightened Congress passed the funding for the expansion of American military power urged by the authors of NSC-68. The Korean War was not the cause of all that followed, but it was where the cold war became deadly earnest.

Before Korea containment was viewed as a problem that concerned Europe; after Korea it went global. This had been presaged by the Truman Doctrine of 1947, but it became activated only after the United States entered the fighting in Korea. The Korean War enlisted the entire nation behind a program that had hitherto been confined to policymakers and strategists. It created the cold war consensus. This is why Acheson believed that Korea "came along and saved us." Korea was a tragedy in one sense, but an opportunity in another.

Not only did it make American foreign policy far more expansive—and expensive—but it also stimulated the creation of what Eisenhower called a "military-industrial complex" with needs and ambitions of its own. And in making foreign policy the highest priority it changed the nature of the American government. Putting the nation on a semi-war footing helped turn a largely decentralized state into a far more centralized one. Economic and political decisions once made in regional capitals gravitated increasingly to Washington. Just as the federal government became the nation's largest employer, so it also provided subsidies for a variety of industries and created the military-industrial configuration that troubled Eisenhower.

The growth of government and the centralization of power enhanced the authority of the presidency. Whereas war once required a declaration by Congress, the United States fought two major wars, in Korea and Vietnam, by presidential order. The expansion of American power in the name of containment meant that millions of Americans served in distant foreign outposts—whether in the armed forces or in one of the quasi-civilian agencies in the related fields of economic development, intelligence-gathering, professional training, cultural influence, labor organization, and police instruction. It was not until the end of the cold war, and the decrease of the military budget, that the relative authority of the executive branch over the legislature began to narrow.

The increasing strains put on the American imperium from the time of the Vietnam War until the collapse of the Soviet Union should not obscure its remarkable success. Its primary declared objective, the containment of Soviet and Chinese power, was achieved. The ideology promulgated by those states was also largely contained, with a few exceptions in the Third World. In Western Europe and Japan communism was defeated and discredited as a political force. The Soviets were confined to a part of Europe of minor economic significance that ultimately became a drain on their own economy. Furthermore, they were excluded from the vital oil-producing areas of the Middle East.

In Asia the success of the American imperium was equally great. Japan, Korea, Indonesia, and India became integrated into the global capitalist system. The communist movement in Indonesia was destroyed in a quasi-genocidal bloodbath. Indigenous revolutionary movements in Malaya and the Philippines were wiped out, with American assistance. Even China, although remaining Stalinist, broke from the Soviet Union and became bitterly inimical—despite the refusal of American officialdom for years to recognize this growing enmity. Only in Vietnam were American objectives seriously defeated. But that was self-inflicted, caused by Washington's own ideological blinders. Had the United States allowed Ho Chi Minh to unify the country, it would not only have spared itself a futile and self-destructive war, but would have advanced its own larger objective of containing China.

While the cold war raged throughout key areas of the Third World, the major players—America, Russia, and China, after the stalemate in Korea—went to considerable pains to avoid direct confrontation. American policymakers were intent on preventing a repetition of Korea during the Vietnam War and studiously avoided the provocation of China. Similarly, after the near-miss of the Cuban missile crisis of 1962, the United States and the Soviet Union assiduously confined their interventions to the periphery. Nowhere did they directly challenge one another's sphere of influence, as was demonstrated by Washington's acceptance of Moscow's repression in Budapest and Prague.

This raises an interesting question about the cold war: not why it finally ended when it did, but why it lasted so long. Why did America and Russia, having worked out the ground rules of détente in the 1970s, not simply call a truce that would have spared them costly proxy wars and the virtually uncontrolled expenses of the arms race? Why did the West Europeans and the Japanese not use their newly acquired and constantly growing wealth to gain military independence from their protector? Why did the Americans not support, rather than struggle at great cost to prevent, the unification of Vietnam in order to constrain China? Why did Washington, as the balance of payments steadily worsened and the burden increased, continue to insist on paying much of the cost of its allies' defense?

There are a number of reasons. First, the cold war allies were largely content with American military protection, and the burdens of such hegemony were relatively light. They did not have to fight in America's colonial wars, and were able to concentrate on what concerned them most: the growth and enrichment of their own societies. America's "free defense" paid for the European welfare state.

Second, both Washington and Moscow found advantages in the cold war system. For the Soviets it ensured a free hand in Eastern Europe, kept the Germans divided and in check, and allowed a country with a backward civilian economy to pose as an equal to the United States. Only in the currency of the cold war—weapons of mass destruction—could one speak of "twin superpowers," and naturally this was congenial to the Kremlin. For the Americans the cold war offered military dominance and political leverage over such potential rivals as Germany and Japan, fortified the domestic economy with infusions of Pentagon dollars, and sustained the integrated global market economy that was the goal of America's postwar planners.

Third, empire is rewarding psychologically, even when costly economically. The commissars in the Kremlin found great satisfaction in dominating a ring of obedient European satellites, in possessing client states on every continent, in making the haughty Americans take notice and on occasion even tremble. Similarly, the Americans, in thrall to their own ideology of market democracy, enjoyed reaping the rewards of leadership of what they called the Free World.

Finally, empire, once acquired, is not easy to relinquish. It has its own logic. To lose control over clients, dependencies, or allies—however defined—means the loss of power and status. It means that one has to accommodate to the will of others. An empire requires both power and determination. Once the imperial power weakens, once it loses the power to impose its will, the entire structure is at risk. Consider the collapse of the British empire following 1945, or the even more stunning disintegration of the Soviet empire, and even the Soviet state, after 1989. To whittle away at an empire is to risk losing it all.

Was there in fact truly an American empire? Certainly millions of others thought so, even if Americans, who have an aversion to the term, did not. In the absolute power it commanded over allies, and with which it threatened foes, it was greater than any empire of the past. Excluding the territories of Russia and China, it knew virtually no limits, dominating not only poor and weak states but also the proud nations that preceded it in imperial grandeur.

Some, who take exception to the term, argue that the United States was not an imperial power, but merely a great power. The line is not always clear. Raymond Aron maintained that there is "only a difference of degree between an 'imperial' diplomacy and a great power diplomacy," and that every great power tends toward an imperial diplomacy with regards to its allies. Aron also observed, quite correctly, that the American empire came about in part by invitation: at least as concerns Western Europe and Japan. With regard to these nations, the empire, or imperium, was—and is—certainly a voluntary one. The allies received protection in return for a certain degree of acquiescence to American political and military objectives. These were never so demanding as to tempt them seriously to rebel.

One could argue that the relationship between Washington and these advanced industrial states has a parallel in the early years of the Delian League, with the United States as leader of "autonomous allies who took part in common synods." One should also note, however, that withdrawal from this cold war league was not easily tolerated. When in 1948 it seemed as though the Italians might vote a communist government into power, the United States—through bribery, coercion, and even the threat of a military coup—took steps to ensure that this would not happen.

The effort to maintain friendly governments in the Third World was, however, far more difficult. On numerous occasions during the cold war the American government subverted, intimidated, and overthrew governments not to its liking. The list includes Iran, Guatemala, South Vietnam, Grenada, Nicaragua, Guyana, and Chile. When Henry Kissinger observed, with regard to the electoral success of Marxists in Chile, that the United States should not be expected to sit back and let other countries behave irresponsibly, he was expressing the logic of imperialism.

Where this leaves us is with the conclusion that an imperial power may behave with magnanimity, just as a great power may behave imperially. Often the decision is governed by the stakes involved. The terms are not mutually exclusive. If a great power can dictate the domestic and foreign policy of an allied or client state—either by withholding support or through active subversion—it is in an imperial relationship with that weaker state. This has been the relationship of the United States with a number of Third World countries, and even such European ones as Greece, throughout the cold war. In this sense we can speak of an American "empire" or imperium.

It is true that the United States has behaved with considerable forbearance toward its stronger allies. It has made trade and economic concessions that have been very costly to the American people. But the purpose of these was to make certain that such states would not try to withdraw from the political and military alliance. Washington would protect their interests for them, thereby making it unnecessary for them to be fully independent militarily and diplomatically. Maintaining a voluntary empire can be extremely costly—as can be seen in the periodic quarrels with Japan over trade balances and market access. Whenever there has been a showdown, American presidents have given in, preferring to placate Japan at America's expense for fear of stimulating a drive toward Japanese military independence. Once an imperial relationship is established, it is difficult to relinquish it.

What is particular about the American empire is its avowedly democratic vocabulary. The notion of an "empire of liberty" has remained a powerful one in American history and rhetoric. While the dedication to democracy has been honored in the breach as well as in the observance (the interventions in Iran and Guatemala come easily to mind), there has been a desire—when sphere of influence and great power factors are set aside—to encourage friendly democratic governments. The cold war interventions, like those that preceded and followed them, have always been justified in terms of a higher moral purpose. That purpose is usually described as the promulgation of democracy, or liberty.

It is striking that during the very time that the United States government was engaged in a distant and unpopular war to maintain an American outpost in Southeast Asia, Americans displayed an extraordinary degree of freedom and opposition. The struggle to maintain the empire did not, during the tempestuous 1960s, imperil the liberties of Americans. Whatever policies the American government pursued abroad in this period, it respected democracy and dissent at home. This is in sharp contrast to the political hysteria of the McCarthy period, which was synonymous with the Korean War. By the time of the second war of hegemonic empire, the effort to maintain the protectorate in South Vietnam, dissent was legitimatized and no longer equated with treason.

The question raised in Vietnam for dissenters was whether the government, in an effort to maintain the hegemonic empire, had betrayed the larger interests of the state, and even the deeper values of the society. Ultimately the government had to readjust its objectives to the level of public tolerance. In Korea, by contrast, dissent was mainly from those who felt that the war was not being fought with sufficient ardor. Thus the Korean War—which won congressional and public support for the creation of a national security state of global dimensions—was the forge of America's informal empire. For this reason it is of critical importance to an understanding of all that followed in its wake.

Notes

1. Liska in Raymond Aron, *The Imperial Republic* (Englewood Cliffs: Prentice-Hall, 1974), p. 303.
2. Ball, "The Dangers of Nostalgia," *Department of State Bulletin*, April 12, 1965, pp. 535–36.
3. Dulles in Ronald Pruessen, *John Foster Dulles: The Road to Power* (New York: Free Press, 1982), p. 200.
4. Kennedy, from an undelivered address, Dallas, November 22, 1963.
5. Marx in Richard Barnet, *The Rocket's Red Glare* (New York: Simon and Schuster, 1990), p. 112.
6. Tony Smith, *America's Mission: The United States and the Worldwide Struggle for Democracy in the 20th Century* (Princeton: Princeton University Press, 1994), p. 348.
7. Johnson, in Ronald Steel, *Pax Americana* (New York: Viking Press, 1967), p. 6.
8. Schlesinger in Barnet, *Rocket's Red Glare*, p. 114.
9. Roosevelt in Barnet, *Rocket's Red Glare*, p. 114.
10. James in Barnet, *Rocket's Red Glare*, p. 137.
11. Wilson, address to Congress, April 2, 1917.
12. Cheney, "The Military We Need in the Future," *Vital Speeches of the Day* 59, no. 1 (October 12, 1992), p. 3.
13. Patrick E. Tyler, "U.S. Strategy Calls for Insuring No Rivals Develop," *New York Times*, March 8, 1992.
14. Lippmann, March 15, 1947, in Ronald Steel, *Walter Lippmann and the American Century* (Boston: Little, Brown, 1980), p. 348.
15. *Foreign Relations of the United States 1977*, vol. 1, pp. 126–492.
16. Acheson, *Present at the Creation: My Years in the State Department* (New York: Norton, 1969), pp. 374–75.
17. Bundy, "The Korean War 40 Years Later," *New York Times*, June 25, 1990.
18. On the domestic political factors see Snyder, *Myths*, pp. 293–96; and Bruce Cumings, *The Origins of the Korean War*, vol. 2, "The Roaring of the Cataract, 1947–1950" (Princeton: Princeton University Press, 1990), ch. 13.
19. See "The Cold War in Asia," *Cold War International History Project*, issues 6–7,

winter 1995/1996, pp. 30–125. Woodrow Wilson International Center for Scholars, Washington, D.C.

20. Kennan, *Memoirs 1925–1950* (Boston: Little, Brown, 1967), p. 498. For a full account of the war see Jon Halliday and Bruce Cumings, *Korea: The Unknown War* (New York: Pantheon Books, 1988).

3

The American Empire: A Case of Mistaken Identity

Robert Kagan

Of the many interesting points raised in Professor Ronald Steel's paper, I found the most intriguing in the very first paragraph. For there he suggests that the American empire, for so long only dimly perceived by the American people, has now ceased to exist or at least has changed its nature and become not an empire but a hegemony. It is fair to say that Steel has come not to praise the American empire but to bury it, and if he is not quite dancing on its grave, he is also not mourning its departure.[1]

In the case of the untimely demise of the American empire, I am led to ask, in the spirit of criminal forensics, what were the circumstances of death? For in the answer to that question, I believe, lies the answer to the more fundamental question: What exactly was it that has died? Was it an empire or was it something else?

Immediately, of course, we run into the problem of definitions. For we often have a hard time determining just what constitutes an empire when the usual model—of a metropolitan power controlling conquered lands through the stationing of occupying forces and the operation of a colonial administration—is not applicable. And such a model surely has not applied consistently to the United States. The United States was an empire in this technical sense for some years after the Spanish-American War, but it actually exercised less global control then, when it clearly was an empire, than it did after World War II, when it was less obviously so.

Professor Steel, himself, employs a variety of terms to describe the United States. He refers to it variously as an empire, a global empire, an "imperial power, possessed of an informal empire," a "hegemon," and a "great power." And he uses an equally varied set of terms to describe the other powers in America's orbit, calling them "colonies," "dependent allies," "ostensible allies," "military allies," and "cold war allies." Despite his admirably dogged insistence that the United States was "certainly an imperial state," therefore, one senses an uncertainty at the core of Steel's argument and a recognition that matters may be more complicated.

In order to get to the heart of the problem, it is worth recalling Raymond Aron's point that there is "only a difference of degree between an 'imperial' diplomacy and a great power diplomacy, and that every great power tends toward an imperial diplomacy with regard to its allies." Perhaps one reason for Professor Steel's confusion is that a great power can, indeed, behave very like an imperial power.

Thus Steel asserts that the United States has used "classical imperial methods" in its foreign policy. It has established military garrisons in distant outposts, applied sanctions and military force against recalcitrant states, employed "a veritable army of colonial administrators" (by which he means diplomats of the foreign service, AID, Peace Corps), sustained client governments and their leaders. But which if any of these methods belong exclusively to empires? All countries have diplomats overseas. Many provide economic aid and assist the development of less developed countries. Most countries apply sanctions. Almost all great powers throughout history have used military force against states they deemed recalcitrant, and quite a few have for a variety of geostrategic reasons maintained troops on foreign territories. As for sustaining client governments and their leaders, this too has been a practice of great powers throughout the centuries. Steel's definitions, in short, take us nowhere.

The most useful distinction to be made between a great power, or even a superpower, and an imperial power is the relationship between the alleged imperial master and its alleged subjects. The historical analogy drawn in this conference is particularly useful in illustrating this point, for one of the powers we have discussed managed at different times to be a great power and then an imperial power. And it was easy to tell which was which, for both subjects and master.

In the early years of the Delian League, Thucydides tells us, the Athenians were "leaders of autonomous allies who took part in common synods." But over time, as one scholar recounts, the Delian League was transformed from "voluntary league to involuntary empire ruled by Athens"—a transformation symbolized by the very significant movement of the League's treasury from Delos to the Acropolis in Athens in 454/453. By the time the Peloponnesian War broke out, Thucydides reports, "Good will was thoroughly on the side of the Spartans, especially since they proclaimed that they were liberating Greece. . . . So great was the anger of the majority against the Athenians, some wanting to be liberated from their rule, the others fearing that they would come under it" (Thuc. 2.8). The Athenians were aware of the problem. Pericles warned his fellow citizens in 430 of the potential loss of empire and of "the danger from those in the empire who hate us. . . . For by now you hold this empire as a tyranny, which it may have been wrong to acquire but is too dangerous to let go" (Thuc. 2.63).

The comparison with Athens leads us to ask, Which sort of relationship existed between the United States and the nations in its orbit? Was it more like the early years of the Delian League, that is, the United States as leader of "autonomous allies who took part in common synods"? Or was it more like the later involuntary empire, in which Athens ruled over those who despised its rule? A few quotations from Steel's own paper provide a clear answer to these questions. For although in one breath Steel declares that "In the *absolute power* it commanded [the United States] was far greater than any empire of the past" (emphasis added), and that for a time "it virtually ruled the world," in many subsequent breaths he completely undercuts this confident assertion.

The story Steel tells of the empire's demise, in fact, is one in which the alleged "subjects" routinely appear more in control of their destiny, more capable of freely choosing the policies they wish to pursue, than the alleged "master," which in turn appears locked on a self-destructive course without the freedom to deviate even enough to save itself. Thus Steel writes, for instance, that when America's relative economic strength—"the essential foundation of hegemonic power"—began to erode in the late 1960s, the "full effects of this were masked by the military dependency of client states—*a dependency that the Europeans and Japanese perpetuated because they found it economically and politically advantageous to do so*" (emphasis added). Steel then correctly asks, "Why did the West Europeans and the Japanese not use their newly acquired and constantly growing wealth to gain military independence from their protector?" And he provides the correct answer: the cold war allies *did not want to take over their own defense burden.* They wanted to rebuild their economies to emulate, and one day overcome and challenge, the U.S. And, indeed, Steel argues that at the end of the Cold War, America's allies were the "real victors" and the United States the real loser.

The American response to the allies' strategy to overthrow its alleged imperial rule, meanwhile, was not merely to allow the allies to grow strong but to encourage them to do so. The United States, Steel reports, "behaved with remarkable forebearance toward its dependent allies, enduring slights and inconveniences of the sort that earlier seats of empire found intolerable and punished accordingly." In the process, according to Steel, the United States bought victory not for itself but for the capitalist system in which it could no longer be the dominant member. And it did so "at the price of its own long-term hegemony." It does not occur to Steel that the United States he has described is not an Athens as imperial master, but an Athens as a "leader of autonomous allies."

If one takes Steel's analysis seriously, however, then in determining the circumstances of death in the case of the American empire's demise, we come to some very interesting conclusions.

The *cause of death*, it seems, was the rise of allies, who were autonomous enough to choose and carry out their strategy for one day overcoming U.S. power. The victim put up no struggle, however. The United States did not adjust to dangerously changing circumstances by asserting its dominance. As Steel rightly points out, "An effective hegemon must maintain military superiority against its rivals and political dominance over the other states within the system, especially its ostensible allies." He suggests that the United States was not able to maintain this dominance, but he provides sufficient evidence to show that, in fact, the United States did not even try.

Indeed, by this interpretation, the cause of death might be described most accurately as suicide, since it was the very heart of U.S. Cold War strategy to encourage this economic growth and the ability of America's allies to stand on their own two feet. When they did stand on their feet and carry out their alleged strategy to overleap American power, the American response was encouragement during the Cold War and indifference afterward. The end of the Cold War was attended in the United States by a cry for a "peace dividend" consisting of vast reductions in defense budgets, and by a renewed tendency toward isolationism and retrenchment and "burden-sharing" by allies.

The *time of death* Steel identifies as the collapse of the Soviet Union. This, too, ought to be a strong indicator of what the U.S. relationship to its allies was all about. But Steel is as confused on this important question as he is on many others. He frequently states in his paper that anti-communism was merely America's justification for empire-building. He cites the Korean War as the turning point in American imperial ambitions, arguing with the revisionists that the war was really only a trick by Paul Nitze and others to force the implementation of the huge defense budget increases called for in NSC-68, which was a recipe for empire for its own sake. In the first section of his paper Steel writes that the American empire was "built, like most empires, on profit, power, and order. But it was cemented together during the cold war . . . by anti-communism." That in turn was the "justification" for the complex of alliances.

Later in the paper, however, Steel contradicts himself again, for he writes that the "primary objective" of the empire was "the containment of communism," and in this, he declares, it was attended by success. And it was precisely the success of the endeavor, "the collapse of [the alliance's] great rival," that brought about the empire's demise by removing its purpose. In short, the alliance, or the "empire," as Steel wants to call it, was not merely justified by anti-communism. It was in fact *about* anti-communism.

The alleged demise of the alleged empire came only after "the disappearance of the security threat that *persuaded* allied states to *grant the U.S. such military and economic dominance* over them" (emphasis added). Once again, in the attempt to show an empire at work, Steel winds up describing instead another Delian League in its early "voluntary" phase. Just as the Delian League members voluntarily submitted to Athenian leadership in the struggle against Persia, so America's allies "grant[ed]" the U.S. dominance over them in the fight against the Soviet Union.

If you put together these two bits of evidence concerning this crime—the cause of death and the time of death—you are left with one conclusion: that the victim has been misidentified by the coroner. What died in the aftermath of communism's demise was not an empire but a voluntary league of autonomous states led by the strongest of them aimed at a specific purpose. When that league's purpose disappeared, the allies voluntarily took their own course.

As with all entertaining murder mysteries, however, even that conclusion is wrong and must be replaced by a still more startling realization: There was no murderer. There was, in fact, no murder. And to top it all off, there was no death.

Despite Steel's hope to dance on the grave of the American "empire," it is alive and well and living under another name. At the end of the Cold War, the United States remains the world's most powerful nation, the sole surviving superpower. Its allies remain allies. They seek America's continued involvement in the world and fear only its withdrawal, in stark contrast with Athens' problem at the outbreak of the Peloponnesian War. They want it to maintain its military garrisons (in Europe and in Asia), and its colonial administrators (in Latin America and Africa). They want it to exercise sanctions, both economic and military, against recalcitrants in Haiti and the Persian Gulf. They want it to support clients, in Sarajevo.

And where are the most angry opponents of continued American predominance? Not in Bonn, Paris, London, Rome, or Tokyo. Not in Seoul or even in Managua. Some can be found in Moscow, more in Beijing and Pyongyang, Baghdad and Teheran.

But, of course, the largest number of opponents of continued American predominance can be found right here, where the calls for a return to isolationism come from Republican isolationists and liberal academics alike. The worry of Athenians was that everyone wanted to escape their rule or avoid being swept into their empire. The worry of many Americans is that everyone wants to join their so-called empire, especially those just recently liberated from the genuine empire of the latter half of the twentieth century.

Let us not quibble over words. Referring to the American "empire" may indeed be useful as a way of understanding how the American system of global influence works. In 1983 the historian John Lewis Gaddis suggested that comparative studies of past empires might be helpful in understanding and explaining American behavior.[2] Such an examination is worthy, as long as we realize that the comparison may also be distorting and misleading.

The United States has, since its earliest days, sought to shape the world around it in ways congenial to the preservation of its security, which it has defined broadly, and to its philosophical proclivities. This, I would argue, does not set it apart from most nations or city-states anywhere in the world today or in history. Where it differs from other nations is in the economic, military, and cultural power it has amassed, only partly by design, and in the immense force of the ideals it stands for. That combined power puts it in a special class in history. It can be compared to only a handful of other great powers the world has known, and most if not all of these were empires. And that is why, in trying to understand the behavior of the United States in the world, it has been instructive to compare it to those empires. It shares much in common with them.

But if our goal is analysis and understanding, as opposed to merely advocacy and name-calling, we must be careful not to stretch the comparison too far. The United States, while wielding immense power, has done so in its own way. It has had neither the need nor the proclivity to behave as past empires. And in fact the differences may be more important than the similarities. The historian Paul Kennedy looked at the world in the late 1980s and believed he saw two overstretched empires. He claimed he could not tell which of them would fail first. Perhaps his error was in not seeing that the Soviet Union possessed an empire in the traditional sense, but the United States did not.[3]

Careful comparisons of the United States and previous empires would yield at least one important difference. Empires' strategies and focus of attention tend to be much more fixed by their colonial possessions than is America's global hegemony. America has famously turned its attention on and off in various parts of the world. Perhaps the United States has proven uniquely capable of responding so effectively to a variety of different challenges precisely because it is not bogged down by the need to preserve its empire. And this is how it has avoided imperial overstretch. Indeed, although it is still too early to tell, the United States may have stumbled upon a means of preserving its power and influence that is more successful than any empire. That, at least, would be my prediction.

And I believe it is that power and the exercise of it to which Steel objects. In so objecting he represents that part of the American soul which has always been suspicious of power, even when wielded by the United States itself, and for some, especially when wielded by America. That is the subject for another discussion. I would like to close this discussion with a quotation from Walter Lippmann, the man whose ideas about American foreign policy Steel so brilliantly sketched for us in his biography of Lippmann, *Walter Lippmann and The American Century*. Writing in the 1920s on the tortuous debate then occurring in the U.S. about America's proper role in the world, Lippmann castigated both sides: those who raised a "great outcry about imperialism," and those "hard-headed" types who sought clear and narrow definitions of American self-interest. "There can be no remedy for this," Lippmann wrote, "until Americans make up their minds to recognize the fact that they are no longer a virginal republic in a wicked world, but they are themselves a world power, and one of the most portentous which has appeared in the history of mankind. When they have let that truth sink in, have digested it, and appraised it, they will cast aside the old phrases which conceal the reality, and as a fully adult nation, they will begin to prepare themselves for the part that their power and their position compel them to play."[4]

Notes

1. Delivered at the conference, this paper responds to an earlier draft of Ronald Steel's "The American Imperium." Although Steel has responded to some of my criticisms in his chapter herein, the fundamental difference between the two of us remains: Steel considers America an empire while I do not.

2. John Lewis Gaddis, "The Emerging Post-Revisionist Synthesis on the Origins of the Cold War," *Diplomatic History* 7 (Summer 1983): 182–83.

3. Paul M. Kennedy, *The Rise and Fall of the Great Powers: Economic Change and Military Conflict from 1500 to 2000* (New York: Random House, 1987).

4. Ronald Steel, *Walter Lippmann and the American Century* (Boston: Little, Brown, 1980).

Part II

Categorizing Wars: Civil or Hegemonic, Decisive or Cyclical?

4

When Sparta Is Sparta but Athens Isn't Athens: Democracy and the Korean War

Bruce Cumings

What Is Democracy?

We live in impoverished times for democracy. Political scientists like Adam Przeworski have narrowed the definition to say that democracy exists if another party can win an election; nothing is said about the quality of that democracy, or the character of the citizens the process depends upon and produces. Joseph Schumpeter also preferred a limited form of democracy, yielding a circulation of party elites, and little participation for the electorate—just some occasional she-nanigans by the politicians to keep the masses entertained. The premier demo-cratic theorist in the United States, Robert Dahl, would not disagree, although he places critical emphasis on political equality as a key requisite for democracy, or what he calls polyarchy. But Dahl's account, through its many iterations over the years, is deeply imbued with the implicit idea that democracy is what you get in the West, but not elsewhere for the most part. This form of democracy is passively satisfied with the outcomes of procedural justice, and unconcerned with the citi-zen who is passive, that is, a non-participant. In Seymour Martin Lipset's eyes, such passivity was salutary for a democracy's stability.[1]

C.B. Macpherson's work was, on the contrary, imbued with the idea of *sub-stantive justice*: democracy is that system that empowers a conception of man as a maker, as a fully realized human being. A truly democratic system must encourage the manifold development of human capacities in all people. From Macpherson's standpoint, the civic virtue of Athenian citizens, as related by Thucydides, seems superior to our own in present-day America: he has Pericles say this, for example: "Here each individual is interested not only in his own affairs but in the affairs of the state as well . . . we do not say that a man who takes no interest in politics is a man who minds his own business: we say that he has no business here at all."[2]

Secondly for Macpherson, democracy in the world of inegalitarian distribution that we all live in is inseparable from "the cry of the oppressed." Democracy is and must be the means for a redress of human inequality, especially economic inequality. Without these two things, political equality is often rendered meaningless.[3]

If we look at Korea in this light, South Korea achieved in 1992 a democracy that allows a circulation of elites, and a limited form of pluralism. This came after nearly fifty years of democratization struggles, and remains a partial victory. Even after the inauguration in 1998 of former dissident Kim Dae Jung, the infamous national security law remains on the books, a law that makes hash of civil liberties if the issues at hand have anything to do with North Korea, or the left side of the political spectrum.

We can examine the South Korean case, particularly the relationship between economic development (as usually conceived) and democracy, in an excellent new book, entitled *Capitalist Development and Democracy*. Capitalist development, according to the authors of this book (Dietrich Rueschemeyer, John Stephens, Evelyn Huber Stephens), is associated with democracy because it transforms class structures, undermining old ones and creating new ones. The new middle classes, however, will fight to the point of their own democratic representation, but not beyond: after that, they will seek to restrict working-class representation. And here we have a nice explanation for the absence of labor representation in contemporary South Korea. Perhaps this means that the middle class will also seek to disestablish working-class representation, an explanation of the rise of "neo-Democrats" in the Clinton administration.[4]

What Is War?

This seems to be an easier question to answer. Yet "war" is a fungible term. We hear about "trade wars" over automobiles, or "culture wars" about differing conceptions of morality, or the wars in Bosnia and Kosovo that often consist of random mayhem. The Korean War was clearly a war, but of what kind? A conventional war of aggression was the answer in the 1950s, "another Munich"—and this was still the answer that Michael Walzer accepted in his discussion of Korea in *Just and Unjust Wars*. This is the most important American text on just war doctrines, but Walzer's usually impeccable logic falls down when he briefly examines American thinking about the 38th parallel. An inviolable line when the North Koreans crossed it in June 1950, it becomes for Warren Austin (the American representative to the United Nations) a newly violable one ("an imaginary line" he called it) in September 1950 when the United States was poised to march north. Walzer comments, "I will leave aside the odd notion that the 38th parallel was an imaginary line (how then did we recognize the initial aggression?)."

Walzer leaves this mouthful without further thought, yet it is basic to his argument on the war.[5] Had he probed its meaning, had he examined the origin of this "imaginary line," he might have come to the critical paradox in the Truman administration's (and Walzer's) definition of this war, which we may simply state: "June 25, 1950: Koreans invade Korea." In the midst of the December 1950 crisis that ineluctably followed upon the debacle of the American march across the 38th parallel to the border with China at the Yalu River, Richard Stokes (British Minister of Works), intuited this paradox. The 38th parallel decision was "the invitation to such a conflict as has in fact arisen":

> In the American Civil War the Americans would never have tolerated for a single moment the setting up of an imaginery [*sic*] line between the forces of North and South, and there can be no doubt as to what would have been their re-action if the British had intervened in force on behalf of the South. This parallel is a close one because in America the conflict was not merely between two groups of Americans, but was between two conflicting economic systems as is the case in Korea.[6]

Ever since 1950, Stokes's dreaded "civil war" conception has been like a Rumplestiltskin for the official American view: say it and the logic collapses, the interpretation loses its power. But Stokes carried his argument one step further: not just a civil war, but a war between two conflicting economic systems. It is precisely *that* Korean war which continues today, with the United States using every resource at its command to support the economic system of the South since 1950—even if in somewhat altered form after the $60-billion bailout in late 1997.[7]

The Korean War was (and is) a civil war; only this conception can account for the 100,000 lives lost in the South before June 1950 (at least 30,000 of which came amid the 1948–49 rebellion on Cheju Island—as far away as you can get from the North and still be in Korea), and the continuance of the conflict down to the present, in spite of claims that this was really "Stalin's war," or that Moscow's puppets in Pyongyang would surely collapse after the USSR itself met oblivion in 1991. It is therefore instructive to see what Thucydides had to say about civil war. Perhaps the most famous line from his book, "war is a stern teacher," comes from the civil war in Corcyra:

> War is a stern teacher. So revolutions broke out in city after city. . . . What used to be described as a thoughtless act of aggression was now regarded as the courage one would expect to find in a party member; to think of the future and wait was merely another way of saying one was a coward; any idea of moderation was just an attempt to disguise one's unmanly character; ability to understand a question from all sides meant that one was totally unfitted for action. Fanatical enthusiasm was the mark of a real man, and to plot against an enemy behind his back was perfectly legitimate self-defense. Anyone who held violent opinions could always be trusted, and anyone who objected to them became a suspect. . . .[8]

This a mnemonic for "Korea." The passage fits the Korean civil war with no necessity to dot "i's" or cross "t's," and it explains the continuing blight on the

Korean mind drawn by that war, just like a doctor drawing blood: to understand the Korean War "from all sides" is still to go to jail in either North or South. It also fits the American civil war, by far the most devastating of all American wars to Americans, but one that happened long enough ago that most Americans have no idea what it means to have warfare sweeping back and forth across the national territory, or to have brother pitted against brother.

The Two Korean Sides: Democracy? War?

For the past fifty years all the special pleading and extenuated rationalization in the United States has gone to the southern side in this war, so it is important to begin with truths that lay behind that special pleading, well known to American officials of the time from formerly classified documentation. South Korea at the time of the Korean War was far from a democracy, if always touted as such by American spokesmen. Internal CIA analyses in the late 1940s painted a devastating picture of what many observers took to be the worst police state in Asia.[9] South Korean political life, one CIA study said, was "dominated by a rivalry between Rightists and the remnants of the Left Wing People's Committees," the latter persisting doggedly in demands for American recognition—which the Americans would never give, of course, even though the CIA recognized the imprimatur of this movement: a "grass-roots independence movement which found expression in the establishment of the People's Committees throughout Korea in August 1945," led by "Communists" who based their right to rule on the resistance to the Japanese.[10] CIA studies also drew the connection between police state politics and the small but powerful landed class in southern Korea—a class with all the accumulated history and feudal arrogance of southern planters in the United States:

> The leadership of the Right [*sic*] . . . is provided by that numerically small class which virtually monopolizes the native wealth and education of the country. Since it fears that an equalitarian distribution of the vested Japanese assets [i.e., colonial capital] would serve as a precedent for the confiscation of concentrated Korean-owned wealth, it has been brought into basic opposition with the Left. Since this class could not have acquired and maintained its favored position under Japanese rule without a certain minimum of "collaboration," it has experienced difficulty in finding acceptable candidates for political office and has been forced to support imported expatriate politicians such as Rhee Syngman and Kim Koo. These, while they have no pro-Japanese taint, are essentially demagogues bent on autocratic rule.

Thus, "the extreme Rightists control the overt political structure in the US zone," mainly through the agency of the National Police, which had been "ruthlessly brutal in suppressing disorder."

General Albert Wedemeyer, a staunch supporter of Chiang Kai-shek, reported much the same evidence during his tour of Korea in late 1947.[11] Wedemeyer found in conversations with Koreans that many had turned to the Left because they could not stomach the pro-Japanese collaborators in power, not because they were communists. The litterateur Chŏng In-bo told the general that communists had their hold on people not because of northern intrigue, but because of the lingering memory of their anti-Japanese patriotism: "communism here has been nurtured with the fertilizer of nationalism." Furthermore, Chŏng observed pointedly, for decades "only Russia, contiguous to us, shared our enmity against Japan." The novelist Younghill Kang, also an anti-communist, wrote to Wedemeyer that "Korea was one of the worst police states in the world"; the struggle in Korea, he said, was "a fight between the few well-fed landed and the hungry landless. These few today control [the government] and the mass of people want to rectify these ancient wrongs."

Quickly some new issues are on the table: not just a civil war, but a struggle for economic justice. Not just a civil war, but a heavy dose of anti-colonial nationalism. Not just a civil war, but one *against* the hated legacy of Japanese totalitarianism. In 1949 Syngman Rhee ran something he called the National Guidance Alliance, using prewar Japanese methods of interrogation and torture to "convert" leftists and communists. The alliance frequently made claims that it was converting as many as 3,000 people per week, as many as 10,000 converts per province. Its chief, Pak U-ch'ŏn, explained to a U.S. Embassy officer how the process worked:

> In order to be sure that conversions are sincere and complete, each individual upon surrendering himself to the Alliance is required to prepare a complete written confession. . . . Most important, he must set down the names of all individuals who served in the same cell. . . . For a period of one year confessions are subject to constant recheck, largely by matching name lists. If a confession proves false or deficient at any time during the year, the person who made it becomes liable to the full legal penalty for his action and for his leftist affiliations.

South Korea had many "unconverted" political prisoners from the Korean War era in its jails until the 1990s, when they began to be released, slowly. But the longest-serving political prisoner in the world is still in those jails as of this writing, a man named U Yŏng-gak, who remains imprisoned because he refuses to renounce his allegiance to the North. Meanwhile, the longtime independent political analyst Chŏng Kyŏng-mo told a Tokyo conference in March 1998 that the National Guidance Alliance slaughtered 100,000 people for political reasons, before the ostensible Korean War began.[12]

What about North Korea? Can we speak of democratic development in such a totalized political system? It is heresy in the American context even to broach this question, but we can do so in Macpherson's sense of a redress of human inequality. Captured documents from the time, which include major collections of biographies of party activists, show that huge numbers of poor peasants were given political responsibilities in the new northern regime: in local people's committees, in the comparatively large "mass party" known as the Korean Workers Party, and in any number of women's, peasants', workers' and youth organizations. Mass literacy campaigns accompanied this empowerment of people who never before had been asked the time of day when it came to politics. The class structure was turned upside down in the late 1940s, with a profound and polarizing effect on South Korean politics.[13]

The quality of poor peasant political participation in North Korean affairs would not impress a Western democrat, but there is no question that poor peasants at the village level acquired new voice in decisions affecting the fundamentals of their daily life, namely, who should own what land and under what conditions, and what class of people should occupy positions of village authority. Even deeply ingrained marriage patterns changed overnight; now a mother sought to marry her daughter off to a person of poor peasant background, shunning the sons of landlords who would have been matrimonial prizes at any previous point in Korean history.[14] At higher levels, of course, all power was monopolized by Kim Il Sung and other guerrillas who had fought in Manchuria in the 1930s, and voting for provincial and national representatives was closely monitored and restricted. With the consolidation of Kim's rule after the Korean War, North Korea came to resemble a modern dystopia, with every Platonic rationalization used to ensure conformity, hierarchy, and discipline—above all, the myth that the *suryǒng,* or "leader," was the fount of all good ideas, a contemporary philosopher-king whose own individual character determined the nature of the state.[15] But that was after the war, and in some ways a result of the war.

Sparta Occupies Athens

For Plato, a more sober analyst of war than Thucydides because he lived after the Athenian defeat and the collapse of empire, civil wars were the worst wars, to be prevented by all means; but they were also supreme tests of human character and virtue.[16] Thus the best place to examine the question of war and democracy in Korea is to look at the behavior of both regimes in the most extreme circumstances, namely, their respective occupations of the other side during the Korean War.

The Korean people had two regimes to choose from, led by Kim Il Sung and Syngman Rhee, not by Karl Marx and Thomas Jefferson. North Korea might have been Sparta, true, but Seoul was a far cry from Athens. In 1950 (not in 1910 or 2000), in the now-defunct era of anti-colonialism and national liberation that swept the world, the regime of the "people's republic" offered more of what Koreans wanted from their politics, and to more Koreans, than did the Rhee regime. This was the basic source of the DPRK's strength. It is why American intelligence found, during the war for the South, that a "very small percentage of South Koreans in Communist held areas has fled south to friendly territory."[17] The North also had a civil war agenda very different from the South: it wanted not just to conquer the other side, but to carry out a revolution. It did so in the summer of 1950, during a three-month occupation. Let us briefly look at "democracy in wartime," North Korean–style.

Within days of the opening of the war on June 25, the northerners revived and then sponsored elections for local people's committees throughout the South: always with an attempt at control from above, but something that was thwarted from below more often than one might imagine. Top secret instructions to political cadres on how to run elections in Anyang illustrate the process. Kang Yong-su, chief of the interior department for Sihŭng County, ordered the establishment of a broad patrolling network (kyŏngbi-mang) to guard elections held in wartime conditions during a ten-day period, July 20–30, 1950. The elections began at the village level on July 25, moving up to the township level two days later, and to the county level on July 28. Self-defense units were to guard election points against "impure elements, wreckers, and arsonists," and to send two or three people around to gather up public opinion and to deal with "evil" ideas and "impure" plots. Daily reports on such activity were to be dispatched to higher levels. Armed force was not to be displayed at the election places; otherwise, "it will look like the forcible elections held under the Rhee puppet regime." It was thought important "to raise [people's] political awareness to a high point."

People who could not vote included the following, in descending order: "pro-Americans," defined as members of the Republic of Korea (ROK) government or the National Assembly or "reactionary organizations"; "national traitors," members of terrorist groups, and people "who actively helped American imperialism through economic aid"; the "pro-Japanese clique," including central, provincial, and county administrators and police during the colonial period, plus "people who actively helped Japanese imperialism through economic aid." Political and social organizations should be examined by party cadres, according to these documents, singling out "the nominees for village, township and county committees to see that individuals are recommended who are truly capable of serving the people."[18]

A top secret July 16, 1950, directive to cadres at the township level from the Sihŭng County police read as follows:

1. Rapidly, after a few days of concrete examination of the local situation, organize an intelligence network, and widely begin the work of voluntary confessions [chasu]; get reactionaries voluntarily to surrender hidden weapons; arrest the ringleaders of reactionary political parties, social organizations, and police organs, giving them no opportunities [for resistance].
2. Collect important documents from the reactionary police, study and analyze them for help with your work.
3. Punish spies who are from the North first.
4. All government officials, spies for south Korea and the U.S., officials of the National Guidance Alliance, cadres of the Korean National Youth, responsible officials of political parties led by Rhee, Kim Sŏng-su, Shin Sŏng-mo, and Yi Pŏm-sŏk should be purged and arrested; then get in touch with this office [the county police].
5. For the lower-ranking officials of the above groups, use voluntary confessions to find out about the local situation.

6. In investigating crimes, never forget political awareness and an attitude of vigilance.
7. Enemy property is to be confiscated and reported according to the DPRK Constitution; keep a precise accounting; absolutely no confiscated property may be used just as you please.

The township authorities were also told to organize small numbers of youths for village security, and to establish "communication networks" of four or five people and "defense committees" to repair war damage and the like.[19]

An American social science/intelligence study of the occupation of the South found that village-level committees would have a slate of three or four candidates, usually chosen from among younger members of very poor families; mass meetings would precede the voting, with Korean Workers Party (KWP) cadres getting people to discuss the candidates and urging everyone to vote. Usually there was one candidate whom the KWP favored, and occasionally peasants nonetheless missed the point and elected the wrong person, so the election would be held again. Local records show that KWP members made up one-third to one-half of village and township committees.[20] None of this is surprising, especially for elections held during a vicious war, but it does indicate that people's committee elections were unlikely to spawn the complex social composition found in 1945 when they first emerged in South Korea.

Many committee members were women, and in the countryside almost all were poor peasants; they were also remarkably young. In elections in nine villages in Koyang County, for example, 86 percent of the eligible voters participated (a comparatively low count), electing 57 people of whom 10 were women, 19 were workers, and 29 were poor peasants. In a nearby township election, of 17 people elected, 10 were under thirty years of age; only 2 were over forty. In Poŭn County, for which there are rather complete data, nearly all the committee members were poor peasants; in some case as many as half were KWP members, and the vast majority were under forty years of age. Elections for the 41-member people's committee in Yangju County brought forth 10 women, 27 poor peasants, and 9 workers. Of some 6,100 village committee members elected by August 6 in Kyŏnggi Province, about 5,800 were peasants and workers; the remainder were *samuwŏn* (a vague category for people with some education), students, intellectuals, and businessmen.[21]

The U.S. social science/intelligence study of the northern occupation examined two communities near the city of Taejŏn: Kŭmnam, a township of 14,000 made up mostly of smallholding peasants, few possessing more than three *chŏngbo*, engaged in irrigated rice farming; and Kach'ang, a village of 600 surrounded by very productive land, about half of which had been owned by absentee Japanese landlords. Most residents were very poor tenants. Within Kŭmnam the researchers studied one neighborhood village, run by a single—but very divided—clan.[22]

North Korean regular units who had fought with the Eighth Route Army in China passed quickly through the area in the early stages of the war and garrisoned it, followed by civil officials who arrived in late July. All the North Koreans emphasized that they were fighting the war with their own resources, with no help from the Soviets; the Soviet role in fact was "passed over with almost no mention." The Americans were also "dismissed very lightly." Instead, great emphasis was placed on "the reputed self-sufficiency of the North Koreans," along with the three major themes of the occupation: reunification, land reform, and the restoration of the people's committees.

The communists retained the basic administrative divisions of local government, while replacing the existing offices with people's committees as "the sole governing bodies." All local committee officials were selected through elections, which the American authors noted was unprecedented in previous local experience (local officials were appointed from Seoul until the 1990s). Mass organizations of women, peasants, workers, and youths were formed, and pre-1948 left-wing unions were reactivated in local factories. All sorts of other groups came into being, exemplifying the remarkable North Korean emphasis on placing everyone in an organization: self-defense groups, construction units, crop-estimation teams. The American social scientists, who did not want to find this result but honestly reported it anyway, stated that the result of all these North Korean measures at the local level was "a nearly autonomous administration, answerable to the community through elections," an administration that provided services "on a scale never attempted before."

Responsible police officers were all North Koreans, but were small in number and were limited mostly to the county and township level; they rarely appeared in the villages. Instead, an informal communist system existed at the village level, with county-level specialists coming down to set various kinds of organizations in motion. A KWP branch was organized at the neighborhood and village level, but citizens were less aware of it than an informal grouping of leaders within the party: "the villager tended to see power resident in persons," less than in formal institutions like the KWP. In Kŭmnam, about nine people were thought to be powerful: seven outsiders, including northerners and southerners, and two villagers, the chiefs of the people's committee and the KWP branch. Top authority was thought to reside in a North Korean who remained behind the scenes and kept aloof (another long-standing pattern of North Korean politics). The authorities sought to keep old administrators in their positions, if they were not top-level people, and reemployed neighborhood chiefs who had not fled before the People's Army.

The most interesting pattern that emerged in this village was a parallelism between subordinated clans and communist politics. Kinship cleavages can be very deep in Korean villages, and here they recapitulated distinctions between those who supported the North and those who remained passive, or pro-South. Kŭmnam was a one-clan village, but the clan was split between the richer descendants of the founder's first wife, and the disadvantaged descendants of his second wife. The latter enthusiastically supported the new regime, while the leader of the advantaged side of the clan was executed as an avidly pro-Rhee reactionary (the single person to be executed in Kŭmnam). The head of the disadvantaged lineage became the village committee leader. A man of aristocratic descent then in his late thirties and with but a primary school education, he had won his position because his infant son had died several years before, when he and his wife were being beaten by Rhee's police for allegedly allowing a man who had attacked a police station during the autumn 1946 uprisings to take refuge in their home. Therefore this man was a "martyr" of the Rhee regime.

The other village studied by the Americans did not have the profound cleavages found in Kŭmnam, and traditional leaders provided a buffer between the new regime and the people. The village people's committee "was in every particular directed by the traditional leaders of the community as a superficial compromise necessary in order to insure the security of the settlement." People's committee leaders included several prominent farmers who worked their own land, and their leader was the owner of the local rice-cleaning mill. This same group had always selected the village chief in the past; they instructed villagers to cooperate with the outsiders from the North. The American study concluded that this "evasion" was "completely successful," but was highly dependent on village solidarity. Had there been one dissenter, the villagers thought, the evasion would not have worked. In the end, the communists "found no way to gain control over the community."

There is much more to be said about the northern occupation of the South, but based on unimpeachable primary evidence from the time and careful research done just after the fact, it seems to have been a vintage example of North Korean politics: highly organized, carefully controlled empowerment of the historically disenfranchised mass of the Korean people, unlettered peasants and workers. The Spartans delivered on their promises, but always in good Spartan fashion: regimentation from the top down, yielding imperfect results from the standpoint of the center: too many villages succeeded in staying aloof from the program.

Athens Occupies Sparta?

What about the southern occupation of the North, from October 1 to mid-December in 1950? On the day before southern army units crossed over to the North, Secretary of State Dean Acheson said that the 38th parallel no longer counted: "Korea will be used as a stage to prove what Western Democracy can do to help the underprivileged countries of the world."[23] American ideas for the occupation of the North called for the "supreme authority" to be the United Nations, not the Republic of Korea; in pursuit of instituting UN authority, Americans had made plans for a temporary trusteeship or an American military government in the North. On October 12, nearly two weeks after ROK units crossed the 38th parallel, the UN resolved to restrict ROK authority to the South for an interim period. In the meantime the existing North Korean provincial administration would be utilized, with no reprisals against individuals merely for having served in middle- or low-level positions in the DPRK government, in political parties, or in the military.

DPRK land reform and other social reforms were to be honored; meanwhile, an extensive "re-education and re-orientation program" would show Koreans in the North the virtues of a democratic way of life.[24]

The leaders of the Republic of Korea (ROK) saw things very differently. Like the North, they considered theirs to be "the only legal government in Korea," signaling their intention to incorporate northern Korea under the ROK aegis on the basis of the 1948 Constitution (in which the North would get but 100 seats in the National Assembly, about one-third of the total). The United Nations, however, had opposed extending the ROK mandate into the North at least for the time being, and the British and French were positively opposed to the idea. The British Foreign Office termed the Rhee regime one of "black reaction, brutality, and extreme incompetence," and had grave reservations about letting it run the North; indeed at one point it even suggested that ROK weakness and corruption, and the possibility that it might "provoke a widespread terror," raised questions about whether it should be allowed to reoccupy *the South*.[25]

The effective politics of the southern occupation, however, consisted mostly of the Korean National Police (KNP, the colonial police agency now populated by Koreans) and the rightist youth corps that shadowed it and provided recruits since 1945. Cho Pyŏng-ok, former director of the National Police but by then Home Minister, announced on October 10 that the KNP controlled nine towns north of the parallel, with a special force of 30,000 in recruitment for occupation duty. Shortly, the U.S. representative, Everett Drumwright, told Washington that its idea that there should be only a minimum of ROK personnel in the North was "already outmoded by events." Some 2,000 police had already crossed the parallel, but Drumwright thought some local responsibility might result if police who originally came from the North could be utilized. (Thousands of police who had served the Japanese in northern Korea had fled South in 1945, and Rhee had always seen them as the vanguard of his plans for a "northern expedition" in the same way that he placed former Japanese army officers from the North in the top ranks of the ROK army.) By October 20 if not earlier, An Ho-sang (a particularly virulent southern fascist who modeled his youth corps on the Hitler *Jugend*), had his people conducting "political indoctrination" in the North.[26]

Shortly after the northern capital at Pyongyang was occupied, a state of anarchy existed with no effective administration, police reprisals against any suspected collaborators (including many assassinations), and rightist youths roaming the streets, looting what little property was left. Stores were shuttered and much of the population had departed (citizens who remained, however, appeared in "much better condition than did those in Seoul"). The new city council included older businessmen who had remained in Pyongyang since 1945: U Che-sŏn, a former tungsten mine owner who ran a large soy sauce factory until 1949; O Chin-hwan, who had owned two firms in Manchukuo before 1945 and who still retained two large homes in Pyongyang; and Yun To-sŏng, a banker with the Shokusan Ginko (the colonial industrial bank) who ran textile and leather firms in the DPRK. They were fig leafs with no power, of course, but it would appear that some members of the "class enemy" fared rather well in the North until 1950. American Civil Affairs officers on the scene were "pathetically few," and barely experienced:

> The recruitment of a provisional city council for Pyongyang would have been farcical, if the implications were not so obviously tragic. It was rather like watching an Army sergeant selecting men for fatigue duty. As a result, weeks after the fall of the city there were no public utilities, law and order was evident only on the main streets during the hours of daylight, and the food shortage due to indifferent transport and distribution had assumed serious proportions.[27]

People in the North tended to remain close to their homes, according to UN observers, because of the "pillaging and looting and general violence of ROK troops." Troops entering a village were witnessed demanding "the first thousand bags of rice" from the fall crop; officers confiscated houses "without any authority." Reprisals were threatened against any citizen reporting such activity to Americans.[28] If this account of the occupation seems one-sided, the evidence does not support an alternative: nary a good word is spoken about ROK behavior in primary materials from the time. And it gets worse.

According to internal intelligence reports and the highest levels of the American government, the ROK perpetrated a nauseating reign of terror and called it liberation, in the name of democracy and the United Nations. State Department officials had sought some mechanism for supervision of the political aspects of the rollback into the North "to insure that a 'bloodbath' would not result. In other words . . . the Korean forces should be kept under control."[29] But occupation forces in the North were under no one's control. The social base of the DPRK was broad, enrolling the majority poor peasantry, so potentially almost any northerner could be a target. Furthermore, the South's definition of "collaboration with communism" was utterly incontinent, spilling over even to old women caught washing the clothes of People's Army soldiers.[30]

The British had evidence by the end of October 1950 that the ROK as a matter of official policy sought to "hunt out and destroy communists and collaborators"; the facts confirmed "what is now becoming pretty notorious, namely that the restored civil administration in Korea bids fair to become an international scandal of a major kind." The Foreign Office urged that immediate representations be made in Washington, because this was "a war for men's minds" in which the political counted almost as much as the military. Ambassador Oliver Franks accordingly brought the matter up with Dean Rusk (then the undersecretary for Far Eastern Affairs) on October 30, getting this response: "Rusk agrees that there have regrettably been many cases of atrocities" by the ROK authorities, and promised to have American military officers seek to control the situation.[31] Cho Pyŏng-ok announced in mid-November that 55,909 "vicious red-hot collaborators and traitors" had been arrested by that date, a total that was probably understated. Internal American documents show considerable awareness of ROK atrocities; for example, Korean

Military Advisory Group (KMAG) officers said the entire North might be put off limits to ROK authorities if they continue the violence, and in one documented instance in the town of Sunch'ŏn, the Americans replaced marauding South Korean forces with American First Cavalry elements.[32]

Once the Chinese came into the war and the retreat from the North began, newspapers all over the world reported eyewitness accounts of ROK executions of people under detention. United Press International estimated that 800 people were executed from December 11 to 16 and buried in mass graves; these included "many women, some children," executed because they were family members of Reds. American and British soldiers witnessed "truckloads [of] old men[,] women[,] youths[,] several children lined before graves and shot down." A British soldier on December 20 saw about forty "emaciated and very subdued Koreans" being shot by ROK military police, their hands tied behind their backs and rifle butts cracked on their heads if they protested. The incident was a blow to his morale, he said, because three fusiliers had just returned from North Korean captivity and had reported good treatment. Elsewhere British troops intervened to stop the killings, and in one case opened a mass grave for 100 people, finding bodies of men and women, but in this case no children. There were many similar reports at the time, from soldiers and reporters. The British representative in northern Korea said that most of the executions occurred when KNP officials sought to move some 3,000 political prisoners to the South:

> As threat to Seoul developed, and owing to the destruction of the death-house, the authorities resorted to these hurried mass executions by shooting in order to avoid the transfer of condemned prisoners South, or leaving them behind to be liberated by the Communists. However deplorable their methods one can readily grasp the problem.[33]

President Rhee defended the killings, saying "we have to take measures," and arguing that "all [death] sentences [were] passed after due process of law." U.S. Ambassador John Muccio generally backed him up, defending the ROK against the atrocity charges. Muccio was aware of ROK intentions by October 20 at the latest, cabling Washington that ROK officials would give death sentences to anyone who "rejoined enemy organizations or otherwise cooperated with the enemy," and said that the "legal basis" would be the ROK National Security Law and an unspecified "special decree" promulgated in Japan in 1950 for emergency situations—something that may indicate the involvement of MacArthur's command in the executions. CIA sources commented with bloodcurdling aplomb that ROK officials had pointed out to UN officials that "the executions all followed legal trials," and that MacArthur's UN Command "has regarded the trial and punishment of collaborators and other political offenders as an internal matter for the ROK." The legal instrument the CIA had in mind was "Emergency Law Number One," a special decree that, the CIA seemed to suggest, justified the murder of POWs and political criminals (after field trials, of course). Many of the murders witnessed by foreign representatives in fact had no legal procedure whatsoever, and were carried out not just by police but by vigilante youth squads. A Japanese source, cited by a conservative scholar, estimated that the Rhee regime executed or kidnapped some 150,000 people in the political violence of the North's "liberation."[34]

A secret account by North Korean authorities, for internal consumption, detailed South Korean atrocities committed in Seoul after it was retaken by the North in early 1951: nearly 29,000 people were said to have been "shot" by ROK authorities, with 21,000 executions occurring in prisons and the rest perpetrated by police and "reactionary" organizations. Entire families of people's committee leaders were slaughtered. The document accused the ROK and the United States of "slave labor" treatment of those collaborators with North Korea (and their families) who were not executed; they were not allowed to carry ROK citizenship cards, and were used for various corvée labor projects. The report detailed gruesome tortures, and alleged that 300 female communists and collaboraters were placed in brothels where they were raped continuously ("day and night") by South Korean and American soldiers. This report may of course be false, but then why would DPRK officials lie to their superiors in secret internal materials that only saw the light of day in 1977?[35]

American Athenians?

War is indeed a stern teacher. Civil wars, not to mention revolutions, have witnessed comparable political violence everywhere in the world. The forces of order in the ROK were mostly trained in Japanese colonial institutions; collaborators with Japan in the ROK police and military lacked legitimacy, and knew that fact better than anyone else. In such circumstances people have little recourse but to use force to maintain their rule. Far more troubling is the complicity and involvement of Americans in the atrocious character of the northern occupation, going beyond the shameless justifications for ROK behavior discussed above.

We find chilling American instructions to political affairs officers on the ground in the North, at odds with the benign, magnanimous occupation envisioned by the State Department, or Acheson's messianic call to show the world the democratic way. Counterintelligence personnel attached to the U.S. Tenth Corps were ordered to "liquidate the North Korean Labor Party and North Korean intelligence agencies," and to forbid any political organizations that might constitute "a security threat to X Corps." "The destruction of the North Korean Labor Party and the government" was to be accomplished by the arrest and internment of the following categories of people: all police, all security service personnel, all officials of government, and all current and former members of the KWP and the South Korean Workers Party. The compilation of "black lists" would follow, the purpose of which was unstated. These orders are repeated in other Tenth Corps documents, with the added authorization that agents were to suspend all types of civilian communications, impound all radio transmitters, even to destroy "[carrier] pigeon lofts and their contents."[36] The KWP was a mass party with as much as 14 percent of the entire population on its rolls; such instructions implied the arrest and internment of upwards of one-third of North Korean adults. Perhaps for this reason the Americans found that virtually all DPRK officials, down to local government levels, had fled before the onrushing troops.[37]

Other internal materials document odious American connivance in South Korean atrocities. U.S. Counterintelligence Corps detachments, accompanied by KNP investigators, were instructed from the occupation of Inchon (September 15, 1950) onward to draw up "white" and "black" lists of Koreans; members of people's committees were particular targets for blacklisting. "Team agents [also] made use of rightist organization[s]," including getting them "to assist in establishing order." American counterintelligence personnel were present with many ROK police and intelligence units in the North. KMAG advisors also accompanied ROK military units, and stated on October 2 that the KNP would be used "to control civil population and maintain order as soon as possible after liberation." Perhaps anticipating what would happen next, KMAG recommended a "method [of] silencing" reporters following in the ROK wake.[38]

Original ROK blacklists exist in the U.S. collection of captured North Korean documents; apparently American investigators thought they were North Korean blacklists. One shows that the South Korean "White Tiger" unit listed sixty-nine residents of Hyesan County, N. Ham'gyŏng Province as "reactionaries" [pandongja], of which nine were designated as "spies" [milchŏng], nineteen as "incorrigibly bad" [akjil], and twelve as "both"; the rest had no designation other than reactionary. The charges against them included cooperating with the army, being members of the party, jailing "patriots," and spying on ROK units or agencies.[39]

During firefights with guerrillas in the North in October 1950, a memorandum from an Army intelligence officer named McCaffrey to Major General Clark Ruffner suggested that, if necessary, the Americans could organize "assassination squads to carry out death sentences passed by ROK Government in 'absentia' trials to guerrilla leaders," and went on to say, "if necessary clear the areas of civilians in which the guerrillas operate," and "inflame the local population against the guerrillas by every propaganda device possible." In the aftermath of the Chinese intervention, a staff conference with Generals Ridgway, Almond, Coulter, and others in attendance brought up the issue of the "enemy in civilian clothing." Someone said, "we cannot execute them but they can be shot before they become prisoners." To which Coulter replied, "We just turn them over to the ROK's and they take care of them."[40]

This important element of the Korean War has been lost from the collective memory, as if Vietnam were the only intervention where Americans lost the distinction between just and unjust war, the only one where a My Lai occurred. One might think that this is because the information just adduced came from classified sources. Not so: in 1950, the guerrillas in "white pajamas," as the Americans called traditional Korean clothing, and what they provoked in Americans was as accessible as the neighborhood barbershop reading table. Military historian Walter Karig, writing in *Collier's*, likened the fighting to "the days of Indian warfare" (a common analogy); he thought Korea might be like Spain—a testing ground for a new type of conflict, which might occur later in places like Indochina and the Middle East. "Our Red foe scorns all rules of civilized warfare," Karig wrote, "hid[ing] behind women's skirts"; he then presented the following colloquy:

> The young pilot drained his cup of coffee and said, "Hell's fire, you can't shoot people when they stand there waving at you." "Shoot 'em," he was told firmly. "They're troops." "But, hell, they've all got on those white pajama things and they're straggling down the road." . . . "See any women or children?" "Women? I wouldn't know." "The women wear pants, too, don't they?" "But no kids, no, sir." "They're troops. Shoot 'em."[41]

John Osborne told the readers of *Life* that G.I.s were ordered to fire into clusters of civilians by their officers, quoting one of them: "it's gone too far when we are shooting children." This was a new kind of war, he said, "blotting out of villages where the enemy *may* be hiding; the shelling of refugees who *may* include North Koreans."[42]

Charles Grutzner, who reported the war for the *New York Times*, said that in the early going, "fear of infiltrators led to the slaughter of hundreds of South Korean civilians, women as well as men, by some U.S. troops and police of the Republic." He quoted a high-ranking U.S. officer who told him of an American regiment that panicked in July 1950 and shot "many civilians."[43]

Veteran correspondent Keyes Beech wrote in the *Newark Star-Ledger*, "It is not the time to be a Korean, for the Yankees are shooting them all . . . nervous American troops are ready to fire at any Korean."[44] Reginald Thompson, an Englishman, authored *Cry Korea*—a fine, honest eyewitness account of the first year of the war. War correspondents found the campaign for the South "strangely disturbing," he wrote, different from World War II in its guerrilla and popular aspect. "There were few who dared to write the truth of things as they saw them." G.I.s "never spoke of the enemy as though they were people, but as one might speak of apes." Even among correspondents, "every man's dearest wish was to kill a Korean. 'Today,' . . . 'I'll get me a gook.'" Americans called Koreans gooks, he thought, because "otherwise these essentially kind and generous Americans would not have been able to kill them indiscriminately or smash up their homes and poor belongings."[45]

All this cannot be blamed on the benighted views of the American military in 1950. Consider the judgment of the venerable military correspondent and editor of the *New York Times*, Hanson Baldwin, three weeks into the war:

> We are facing an army of barbarians in Korea, but they are barbarians as trained, as relentless, as reckless of life, and as skilled in the tactics of the kind of war they fight as the hordes of Genghis Khan. . . . They have taken a leaf from the Nazi book of blitzkreig and are employing all the weapons of fear and terror. . . .

Chinese communists were reported to have joined the fighting (he erred in saying), and not far behind them might be "Mongolians, Soviet Asiatics and a variety of races"—some of "the most primitive of peoples." Elsewhere Baldwin likened the North Koreans to invading locusts; he ended by recommending that Americans be given "more realistic training to meet the barbarian discipline of the armored horde."[46]

Then there was another troubling fact about the war: our Koreans would not fight. In 1950 a puzzled John Foster Dulles found the North Koreans "fighting and dying, and indeed ruining the whole country, to the end that Russia may achieve its Czarist ambitions." Rusk thought it important to find out how the Russians get the satellites "to fight their actions" for them—"here was a technique which had been very effective and it was not obvious how the success had been achieved." There appeared to be a "nationalist impetus," too, so it would also be well to figure out how the Russians "stimulate this enthusiasm."[47] As late as 1969 General Ridgway was still vexed by this conundrum, even though, as he said, "My acquaintance with Orientals goes back to the mid-1920s." (He might have added that his experience included chasing Sandino in Nicaragua.)[48] The North Koreans were more "fanatical" fighters than the Chinese, he said, yet the South Koreans were not good soldiers: "I couldn't help asking why. Why such a difference between the two when they were the same otherwise." He speculated that perhaps the KPA was using "dope," but never found evidence of it.[49]

The same American society that fought for freedom in Korea prohibited Koreans from entering the country in 1950 under existing racial quotas, and denied naturalization to 3,000 Koreans who came to the United States before 1924. Fifteen American states prevented Korean-Caucasian marriages, eleven states refused to allow Koreans to buy or own land, and twenty-seven occupations in New York City were proscribed to Koreans.[50] Korea, in short, was an example not of the best but of the worst that Americans had to show for themselves in the Third World: it was the harbinger of experiences that would later unfold in Vietnam, Guatemala, Nicaragua, El Salvador, and many other places where Americans worked hand in glove with some of the most repressive of twentieth century regimes, their one virtue being that they were anti-communist. The wartime experience also, of course, explains the North Korean contempt for American and South Korean claims to democracy ever since.

American Athens? Democracy and the Korean War in the United States

Other articles in this volume address the prime threat to American democracy during the Korean War, namely, McCarthyism. I would like to say something about that as well, since in my view the phenomenon is still poorly understood (particularly its connections to East Asian policy, the Korean War, and the "who lost China" issue in the 1950s), and because it helps to explain how this war later became "forgotten": it could not be known in the first place, because of censorship by MacArthur's command and repression in the United States. First, however, it is important to examine the process by which American power was committed to the Korean War, one that demolished constitutional procedure and set a key precedent for later military interventions by the executive branch and the "imperial presidency."

Acheson and Rusk were virtually the only high officials in Washington when the war broke out on Saturday evening, American time; President Truman had left for his home in Independence, Missouri, that morning. In succeeding days Acheson dominated the decision-making that soon committed American air and ground forces to the fight. Acheson (together with Rusk) also made the decision to take the Korean question to the United Nations, even before he had notified Truman of the fighting; he then told Truman there was no need for him to return to Washington until the next day. At the famous Blair House meetings on Sunday evening (June 25) Acheson argued for increased military aid to the ROK, American air cover for evacuating U.S. citizens, and the interposition of the Seventh Fleet between Taiwan and the mainland. On the afternoon of June 26 Acheson labored alone on the fundamental decisions committing American air and naval power to the Korean War, ones approved that evening at another Blair House conference. Thus the decision to intervene was Acheson's decision, supported by the president but taken before United Nations, Pentagon, or congressional approval.[51]

Acheson later acknowledged that these decisions were taken not only before congressional consultations, but that the United Nations was used for after-the-fact ratification of his decisions ("it wasn't until 3:00 in the afternoon [on June 27] that the United Nations asked us to do what we said we were going to . . . in the morning").[52] Acheson's first recourse was to the United Nations, not to the United States Congress, because he knew he would get a better vote in that legislature—and that in 1950 the United Nations was a legislature with no clout unless Washington backed its decisions. The United States thus acted first and got the United Nations to ratify its decisions later, just as Acheson acted first in calling the Security Council together and told Truman later. Had the United Nations not backed up their decisions, Acheson and George Kennan later agreed, the United States would have gone ahead anyway, but would have had more problems with that residual, nagging irritant—public opinion.[53]

The war powers provisions of the Constitution did not cause Acheson a single caution. He did not consult Congress on June 24–26, he later said, because "you might have completely muddied up the situation which seemed to be very clear at the time," and because congressmen might have "circumscribed the President's prerogatives" (no problem there from the United Nations). He later expressed some astonishment when told that as early as June 28, 1950, Senator Taft referred to the Acheson–Truman decisions as "a complete usurpation by the President of the authority to use armed forces."[54] It had not occurred to him, perhaps, that war-making is the Congress' prerogative.

Acheson was a classic elitist, with a confident, easy contempt for the opinions of those who disagreed with him, whether in the military, the Congress, or the public at large (public opinion was for him a cranky constraint on his autonomy of decision). He preferred the autonomy of private decisions taken mostly with his own counsel, the only one he really trusted, and then presented to an inexperienced president who so often deferred to Acheson. Truman with all his limitations was a democrat with an abiding belief in the judgment of the common man, something that he himself exemplified. As I.F. Stone once put it, Truman was "as honorable and decent a specimen of that excellent breed, the plain small-town American, as one could find anywhere in the U.S.A."[55]

American military leaders were more sober and measured about the limits to American power, and reticent about committing ground forces to the war in Korea, as the Blair House records make clear.[56] But Acheson had little use for their judgment, either. The Joint Chiefs of Staff, he said later, "do not know what they think until they hear what they say." But once they have spoken, "the Pope has spoken, and they are infallible." In NSC meetings, according to Acheson, the JCS would present their viewpoint in an involved paper that no one usually read, "then a discussion, and then—in my experience, always—the president deciding in favor of what I thought was the sound view, which was the one I presented to him." If a controversy existed between Defense and State, the president almost always followed Acheson's position, "not because I presented it, but because the other view was so silly. There wasn't any sense to it. It hadn't been thought through."[57] Acheson had the brains of a great man and the hubris to use them as he chose, in a raucous democracy that knew next to nothing about his agenda. But his decisions had not been quite thought through, either.

Acheson committed the United States to a war whose dynamics he could not master—in the field or in the American body politic. As the war lurched into its own dialectic, taking American forces into a new war with China, within six months it brought on a debacle threatening the peace of the world, and the only dog barking in Washington was the voice of principled Republican conservatism—a voice soon silenced for good by a bipartisan centrist coalition dedicated to supporting the interests of the national security state.

When Soviet ambassador to the United Nations Jacob Malik returned to his vacated seat, on August 1, he declared that the resolutions on Korea had been illegal and should be rescinded. The *New York Times* called this "the ultimate arrogance of a power-maddened despotism which mistakes the UN for one of its zombie soviets and proposes to make it kowtow to the great Khan in Moscow."[58] The *Times* editorial aptly reflected the political atmosphere in the United States in mid-1950, to which we now turn.

Barbarian in Athens?: McCarthyism and the Korean War

The case of Owen Lattimore says much about McCarthyism, the China Lobby, and its relationship to Korea. It is usually forgotten that McCarthy began his attacks well before the Korean War, that Lattimore's views on Korea were one of McCarthy's central subjects, and that by June 1950 McCarthyism seemed to be losing its momentum—its capacity to establish "China" as an issue in American politics. McCarthy first attacked Lattimore indirectly on March 13, 1950, then alleged that he had found a "chief Russian spy" on March 21, and finally named Lattimore when information leaked from his committee. Beyond Lattimore he was after Under-Secretary of State Philip Jessup, "a dangerously efficient Lattimore front," but ultimately his object was the secretary of state himself, whom McCarthy termed "the voice for the mind of Lattimore."[59] Acheson was his final target: why? In part it was because, by the spring of 1950, he was the last high official, besides Truman himself, standing between Chiang Kai-shek and the American backing he desperately needed to survive an impending communist invasion of Taiwan.

In early April McCarthy claimed to have a document incriminating Lattimore as a Soviet agent, prompting Lattimore to release it to the press—a memorandum he wrote in August 1949, arguing that "the U.S. should disembarrass itself as quickly as possible of its entanglements in South Korea." Lattimore saw Korea as "little China," and Rhee as another Chiang: if we could not win with Chiang, he said, how could we win with "a scattering of 'little Chiang Kai-sheks' in China or elsewhere in Asia." The argument was cogent, and hardly treasonable: this was a common theme of anti-Chiang liberals from 1945 on. Of greater moment, Lattimore's memo also implicitly criticized the developing bureaucratic momentum for some kind of armed riposte to the communists in the summer of 1949:

> It certainly cannot yet be said . . . that armed warfare against communism in the Far East . . . has become either unavoidable or positively desirable. Nor can it be said with any assurance that . . . the Far East would be the optimum field of operation. There are still alternatives before us—a relatively long peace, or a rapid approach toward war. If there is to be war, it can only be won by defeating Russia—not northern Korea, or Viet Nam, or even China.[60]

This document was part of a general reassessment of East Asian policy, bringing in outside consultants such as Lattimore for their views. It reads as if Lattimore were aware of the developing dialectic between strategies of containment and "liberation" or rollback in the Far East.[61]

Lattimore's fuller views on Korea were given in the fall of 1949 when the State Department called in experts to consult with them on the new Asian policy. Generally speaking, liberal scholars such as Lattimore, Cora DuBois, and John K. Fairbank sought merely to point out that the revolution sweeping much of East Asia was indigenous, the culmination of a century of Western impact. Conservative scholars like Phillip Taylor, William Colegrove, and Bernard Brodie sought instead to argue that Soviet machinations were behind Asian revolution; Taylor and Colegrove were particularly anxious to reestablish Japan's position in East Asia, Taylor saying that "we have got to face it head on. We have to get Japan back into, I am afraid, the old co-prosperity sphere . . . and include India in it." The liberal wing was dominant within scholarly circles, however, and in these meetings a consensus emerged looking forward to the establishment of relations with the PRC. Taylor and the others were in the distinct minority.

Lattimore's views on Korea during these sessions were prescient:

> Korea appears to be of such minor importance that it tends to get overlooked, but Korea may turn out to be a country that has more effect upon the situation than its apparent weight would indicate.

After this prophetic mouthful, he went on to say accurately that the ROK politically was "an increasing embarrassment," an "extremely unsavory police state" where the

> chief power is concentrated in the hands of people who were collaborators of Japan. . . . Southern Korea, under the present regime, could not resume close economic relations with Japan without a complete reinfiltration of the old Japanese control and associations . . . the kind of regime that exists in southern Korea is a terrible discouragement to would-be democrats throughout Asia. . . . Korea stands as a terrible warning of what can happen.

Once the war began, however, Lattimore expressed his support for the American intervention.[62]

In mid-May 1950 McCarthy again attacked the "Acheson Lattimore axis" (or, the "pied pipers of the Politburo") on Korea policy, saying Lattimore's plans for Korea would deliver millions to "Communist slavery." Taking direct aim at Taiwan's principal antagonist, Acheson, he blared, "fire the headmaster who betrays us in Asia."[63] In spite of the obviously political and mendacious nature of McCarthy's witch hunt against Lattimore, within a few weeks liberal organs of opinion were already giving the classic formulation that enabled them to escape McCarthy's gun sights: supporting Lattimore's right to his opinions, but condemning them as irresponsible or extreme. In mid-April 1950 the *New York Times* singled out his "unsound" position on Korea; it found Lattimore's view "quite shocking," saying that the State Department had "rejected flatly Mr. Lattimore's advice to cut and run in Korea."[64] McCarthy's assault on Lattimore drew precisely the new boundaries of acceptability: a left-liberal China scholar was out, Red-baiting pamphleteer Freda Utley was in.[65]

McCarthy (and later, Senator Patrick McCarran) also had critically placed allies inside the government. They both were supplied documentation on alleged subversives, most of it classified, by J. Edgar Hoover, Willoughby and Whitney of MacArthur's staff, and even Walter Bedell Smith of the CIA. In 1953 the Justice Department went so far as to work with Willoughby, Ho Shih-lai, and Chiang Ching-kuo on the cases of Owen Lattimore and John Paton Davies—Chiang, of course, being the son of Chiang Kai-shek, with long experience in the KMT secret police. Perhaps most shocking, it is now clear that several of the cases that came out of such investigations were faked.[66]

Tail-gunner Joe was a good marksman: he left a generation of liberals looking over their shoulder to the right, fearing yet another case of mistaken identity. This is the real source of liberal outrage against McCarthyism: not that thousands of

American leftists and communists were unfairly and unconstitutionally persecuted, but that liberal innocents got caught in his gun sights. In the atmosphere of
McCarthyism, Godfrey Hodgson was right to say, "liberals were almost always
more concerned about distinguishing themselves from the Left than about distinguishing themselves from conservatives." Thus they joined "the citadel of . . . a
conservative liberalism." Hodgson's explanation of this persistence is also a good
one: "if the fear of being investigated had shown the intellectuals the stick" in the
early 1950s, "the hope of being consulted had shown them the carrot" thereafter.
Being an influential client meant accepting the confines of one's patronage.[67]

McCarthyism's success by the summer of 1950 meant that about the only place
one could go in the American media to find a principled, independent critical
stance, and a sincere inquiry into what really had happened in Korea, was to three
periodicals of miniscule circulation: *Monthly Review*, Scott Nearing's *World Events*,
and George Seldes's *In Fact*. *Monthly Review* called Korea a civil war, and implied
that the utter collapse of the Rhee regime made it rather a people's war as well. "It
is pretty clear that responsibility for the outbreak of full-scale warfare rests on the
North," it said, but it saw no moral difference between this and Lincoln's offensive
after the South shelled Fort Sumter. Alone among all American commentary on
Korea that I have seen, it argued that the real issue was not who attacked first. It
predicted that the United States would suffer "a disastrous defeat" in Korea.[68]

Scott Nearing was an independent social democrat who cranked out his little
magazine from a farm in New England, offering the best critical commentary on
American foreign policy in 1950. In the spring of 1950 he listed Korea with Indochina
as places where wars were "in progress." The Korean War, he said later, was a civil
war, but also a war for national independence; it had been a testing ground in the
Cold War since 1945, not since June 1950. George Seldes was also a voice of reasoned dissent until his magazine folded for lack of funds.[69] It is not simply that such
views were accurate, in the light of history, more daunting is that they recapitulated
the private opinions of the Americans most knowledgeable about Korea.

George McCune's fine book, *Korea Today*, a sincere account by a principled
liberal who was America's leading expert on Korea at the time, was published by
Harvard University Press just as the war began. John Foster Dulles took time from
his duties to write William Holland, secretary-general of the Institute for Pacific
Relations (which had sponsored the book), with this chilling admonition: "I question whether its publication at this time will serve to promote real insight into the
issues which today so deeply engage our nation."[70]

The first demonstrations against the war occurred mostly in New York, and
drew mostly communist support. Paul Robeson spoke at one rally in Harlem in
early July. The Fellowship of Reconciliation offered a mild protest shortly thereafter. In early August some peace demonstrators sought to march in New York,
but police broke it up immediately; the *New York Times* was hostile to the very
idea of a march.[71]

The United States during this period is not to be compared with authoritarian states like prewar Japan or Germany, or the Soviet Union. It remained open, over the long term, to a reversal of some of the worst excesses of 1950 (although by no means all of them); the press was not muzzled and dissenters were not confined, unless they were the leaders of the Communist Party, and the Supreme Court later overturned their convictions under the Smith Act. But this is not really the point. Judged by the ideals America established for itself, and its presumed fight for freedom on a world scale, the early 1950s were a dark period indeed, a maximization of the potential for absolutist conformity that Louis Hartz had the courage to explore in his 1955 book. If critics were not shot or tortured, they nonetheless suffered loss of career, ostracism, intense psychological pressure, and admonitions to change their thoughts or be excluded from the spectrum of political acceptability.[72]

An epitaph for this period—and for the 38th parallel—came from the fearless independent historian Harry Elmer Barnes. In *Perpetual War for Perpetual Peace*, Barnes wrote,

> Fantastic political boundaries are set up carelessly and arbitrarily, but once they are established . . . they take on some mysterious sanctity. . . . Every border war becomes a world war, and world peace disappears from the scene. By this absurd policy, internationalism and interventionism invite and insure "perpetual war for perpetual peace," since any move which threatens petty nations and these mystical boundaries becomes an "aggressive war" which must not be tolerated, even though to oppose it may break the back of the world.[73]

Conclusion

In this chapter I have not treated the effects of the Korean War on democracy in post-1953 Korea. Some suggestive indications of the war's profound effect on the possibilities of democracy would be (1) the institution of a total garrison state in the North, a vintage "Sparta,"[74] with every citizen required to enter the military and to join a militia after military service—and that merely begins a long list of garrison-state features, including huge efforts to build all kinds of installations deep underground, and above all a tense posture of continuous readiness for war; (2) the rise of a huge and strong military in the South, which provided military training to the whole male population through successive generations, something salutary for the disciplines of economic development but not for democracy—an effect of the war better exampled by the intervention of the military in politics from 1961 to 1992; (3) the final accomplishment of land redistribution in the South in 1951, courtesy of the communist land reform in the preceding summer, which cleared away Korea's historic landed class, and left widespread egalitarian distribution of wealth among the majority peasant population (thus yielding most of the "income equality" that ahistorical economists have found in Korea's developmental model); (4) a permanent national security state in both Koreas, with a National Security Law in the South and various codes in the North making any unofficial attempts at reunification or even praise for the other side treasonable.

With the last point we begin to understand the post-1953 completion of a "division system" (using Paik Nak-chung's term)[75] in which "containment" strategies by both sides built enormous walls and dykes against any serious accommodation between North and South, with those people who police the division getting the highest rewards—often, Koreans from the North running ROK intelligence groupings, and Koreans from the South running DPRK security bureaus. This is perhaps the ultimate proof that it was, is, and (if conflict comes again, perish the thought) will be a civil war. Neither Korea can be a fulfilled democracy until it finally comes to an end.

Notes

1. See Dahl, *Democracy and Its Critics* (New Haven: Yale University Press, 1989), pp. 322–23 (on political equality), and p. 264, where Dahl lists a number of conditions for democracy (or polyarchy in his terms). The conditions listed barely go beyond the categories elaborated by Gabriel Almond and Sidney Verba in their early 1960s work on civic culture. Lipset's views are available in his classic work, *Political Man* (New York, 1960).

2. Thucydides, *History of the Peloponnesian War*, trans. Rex Warner (New York: Penguin Books, 1954), p. 147.

3. C.B. Macpherson, *Democratic Theory: Essays in Retrieval* (New York: Oxford University Press, 1973), pp. 3–8, 78–90. On page 78 Macpherson links Schumpeter's conception of democracy with Dahl's.

4. Dietrich Rueschemeyer, Evelyne Huber Stephens, and John D. Stephens, *Capitalist Development and Democracy* (Chicago: University of Chicago Press, 1992). In the United States the Democratic Party functioned as a business/labor party from 1932 into the 1980s, but the "neo-democrats" now in power in Washington think they must attend to the interests of the middle classes, to the detriment of the old Democratic coalition with labor—just as this book would predict.

5. Michael Walzer, *Just and Unjust Wars: A Moral Argument with Historical Illustrations* (New York: Basic Books, 1977), pp. 117–23.

6. Public Record Office, London, Foreign Office file 317 (hereafter FO317), piece no. 83008, Stokes to Bevin, December 2, 1950.

7. Imagine, if only for purposes of argument, that the $60 billion had gone to North Korea instead. Or imagine that it was the North rather than the South that got an average largess of $600 per person annually from the United States, as the South did from 1945–65.

8. Thucydides, *Peloponnesian War*, pp. 242–43.

9. See the following CIA studies: "Korea," SR-2, summer 1947; "The Current Situation in Korea," ORE 15–48, March 18, 1948; "Communist Capabilities in Korea," ORE 32–48, February 21, 1949; Carrollton Press, Retrospective Collection (CRC), 1981, items 137 B, C, D, 138 A to E, 139 A to C, "National Intelligence Survey, Korea," NIS 41 (compiled in 1950 and 1952).

10. I cover the rise and demise of the people's committees in southern Korea in *The Origins of the Korean War: Liberation and the Emergence of Separate Regimes, 1945–1947* (Princeton: Princeton University Press, 1981).

11. RG59, Lot File 55 D150, "Records of the Wedemeyer Mission to China," various memoranda in box 3 and box 10; see especially box 3, "Korean Interim Government Briefing."

12. I spoke at this conference, held on March 10, 1998. Mr. Chŏng also said that leaders of the Alliance remain in the South, never prosecuted for their political murders.

13. I used this huge documentary collection, which was declassified in 1977 by the U.S. National Archives, in my *Origins of the Korean War, II: The Roaring of the Cataract, 1947–1950* (Princeton: Princeton University Press, 1990).

14. A point well made by Lee Mun Woong, in *Rural North Korea Under Communism: A Study of Sociocultural Change* (Houston: Rice University Studies, 1976).

15. In using the term "Platonic" I draw on Alvin W. Gouldner's discussion of the *Republic* in *Enter Plato: Classical Greece and the Origins of Social Theory* (New York: Basic Books, 1965), pp. 58–61. Gouldner's discussion of military regimentation in Plato's *Laws* is also redolent of the garrison state in post–Korean War North Korea (pp. 72–73).

16. Gouldner, *Enter Plato*, p. 70.

17. Harry Truman Presidential Library (hereafter HST), Presidential Secretaries File (hereafter PSF), "Selected Records Relating to the Korean War," box 3, Office of Intelligence Research, (hereafter, OIR file), report no. 5299.1, June 30–July 1, 1950. This was several days after the occupation of Seoul, and continued to be true as the KPA moved southward. The received wisdom that millions of Koreans fled before communist armies is an artifact of early 1951 and the second seizure of Seoul.

18. Record Group 242, "Captured Enemy Documents," National Records Center (hereafter RG 242), SA2010, item 5/121, top secret instructions to lower levels on PC elections, signed by Kang Yong-su, chief of the Sihŭng County interior department, July 20, 1950.

19. RG242, SA2010, item 5/121, Sihŭng County police memo to township police, top secret, July 16, 1950; also documents signed by the Sihŭng County interior department chief, Kang Yong-su, August 22, 1950.

20. Maxwell Air Force Base, "A Preliminary Study of the Impact of Communism" (1951) part III, pp. 159–60.

21. *Haebang ilbo*, July 29, 31, August 9, 1950. RG242, SA2010, item 4/74, table on the political affiliations and class backgrounds of PC members in Poŭn County, no date but probably August 1950.

22. "Preliminary Study of the Impact of Communism," part III, pp. 106–85. The primary writers of this portion were John Pelzel and Clarence Weems; the study was under the overall direction of Wilbur Schramm.

23. HST, Matthew Connelly Papers, box 1, Acheson remarks in cabinet meeting minutes for September 29, 1950.

24. FO317, piece no. 84072, Washington Embassy to FO, November 10, 1950, enclosing a State Department paper on the occupation. Allison told the British that Foreign Minister Ben Limb's claim that the ROK government was "the only legitimate government of all Korea" was "in direct conflict with the position taken by the US Government" and by the United Nations, both of which saw the ROK as having jurisdiction only in those areas where UNCOK observed elections. National Archives, diplomatic branch, 795.00 file, box 4268, Allison to Austin, September 27, 1950 (hereafter, 795.00 file). On the UN resolution, see no. 602, p. 8 of notes, also *London Times*, November 16, 1950.

25. FO317, piece no. 84093, Tomlinson memo of July 9, 1950; Moscow Embassy to FO, July 26, 1950; piece no. 84100, John M. Chang to Acheson, September 21, 1950, relayed to the FO by the State Department; see also *Foreign Relations of the United States* (hereafter *FR*) 1950, 3: pp. 1154–58, minutes of preliminary meetings for the September Foreign Minister's Conference, August 30, 1950; also FO317, piece no. 84099, "Korea: The 38° Parallel," October 2, 1950; also various documents in piece no. 84097, 84098, and 84099, autumn 1950. See also 795.00 file, box 4265, Feis memo attached to Dulles's July 14, 1950, memo.

26. 795.00 file, box 4268, Acheson to Muccio, October 12, 1950. Acheson wanted Muccio to assure that the KNP would operate under the UN Command. See also box 4299, Drumwright to State, October 14, 1950; *New York Times*, October 20, 1950.

27. *London Times*, November 16, 1950.

28. H.W. Bullock, UNCURK Memo no. 2, "Conditions in Pyongyang," November 16, 1950, courtesy Gavan McCormack.

29. 795.00 file, box 4268, Durward V. Sandifer to John Hickerson, August, 1950, top secret.

30. British sources encountered the elderly woman charged with washing soldier's clothes in late November; she was among knots of "emaciated, dirty, miserably clothed" people tied in ropes and being herded through the streets. F0317, piece no. 84073, Korea to FO, November 23, 1950.

31. Ibid., handwritten FO notes on FK1015/303, U.S. Embassy press translations for November 1, 1950; piece no. 84125, FO memo by R. Murray, October 26, 1950; piece no. 84102, Franks memo of discussion with Rusk, October 30, 1950; Heron in *London Times*, October 25, 1950.

32. *Manchester Guardian*, December 4, 1950; RG338, KMAG file, box 5418, KMAG journal, entries for November 5, 24, 25, 30, 1950.

33. 795.00 file, box 4270, carrying UPI and AP dispatches dated December 16, 17, 18, 1950; FO317, piece no. 92847, original letter from Private Duncan, January 4, 1951; Adams to FO, January 8, 1951; UNCURK reports cited in HST, PSF, CIA file, box 248, daily summary, December 19, 1950. See also *London Times*, December 18, 21, 22, 1950.

34. *London Times*, UPI December 16, 1950; 795.00 file, box 4299, Muccio to State, October 20, 1950; CIA file, ibid., daily summaries for December 19, 20, 21, 1950. The CIA also reported that UNC officials had made representations to ROK officials about the atrocities, but "appear to have had little effect." The *Manchester Guardian* reported that American infantry elements saved one woman after arriving in the midst of executions carried out by members of An Ho-sang's Korean Youth Defense League; twenty-six others, including three women, a nine-year-old boy, and a thirteen-year-old girl, were already murdered: "when they grow up, they too would be Communists," the murderers said (December 18, 1950). The Japanese figure is in Nam, *North Korean Leadership*, p. 89.

35. RG242, SA2012, item 5/18, *Sŭl Si wa kŭ chubyŏn chidae esŭi chŏkdŭl ŭi manhaeng* [Enemy atrocities in Seoul city and its vicinity], two secret reports compiled by the Seoul branch of the KWP after the second capture of Seoul, no date but early 1951.

36. Carlisle Military Barracks, Almond Papers, General Files, X Corps, "Appendix 3 Counterintelligence," November 25, 1950; William V. Quinn Papers, box 3, X Corps periodic intelligence report dated November 11, 1950 (Quinn was the X Corps G-2 chief) [emphasis added].

37. FO317, piece no. 84073, Tokyo to FO, November 21, 1950.

38. MacArthur Archives, RG6, box 61, intelligence summary no. 3006, December 2, 1950; this document refers back to operations in Inchon in September, and suggests that such methods were standard. See also RG338, KMAG file, box 5418, KMAG journal, entry for October 2, 1950.

39. RG242, SA2010, item 2/99, *pandongja myŏngbu* [list of reactionaries], no date, but autumn 1950.

40. Carlisle Military Barracks, William V. Quinn Papers, box 3, X Corps HQ, McCaffrey to Ruffner, October 30, 1950; Ridgway Papers, box 20, highlights of a staff conference, with Ridgway and Almond present, January 8, 1951.

41. Walter Karig, "Korea—Tougher than Okinawa," *Collier's* (September 23, 1950), p. 24–26. Gen. Lawton Collins remarked that Korea saw "a reversion to old-style fighting—more comparable to that of our own Indian frontier days than to modern war" (*New York Times*, December 27, 1950).

42. John Osborne, "Report from the Orient—Guns Are Not Enough," *Life* (August 21, 1950), pp. 74–84.

43. *New York Times*, September 30, 1950.

44. Keyes Beech, *Newark Star-Ledger*, July 23, 1950.

45. Thompson, *Cry Korea*, pp. 39, 44, 84, 114.

46. *New York Times*, July 14, 1950.

47. *FR* (1950), 6: pp. 128–30, Dulles to Acheson, August 4, 1950; FO317, piece no. 83014, notes on talk between Dening and Rusk, July 22, 1950.

48. Thomas McPhail, KMAG advisor who finished his career as head of the U.S. Military Advisory group to Nicaragua under Somoza, wrote to Ridgway in 1965, "the old Guardia [National Guard] members who fought with the Marines against Sandino still talk about General Ridgway" (Ridgway Papers, box 19, Thomas D. McPhail to Ridgway, April 15, 1965).

49. Ridgway Papers, oral interview, August 29, 1969. His interviewer, a Vietnam veteran, told him the North Koreans sounded "about the same" as the Vietcong.

50. *The Nation*, August 26, 1950.

51. Acheson says that he instructed Hickerson at 10:30 P.M. to call the Security Council together, which would be almost an hour before he called Truman; he is most explicit in saying Hickerson was told he should "proceed at once," and "if the President had a different idea, it would be perfectly possible to change what [Hickerson] was doing." When he called Truman, he told him "what I had authorized Hickerson to do," and the president "approved." Acheson also related that on July 19, 1950, Truman sent him a note saying in part that Acheson's initiative in "immediately calling the Security Council of the U.N. on Saturday night and notifying me was the key to what developed afterwards. Had you not acted promptly in that direction, we would have had to go into Korea alone." Acheson also says that Truman had wanted to return at once, but Acheson suggested he wait until the next day. See Acheson's account in Acheson Seminars, February 13–14, 1954.

52. Acheson Seminars, transcript of February 13–14, 1954.

53. Ibid.

54. Ibid.

55. I.F. Stone, *The Hidden History of the Korean War* (New York: Monthly Review Press, 1952), p. 105.

56. General Bradley supported Achesonian containment at the first Blair House meeting, remarking that "we must draw the line somewhere." But he questioned "the advisability" of introducing American ground troops in large numbers, as did Frank Pace and Louis Johnson. At the second meeting on June 26, Generals Bradley and Collins again expressed the view that committing ground troops would stretch American combat troop limits, unless a general mobilization were undertaken. Louis Johnson now supported Acheson, however, while falsely leaking to the press that he, not Acheson, had advocated a defense of Taiwan. *FR* [1950], 7: pp. 157–61 and 178–83.

57. Acheson Seminars, see note 52 above.

58. *New York Times* editorial, August 30, 1950.

59. *New York Times,* March 14, 22, 27, and 31, 1950. For a good account of the Lattimore case see Stanley I. Kutler, *The American Inquisition: Justice and Injustice in the Cold War* (New York: Hill and Wang, 1982), pp. 183–214.

60. *New York Times,* April 4, 1950.

61. A phenomenon which I discuss at length in *Origins,* vol. 2.

62. "Transcript of Round Table Discussion on American Policy Toward China," State Department, October 6–8, 1949, CRC 1977, item 316B. On Lattimore's support for the U.S. role in the Korean War, see *New York Times*, August 1, 1950.

63. *New York Times*, May 16, 1950.

64. *New York Times* editorials, April 5 and 19, 1950. Other responsible officials who held this "shocking view" were, for example, most of the high Army Department officials in 1948–49, who were ready to write off the ROK even if it meant a communist takeover; Gen. Lawton Collins told the MacArthur Hearings in testimony deleted at the time, that Korea "has no particular military significance," and if the Soviets were fully to occupy the peninsula, Japan would be in little greater jeopardy than it already was from Vladivostok and the Shantung Peninsula.

65. At least in business circles. *American Affairs,* the journal of the National Industrial Conference Board, ran Utley's laudatory review of Burnham's *Coming Defeat of Commu-*

nism (12/2 [April 1950], pp. 121–24), also another article entitled "The Soviet Worm in Our School Libraries." Yet this was a centrist journal of opinion.

66. On Hoover, Willoughby, Whitney, and Smith helping McCarthy, see Thomas C. Reeves, *The Life and Times of Joe McCarthy. A Biography* (New York: Stein and Day, 1982), pp. 318, 502; for the 1953 episode see Willoughby Papers, box 23, John W. Jackson letters, written on Justice Department stationery to Willoughby and to Ho Shih-lai, both dated October 16, 1953. The faked files (on Lattimore, John Service, and others) are discussed in Robert P. Newman, "Clandestine Chinese Nationalist Efforts to Punish Their American Detractors," *Diplomatic History* 7, no. 3 (Summer 1983): 205–22.

67. Godfrey Hodgson, *America in Our Time* (New York: Doubleday, 1976), pp. 89, 97.

68. Lead editorial, *Monthly Review* 2, no. 4 (August 1950): 110–17.

69. *World Events* 7, no. 2 (Spring 1950); 7, no. 4 (Fall 1950).

70. Dulles Papers, box 48, Dulles to Holland, August 17, 1950. Dulles slandered McCune, who died before his book appeared, by saying he was one of those "who contrast the perfection of communist words with the inevitable imperfection of our fallible deeds."

71. *New York Times*, July 4, 10, August 3, 1950.

72. Among the best accounts along these lines is Victor Navasky, *Naming Names* (New York: Viking Press, 1980).

73. Barnes, ed., *Perpetual War for Perpetual Peace* (Caldwell, ID: Caxton, 1953), p. 657.

74. The late Allan Bloom would not like me to say so, but his discussion of "spiritedness" among the guardians in *The Republic* reminds me of North Korea. See Bloom, "Interpretive Essay," in Bloom, trans., *The Republic of Plato* (New York: Basic Books, 1968), pp. 348–50.

75. Paik Nak-chung, *Pundanŭi Cheje* [Division System] (Seoul: Ch'angbi-sa, 1992).

5

Stalin and the Decision for War in Korea

Kathryn Weathersby

From the earliest years of the cold war, scholars and policymakers in the West have turned to Thucydides' history of the Peloponnesian War to understand as fully as possible the perilous great-power struggle of recent decades. Indeed, no less a cold war personage than General George C. Marshall declared in a speech delivered at Princeton University in 1947 that he doubted "whether a man can think with full wisdom and with deep convictions regarding certain of the basic international issues of today who has not at least reviewed in his mind the period of the Peloponnesian War and the fall of Athens."[1] Statesmen and scholars of the post–World War I period also drew parallels between the European Great War and the ancient struggle between Athens and Sparta, but the analogy to the American/Soviet conflict of the post–World War II period became particularly potent as the existence of nuclear weapons and an unusually malignant foe led many Westerners to fear that without proper vigilance America might, like ancient Athens, fall before her powerful authoritarian opponent.

At first glance, the similarities between the Peloponnesian War and the cold war are striking. Sparta, like the Soviet Union, was an authoritarian, autarchic, land power. Athens, like the United States, was a democratic, commercial, sea power. As the Soviet Communist Party placed highest priority on preserving its hold on power and consequently maintained a vast system of domestic control, the Spartan elite similarly sought to maintain a static, isolated society in order to prevent a revolt by its numerous slaves. Athens, by contrast, like postwar America, was a dynamic, expansionistic society dominated by a new commercial elite concerned above all with amassing wealth and power. Furthermore, Sparta and the Soviet Union both fell behind their Athenian/American rival in wealth and power as their preoccupation with internal stability inhibited economic and technological innovation.

When we examine the outbreak of war between the ancient and the modern rivals, however, the analogy between the two conflicts becomes more problematic. Thucydides concluded that the expansion of Athenian power and the fear in both Sparta and Athens that the other would gain power at their expense drove the two hegemons into a war that neither desired. The Soviet Union and the United States similarly feared an expansion of the military power of the other, but this "security dilemma" did not drive the two superpowers into full-scale war. On the contrary, although both sides continued to fear strategic disadvantage, war between the Soviet Union and the United States became less likely the longer the cold war continued; the most dangerous period was the early years, from the Berlin blockade through the Cuban missile crisis.

While the cold war hegemons avoided military conflict on the scale of the Peloponnesian War, they did, nonetheless, in one case engage one another's armed forces on a large scale over a protracted period of time. During the Korean War of 1950–53, the air forces of the Soviet Union and the United States fought a ferocious and costly war in the skies over North Korea, a conflict both sides succeeded in keeping largely hidden from public knowledge. Soviet ground forces were also heavily involved in the Korean War as trainers, advisors, and suppliers of the Chinese and North Korean armies. Furthermore, during the first months of the war, prior to the entry of Chinese troops in November 1950, Soviet military officers were responsible for planning and overseeing the execution of the war. After Chinese troops took over responsibility of the ground fighting from the North Koreans in November 1950, Chinese commanders directed the war on a day-to-day basis. However, until his death in March 1953 Joseph Stalin continued to exercise the final voice in decision making on the communist side. As the Chinese and North Koreans were dependent on the Soviet Union for military supplies and technical expertise in fighting a conventional war against a formidably equipped adversary, Mao Zedong and Kim Il Sung appealed to Stalin for advice and approval before taking any significant action. The Soviet leader also intervened with military instructions whenever he saw fit.[2]

Unlike the Peloponnesian War, the Korean War did not lead to the destruction of either great power.[3] The modern peninsular conflict did, however, significantly alter the nature and course of the cold war. The war in Korea had the immediate effect of militarizing the Soviet/American conflict. Because the United States and its allies feared that the North Korean attack on South Korea was the opening salvo in a new wave of Soviet aggressiveness, Washington began to view containment in primarily military rather than political terms. The war spurred the U.S. Congress to approve massive rearmament and prompted NATO members to solidify their alliance, increase their armed forces, accept the semipermanent stationing of American troops in Europe, and move toward rearming Germany. In Asia, the outbreak of war in Korea led the United States to conclude a separate peace treaty with Japan, support the French position in Indochina, defend the Nationalist regime in Taiwan, and station troops indefinitely in South Korea, Japan, and the Philippines. On the communist side, the Korean War greatly accelerated the development of modern military forces in China and North Korea; during the war Moscow and Beijing devoted particular attention to creating a navy and air force for the People's Republic of China.[4] Although the Soviet Union was already highly militarized, the war in Korea forced Moscow to devote a still larger proportion of its resources to weapons development as the prolonged combat against American forces revealed with painful clarity the technological and productive might of the United States.[5]

With these thematic resonances between the Korean and the Greek conflicts, the Korean War is an appropriate place to explore the important question raised by the oft-cited analogy between the Peloponnesian War and the cold war. The Korean War is also at present a particularly fruitful focus of examination because, with the recent release of a substantial portion of the vast holdings on the war in Russian archives and the recent publication of a limited but significant collection of high-level documents on the war from Chinese archives, we can at last begin to examine this complex and pivotal conflict from both the "Spartan" and "Athenian" sides, a level of analysis not previously possible.[6] The new documentary evidence on the Korean War from the communist side illuminates many long-standing questions about the war and raises new ones, ranging from the causes of the war to the nature of the alliance on the communist side, the complex dynamics of the armistice negotiations, and the effect of the war on postwar Soviet, Chinese, and North Korean foreign relations. This chapter will examine the aspect of the Korean War that perhaps most directly bears on a comparison of it with the Peloponnesian War—the central, long-contentious question of the cause of the outbreak of war in Korea.

Thucydides' conclusion that Spartan fear of the growth of Athenian power made war between them inevitable has spurred a discussion that echoes the debate over the roles of American and Soviet aggressiveness in causing the cold war. Thucydides' analysis of the Peloponnesian War has been challenged from many sides, from assertions that Athens' power had not in fact increased prior to the outbreak of the war[7] to arguments that Thucydides' own narrative contradicts his analysis, indicating instead that the war had complex causes that operated on several levels.[8] Historians of the ancient world also dispute whether the Greek international system was bipolar or multipolar, with attendant disagreement over the role smaller powers such as Corinth, Sicily, and Corcyra played in the outbreak and course of the war. The importance of culture in the formulation of Athenian and Spartan foreign policy is also contested, as is the role played by individual statesmen, with some scholars arguing, for example, that Pericles' poor judgment was a primary cause of the war.

This essay will argue that the new documentary evidence from Soviet archives indicates that the "security dilemma" identified by Thucydides was indeed the central factor in the outbreak of the Korean War. The existence of this dilemma did not, however, make war between the two Koreas or between the Soviet and American alliances inevitable. Instead, the war came about because one hegemon erroneously concluded that it would be able to succeed in gaining strategic advantage by allowing its client state in Korea to seize quickly the other half of the peninsula before the opposing hegemon could or would intervene in the conflict. The impetus for war came from the two Korean states, but the decision for war was made by the communist superpower. As Soviet foreign policy in the postwar Stalin years was to an extraordinary degree made by Stalin himself, the role of the individual statesman is not in dispute. The role of Soviet political culture in the decision for war in Korea is, on the other hand, more difficult to identify with precision. This essay will suggest that the Bolshevik assumption that the "capitalist " powers would inevitably wage war against the "socialist" world led Stalin to exaggerate the threat the Soviet Union faced in East Asia and thus to risk the fateful action in Korea. At the same time, however, Stalin's ideological blinkers did not make his decision on Korea inevitable. Had the United States let it be known that it would commit American forces to defend the Republic of Korea, the Soviet leader would never have approved a North Korean attack on South Korea.

The Respective Rules of North Korea and the Soviet Union

Since the early 1980s much of the debate over the Korean War has focused on the argument that the conflict was a civil war mistakenly viewed by the Western allies as a manifestation of the superpower struggle between the United States and the Soviet Union.[9] In this interpretation Korea was not as much a Melos, caught in the imperial aspirations of superpowers, as it was the locus of both Sparta and Athens. The war of 1950–53 should therefore be seen as comparable to the Peloponnesian War itself rather than as one aspect of the larger struggle between Athens/America and Sparta/Soviet Union.

On most counts, the Russian documentary sources contradict the civil war thesis quite sharply. They reveal that the outbreak of full-scale fighting along the 38th parallel on June 25, 1950, was not simply an escalation of the border skirmishes that had been occurring along the 38th parallel since the summer of 1949, but was instead a conventional offensive campaign prepared by North Korea and the Soviet Union over a period of several months. Most importantly, though Kim Il Sung pressed Stalin for permission to attack South Korea, the decision to undertake the campaign to seize control over southern Korea was made by Joseph Stalin, not by the North Korean leadership.

In March 1949 while Kim Il Sung was in Moscow concluding the initial series of economic, cultural, and military agreements between the newly established Democratic People's Republic of Korea (DPRK) and the Soviet Union, he asked Stalin about the possibility of reunifying Korea by military means. According to Kim's account of the conversation, Stalin refused this request, saying that it was "not necessary" to attack the South, that the North Korean army could move across the 38th parallel only as a counterattack to an assault by South Korean forces.[10]

After heavy but inconclusive fighting along the 38th parallel in the summer of 1949, Kim Il Sung again requested permission to attack South Korea. After reporting to a Soviet official in Pyongyang on August 12 and September 3 that South Korea was preparing to attack the territory of the DPRK, in keeping with Stalin's guidelines, Kim requested permission to make a roughly equivalent counterattack. He added that "if the international situation permits," which was no doubt a reference to possible American actions, the Korean People's Army could easily seize control of the remainder of the peninsula.[11]

This time, with American forces having withdrawn from South Korea in June, Stalin was ready to entertain Kim's request. On September 11 G.I. Tunkin, political advisor at the Soviet Embassy, was instructed to ask Kim Il Sung for specific information about the size and fighting capacity of the military forces of North and South, the condition of the partisan movement in the South, the likely attitude of the population to an attack by northerners and possible intervention by the Americans. Kim Il Sung and Pak Hŏn-yŏng replied to these questions in conversations with Tunkin over the next two days. Their answers were not persuasive, however, for on September 24 the Soviet leadership decided that an attack on South Korea was inadvisable at that time. The Politburo instructed Ambassador Shtykov to communicate to Kim Il Sung that "from the military side it is impossible to consider that the People's Army is prepared for such an attack" since "North Korea does not have the necessary superiority of military forces in comparison with South Korea." Likewise, "from a political side, a military attack on the South by you is also not prepared for" since "very little has been done to develop the partisan movement and prepare for a general uprising in South Korea." The Politburo explained to Kim that "it is necessary to consider that if military actions begin at the initiative of the North and acquire a prolonged character, then this can give to the Americans cause for any kind of interference in Korean affairs." Consequently, while supporting the goal of unifying Korea by "liberating" the South, the Soviet leadership ruled that at the present time the North Koreans must focus their efforts on developing the partisan movement in the South, preparing for a general armed uprising against the Syngman Rhee regime, and strengthening the People's Army.[12]

Ambassador Shtykov reported to Stalin on October 4 that Kim Il Sung and Pak Hŏn-yŏng received the Politburo directive "in a reserved manner." Kim was clearly disappointed, responding only "Very well." Pak was more expressive, however, stating that the decision was correct, that they must develop the partisan movement more widely. Shtykov added that Kim and Pak subsequently reported to him that they had sent approximately 800 persons to the South to lead the partisans and that the movement was growing.[13]

The Politburo decision of September 24 ended discussion of a military campaign against South Korea for the remainder of 1949. However, on January 19, 1950, while Mao Zedong was in Moscow negotiating the terms of the alliance between the Soviet Union and the nearly established People's Republic of China,

Kim Il Sung again raised the issue, this time with increased urgency. Speaking to Soviet advisors after a reception at the new Chinese Embassy in Pyongyang, Kim "in an excited manner began to speak about how now, when China is contemplating its liberation, the liberation of the Korean people in the south of the country is next in line." Ambassador Shtykov's account of Kim's appeal is worth quoting at length. In the account, Kim continued by stating:

"The people of the southern portion of Korea rely on our armed might. Partisans will not decide the question. The people of the south know that we have a good army. Lately I do not sleep at night, thinking about how to resolve the question of the unification of the whole country. If the matter of the liberation of the people of the southern portion of Korea and the unification of the country is drawn out, then I can lose the trust of the people of Korea." Further, Kim stated that when he was in Moscow, Comrade Stalin said to him that it was not necessary to attack the south, in case of an attack on the north of the country by the army of Rhee Syngmann, then it is possible to go on the counteroffensive to the south of Korea. But since Rhee Syngmann is still not instigating an attack, it means that the liberation of the people of the southern part of the country, and the unification of the country are being drawn out, that he (Kim Il Sung) thinks that he needs again to visit Comrade Stalin and receive an order and permission for offensive action by the People's Army for the purpose of the liberation of the people of Southern Korea. Further Kim said that he himself cannot begin an attack, because he is a communist, a disciplined person and for him the order of Comrade Stalin is law. Then he stated that if it is not possible to meet with Comrade Stalin, then he will try to meet with Mao Zedong, after his return from Moscow. Kim underscored that Mao Zedong promised to render him assistance after the conclusion of the war in China. (Apparently Kim Il Sung has in mind the conversation of his representative Kim Il with Mao Zedong in June 1940, about which I reported by ciphered telegram).

The advisors of the embassy, Ignatiev and Pelishenko, avoiding discussing these questions, tried to switch the discussion to a general theme, then Kim Il Sung came toward me, took me aside and began the following conversation: can he meet with Comrade Stalin and discuss the question of the position in the south and the question of aggressive actions against the army of Rhee Syngmann, that their people's army now is significantly stronger than the army of Rhee Syngmann. Here he stated that if it is impossible to meet with Comrade Stalin, then he wants to meet with Mao Zedong, since Mao after his visit to Moscow will have orders on all questions.

Then Kim Il Sung placed before me the question, why don't I allow him to attack the Ongjin peninsula, which the people's army could take in three days, and with a general attack the people's army could be in Seoul in several days.

I answered Kim that he has not raised the question of a meeting with Comrade Stalin and if he raises such a question, then it is possible that Comrade Stalin will receive him. On the question of an attack on the Ongjin peninsula I answered him that it is impossible to do this. Then I tried to conclude the conversation on these questions and, alluding to a later time, proposed to go home. With that the conversation was concluded.

After the luncheon Kim Il Sung was in a mood of some intoxication. It was obvious that he began this conversation not accidentally, but had thought it out earlier, with the goal of laying out his frame of mind and elucidating our attitude to these questions.

In the process of this conversation Kim Il Sung repeatedly underscored his wish to get the advice of Comrade Stalin on the question of the situation in the south of Korea, since (Kim Il Sung) is constantly nurturing his idea about an attack.[14]

Stalin replied to this fervent appeal from Kim Il Sung with a succinct telegram to Ambassador Shtykov on January 30, 1950. The Soviet leader wrote that he understands the dissatisfaction of Comrade Kim Il Sung, but Kim "must understand that such a large matter in regard to South Korea such as he wants to undertake needs large preparation. The matter must be organized so that there would not be too great a risk. If he wants to discuss this matter with me then I will always be ready to receive him and discuss with him." Stalin instructed Shtykov to "transmit all this to Kim Il Sung and tell him that I am ready to help him in this matter."

Stalin's telegram continued with a second brief message, apparently presenting the price the North Koreans would have to pay for Soviet assistance in reunification.

> I have a request for Comrade Kim Il Sung. The Soviet Union is experiencing a great insufficiency in lead. We would like to receive from Korea a yearly minimum of 25,000 tons of lead. Korea would render us a great assistance if it could yearly send to the Soviet Union the indicated amount of lead. I hope that Kim Il Sung will not refuse us in this. It is possible that Kim Il Sung needs our technical assistance and some number of Soviet specialists. We are ready to render this assistance. Transmit this request of mine to Comrade Kim Il Sung and ask him for me, to communicate to me his consideration on this matter.[15]

Ambassador Shtykov reported to Stalin the following day that Kim received his message with great satisfaction. "Your agreement to receive him and your readiness to assist him in this matter made an especially strong impression. Kim Il Sung, apparently wishing once more to reassure himself, asked me if this means that it is possible to meet with Comrade Stalin on this question. I answered that from this communication it follows that Comrade Stalin is ready to receive you. Kim Il Sung further stated that he will prepare himself for the meeting." Kim also, needless to say, assured Shtykov than he would "take all necessary measures" to secure delivery to the Soviet Union of the requested quantity of lead.[16]

Kim Il Sung and Foreign Minister Pak Hŏn-yŏng traveled to Moscow in late March to discuss the issue with Stalin in greater detail. They remained in Moscow for the month of April, apparently working with Soviet officers to plan the campaign. Documentation on the April deliberations has unfortunately not been released,[17] but we do know that Stalin made final approval of the operation contingent on Kim's securing the consent of Mao Zedong. Consequently, following his return to Pyongyang from Moscow, Kim Il Sung traveled to Beijing to present the plan to Mao. Given his dependence on military and economic support from Moscow, Mao

Zedong had little choice but to consent to the Korean operation, but he apparently resented being presented with this fait accompli.[18] In April and May the large quantities of weaponry and supplies needed for a full-scale offensive operation were shipped from the Soviet Union to North Korea via Manchuria, and Soviet advisors with experience in the war against Germany were dispatched to Pyongyang to draw up the plan of battle.[19]

The Russian documentary record thus reveals quite clearly that while the North Korean leadership fervently wished to launch a military campaign to gain control over southern Korea, the decision to undertake such an operation was made by Joseph Stalin. It is important to emphasize that Russian archival records also reveal that it would have been completely impossible for the North Korean leadership to act alone on a matter of such seriousness. As the hundreds of files on Korea in the Central Committee and Foreign Ministry archives reveal in exhaustive detail, on matters of concern to Moscow, the Soviet Union maintained tight control over its client state in Korea.

The general pattern was that strictly internal issues with no foreseeable impact on the Soviet Union were resolved by the North Korean party and government bodies themselves, although Soviet advisors attended the meetings and reported the proceedings to Moscow. On internal matters of greater importance, North Korean party and government leaders cleared their proposals with Moscow before taking action. In a surprising number of cases Stalin himself reviewed the questions concerning North Korea. For example, on February 4, 1960, Kim Il Sung requested a meeting with Ambassador Shtykov to report the proposed agenda for a meeting of the Supreme People's Assembly. Shtykov relayed the agenda to Stalin, Minister of Foreign Affairs Vyshinsky, and six other high Soviet officials, and Stalin informed Shtykov that he had no objections to the proposed agenda. At the same meeting on February 4, Kim Il Sung also requested a Soviet decision regarding whether the DPRK could issue a bond. Stalin gave an answer to that question as well, ruling "it is possible."[20]

The extent and nature of Soviet control over North Korea was not noticeably altered by the withdrawal of Soviet troops in late 1948. The pattern of supervision in 1949 and 1950 was essentially the same as it had been during the occupation period. For example, on January 20, 1948, Shtykov reported to Foreign Minister Molotov about plans to hold a session of the People's Assembly. He asked Molotov to approve both the agenda for the Assembly and the text of a resolution on creation of a provisional constitution for Korea to be adopted by the Assembly. Deputy Foreign Minister Vyshinsky forwarded Shtykov's letter to Stalin on January 24 recommending that Shtykov's proposal be approved and attaching a draft resolution of the Central Committee to that effect.[21] On April 23, 1948, L. Baranov of the Central Committee sent to Deputy Foreign Minister Malik detailed notes on changes to be made to the draft of a provisional constitution for the DPRK, recommending that it be reviewed a second time after the necessary changes had been made. The following day Baranov's comments were sent to Stalin for approval along with a draft Central Committee resolution ordering Shtykov, in case separate elections were held in South Korea, to recommend to Kim Il Sung that he call an extraordinary session of the People's Assembly of North Korea to adopt the attached resolution.[22]

Matters involving contact with other countries, whether political or purely economic, were closely monitored by Soviet officials in Pyongyang. For example, on August 27, 1949, DPRK foreign minister Pak Hŏn-yŏng met with political advisor of the Soviet Embassy G.I. Tunkin to inform him that the Chinese had asked the DPRK to send an additional eight to ten kilowatts of electricity from the Supun power plant. Tunkin recommended that the DPRK satisfy the Chinese request and Pak replied that he would communicate this recommendation to Kim Il Sung and draft a resolution to this effect.[23] The North Koreans also had to secure Soviet permission before joining the International Red Cross. On September 14, 1949, N. Fedorenko at the Soviet Foreign Ministry sent to Andrei Gromyko, for forwarding to Stalin, his recommendations about the advisability of allowing the DPRK to apply for membership in the International Red Cross.[24]

On another political issue, on June 6, 1949, Shtykov informed Moscow of the plans of the DPRK to create a United Democratic Fatherland Front, in the name of which to proclaim a plan to hold general elections in both South and North and to create a single organ of power for all of Korea according to the results of this election. A decision was delayed until June 24, when (now) Foreign Minister Vyshinsky sent a telegram to Shtykov approving the proposal.[25] Similarly, the text of the appeal for peaceful unification issued by the Presidium of the Supreme People's Assembly of the DPRK to the National Assembly of South Korea on June 19, 1950, was first sent to Moscow for approval.[26] Given this level of Soviet control over North Korean affairs, it would have been absolutely unthinkable for the North Korean leadership to attempt a military campaign against South Korea without Soviet approval.

The DPRK was also physically incapable of launching an attack on South Korea on its own since it lacked the necessary weapons, supplies, and military expertise to conduct such an operation. Those essential ingredients were shipped to North Korea from the Soviet Union after Stalin approved the campaign. The DPRK's dependence on Soviet resources and expertise must have been the primary reason why the nationalistic and strong-willed Koreans in power in Pyongyang tolerated a high level of Soviet control over their affairs. The files on Korea in the Soviet Foreign Ministry archive reveal that North Korea was heavily dependent on the Soviet Union for the material resources and expertise needed to construct the new socialist state. Due to Soviet occupation policy and the civil war in China, from 1945 to 1949 North Korea was cut off from its former economic ties with southern Korea, Japan, and Manchuria. Except for very limited trade with Hong Kong and two Manchurian ports, the Soviet Union was the only source of manufactured goods and raw materials not produced internally and the only market for North Korean goods. The DPRK also apparently did not have its own supplies of hard currency, and therefore could not conduct foreign trade on anything other than a barter basis. In 1949 when North Korean delegations attended a youth festival in Budapest, a peace conference in Paris, and a trade union congress in Milan, the DPRK had to appeal to the Soviet Union to provide the delegations with the necessary foreign currency.[27] Furthermore, to an unusual degree, North Korea was dependent on the Soviet Union for technical expertise.[28] Japanese colonial policy had permitted only a small number of Koreans to gain higher education or management experience, and the politics of the Soviet/American occupation prompted most northerners who possessed such skills to flee to the South. Because of these economic and demographic circumstances, the DPRK was much more fully subordinate to the Soviet Union than were the East European states that came under Soviet control.

The North Korean leadership was also subordinate to Moscow for political reasons. According to Shtykov, Kim Il Sung stated in his final appeal for approval to attack South Korea that "he himself cannot begin an attack, because he is a communist, a disciplined person and for him the order of Comrade Stalin is law."[29] In Comintern circles Korean communists had long been infamous for their nationalism, factionalism, and general willfulness,[30] but the Korean communists who rose to power under the Soviet occupation had primary allegiance to the Soviet Communist Party, rather than to the Chinese party or to domestic leaders who had remained in Korea. This deference to the Communist Party of the Soviet Union was certainly not unusual in the early postwar period; even the Chinese communists were remarkably obedient to Moscow from 1945 to 1949.[31] Pyongyang's deference to Moscow, however, was strengthened even further by the circumstances in which the North Korean communists found themselves. Like their rightist counterparts in South Korea, they had been placed in power by the occupation force controlling their half of Korea; they had not, like the Yugoslav or Chinese party, risen to power on their own. Although they seem to have faced little opposition from the population that remained in the North, they nonetheless faced the implacable hostility of the rightist government in Seoul that was backed by American money and expertise.

Furthermore, the North Korean communists were experienced only in guerrilla fighting and underground resistance. As they undertook the massive task of constructing a new socialist state, the only model to which they could turn was the Soviet Union.[32] Not only did they need assistance in running factories, railroads, banks, and so forth, but they also needed to learn how to organize their matters in a proper socialist way. Prior to 1950, the only place to learn socialist state-building was Moscow. The North Korean communists therefore had their own reasons for subordinating themselves to Moscow's superior knowledge and power. After Stalin's death and the weakening of Soviet prestige that followed Khrushchev's de-Stalinization campaign of the late fifties Kim Il Sung was able to develop a distinctly Korean ideology and to maintain a remarkable level of national autonomy within the communist world. In 1949 and 1950, however, his circumstances were sharply different. At the time of the outbreak of the Korean War, the Korean communists were in no position to act independently of Moscow.

To summarize, the role North Korea played in the decision to launch a war against South Korea was to raise the issue. They presented the Soviet leader with the basic ingredients—an army and government willing and eager to seize control of South Korea—and pressed the option. The "civil war" interpretation is thus correct in emphasizing that the leadership of both North and South Korea fervently wished to end the division of their country and to extend their own authority over the other half.[33] Stalin did not devise this plan out of whole cloth and then order North Koreans to attack South Korea. However, while both Korean governments were willing to use military force to bring about reunification, neither was able to do so on its own. Because of the political, economic, and military dependence of both North and South, the decision to wage war for reunification lay not with the Koreans themselves but with their great-power patrons. The war came about because the Soviet Union eventually approved the request of its Korean client while the United States did not.

Stalin's Decision for War in Korea

Having established the relative roles of hegemon and small state in the outbreak of war in Korea, we must now turn to the question of why Stalin decided in early 1950 to allow North Korea to launch a military campaign against South Korea. Unfortunately, while the Russian archival documents provide clear evidence of the locus of decision making, they provide only indirect evidence of the reasoning behind Stalin's decision. The only written explanation Stalin gave for his decision was in a telegram to Mao Zedong on May 14, 1950, in response to Mao's request for confirmation of Kim Il Sung's report that Stalin had approved an attack on South Korea. Stalin's reply verified Kim's account of his meetings in Moscow, explaining that due to the "changed international situation" it was now possible to support the request of the Korean comrades.[34] What Stalin meant by "changed international situation" has thus been the central interpretive question regarding the reasons for the outbreak of war in Korea.

The Russian scholar and diplomat Valeri Denissov emphasizes the importance of the saber-rattling by South Korean leaders in 1949 and 1950 in convincing Moscow to approve an offensive campaign against the South. Denissov, a Korea specialist who served for fifteen years in the Soviet Embassy in Pyongyang and participated in declassifying the Soviet documents on Korea, concludes that the Soviet leadership viewed the withdrawal of American troops from South Korea in 1949 as having unleashed Syngman Rhee, enabling him to attempt reunification by military means. The war-mongering speeches by Syngman Rhee and others in Seoul were taken seriously in Moscow and provided Kim Il Sung with a persuasive argument in favor of a preemptive strike by North Korea.[35]

Another Korea specialist in the Russian Foreign Ministry familiar with the newly declassified documents, Evgenii Bazhanov, argues that because the cold war was in full swing by 1950, a war in Korea had become admissible. Stalin saw the creation of NATO as a serious danger to the Soviet Union; control of all of Korea could offset American control over Japan. In addition, the victory of the Communist Party in China made it seem possible that the North Korean party might also succeed, especially since China could now assist them. Stalin was also influenced, Bazhanov argues, by his acquisition of nuclear weapons and by the Americans' abandonment of the Nationalists in China.[36]

While each of the factors mentioned above undoubtedly played a role in Stalin's thinking, I would argue that the decision for war in Korea can best be understood as part of a general reformulation of strategy toward East Asia that Stalin made in December 1949 and January 1950, as he resolved the thorny question of what kind of alliance to establish with the People's Republic of China. Transcripts of two conversations between Stalin and Mao Zedong released by the Presidential Archive in Moscow provide good evidence of the evolution of Stalin's thinking.

Two months after declaring the establishment of the People's Republic of China in Beijing in October 1949, Mao Zedong traveled to Moscow to secure from Stalin desperately needed economic and military support and to resolve the difficult issue of whether the 1945 treaty between the Nationalist government of China and the Soviet Union would remain in force. This treaty recognized Soviet control over Outer Mongolia and gave the Soviet Union joint control over the Russian-built Manchurian railroad as well as control over the important Manchurian ports of Dalny and Port Arthur, and wide commercial influence in Sinkiang. The Chinese thus regarded the treaty as an infringement of Chinese sovereignty, an insult that the Chinese communists, despite their ideological kinship with Moscow, wished to remedy.[37]

In the initial meetings between the two communist leaders, on December 16, Stalin stated rather quickly that they "must ascertain whether to declare the continuation of the current 1945 treaty of alliance and friendship between the USSR and China, to announce impending changes in the future, or to make these changes right now." The main difficulty with changing the 1945 treaty, Stalin explained, was that it was concluded as part of the Yalta agreement with the United States and Great Britain. Since this agreement provided for the territorial gains in East Asia that the Soviet Union received in exchange for its participation in the war against Japan, "we, within our inner circle have decided not to modify any of the points of this treaty for now, since a change in even one point could give America and England the legal grounds to raise questions about modifying also the treaty's provisions concerning the Kurile Islands, South Sakhalin, etc."[38]

Stalin suggested instead that they keep the treaty formally in force while in practice modifying those conditions that the Chinese wished to alter. For example, Moscow could formally maintain "the Soviet Union's right to station its troops at Port Arthur while, at the request of the Chinese government, actually withdrawing the Soviet armed forces currently stationed there." Such a resolution was not what

Mao wished, but he nonetheless agreed with Stalin's proposal, stating that "in discussing the treaty in China we had not taken into account the American and English positions regarding the Yalta agreement. We must act in a way that is best for the common cause. This question merits further consideration. However, it is already becoming clear that the treaty should not be modified at the present time, nor should one rush to withdraw troops from Port Arthur." The two leaders then spent the remainder of the conversation discussing economic questions.[39]

On January 2 Vyacheslav Molotov informed Mao Zedong that Stalin had decided to abandon the 1945 Sino-Soviet treaty.[40] Stalin then confirmed this policy change on January 22, when after a tense delay of more than a month, he had a second conversation with Mao Zedong. At this meeting Stalin immediately raised the issue of the 1945 treaty, informing Mao that "we believe that these agreements need to be changed, though earlier we had thought that they could be left intact." Mao agreed, pointing out that while the 1945 treaty had spoken of cooperation in the war against Japan, attention must now be turned to preventing future Japanese aggression. As the two leaders discussed specific provisions for a new treaty, Stalin stated that the present agreement on Port Arthur was inequitable. When Mao responded that "changing this agreement goes against the decisions of the Yalta Conference" Stalin replied, "True, it does—and to hell with it! Once we have taken up the position that the treaties must be changed, we must go all the way. It is true that this entails certain inconveniences, and we will have to struggle against the Americans. But we are already reconciled to that."[41]

Stalin and Mao went on to discuss general terms for settlements on Port Arthur, Dalny, and the Manchurian railroad, and for military and economic cooperation between the two countries, terms that eventually formed the basis for the Sino-Soviet treaties signed in Moscow in February.

These pivotal conversations between Stalin and Mao Zedong reveal that a significant change occurred in Stalin's thinking from the time of his initial meeting with Mao on December 16, 1949, to their second meeting on January 22, 1950. In December Stalin's primary concern was to avoid losing the territorial gains secured by the Yalta agreement, a concern that had guided his foreign policy since 1945.[42] Two weeks later, however, he was willing to throw away the security Yalta had provided, a dramatic step that would necessarily affect the Soviet position in Europe as well as in East Asia.

Why did Stalin make such a radical change in his security strategy? Possession of an atomic bomb and the victory of the Chinese Communist Party (CCP) cannot account for this shift, since those significant developments occurred in July and October 1949. Given the timing and the logic of the situation Stalin faced, the most likely cause for this strategic shift was the Truman administration's adoption of a new Far Eastern strategy, NSC-48/2, on December 30, 1949. For military, economic, and political reasons, the United States decided to avoid intervention on the Asian mainland and instead to maintain control over island possessions in the Pacific, particularly Japan and the Philippines. President Truman referred to the new strategy in a press conference on January 5 and Secretary of State Dean Acheson outlined the new policy in his infamous "defense perimeter" speech to the National Press Club on January 12. It is likely, however, that Stalin knew of the change in American strategy before January 2, through British spies operating in Washington and London. At any rate, he clearly was aware of the new American policy by January 7, when Molotov discussed Acheson's speech with Mao.[43]

From Stalin's viewpoint, the American withdrawal from the Asian mainland in the wake of the communist victory in China must have implied U.S. acceptance of a new Sino-Soviet alliance. The Yalta system in East Asia was thus no longer operative; Stalin would have to secure Soviet gains through different means. Complicating the situation was that Stalin knew that some members of the Chinese communist leadership, particularly Zhou Enlai, had spoken of the advantages to China of balancing Soviet influence in China with ties to the United States.[44] Aware that the Chinese communists had good reason to doubt Soviet concern for their welfare, given the troubled history of CCP relations with Moscow since 1927, Stalin had to offer the Chinese terms that would tie the huge new communist state firmly to the Soviet Union. Molotov's conversation with Mao Zedong on January 17 reveals considerable anxiety about American and British attempts to drive a wedge between the two communist states. Stalin's chief deputy visited Mao personally in order to formulate a joint response to what he termed the "slanderous" charge of Secretary of State Acheson that Moscow was attempting to separate the northern portions of China from Beijing's control.

In light of the dramatically altered configuration of power among China, the Soviet Union, and the United States, Stalin decided to "go all the way" in revising the treaties regulating international relations in East Asia. Abandoning the outmoded Yalta system, he would secure Soviet strategic interests through an alliance with the People's Republic of China that offered mutual defense and Soviet economic assistance in exchange for continued Soviet control over the Manchurian ports and railroad, the maintenance of Soviet control over Outer Mongolia, and the exclusion of other powers from Xinjiang. Anticipating that the United States would revive Japanese military power, the Sino-Soviet treaty provided mutual defense in case of war against Japan or any state allied with Japan.

Having secured Soviet strategic interests in northern China, Stalin then moved quickly to eliminate the danger that the Korean peninsula could again be used as a staging ground for Japanese (now Japanese/American) aggression against China or the Soviet Union. In January 1950 the United States declared that South Korea lay outside its defense perimeter, Syngman Rhee's regime was weak, and the leadership of North Korea was eager to extend its control over the remainder of the peninsula. Stalin was determined above all to avoid military conflict with the United States until the Soviet Union could recover sufficiently from the devastation of World War II, and therefore feared that despite its new strategic policy the United States might nonetheless intervene in defense of South Korea. The objective conditions indicated, however, that if an offensive against South Korea were carried out quickly and efficiently the Soviet Union would be able to eliminate the security danger posed by an American-allied South Korea with minimal risk of an expanded conflict. Stalin therefore cabled his ambassador in Pyongyang on January 30 to inform Kim Il Sung that he was ready to "assist him in this matter," adding that the operation "must be organized so that there would not be too great a risk."

As Soviet and North Korean military officers prepared the campaign in the weeks following Stalin's January decision, the United States gave little indication that it would change its policy toward the defense of South Korea. Early in May, the chairman of the Senate Foreign Relations Committee, Tom Connally, stated that South Korea would probably be overrun by the communists "whether we want it to or not." Responding to Connally's statement, Secretary of State Acheson refused to commit the United States to the use of force to prevent such an occurrence.[45] Nonetheless, Stalin remained anxious to minimize the risk that the military operation in Korea might pull the Soviet Union into armed conflict with the Americans. Only days before the attack he withdrew Soviet military advisors from the front line[46] and refused Kim Il Sung's request that Soviet advisors man naval vessels for the assault.[47] One cannot avoid the conclusion that if the United States had made it clear that it would not tolerate a North Korean attack on South Korea, Stalin would never have approved the operation.

Thucydides and Stalin's Korean Decision

What can we conclude from the Soviet documentary record about why war broke out in Korea in June 1950? Did the increased power of one hegemon, like that of ancient Athens, make war between the two great powers inevitable? What were the roles of great power and small state and of individual statesmen? Finally, how did the political culture of the large and small states contribute to the coming of war?

The evidence is most convincing regarding the relative roles of hegemon and small state in the outbreak of the Korean War. The eagerness of the North Korean leadership to end the division of their country through a military campaign against the South Korean regime gave the Soviet Union a convenient tool with which to pursue its strategic interests. It must be emphasized, however, that Kim Il Sung's persistent appeals did not pressure Stalin into eventually approving the campaign. The record of Stalin's dealings with political leaders in territories under his control makes it all too clear that the Soviet ruler could have and would have eliminated Kim at any moment if he had concluded that Soviet interests would be served by doing so. The division of Korea, the extreme polarization of Korean political forces, and the formal commitment of the great powers to the eventual reunification of the country created circumstances that facilitated a military reunification by one side. But conditions within Korea did not make it inevitable that one of the hegemons would eventually unleash its client. As the post-1953 period has shown, the division of the peninsula could have gone on indefinitely if both great powers had considered it in their interests to ensure that the stalemate in Korea continue.

The decision for war in Korea was clearly made by Stalin, but was this decision inevitable? Did the growth of American power, like that of ancient Athens, force Stalin to make this fateful strike against the remaining American outpost on the Asian mainland? The Soviet documentary record suggests that Stalin's decision regarding the advisability of a military campaign against South Korea was in fact based on his perception of a "security dilemma." Fearing a revived Japan allied with the United States, he decided it was worth incurring some risk in order to deny Japan/America the use of the southern portion of Korea as a staging ground for offensive operations against the Asian mainland.

Stalin's fear, however, was based on the assumptions underlying his worldview, not on "objective reality," as he would phrase it. In January 1950 Japan and the United States had not rearmed, NATO was an alliance with little military substance, and all indicators pointed to a long-term withdrawal of America presence from the East Asian mainland. Stalin took the fateful step in Korea that led to a massive expansion of American military power and commitment because his view of the world blinded him to the reality of the situation. As an Old Bolshevik from the revolutionary era he assumed that the "capitalist" powers, in their death throes, would eventually wage war against the "socialist" world. It was only a matter of time before the Soviet Union would face this final test; it was therefore his responsibility to secure Soviet defenses as firmly as possible in preparation for the inevitable conflict. Stalin could compromise with Western powers on specific issues when the situation required it, but he was unable to imagine a long-term, peaceful balance of power between the Soviet Union and the United States and its allies. Stalin's ideological blinders did not, however, make war in Korea or elsewhere inevitable. The Soviet tyrant was a cautious, pragmatic statesman; in both domestic and foreign affairs, he moved only when the ground had been prepared and conditions appeared favorable. He was nothing if not constantly and keenly aware of the configuration of power. Thus, if the United States had let it be known that it would defend South Korea against attack from the North, the war in Korea would never have begun. The messiness of foreign policymaking in a democracy thus combined with the blinkered foreign policy of a communist state to produce the unexpected and unwanted war in Korea.

Notes

1. Quoted in Robert Gilpin, "Peloponnesian War and Cold War," in Richard Ned Lebow and Barry S. Strauss, eds., *Hegemonic Rivalry: From Thucydides to the Nuclear Age* (Boulder: Westview Press, 1991), 31.

2. This analysis is based on the large collection of high-level documents on the Korean War released by the Archive of the President of the Russian Federation. For translations into English of a sizable portion of these documents see *Cold War International History Project Bulletin*, Issue 6–7 (Winter 1995/96): 30–84.

3. The war brought immense destruction to Korea, however, particularly to the northern half, which was subjected to massive, sustained saturation bombing by American planes.

4. For documentation on the creation of the Chinese navy and air force see *Cold War International History Project Bulletin,* Issue 6–7 (Winter 1995/96): 30–84.

5. During the Korean War, particularly in 1952, the Soviet Union carried out a systematic effort to locate downed American equipment and transport it to military institutes in Moscow. Based on interviews and archival research in Russia, Paul Lashmar has argued persuasively that access to the latest American military technology, particularly to the F-88 fighter jet, played an important role in the subsequent development of Soviet military capability. See P. Lashmar, "POW's, Soviet Intelligence and the MIA Question," unpublished paper presented at a conference organized by the Korea Society of Washington titled "The Korean War: An Assessment of the Historical Record," held at Georgetown University, July 24–25, 1995. See also the BBC film directed by Mr. Lashmar titled *Korea, Russia's Secret War,* which was broadcast in the United Kingdom in January 1996.

6. For detailed examinations of the Chinese role in the Korean War based on recently released documentary material from China see Chen Jian, *China's Road to the Korean War: The Making of the Sino-American Confrontation* (New York: Columbia University Press, 1994); and Zhang Shu-Guang, *Mao's Military Romanticism: China and the Korean War 1950–1953* (Lawrence: University of Kansas Press, 1995). Since 1991, the archives of the Soviet Foreign Ministry, the Central Committee of the Communist Party of the Soviet Union, and the General Staff of the Armed Forces of the Soviet Union have gradually made a large portion of their holdings on Korea accessible to scholars. In addition, in 1995 the Presidential Archive (the Kremlin repository that holds documents of the greatest sensitivity) released 1,200 pages of high-level documents on the Korean War. For discussion of the new Russian sources see articles by this author in the *Cold War International History Project Bulletin,* Issue 5 (Spring 1995), and Issue 6/7 (Winter 1995/96).

7. See Donald Kagan, *The Outbreak of the Peloponnesian War* (Ithaca, N.Y.: Cornell University Press, 1969).

8. Richard Ned Lebow, "Thucydides, Power Transition Theory and the Causes of War," in Lebow and Strauss, *Hegemonic Rivalry,* 125–165.

9. For the most substantial argument that the war should be classified as a civil war, see Bruce Cumings, *The Origins of the Korean War: Liberation and the Emergence of Separate Regimes, 1945–1947* (Princeton: Princeton University press, 1981), and *Origins of the Korea War II: The Roaring of the Cataract* (Princeton: Princeton University Press, 1990).

10. Ciphered telegram from Soviet ambassador to the DPRK Terentii F. Shtykov to Soviet foreign minister Andrei Vyshinsky. Archive of the Foreign Policy of the Russian Federation, Fond 059a, Opis 5a, Delo 3, Papka 11, Listy 87–91.

11. Ciphered telegram from Shtykov to Vyshinsky, September 3, 1949. Archive of the Foreign Policy of the Russian Federation, Fond 059a, Opis 5a, Delo 4, Papka 11, Listy 136–138. Ciphered telegram from Andrei Gromyko at the Soviet Foreign Ministry to G.I. Tunkin, political advisor at the Soviet Embassy in Pyongyang. Archive of the Foreign Policy of the Russian Federation, Fond 059a, Opis 5a, Delo 3, Papka 11, Listy 45.

12. Politburo directive to the Soviet ambassador in Pyongyang. AVPRF (Russian Foreign Ministry Archives, Moscow), Fond 059a Opis 5a, Delo 3, Papka 11, Listy 75–77.

13. Ciphered telegram from Shtykov to Stalin, October 4, 1949. AVPRF, Fond 059a, Opis 5a, Delo 3, Papka 11, Listy 78.

14. Ciphered telegram from Shtykov to Vyshinsky, January 19, 1950. AVPRF, Fond 059a, Opis 5a, Delo 3, Papka 11, Listy 87–91.

15. Ciphered telegram from Stalin to Shtykov January 30, 1950. AVPRF, Fond 059a, Opis 5a, Delo 3, Papka 11, Listy 92.

16. Ciphered telegram from Shtykov to Stalin January 31, 1950. AVPRF, Fond 059a, Opis 5a, Delo 3, Papka 11, Listy 92–33.

17. Korea specialists in Moscow who served on the declassification committee have all stated that they have been unable to locate the records of the April meetings. It is possible

that these records are stored only with military documents and not in the Foreign Ministry archive or the Presidential Archive. At this point the question of their whereabouts remains unresolved.

18. In a conversation with the Soviet ambassador to Beijing on March 31, 1956, Mao stated that on the question of the Korean War, "we were not sufficiently consulted." Ciphered telegram to Moscow from P. Yudin, the Soviet ambassador to Beijing. Archive of the President of the Russian Federation, List 157, Fond, Opis, and Delo not given. From the collection obtained by the Cold War International History Project and the Center for Korean Research of Columbia University, which has been deposited at the National Security Archives in Washington, D.C.

19. For documentation of these statements and for the full texts of documents quoted earlier see the *Cold War International History Project Bulletin,* Issue 3 (Fall 1993), Issue 5 (Spring 1995), and Issue 6/7 (Winter 1995/96), Woodrow Wilson International Center for Scholars, Washington, D.C.

20. Ciphered telegram of February 7, 1950, from T.F. Shtykov to Soviet foreign minister Andrei Vyshinsky reporting a meeting with Kim Il Sung, held at the latter's request on February 4, (Archive of the Foreign Policy of the Russian Federation, Fond 059a, Opis 5a, Delo 4, Papka 11, Listy 145–146). For a translation into English of the text of the document see the *Cold War International History Project Bulletin* Issue 6/7 (Winter 1995/96): 36.

21. AVPRF, Fond 07, Opis 21, Delo 316, Papka 22, Listy 1-4.

22. Russian Center for the Preservation and Study of Documents of Recent History (the former Central Party Archive), Fond 17, Opis 128, Delo 1173, Listy 47–51.

23. Tunkin to Gromyko and Vyshinsky at the Foreign Ministry, August 27, 1949 (Archive of the Foreign Policy of the Russian Federation, Fond 7, Opis 22, Delo 232, Papka 37).

24. AVPRF. Fond 0102, Opis 5, Delo 71, Papka 16, Listy 14–15.

25. Chronology of events in Korea prepared by the Russian Foreign Ministry's declassification committee. It has no archival location, but a copy has been deposited at the National Security Archives in Washington, D.C., where it is available to all researchers.

26. Archive of the Foreign Policy of the Russian Federation, Fond 07, Opis 23a, Delo 25c, Papka 20.

27. AVPRF, Fond 0102, Opis 5, Delo 20, Papka 12, Listy 1–5; Delo 21, Listy 2, 4; and Delo 22, Listy 1. It seems likely that Soviet occupation authorities shipped to the Soviet Union any supply of hard currency found in North Korean banks or enterprises.

28. A major portion of the records on Korea in the Foreign Ministry archive in Moscow are requests from North Korea for assistance in training workers in virtually every branch of economic and cultural activity and Soviet arrangements for fulfilling these requests. The level of technological dependency of North Korea is one of the most significant ways in which DPRK relations with Moscow differed from Soviet relations with its satellite states in Eastern Europe.

29. Ciphered telegram from Ambassador T.F. Shtyhov to Foreign Minister Andrei Vyshinsky, January 19, 1950 (Archive of the Foreign Policy of the Russian Federation, Fond 059a, Opis 5a, Delo 3, Papka 11, Listy 87–91).

30. The most influential criticism of the Korean communists came in an article published in *Revoliutsionnii vostok* (Revolutionary East) in 1931 by Otto Kuusinen, a member of the Comintern Executive Committee with a special interest in the Korean movement. Kuusinen excoriated the Koreans for ideological deviationism and factionalism. Dae-Sook Suh, ed., *Documents of Korean Communism, 1918–1948* (Princeton: Princeton University Press, 1970), 257–282.

31. For persuasive evidence of Chinese communist deference to Moscow based on recently released documentary sources from Nanjing and Taipei, see Odd Arne Westad, *Cold War and Revolution: Soviet-American Rivalry and the Origins of the Chinese Civil War, 1944–1946* (New York: Columbia University Press, 1993).

32. The Foreign Ministry and Central Committee archives include numerous records of visits to Soviet ministries by official delegations from the DPRK, who always brought with them long lists of practical questions about how to organize and manage schools, hospitals, youth organizations, industrial enterprises, parks, and so forth.

33. For an account of initiatives for war from the southern leadership, see Bruce Cumings, *The Origins of the Korean War*, vol. 2, *The Roaring of the Cataract* (Princeton: Princeton University Press, 1990).

34. Ciphered telegram from Vyshinsky to the Soviet Ambassador in Beijing, sending the text of a message from Stalin to Mao Zedong (Archive of the Foreign Policy of the Russian Federation, Fond 059a, Opis 5a, Delo 3, Papka 11, Listy 106.

35. Valeri Denissov, "The Korean War of 1950–1953: Thoughts About the Conflict's Causes and Actors," unpublished paper presented at Georgetown University, July 24, 1995, at a conference organized by the Korea Society of Washington, D.C.

36. Evgenii Bazhanov, "Assessing the Politics of the Korean War, 1949–51," *Cold War International History Project Bulletin*, Issue 6–7 (Winter 1995/96): 54, 87–91.

37. For an extensive examination of the tension between Stalin and Mao over the territorial issues see Sergei N. Goncharov, John Lewis, and Xue Litai, *Uncertain Partners: Stalin, Mao and the Korean War* (Stanford: Stanford University Press, 1993).

38. Record of Conversation between Comrade I.V. Stalin and Chairman of the Central People's Government of the People's Republic of China Mao Zedong on December 16, 1949. Archive of the President of the Russian Federation, Fond 45, Opis 1, Delo 329, Listy 9–17. Quotations are from the translation by Danny Rozas, *Cold War International History Project Bulletin*, Issue 6–7 (Winter 1995/96): 5–7. It should be noted that according to the translator of the memoirs of Shi Zhe, Mao Zedong's translator at his meetings in Moscow, the Russian documents agree with Shi's account, though they are more detailed. See Chen Jian, "Comparing Russian and Chinese Sources: A New Point of Departure for Cold War History," *Cold War International History Project Bulletin*, Issue 6–7 (Winter 1995/96): 20–21.

39. *Cold War International History Project Bulletin* 6–7 (Winter 1995/96): 5–7.

40. Mao jubilantly cabled this news to Beijing on January 2. Goncharov, Lewis, and Xue, *Uncertain Partners*, p. 242.

41. Record of Conversation between Comrade I.V. Stalin and Chairman of the Central People's Government of the People's Republic of China Mao Zedong January 22, 1950. Archive of the President of the Russian Federation, Fond 45, Opis 1, Delo 329, Listy 29–38. Quotations are from the translation by Danny Rozas, *Cold War International History Project Bulletin*, Issue 6–7 (Winter 1995/96): 7–9.

42. For a detailed and persuasive argument that Stalin considered the Yalta agreement to be his crowning achievement as a statesman and the foundation of the USSR's international legitimacy see Vladislav Zubok and Constantine Pleshakov, *Inside the Kremlin's Cold War: From Stalin to Khrushchev* (Cambridge: Harvard University Press, 1996).

43. Memorandum of Conversation between Molotov and Mao, AVPRF, Fond 07, Opis 23a, Papka 18, Delo 234, Listy 1–7.

44. See Vladislav Zubok, "To Hell with Yalta! Stalin Opts for a New Status Quo," *Cold War International History Project Bulletin* Issue 6–7 (Winter 1995/96): 24–27.

45. William Stueck, *The Korean War, An International History* (Princeton: Princeton University Press, 1995), 16–37.

46. Nikita Khrushchev (Strobe Talbott, ed.), *Khrushchev Remembers* (Boston, 1970), 370. Khrushchev's account is consistent with archival evidence that Kim Il Sung asked Stalin on July 8 to allow the use of Soviet military advisors at the front (ciphered telegram from Shtykov to Stalin, July 8, 1950. AVPRF; Fond 059a, Opis 5a, Delo 4, Papka 11, Listy 151).

47. Dmitrii Volkogonov, "Sleduyet li etogo boyat'sia?" (Should we fear this?), *Ogonyok*, No. 26 (June 1993): 29.

6

The Effects of the Peloponnesian (Athenian) War on Athenian and Spartan Societies

Paul Cartledge

But we were always wrong in these predictions.

—*James Fenton, "Dead Soldiers," from* Memories of War[1]

This chapter is above all an essay in problematization: in problem-posing as opposed to problem-solving.[2] I begin with general problems of three kinds, implicit in all historiography, but each having a particular application to the subject in hand: problems respectively of perspective, of periodization, and of consequentiality (the first section). Next, I consider problems of evidence and its interpretation that are peculiar to what we (merely) conventionally call the Peloponnesian War, with special reference to its aftermath (the second section). Finally, I attempt to draw up a sort of balance sheet, explicitly subjective and moralizing and consciously not confined within the canons of supposed historiographical objectivity (the final section).

This chapter is written in what is—for some of us—a genuinely post-war era, following the ending of the Cold War between the two so-called superpowers.[3] That happy fact, followed as it has been by the fiftieth anniversary of the conclusion of the Second World War and by the passage of well over forty years since the Korean War, has occasioned a paroxysm of retrospection and introspection, or rather a politics of memory—sometimes creative, sometimes destructive, sometimes maybe recovered, but always and inevitably selective.[4]

The problem of viewing in its proper perspective the aftermath of the Second World War, whether globally, nationally, or locally, poses instructive and usually cautionary questions for our present historiographical exercise. For example, has it been the longest period of uninterrupted peace in Western history, thanks to the more or less real and imminent threat of thermonuclear extinction—or has it, rather, been mainly a continuation of that war by other means?[5] Nationally, has the unquestionably unprecedented economic growth also been accompanied by, or indeed caused, unprecedented collective anomie and despair?[6] Locally, even parochially, the aftermath of the Second World War saw the beginning of the mechanized, labor-intensive, heavily subsidized modern era of British farming, yet is not East Anglia, from where I write, the very epicenter now of high tech, "prairie" style overproduction of cereals—and rural unemployment—in England?[7]

In short, the current period of agonized reappraisal would seem to offer an appropriate perspective for reflection on the no less contested aftermath of a major ancient war, the so-called Peloponnesian or Athenian War of 431 to 404 B.C. This is a conflict that, according to its original historian's own optimistic self-assessment, generated a "possession for all time" (Thucydides 1.22.4), but that other commentators both ancient and modern have lamented as representing a (possibly, the) nadir of ancient Greek civilization and culture.

By an unfortunate coincidence (due to the not quite accurate chronography of the sixth-century doxographer Dionysius Exiguus) the Peloponnesian War and its immediate aftermath happen to coincide with what we label the turn of the fifth and fourth centuries B.C. Periodization is both a necessary convenience and an aprioristic, prejudiced, and prejudicial distortion of the seamless flow of past events and processes; it therefore requires constant rethinking, even if that may lead to questioning why we should continue to pay so much attention to ancient Greece and Rome.[8] The intrinsic arbitrariness of the periodization procedure is what causes the problem here. For thanks to our habit of thinking in centuries, a new century tends automatically to be identified with a new era or epoch of human history.[9] In this case, as often, the new has tended to be seen as inferior, and "the fourth century" is thus cast as the Cinderella of the "Classical" Greek epoch. Some of us believe that a time in which there flourished among others Democritus, Eudoxus, Plato, Demosthenes, and Aristotle cannot be all bad, but nevertheless the stigma remains. Though I am no devotee of "what if" historiography, I do therefore like to ask my students to consider what might be the consequences for our periodic conceptualization if the Peloponnesian War had happened to end in (say) 435 B.C.

The issue of consequences, indeed, constitutes the third of our preliminary problems, recalling those faced by utilitarian and other consequentialist philosophies: what, precisely, is a determinate and attributable consequence of what, and when may the consequences of any particular action, or set of actions, be said to end? It is no doubt fairly easy to distinguish abstractly between short-, medium- and long-term consequences, but it is much less easy to do so in practical historiographical actuality, where canons of probability or plausibility rather than certainty are the order of the day.[10]

Ever since Herodotus, evaluating causation has been of the essence of the historian's task.[11] Here, the particular task is to distinguish between, on the one hand, consequences "caused" by the Peloponnesian War in the strongest sense (it was the Peloponnesian War that "made the difference"), and, on the other, consequences that were continuations of processes already under way before the war but significantly affected by it, that is, exacerbated, aggravated, or especially—all "great" wars tend to have this effect—accelerated thereby. A further requirement is to distinguish both those kinds of consequences from changes that "arose from particular post-war needs which had little or nothing to do with the war."[12] However one frames, analyzes, and putatively "solves" the philosophic and/or practical problems, Aristotle's methodological warning that different intellectual projects have different inherent degrees of exactitude merits special regard in this connection. The historian's episteme must in those terms rank pretty low on Aristotle's—or anybody's—scale, notwithstanding the contrary claims of our principal ancient source, Thucydides.

Thucydides was the historian of the War—both definite articles are crucial. The revisionist, one-war theory is his, whereas other contemporary commentators discerned at least two separate wars between 431 and 404; we conventionally call the War "Peloponnesian," not (as the Spartans and their allies saw and called it) "Athenian," thanks to "Thucydides the Athenian," as he styles himself in his opening words; and it is he who has given the (one) War its peculiar shape. In short, "no other war, or for that matter no other historical subject, is so much the product of its reporter."[13] Thucydides is thus the classic instance of a widely received modern historiographical view, that literary historians "make" history as well as—or rather than—merely record the past.[14] Yet we actually know very little about Thucydides. For example, when did he write? As my earlier remarks on perspective tried to emphasize, it makes a great deal of difference interpretatively whether we read Thucydides as if he were writing retrospectively in say the mid-390s, as opposed to either in 404 and immediately thereafter, or (as he seems to claim) as the War was actually happening.[15] In any case, he did not live to complete his manuscript, which was published posthumously—by whom, we know not.

Thucydides' unique status and stature have been duly recognized both by contemporaries, including especially his at least four continuators (who picked up the narrative thread from more or less where his text broke off in mid-sentence), and by moderns—from Guicciardini in the sixteenth century, via Hobbes in the seventeenth, through Macaulay and von Ranke in the nineteenth (the eighteenth had preferred Tacitus) to the army of his modern devotees. Thucydides tends to be the historian's historian, caviare to the modern career general, perhaps, but not to the reflective amateur politician-general of the type that Thucydides himself embodied. Whatever else Thucydides did or did not set out to do and/or achieve, he indubitably has fixed in Western minds the idea of the magnitude of "his" War. Perhaps his History will eventually prove not to have been "a possession for all time," but it is at any rate because of his work that we discuss "the Peloponnesian War" today. Thucydides' History, to conclude, is therefore in a sense a cause as well as a consequence of the Peloponnesian War.

Why, or rather for whom, did Thucydides write? It was, he stated, "for those who wish to have a clear understanding both of events in the past and of those in the future which will, in all human likelihood, happen again in the same or a similar way."[16] Thucydides, however, deceived himself (and many others since) in claiming that he aimed to and did write the truth, all of it, and nothing but it; and he has deceived many others (and perhaps himself) over the degree of exactitude (*akribeia*) attainable, even or especially by a participant observer. Moreover, he construed very narrowly what it was necessary or appropriate for the historian to record, and he took a very great deal for granted. Hence the need for detailed supplementation and exegesis, as well as continuation, of his austerely brilliant text.

Of his continuators and would-be emulators four are known, but none can claim the same historiographical high ground, and of only two does enough survive for us to be able to form a reasoned and reasonable estimate of their outlook and quality. First, there is the "Oxyrhynchus Historian"—the sadly fragmentary remains on papyri from Egypt of a not certainly identifiable (the Athenian Cratippus?) Thucydidean-style pragmatic historian of the earlier fourth century who at least in some crucial respects served as source for the "universal" history of Ephorus (mid-fourth century B.C.). The latter in turn served as the main source for the relevant portion of the extant "Library of History" by Diodorus, a Sicilian Greek writer of the first century B.C.[17] In the Oxyrhynchus Historian–Ephorus–Diodorus tradition we have a serious alternative to the Athenian exile Xenophon, who is the best known of Thucydides' continuators. Xenophon was not perhaps quite as exclusively pro-Spartan as has usually been thought, but he was sufficiently in thrall to his one-time patron, Agesilaus, to publish an unstintingly eulogistic "biography" of the Spartan king on his death in about 360/359. Distinctly undemocratic in outlook, Xenophon made the most of his native Athens' post–Peloponnesian War discomfiture to point an aristocratic moral and adorn an oligarchic tale.[18]

Other literary sources include Aristophanes prominently among the contemporaries, but his fantasy-comedy is hard to interpret in social or political terms despite or because of its pseudo-realism.[19] So too the law court and assembly speeches of contemporary or near-contemporary Attic (Athenian) orators have a clear but ambiguous connection to the realities they purport to exploit or change.[20] In a class of its own is the "Athenian Constitution" treatise attributed to Aristotle or his School: written in the 330s and 320s shortly before the effective demise of democracy at Athens, it presents an untypically upbeat reading of Athens' post-War constitutional development.[21] Of non-contemporaries, Plutarch, who wrote Spartan as well as Athenian lives (the former—Lysander, Agesilaus—perhaps more helpful than the latter), deserves special mention, although his moralizing biography was written in and for an immeasurably different political universe, since by the second century A.D. Greece was but a very minor constituent of the Roman Empire.[22] To the literary sources the documentary inscriptions offer a never less than interesting counterpoint.[23] The non-written—archaeological, art-historical—evidence has its own discourse.[24]

The modern professional historiography of ancient Greece began roughly one hundred and fifty years ago with the publication in 1846 of the first volume of George Grote's unashamedly pro-democratic *History of Greece*.[25] Within this historiography our present problem of the consequences of the Peloponnesian War has provoked not only an abundance of periodical literature but even whole books.[26] Out of this welter there emerge quite sharply etched two opposed scenarios. The first might be called the nightmare or crisis scenario, Apocalypse Then as it were, which echoes the judgment of the second-century A.D. religious traveler Pausanias: "Until that time Greece had walked with her feet well planted on the ground, but that war shook her from her foundations like an earthquake."[27] A few modern illustrations of this scenario will have to suffice.

Among the popularizers an intriguing debate set Edith Hamilton against Herbert Muller. Whereas for Hamilton Athens' defeat in the Peloponnesian War represented nothing less than the defeat of the cause of humanity, because "Greece's contribution to the world was checked and soon ceased," in Muller's view "for the lover of Athens . . . the real tragedy [of her defeat] was . . . that the Greeks made their deepest mark on the world in the centuries following this defeat." Both, in other words, firmly espoused the by then conventional notion of a supposed Golden Age acme of Greek civilization and culture in the fifth century, a notion due ultimately to the fourth-century Athenian public orators (below), but they differed as to their interpretation of the effects of its forcible termination in and by the Peloponnesian War.[28]

Between the popular and the scholarly registers falls Donald Kagan's magisterial *On the Origins of War and the Preservation of Peace*. The author selects five case studies, in order to consider the questions of how and why major wars break out, or (more especially) how and why peace breaks down: for his para-historiographical, political aim is the Thucydidean one of providing practical guidance for the future. The first of Kagan's case studies is the Peloponnesian War, almost inevitably, since this is a subject on which he has written no less than four scholarly books. But although the new book is supposedly about that War's origins, Kagan does not scruple to deliver also and in no uncertain terms his view of its consequences. "The war was a terrible watershed in Greek history, causing enormous destruction of life and property, intensifying factional and class hostility, dividing the Greek states internally and destabilizing their relationship to one another." So far, so bad. But it gets worse, . . . "ultimately weakening the Greek capacity to resist conquest from outside." Finally, for good—or ill—measure: "Sparta's victory also reversed the tendency towards the growth of democracy. When Athens was powerful and successful, its democratic constitution had a magnetic effect on other states, but its defeat was the turning point in the political development of Greece that sent it in the direction of oligarchy instead of democracy." In short, "The war was a tragic event, a great turning point in history, the end of a period of confidence and hope, and the beginning of a darker time."[29]

Exclusively scholarly versions of the nightmare scenario may be divided up according to subject emphasis. One learned French student of Athenian and Greek post-War "ideology" begins his dense, almost 350-page monograph with the ringing pronouncement that "*la cité du IVe siècle est et se sait malade.*" Lévy generalizes, that is, from what he takes to be an unresolved ideological crisis at Athens, consisting in that one city's failure in the fourth century to heal the rifts that developed during especially the latter stages of the Peloponnesian War, to the assumption (shared by many others, not only Kagan) of a general "crisis" (in the loose vernacular sense of practically everything going wrong) of "the" Greek city in the fourth century. Such generalization is literally indefensible, since there were some 700 separate Greek political communities in the Aegean area alone, and well over 1,000 in the Greek world as a whole, so that "the Greek city" is an unacceptably reductionist abstraction. The question remains whether the thesis can be sustained in the weaker form proposed much earlier by Gilbert Murray, who stated that "Greece as a whole felt the tragedy" of Athens' moral as well as physical fall, or whether indeed it is vulnerable even in its specifically Athenian aspect.[30]

Scholars of Athenian and Spartan society and politics tend to be less rhetorical and extreme, if nonetheless astringent. It is hard to point to books or articles that paint a uniformly negative picture of Athens in these respects. On the other hand, as we shall see, her economy (especially the cessation of silver mining), polity (the abolition of democracy and, when democracy was restored, the spate of political trials), culture (the finger is pointed at the demise of great tragic drama), and society (in particular the trial of Socrates on charges of religious abnormality and pedagogical corruption) have all at some time been characterized as at least temporarily disabled or dysfunctional.[31]

Sparta, on the other hand, has usually seemed, paradoxically, to have come out of the Peloponnesian War only too well: that is to say, her greatest military victory proved, remarkably rapidly, to be also her own undoing. Finley's claim that "her greatest military success destroyed the model military state" may perhaps be "altogether too simple," but his former pupil Hodkinson agrees that prolonged external engagement, especially during the Peloponnesian War, did affect the "timing, circumstances, and international repercussions of the Spartiate crisis." And it is not only Sparta's own undoing that is at issue, either: during her paroxysm of imperialist *folie de grandeur* in the first half of the fourth century she managed to wreak untold havoc from Macedon to the Peloponnese and throughout the Aegean world. Viciously narrow oligarchic juntas were imposed by Lysander on Athens, the islands, and Asia Minor, and scarcely less narrow and vicious ones were established by Agesilaus in Boeotia. In addition, by dismembering all Greek multi-state organizations, apart from her own Peloponnesian League, through self-interestedly crooked interpretation of the autonomy clause contained in the King's Peace (also known after its Spartan negotiator as the Peace of Antalcidas), Sparta could be held to have paved the way for the rise of Macedon and thereby the end of the independent Greek city together with its original invention of citizen democracy.[32]

Against these variously negative scenarios must be counterposed a variety of optimistic, even utopian reconstructions. The Peloponnesian War had shown at exhausting length—and precisely by its exhausting length—that no one Greek power could unite politically any significant portion of "Greece" (however defined). Hence the imperious necessity for more sustained and sophisticated intercity diplomacy, which resulted in both a pragmatic quest for neutrality on the part of the small and vulnerable and the expression by the "superpowers" of the seemingly loftier ideals of Common Peace (even if they were too often honored only in the breach), and indeed the first tentative formulation of balance of power theories.[33] Paradoxically, too, Sparta's imperialist promotion of oligarchic regimes throughout its sphere of influence could in retrospect be seen optimistically as having helped to trigger in reaction the most widespread dissemination of various forms of democracy in ancient Greece, indeed the most widespread such dissemination before the modern era. Admittedly, the forms of democracy adopted were usually much less thoroughgoing and radical than the original, Athenian model. But it must be remembered that all ancient Greek political systems, oligarchic no less than democratic, were direct and participatory, in sharp contrast to our modern representative varieties of "democracy"—a term that has indeed been drained of most of its aboriginal content.[34]

So far as Athens specifically is concerned, the defeat was certainly heavy, but she nevertheless can convincingly be argued to have made a remarkably quick post-War economic and political recovery leading to a recrudescence both of population numbers and of the democratic-imperialist impulse.[35] Conversely, Sparta's post-War rush of blood to her (not always very intelligent) head meant that by 360 she had been knocked down and out, for good, as a great Greek power—a considerable blessing for many Greeks, but most especially for the thousands of long-enslaved Greeks in Messenia; and Greece's new balance of power (in fact, weakness) diplomacy was potentially more profitable than any doomed, destructive hegemonial striving on the old models—even if, in practice, the rise of Macedon soon put paid to hopes of genuine long-term independence. Any number of intermediate positions might be espoused on the spectrum between the "Nightmare" and "Utopia" scenarios. In the interests of brevity I mention just one: the soundly based view that "Greek popular morality" (actually, the evidence mainly concerns Athens) remained substantially and consensually stable between—in round figures—430 and 320 B.C.[36]

It is time now to consider in detail the effects of the War on particular facets of Athenian and Spartan society and politics, beginning with Athenian economy. No doubt there was quite widespread impoverishment in the War's immediate aftermath, but to take the *Plutus* ("Wealth") of Aristophanes for unmediated reality, as has been done by Ephraim David, is to mistake comic genius for an official government report. The contrary view developed by Strauss is infinitely more plausible, not least because he is able to make use of Victor Hanson's crucial insight in minimizing the long-term damage inflicted on Athenian agriculture (especially vines and the near-indestructible olives) by Spartan invasion and occupation.[37]

Strauss thus argues effectively against the notions both of a long-lasting general impoverishment and of any serious town-country division in Attica; he does, however, believe that the Athenian thetes (the lowest of the four citizen census groups) suffered twice as many battle casualties as the hoplites (roughly the band from the fiftieth to the ninetieth percentile in wealth), but here his demographics would seem to be questionable.[38]

So far as the Athenians' collective psychology goes, the trial of Socrates in 399 B.C. encapsulates, for some interpreters, a manifestation of collective psychosis or loss of nerve.[39] More telling, probably, is the evidence for serious and unresolved intergenerational conflict.[40] Even a temporary denial of satisfaction of the victory syndrome must have been traumatic, while a powerful case has been mounted for a lasting collective mentality of preoccupation with the military defense of the home territory of Attica.[41] Individually, too, the War will of course have produced emotional as well as physical cripples. A sobering recent study of Greek attitudes toward disability and deformity ends with a telling comparative reference to a Korean War veteran suffering from "post-traumatic stress disorder."[42] Surely there were such severely stressed, as well as more or less severely disabled, war veterans on both sides after the Peloponnesian War, "total" war not being an invention of our century. Yet, for all that War's increased brutality, there was no new technology —such as the artillery shells of the First World War or the atomic and hydrogen bombs of the Second—to provoke exceptional terror of largely unforeseeable and unprecedentedly devastating instruments of war, while war as such—which had been denounced as an evil since Homer—was to remain all too familiar a phenomenon in ancient Greece.

Turning to post-War politics at Athens, we must notice first the overthrow of the Spartan-imposed junta of the Thirty Tyrants; that gain was cemented, moreover, by the General Amnesty of 403, the first such in all known history. Soon, however, the Amnesty was breached in spirit if not in the letter, and indeed throughout the period 403–386 there was an apparent rage at Athens for political trials, the most notorious being that of Socrates. The latter has even been viewed as a breach not merely of the Amnesty but of a supposed fundamental article of Athenian democratic faith, namely freedom of speech and expression.[43] Actually, grossly offensive though these trials are to modern liberal democratic sentiment, there is no reason to see them as extraordinarily rabid or destabilizing, any more than was the spate of judicial executions of top executives (generals, ambassadors). The one truly "McCarthyite" witch-hunt at Athens had already occurred in 415–413, following the manufactured or at least politically manipulated Herms and Mysteries scandals.[44] Moreover, the post-War trials and executions did nothing to prevent, indeed they went hand in hand with, renewed Athenian imperialistic ambition and achievement. This was a consummation desired especially perhaps by the poor majority of citizens, but a widespread imperialist zeal both contributed to, and was premised upon, the restoration of some sort of internal ideological unanimity (*homonoia*).[45]

Likewise furthering civic re-consolidation was the reaffirmation in 403/402 of the Periclean citizenship law of 451, which enjoined that an Athenian citizen must be the legitimate son of two Athenian parents. Indeed, it seems that cohabitation without proper marriage was now (or again?) made illegal, and further buttressing for the institution of citizen marriage was provided by symbolic recourse to the "*hetaira*" figure (literally, and euphemistically, a female comrade or companion, actually a high-class, non-Athenian prostitute) as the antitype or "Other" of the lawfully wedded Athenian citizen wife. If anything, Athenian civic self-identity was thus being circumscribed even more narrowly and exclusively than before.[46] A subsidiary effect, and perhaps motive, of the legislation was to counteract the very small-scale "liberalization" of citizenship access during the War, which had seen the rise to full citizen status of a few ex-slaves and a somewhat freer distribution of honorary Athenian citizenship to foreigners. Either way, the cause of citizen solidarity was fostered, even if it was of a more anxiously policed and tensely self-regarding kind.[47]

According to the author of the Aristotelian "Athenian Constitution," the Athenian constitution after the democratic restoration of 403 became progressively more democratic—that is, gave ever more power to the poor masses of the citizenry; whereas, according to the leading modern scholar of Athenian democratic institutions, there were at least ten points of substance wherein the restored, post-War democracy of the fourth century differed qualitatively from that of the fifth—being more "moderate," or less "radical."[48] Even if we might quarrel with Hansen on individual points (e.g., the introduction of political pay in the 390s also meant that regular attendances at Assembly meetings became larger and so more representatively participatory), we can surely agree that the experience of especially the last phase of the Peloponnesian War and its immediate (tyrannical) aftermath helped pre-empt extremism, promoted "constitutional" over violent means of dispute settlement, and so contributed to prolonging democracy's active life (to 322).[49]

So much, for the time being, for Athens. The best point at which to begin a consideration of post-War Spartan society is the "Cinadon affair" of c. 400. This neatly encapsulates the stresses and strains, exacerbated but not caused by the War, both within the Spartiate full citizen group (known as Homoioi or "Peers") and between the Spartiates and the various (and increasingly numerous) "out"-groups of Spartan society. Within the elite corps of Spartiates, the long-standing gaps between rich and poor citizens and between rich and poor lifestyles were now becoming distinctly more pronounced, as were the political-ideological divisions in attitudes to relations with the outside world. For instance, the acceptance and internal private use of money coined by other cities, which would have both breached the barriers of an ideally "closed" society and created novel sociopolitical tensions, were finally rejected, but only after being seriously and contentiously mooted. The out-groups included conspicuously the following four categories: first, those known collectively as "Inferiors," that is, semi- or ex-Spartiates (some, such as Cinadon himself apparently, falling below the line for economic reasons, others for reasons of misconduct); then, the Helots (the Greek serflike populations, mainly agricultural, of Laconia and Messenia, of whom the Messenians were the really hot property—politically as well as socially, economically, and culturally); next, the ex-Helots (especially the Neodamodeis or "New damos-men," that is, Helots liberated for such specifically military and imperialist purposes as garrison duty in central Greece or Anatolia, but occupying an awkwardly undefined political status between Spartiates and Perioikoi); and, finally, the Perioikoi themselves (free inhabitants of the quasi-autonomous towns and villages of Laconia and Messenia, who were fully integrated in the Spartan army but lacked any corresponding political say in Spartan policymaking).

The post-War imperialism of Lysander and Agesilaus offered to Spartiates unprecedented opportunities for foreign enrichment and power—hence, probably, Cinadon's especial resentment at being not only socially demoted in an acutely hierarchical and highly agonistic society but also prevented thereby from getting what he reckoned as his due share of the imperial bounty. Precisely what he hoped to achieve by his inchoate and abortive revolution (if that is what it was) is, however, unclear: according to the main source, Xenophon, he is said to have wished "not to be inferior to anyone," but that formulation was probably deliberately ambiguous—and anyhow Xenophon is by no means an unbiased witness.[50]

Spartan internal politics were intimately linked to and now more than ever determined by external policy considerations arising out of the new imperialism. For example, the politics of patronage might now be deployed beyond the confines of the traditionally rival royal circles, as was made acutely apparent in the unprecedented career of the well-connected commoner Lysander. He became the first Greek, not just the first Spartan, to be awarded divine honors in his lifetime—though not, needless to add, at home in Sparta, but at Samos, where they were bestowed on him by an extravagantly grateful "decarchy," a narrow junta of just ten men, oligarchic cronies whom he had been instrumental in imposing on a determinedly democratic populace.[51]

Externally, the Spartans had once, according to Thucydides, been traditionally "slow to go to war—unless compelled," as they allegedly were in 432. The force of Thucydides' "compelled" is contested, but it could plausibly be understood to mean that the Spartans typically chose war only when inescapably faced with an alternative adjudged by those in authority (as opposed to the ordinary Spartan masses) to be even worse: thus in 432 not going to war would have involved the serious risk of Sparta's losing control of an organization known to us as the Peloponnesian League, a multi-state military alliance that was as vital to the Spartans defensively as it was for purposes of aggression. They were slow to war in general chiefly because of their endemic domestic Helot problem, and in the 430s they might have been especially slow to embark on a possibly prolonged war that would be sure to place great strain on internal social relationships and, by necessitating unusually extended contact with the outside world, breach an ideally closed society.[52] However, after and as a direct consequence of their victory in the Peloponnesian War the Spartans became, to quote the second-century B.C. Megalopolitan historian

Polybius, "ambitious, eager for supremacy, and acquisitive in the highest degree." But in going for broke, rather than pursuing the quieter and more tolerant imperial policy advocated by King Pausanias (one of Agesilaus' many rivals from the other Spartan royal house), Sparta did in the event break as a great power—with a series of spectacularly loud cracks between 371 and 365.[53]

From Athens and Sparta I widen the angle once more to embrace all "Greece" in the post–Peloponnesian War epoch. This is the most difficult and nebulous terrain, but that difficulty, as we have seen, has not deterred commentators from declaring that the War changed the Greek world and its civilization forever. Such a declaration, however, is at least too sweeping: Spartan and Athenian civilization, though both equally Greek, were far from being precisely identical in all respects.[54] So far as the economic impact of the War is concerned, the majority of scholarship would seem to regard it as firmly negative. For example, the constant availability of an ever-widening pool of mercenary soldiers is usually taken to be a sign of widespread, increasing, and irremediable rural immiseration.[55] On the minority side, though, the fourth century has been viewed positively, as the prerequisite transitional phase between a (too) local Greek and a (suitably) international/inter-regional Middle Eastern economy.[56] A fittingly middle position has been adopted recently by Victor Hanson in the context of a typically challenging and controversial work. On his view, there occurred gradually after 404 the erosion not merely of the individual Greek family farm—for him the "agrarian root of Western civilization"—but of the overall "agrarian polis" that such farms had constituted and sustained. However, as he sagely adds, "If Athens could not reinvent the city-state to meet new social and economic challenges and opportunities, no other Greek community could. But we should not dwell on the end of the autonomous agrarian polis in the fourth century. No state, no political entity exists in perpetuity. . . ."[57]

From international economy I move to international politics. Athens' defeat proved that she could not unite Greece; but this did not deter others even less well placed than she had once been from making a like attempt. The subsequent failures of first Sparta and then Thebes to achieve overall hegemony merely paved the way for the enforced unification of Aegean Greece by Philip and Alexander of Macedon, one of the major costs of which was the termination of Athenian democracy. On the other hand, failure by any one Greek state to achieve a stable and lasting hegemony helped to point the Greeks, somewhat inchoately, in what may be thought the right directions.[58]

Perhaps the most famous, and influential, passage in all Thucydides is his analysis of the stasis (civil war) that broke out in 427 on the island of Corcyra (modern Corfu).[59] As an expression of polarized economic class struggle between rich (oligarchs) and poor (democrats) Greek citizens, Thucydides saw the Corcyra stasis as prototypical, and the evidence of fourth-century writers like Aineas Tacticus, Plato, and Aristotle does nothing to weaken the impression of an endemic and chronic sociopolitical instability throughout the post-War Greek world.[60] The case for assigning a significant measure of responsibility for this to the specific circumstances of the Peloponnesian War is strong. Although class struggle stasis is trace-able at least as far back as the seventh century B.C., the generalized brutality and deceitfulness stimulated by this unusually prolonged and self-destructive war could have weakened significantly any ingrained cultural resistance to doing extreme violence to fellow citizens.[61]

Class war within cities was aggravated, and sometimes caused, by constant warring between cities. In the fourth century that warfare took a more professional turn, literally as well as metaphorically. The art of war became more sophisticated, thanks to such technological inventions as the torsion catapult, the more frequent and sophisticated use of *l'autre guerrier* ("other" in contrast to citizen hoplite militiamen), and the greater emphasis on training and training manuals, special forces, and siegecraft.[62] More and more often, the agents of war were paid profes-sionals, such as the mercenaries resorted to even by such small cities as Kleitor in Arcadia in 370.[63] The military art of Epaminondas, the Theban general who was the architect of Sparta's decisive defeat at Leuctra in 371, "presupposes an intel-lectual shift, or rather, in the terms of C. Castoriadis, a shift in society's image of itself."[64] As Demosthenes despairingly put it, in the course of his unavailing struggles to rouse Athens against Macedon,"I myself believe that, although almost everything has progressed and the present bears no relation to the past, nothing has changed so much as the art of war."[65] There is, however, still room to deny that the Peloponnesian War was directly, solely, or even mainly responsible for all such undoubtedly basic changes—Dionysius of Syracuse and Philip of Macedon, for example, the great military success stories of the fourth century, both had their own, independent trajectories and agendas.

Talk of imaginary social mutations brings us on, or back, to questions of men-tality and ideology. The ratcheting-up of brutality toward hostages and captives as the Peloponnesian War progressed (or regressed), in itself perhaps unsurprising, has been painstakingly and painfully documented.[66] That may, as suggested above, have made a lasting contribution to the prevalence of stasis. At the same time, an increasing use of ruse and deceit in the Peloponnesian War has been detected on as well as off the battlefield, possibly even leading to a positive (re)valuation of de-ceitfulness (*apatê*) in the guise of cunning intelligence (*mêtis*) at least in Athens.[67]

Yet more interpretatively problematic is the visual art of the time. Oscar Wilde's dictum that it is wrongheaded to try to infer the nature or spirit of an age from its art is typically inverted and perverse, but how best to attempt such inference is not obvious. One leading authority has opined that in terms of *"what it expresses"* (italics in original), rather than in virtue of its formal components, the break be-tween "Classical" and post-Classical visual art comes—or should be placed—at about the end of the Peloponnesian War, whereafter the period c. 400–1 B.C. should be viewed as "a continuum."[68] All such art-historical judgments are inevitably somewhat impressionistic and subjective. We may, perhaps, be on safer ground with religion.

The trial of Socrates, staged in 399 in spite of the Amnesty and when memories of active upper-class collaboration with Sparta in Athens' terrible defeat were still too fresh in the popular mind, has been seen by the leading scholar of Greek religion as far more than a purely Athenian affair. For Burkert, it is the most spectacular instance of a general crisis of polis (Greek city) religion brought on by the destructive questioning of "Sophists and Atheists."[69] It is, of course, hard to challenge or impugn Burkert's global knowledge, but one might be inclined to argue that he both overstates and overintellectualizes the supposed "crisis." Far more threatening in contemporary Athenian perception, at any rate, was the miasma emanating from, for example, the Great Plague or the mutilation of the Herms—these were widely taken as veritable manifestations of the anger and retribution of jealously vindictive gods. But even for exceptional traumas like these, normal polis religion could provide satisfactory responses.

A "crisis" view even more extreme than Burkert's was proposed by the distinguished Harvard scholar J.H. Finley. Postulating an "intellectual revolution" by which "conceptual thinking" allegedly replaced the old reliance on myth and which "necessarily shook the brief balance of the great age," he claimed that "[a]s a cause of change, this intellectual revolution far outweighs even the strain of the twenty-seven-year war."[70] That, surely, is interpretative idealism taken to absurd lengths. Far more sober and plausible is the reverse claim that our idea of "the Classical," according to which the era of "Periclean" Athens or Greece as a whole in the fifth century is seen as a golden age of cultural glory, was formulated initially in the fourth century, especially in the improving or self-serving rhetoric of the Attic orators before mass audiences of jurymen or assembly-goers.[71]

To summarize and conclude this second section: The extreme view that (in some way or ways) everything in the Greek world changed abruptly in or around 400 B.C. is, I hope to have shown, untenable. Attention might usefully be drawn once again to the unintended but nonetheless real dangers created by artificial periodization. Moderate alternatives to that extreme view propose, not that there were no significant changes, but that the changes were not always and only changes of kind but also changes of degree and tempo. They properly stress the differential effects of the War on Athens and Sparta, and on the different institutional subsystems of these two, in many ways disparate, Greek communities. They distinguish, above all, between propter hoc radical novelty and post hoc change.

It has been said of the Korean War that, "though it has passed into history," it "has yet to become solely a historical problem."[72] The terms of that distinction are questionable, but let us allow that a problem may become at any rate mainly historical, and that some such problems, the Peloponnesian War for instance, may be relatively less immediately implicated in and complicated by present concerns than others. Hindsight, then, is rightly considered to be one of the historian's principal assets, so long as it is not confused or contaminated with anachronism.[73] But even with perfect hindsight, the question remains "*qu'est-ce que je fais quand je fais de l'histoire?*"[74] Whatever it may be, it is not solely or merely to act as recording angel of *wie es eigentlich gewesen ist*. For, in the memorable phrase of the late Moses Finley, "The dead past never buries its dead."[75] It is we rather who must be the undertakers, the funeral directors, of our own, often very different, even contradictory pasts. So I end with the tentative expression of a very personal view of the after-effects of the Peloponnesian War, couched in the typically ancient Greek form of an antithesis.

On the one hand, without the (or a) Peloponnesian War, there would have been no Thucydides (nor perhaps Euripides, Aristophanes, and Plato, at any rate in their existing forms). That surely would have been a major loss to the heritage of "Western civilization" and "the Western cultural tradition," whether or not one chooses to regard Thucydides as "canonical" or indeed believes that canons can be good as well as usable things.[76] On the other hand, this—by parochial Classical Greek standards—massive and prolonged upheaval brought to the fore for too long, and too influentially, some of the most undesirable aspects of Greek culture and some of their least admirable representatives. To adopt for once the perspective of a moralizing ancient biographer such as Plutarch, for me Alcibiades of Athens and Lysander of Sparta incarnate the most characteristic defects of their respective societies: in the one case, a flamboyantly narcissistic egotism run riot, in the other, an uncontrolled lust for narrowly oligarchic power at all costs.[77]

Despite the best, or worst, efforts of such individuals, however, the succeeding Greek "fourth century" is an epoch by no means entirely contemptible or even merely decadent. To have predicted that it would be, would have been to get it wrong.

Notes

1. Anthologized in Jon Stallworthy, ed., *The Oxford Book of War Poetry* (Oxford: Oxford University Press, 1984), 330.

2. Some heartfelt words of thanks, first to our co-organizers, David McCann and (especially) Barry Strauss; then to our hosts at the Woodrow Wilson Center, especially Jim Morris and Susan Nugent; finally, and not least, to my discussant Kongdan (Katie) Oh.

3. A. Buchan, *The End of the Postwar Era: A New Balance of World Power* (London: Weidenfeld and Nicolson, 1974), looks now to have been somewhat premature.

4. See, e.g., Pierre Vidal-Naquet, *Assassins of Memory. Essays on the Denial of the Holocaust* (New York: Columbia University Press, 1992); cf. Ian Buruma, *Wages of Guilt: Memories of War in Germany and Japan* (London: Jonathan Cape, 1994); Young Whan Kihl, ed., *Korea and the World: Beyond the Cold War* (Boulder, CO: Westview Press, 1994).

5. A.J.P. Taylor, *How Wars End* (London: Hamish Hamilton, 1985), observed acerbically that the patchwork of treaties concluded in the first post–Second World War decade hardly amounted to a "settlement"; cf. S. Dockrill, "Re-ordering Europe After 1945," *International History Review* 16, 4 (1994): 758–74. The title of R. Crockatt, *The Fifty Years' War: The United States and the Soviet Union in World Politics, 1941–1991* (London and New York: Routledge, 1994), speaks for itself; cf. Ellen M. Wood, *Democracy Against Capitalism. Renewing Historical Materialism* (Cambridge: Cambridge University Press, 1995), 265 n. 2: "there has hardly been a major regional conflict anywhere since World War II that has not been initiated, aggravated or prolonged by US intervention."

6. Positive: Kenneth O. Morgan, *The People's Peace. British History 1945–1990* (Oxford: Oxford University Press, 1990). Negative: T. Engelhardt, *The End of Victory Culture*

(London: HarperCollins, 1995); Bernard Bergonzi, *Wartime and Aftermath. English Literature and Its Background, 1939–1960* (Oxford: Oxford University Press, 1993). Generally, C. Coker, *War and the 20th Century: The Impact of War on the Modern Consciousness* (London: Brassey's, 1995).

7. D. Reynolds, *Rich Relations. The American Occupation of Britain 1942–1945* (London: HarperCollins, 1995). Compare Victor D. Hanson, *Fields Without Dreams. Defending the Agrarian Idea* (New York: The Free Press, 1996).

8. Ian M. Morris, "Periodization and the Heroes: Inventing a Dark Age," in Mark Golden and Peter Toohey, eds., *Inventing Ancient Culture: Historicism, Periodization, and the Ancient World* (London and New York: Routledge, 1997), 96–131; yet more relevantly, Barry S. Strauss, "The Problem of Periodization: The Case of the Peloponnesian War," in ibid., 165–75; cf. Morris, ed., *Classical Greece. Ancient Histories and Modern Archaeologies* (Cambridge: Cambridge University Press, 1994), 8–47, on the need for a "new" Classical Greek Archaeology, and what it could and should look like.

9. Marc Bloch, *The Historian's Craft* (Manchester: Manchester University Press, 1954), 182, wryly remarked that France's "eighteenth century" traditionally began in 1714 and ended in 1789. Compare and contrast the invention (and continuing re-invention) of the fifth century B.C. in the Greek world as a "Golden Age."

10. For example, the contention of Jose Harris, *Private Lives, Public Spirit: A Social History of Britain 1870–1914* (Oxford: Clarendon Press, 1994), that the changes inaugurated in Britain in the 1870s were not overwhelmed by the carnage and upheaval of either of this century's world wars but lasted until the 1960s and 1970s, the next great period of social change, is surely not even probable, let alone plausible.

11. Philosophy of causation: Bertrand Russell, *The Analysis of Mind* (London and New York: Routledge, 1921, repr. with intro. by T. Baldwin, 1995), Lecture V; Herbert L.A. Hart and A.M. Honoré, *Causation in the Law*, 2nd ed. (Oxford: Oxford University Press, 1985); John L. Mackie, *The Cement of the Universe. A Study of Causation* (Oxford: Oxford University Press, 1974). Historiography: Christopher Lloyd, *The Structures of History* (Oxford: Blackwell, 1993). The view of Pierre Veyne, *Writing History. Essay on Epistemology* (Manchester: Manchester University Press, 1984 [1971]), 93, that "Every historical account is a plot out of which it would be artificial to cut discrete causes," seems determinedly eccentric.

12. Quotation from Arthur Marwick, *War and Social Change in the Twentieth Century* (London: Macmillan, 1974), a book that raises, but treats unsatisfactorily, the largest and most fundamental methodological questions relevant to the present chapter.

13. Quotation from Moses I. Finley's introduction to the Penguin Classics reprint, *Thucydides, History of the Peloponnesian War,* trans. Rex Warner (Harmondsworth: Penguin, 1972); cf. Strauss, "Problem of Periodization" (op. cit., n. 8). On the necessity of choice and inevitable selectivity of all historiography, cf. Kenneth J. Dover, *Thucydides Greece and Rome*, New Surveys in the Classics 7 (Oxford: Oxford University Press, 1973)— still perhaps the best short treatment in English on Thucydides, though it needs updating in light of, e.g., Simon Hornblower's *Commentary* (Oxford: Oxford University Press, 2 vols. so far, 1991–1997); cf. Kenneth J. Dover, "Thucydides 'as History' and 'as Literature'" (1983), repr. in his *The Greeks and Their Legacy. Collected Papers II* (Oxford: Blackwell, 1988), 53–64.

14. Hayden V. White, "The Historical Text as Literary Artifact," in R.H. Canary and H. Kozicki, eds., *The Writing of History: Literary Form and Historical Understanding* (Madison, WI: University of Madison Press, 1978), 41–72; White's review of Veyne, *Writing History* (op. cit., n. 11), *Times Literary Supplement* (31 January 1986): 109–10; cf. Lloyd S. Kramer, "Literature, Criticism and Historical Imagination: The Literary Challenge of Hayden White and Dominick LaCapra," in Lynne Hunt, ed., *The New Cultural History* (Berkeley: University of California Press, 1989), 97–128. As a commentary on Benedetto

Croce's aphorism that all history is present history, cf. G. Thuillier and J. Tulard, *Le marché de l'histoire* (Paris: Presses Universitaires de France, 1994).

15. Hornblower, *Commentary* (op. cit., n. 13).

16. Thuc. 1.22.1, in the translation of Donald Kagan, *On the Origins of War and the Preservation of Peace* (New York: Doubleday, 1995), 5. Perhaps the most useful English version of Thucydides now available is *The Landmark Thucydides: A Comprehensive Guide to the Peloponnesian War*, ed. Robert B. Strassler (New York: The Free Press, 1996).

17. Paul R. McKechnie and Stephen J. Kern, eds., *Hellenica Oxyrhynchia* (Warminster: Aris and Phillips, 1988); Mortimer H. Chambers, ed., *Hellenica Oxyrhynchia* (Leipzig: Teubner, 1993).

18. A recent general study of Xenophon's historiography, a good one, is John M. Dillery, *Xenophon and the History of His Times* (London and New York: Routledge, 1995); cf. Gerald Proietti, *Xenophon's Sparta* (Leiden: Brill, 1987).

19. Paul A. Cartledge, *Aristophanes and His Theatre of the Absurd* (Bristol: Bristol Classical Press; London: Duckworth, 3rd ed., 1995). See below, p. 110 and n. 37.

20. Michel Nouhaud, *L'utilisation de l'histoire par les orateurs attiques* (Paris: Belles Lettres, 1984); Josiah Ober, *Mass and Elite in Democratic Athens. Rhetoric, Ideology, and the Power of the People* (Princeton: Princeton University Press, 1989); Harvey Yunis, *Taming Democracy: Models of Political Rhetoric in Classical Athens* (Ithaca and London: Cornell University Press, 1996).

21. Peter J. Rhodes, *A Commentary on the Aristotelian Athenaion Politeia* (Oxford: Clarendon Press, 1981).

22. Barbara Scardigli, ed., *Essays on Plutarch's Lives* (Oxford: Clarendon Press, 1995); Judith Mossman, ed., *Plutarch and His Intellectual World* (London: Classical Press of Wales and Duckworth, 1997); Donald Shipley, *Plutarch's Life of Agesilaos. Response to Sources in the Presentation of Character* (Oxford: Oxford University Press, 1997).

23. A selection translated in Phillip Harding, *From the End of the Peloponnesian War to the Battle of Ipsus* (Cambridge: Cambridge University Press, 1985).

24. Archaeology/art-history discourse: Simon Goldhill and Robin Osborne, eds., *Art and Text in Ancient Greek Culture* (Cambridge: Cambridge University Press, 1994).

25. Arnaldo D. Momigliano, "George Grote and the Study of Greek History" (1952), reprinted in *Studies on Modern Scholarship*, Glen W. Bowersock and Timothy J. Cornell, eds. (Berkeley: University of California Press, 1994), 15–31; cf. Jennifer T. Roberts, *Athens on Trial. The Antidemocratic Tradition in Western Thought* (Princeton: Princeton University Press, 1994).

26. Barry S. Strauss, *Athens After the Peloponnesian War. Class, Faction and Policy, 403–386 B.C.* (Ithaca: Cornell University Press; London and Sydney: Routledge, 1986); reviewed by Cartledge, *Hermathena* 144 (Summer 1988): 105–10. Paul A. Cartledge, *Agesilaos and the Crisis of Sparta* (London: Duckworth; Baltimore: Johns Hopkins University Press, 1987).

27. Pausanias 3.7.11, used as epigraph to Cartledge, *Agesilaos*, ch. 4. On Pausanias generally see, e.g., Christian Habicht, *Pausanias* (Berkeley: University of California Press, 1985).

28. Herbert J. Muller, *The Uses of the Past. Profiles of Former Societies* (New York: Oxford University Press, 1952). Despite chapter 5, "The Romantic Glory of Classical Greece," Muller was by no means a starry-eyed romanticizer of the Greeks, though he did culpably denigrate the Hellenistic Age of the Greek Past (roughly the last three centuries B.C.) in contrast with the Classical age—a syndrome from which scholars too are not immune: see esp. Peter Green, *From Alexander to Actium* (Berkeley, Los Angeles, and London: University of California Press, 1990, corr. repr. 1993).

29. Donald Kagan, *Origins* (n. 16), 15–79, at 16.

30. Edmond Lévy, *Athènes devant la défaite de 404. Histoire d'une crise idéologique*

(Paris: Boccard, 1976), 1; but see my review in *Gnomon* 50 (1978): 650–54; and see also n. 45. Gilbert Murray, *Five Stages of Greek Religion* (London: Watts, 1935), 79–80: "The echo of that lamentation [Xen. *Hell.* 2.2.3] seems to ring behind most of the literature of the fourth century, and not the Athenian literature alone." On the evidence for the extent and nature of the ancient Greek polis, the Copenhagen Polis Project, co-directed by Mogens Hansen and Simon Hornblower, aims to be definitive; at the time of writing, five volumes of *Acts of the Copenhagen Polis Centre* (CPC) and four volumes of *CPC Papers* have appeared.

31. Claude Mossé, in *La fin de la démocratie athénienne* (Paris: Presses Universitaires de France, 1962), took a more negative overall view than in her later *Athens in Decline: 404–86 B.C.* (London: Routledge, 1973). The four-volume East German *Hellenische Poleis. Krise-Wirkung-Wandlung*, ed., E.C. Welskopf (Berlin, Academie-Verlag, 1974), was dedicated, unsuccessfully, to proving the ideologically overdetermined assumption of a general "crisis" in Greece, not least at Athens. More balanced views are in Lawrence A. Tritle, ed., *The Greek World in the Fourth Century: From the Fall of the Athenian Empire to the Successors of Alexander* (London and New York: Routledge, 1997).

32. Moses I. Finley, "Sparta and Spartan Society," *Economy and Society in Ancient Greece* (London: Chatto and Windus, 1981), 24–40, at 40. Stephen J. Hodkinson, "The Crisis of Spartiate Society," in John Rich and G. Shipley, eds., *War and Society in the Greek World* (London and New York: Routledge, 1993), 146–76, at 174–75. The case for Sparta's geopolitical responsibility for the end of the independent Greek city is argued at length by Cartledge, *Agesilaos* (op. cit., n. 26).

33. Neutrality: Robert A. Bauslaugh, *The Concept of Neutrality in Classical Greece* (Berkeley: University of California Press, 1991). Common peace/balance of power: M. Jehne, *Koine Eirene. Untersuchungen zu den Befriedungs- und Stabilisierungsbemühungen in der griechischen Staatenwelt des 4 Jahrhunderts v. Chr.* (Stuttgart: Steiner, 1994). This largely supersedes Timothy T.B. Ryder, *Koine Eirene: General Peace and Local Independence in Ancient Greece* (Oxford: Oxford University Press, 1965).

34. Roberts, *Athens on Trial* (op. cit., n. 25).

35. Strauss, *Athens After the Peloponnesian War* (op. cit., n. 26).

36. Kenneth J. Dover, *Greek Popular Morality in the Time of Plato and Aristotle* (Oxford: Blackwell, 1974). Walter Burkert, *Greek Religion. Archaic and Classical* (Oxford: Blackwell, 1985; German original, 1977), 311–17, seems to me to have exaggerated the "crisis" in traditional religion in the later fifth century allegedly represented by "sophists and atheists," but in any case religion and morality were not as mutually reinforcing under the scheme of Greek polytheism as they were to become under the regime of Christian monotheism.

37. Ephraim David, *Aristophanes and Athenian Society of the Early Fourth Century* (Leiden: Brill, 1984). Strauss, *Athens After the Peloponnesian War* (op. cit., n. 26). Victor D. Hanson, *Warfare and Agriculture in Ancient Greece* (Pisa: Giardini, 1983).

38. Strauss, *Athens After the Peloponnesian War* (op. cit., n. 26), ch. 3; but see the review article by Stephen C. Todd, "Factions in Early-Fourth-Century Athens?" *Polis* 7.1 (1987): 32–49, esp. 39–42. See also Mogens H. Hansen, *Demography and Democracy* (Herning: Systime, 1986).

39. Burkert, *Greek Religion*, 317. See, rather, W. Robert Connor, "The Other 399: Religion and the Trial of Socrates," in *Georgica. Greek Studies in Honour of George Cawkwell* (BICS Supp. 58, 1991): 49–56; Robert Garland, *Introducing New Gods. The Politics of Athenian Religion* (London: Duckworth, 1992); Mogens H. Hansen, *The Trial of Sokrates— From the Athenian Point of View* (Copenhagen: Royal Danish Academy of Sciences and Letters, 1995).

40. Barry S. Strauss, *Fathers and Sons. Ideology and Society in the Era of the Peloponnesian War* (London: Routledge, and Ithaca: Cornell University Press, 1993), 131, 199–209 (strong thesis); 214 (understated).

41. Victory syndrome: Michael H. Jameson, "The Ritual of the Athena Nike Parapet," in *Ritual, Finance, Politics. Athenian Democratic Accounts Presented to David Lewis*, ed. Simon Hornblower and Robin Osborne (Oxford: Clarendon Press, 1994), 307–24, at 318 n. 19, quotes Andrew F. Stewart, "History, Myth and Allegory in the Program of the Temple of Athena Nike, Athens," in *Pictorial Narrative in Antiquity and the Middle Ages* (Studies in the History of Art 17; Washington, D.C., 1985), 70, as follows: "in this highly strung world, at war for decades, increasingly unstable both emotionally and politically, mesmerized by Alkibiades, obsessed with conquest and glory, victory needed to be continually present to the eye in all sorts of ways in order to transpire." Collective defensive mentality: Josiah Ober, *Fortress Attica. Defense of the Athenian Land Frontier, 404–322 B.C.* (*Mnemosyne Supp.* 84: Leiden, 1985).

42. Robert Garland, *The Eye of the Beholder. Deformity and Disability in the Graeco-Roman World* (London: Duckworth, 1995), 182, n. 5.

43. Peter Krentz, *The Thirty at Athens* (Ithaca: Cornell University Press, 1982). Freedom of speech: I.F. Stone, *The Trial of Socrates* (London: Cape, 1988). But Stone writes as if Athens had passed the First Amendment; see rather Richard W. Wallace, "Private Lives and Public Enemies: Freedom of Thought in Classical Athens," in Alan L. Boegehold and Adele C. Scafuro, eds., *Athenian Identity and Civic Ideology* (Baltimore: Johns Hopkins University Press, 1994), 127–55.

44. W. Kendrick Pritchett, *The Greek State at War*, vol. 2 (Berkeley: University of California Press, 1974), ch. 1. Richard A. Bauman, *Political Trials in Ancient Greece* (London and New York: Routledge, 1990).

45. Peter Funke, *Homonoia und Arkhe. Athen und die griechische Staatenwelt vom Ende des Peloponnesischen Krieges bis zum Königsfrieden (404/3–387/6 v. Chr.)* (Wiesbaden: Steiner, 1980). Contra: Lévy, *Athènes* (op. cit., n. 30).

46. See generally Nicole Loraux, *The Children of Athena. Athenian Ideas About Citizenship and the Division Between the Sexes* (Princeton: Princeton University Press, 1993 [1984]); Boegehold and Scafuro, *Athenian Identity*. Specifically on the *guné–hetaira* antithesis, see Despina Christodoulou, "Greek Prostitution: A Study of 'the Hetaira'" (Ph.D. thesis, Cambridge University, 1997).

47. Citizenship extension: M.J. Osborne, *Naturalization in Athens,* 4 vols. (Brussels: Royal Belgian Academy, 1981–83). Self-regarding nature, and tighter policing, of citizen identity: see the new chapter (XII), "Athenian Society in the Fourth Century," in John K. Davies, *Democracy and Classical Greece*, 2nd ed. (Glasgow and London: Fontana/HarperCollins, 1993); Virginia J. Hunter, *Policing Athens. Social Control in the Attic Lawsuits, 420–320 B.C.* (Princeton: Princeton University Press, 1994); David Cohen, *Law, Violence, and Community in Classical Athens* (Cambridge: Cambridge University Press, 1995).

48. *Ath. Pol.* 41.2; countered by Mogens Herman Hansen, "Review of R.K. Sinclair, *Democracy and Participation in Athens* (1988)," *Classical Review* 39 (1989): 69–76; "Review of J. Ober, *Mass and Elite* (1989)," *Classical Review* 40 (1990): 348–56; *The Athenian Democracy in the Age of Demosthenes* (Oxford: Blackwell, 1991). But on the impropriety of the modern "moderate" and "radical" terminology see Barry S. Strauss, "Athenian Democracy. Neither Radical, Extreme, nor Moderate," *Ancient History Bulletin* 1 (1987): 127–29.

49. See now Walter Eder, ed., *Die athenische Demokratie im 4. Jahrhundert v. Chr.: Vollendung oder Verfall einer Verfassungsform?* (Stuttgart: Steiner, 1995).

50. Main source on Cinadon: Xen. *Hell.* 3.3.4–11. Spartan society and social change: Stephen J. Hodkinson, "Inheritance, Marriage and Demography: Perspectives upon the Success and Decline of Classical Sparta," in Anton Powell, ed., *Classical Sparta: Techniques Behind Her Success* (London: Routledge, 1989), 79–121. See also Finley and Hodkinson, as cited in n. 32.

51. Lysander and Samos: Cartledge, *Agesilaos*, 82–86.

52. Thucydides 1.118.2, with Geoffrey E.M. de Ste. Croix, *The Origins of the Peloponnesian War* (Ithaca: Cornell University Press; London: Duckworth, 1972), esp. ch. 4.

53. Polybius 6.48; with Cartledge, *Agesilaos*, esp. ch. 6.

54. On the other hand, the gulf between them should not be exaggerated: the polar opposition of Sparta and Athens, which had become a rhetorical trope by—at latest—the real historical date (430 B.C.) of the idealized Funeral Speech placed in the mouth of Pericles by Thucydides (2.25–46), was significantly the product of sectarian political exigencies.

55. Paul R. McKechnie, *Outsiders in the Greek Cities in the Fourth Century* (London and New York: Routledge, 1989); better, Yvon Garlan, *Guerre et économie en Grèce ancienne* (Paris: La Découverte, 1989); cf. Cartledge, *Agesilaos*, 316–17. Other economic factors get at least their fair share of attention in Mossé, *Fin* (op. cit., n. 31) and in her contribution to Edouard Will et al., *Le monde grec et l'orient: le IVe siècle et l'époque hellénistique* (Paris: Presses Universitaires de France, 1975), pt. II.

56. Mikhail I. Rostovtzeff, *Social and Economic History of the Hellenistic World*, 3 vols. (Oxford: Oxford University Press, 1941, corr. impr., 1953). Contrast Green (op. cit., n. 28).

57. Victor D. Hanson, *The Other Greeks. The Family Farm and the Agrarian Roots of Western Civilization* (New York: Free Press, 1995), 390–91. But see Cartledge, *Journal of Peasant Studies* 23.1 (1995): 131–39.

58. See above, pp. 109–110 and n. 33.

59. Thuc. 3.82–3 (Corcyra, 427); cf. 8.47–90 (Athens, 411).

60. David Whitehead, *Aineias the Tactician. How to Survive Under Siege* (Oxford: Clarendon Press, 1990); Alexander Fuks, *Social Conflict in Ancient Greece* (Jerusalem and Leiden: Magnes Press, 1984); Pierre Villard, "Sociétés et armées civiques en Grèce; de l'union à la subversion," *Revue Historique* 266 (1981): 297–310; Hans-Joachim Gehrke, *Stasis. Untersuchungen zu den inneren Kriegen in den griechischen Staaten des 5. und 4. Jahrhunderts v.Chr.* (Munich: Beck, 1985); Paul Cartledge, "La Politica," in Salvatore Settis, ed., *I Greci*, vol. I (Turin: Einaudi, 1996), 39–72, at 60–64; and above all Nicole Loraux, *La cité divisée* (Paris: Payot, 1997).

61. See n. 67.

62. J.K. Anderson, *Military Theory and Practice in the Age of Xenophon* (Berkeley: University of California Press, 1970); François Lissarrague, *L'autre guerrier: archers, peltastes, cavalerie dans l'imagerie attique* (Paris and Rome: La Découverte, 1990).

63. Xen. *Hell.* 5.4.36–7; cf. Cartledge, *Agesilaos*, 323 (op. cit., n. 26). But note that Arcadia had long been a provider of mercenaries, at least since c. 480.

64. Pierre Vidal-Naquet (with Pierre Lévêque), "Epaminondas Pythagoricien," repr. in *Le chasseur noir. Formes de pensée et formes de société dans le monde grec* (Paris: Maspéro, 1981), 117–18.

65. Dem. 9.47–50.

66. Andreas Panagopoulos, *Captives and Hostages in the Peloponnesian War* (Athens: Grigoris, 1978). The post–Peloponnesian War era produced no "Geneva Agreement"—for what that has been worth: Geoffrey Best, *War and Law Since 1945* (Oxford: Oxford University Press, 1994).

67. E. Heza, "Ruse de guerre: Trait caractéristique d'une tactique nouvelle dans l'oeuvre de Thucydide," *Eos* 62 (1974): 227–44; Suzanne Saïd and M. Trédé, "Art de Guerre et Expérience chez Thucydide," *Classica et Mediaevalia* 36 (1985): 65–85; Everett L. Wheeler, *Stratagem and the Vocabulary of Military Trickery* (Leiden: Brill, 1988); Jonathan Hesk, "The Rhetoric of Self-Representation: Deception and the Collective in Classical Athenian Culture" (Ph.D. thesis, Cambridge University, 1998). Contra: David Whitehead, "KLOPE POLEMOU: 'Theft' in Ancient Greek Warfare," *Classica et Mediaevalia* 39 (1988): 43–53.

68. Jerome J. Pollitt, *Art and Experience in Classical Greece* (Cambridge: Cambridge University Press, 1972), 136.

69. Burkert, *Greek Religion* (op. cit., n. 36), 311–17. Plato's *Laws*, for all its theological eccentricity, did not in Burkert's view pose the same sort of challenge as the sophists and atheists allegedly had.

70. John H. Finley, Jr., *Four Stages of Greek Thought* (Stanford: Stanford University Press; London: Oxford University Press, 1966), 107.

71. Nouhaud, *Orateurs* (op. cit., n. 20).

72. J. Cotton, in J. Cotton and I. Neary, eds., *The Korean War in History* (Manchester: Manchester University Press, 1989), 3.

73. "Hindsight" is the title of the final chapter of Max Hastings, *The Korean War* (London: Michael Joseph, 1987), 409–27.

74. François Hartog, "L'art du récit historique," in J. Boutier and D. Julia, eds., *Passés recomposés. Champs et chantiers de l'histoire* (Paris: Autrement, 1995).

75. Moses I. Finley, *Aspects of Antiquity*, 2nd ed. (Harmondsworth: Penguin, 1977), 184.

76. See now J. Peter Euben, "Imploding the Canon: The Reform of Education and the War over Culture," in his *Corrupting Youth: Political Education, Democratic Culture, and Political Theory* (Princeton: Princeton University Press, 1997), 3–31; and Paul Cartledge, "Classics: From Discipline in Crisis to (Multi-) Cultural Capital," pp. 16–28, in Y.L. Too and N. Livingstone, eds., *Pedagogy and Power* (Cambridge: Cambridge University Press, 1998).

77. Barry S. Strauss and Josiah Ober, *The Anatomy of Error. Ancient Military Disasters and Their Lessons for Modern Strategists* (New York: St. Martin's Press, 1990), ch. 2 ("The Alcibiades Syndrome: Why Athens Lost the Peloponnesian War"), and ch. 3 ("Lysander and Agesilaus: The Spartans Who Defeated Themselves").

Part III

Third Forces, or Shrimps Between Whales

7

The Case of Plataea: Small States and the (Re-)Invention of Political Realism

Gregory Crane

Past events lend to current processes their momentum—the mutual slaughter of Bosnian and Serb looks back to the battle of Kosovo; Arab nationalists and Muslim fundamentalists never quite forget, as they contemplate their American and European contemporaries, the short-lived kingdom that the Crusaders established in Palestine; and, even in U.S. culture as a whole (which glories in an obsessive presentism), the seventeenth-century slavers and slave owners not only shaped the institution of slavery but established patterns of behavior and mistrust that continue to shape the ways in which racial groups perceive themselves and each other. The Korean War, and the Cold War which it helped define, had plenty of historical causes. The men who defined American policy, however, looked back not only to the twentieth century but also to the war between Athens and Sparta and especially to Thucydides' brilliant history of those events. Years before war broke out in Korea, George Marshall explicitly pointed out the resemblance between the emerging Cold War and the tensions between Athens and Sparta.[1] His reading of Thucydides shaped his view of the contemporary affairs on which he exercised tremendous influence.

More generally, Thucydides has been cited, and with good reason, as a founding thinker in the Western tradition of political realism: his Mytilenean Debate and Melian Dialogue remain standard fare in anthologies and courses on international relations, law, and political philosophy.[2] The post-war years saw a reaction against the so-called idealism that had guided Western political thinking before the war, and that realism with which Thucydides was so closely identified.[3] Hans Morgenthau's masterful *Politics Among Nations*, first published in 1948, rendered him only the most influential of the new realists who would decisively shape a generation of American foreign policy. Members of this school regularly turned to Thucydides as their earliest member—Robert Gilpin even questioned whether twenty-four centuries have substantially advanced our understanding of how states relate to one another and of why wars occur.[4]

But if Thucydides and the Peloponnesian War influenced the thought of men two thousand years after their own world had vanished, this influence was paradoxically a- (or even anti-) historical, for it was the universalizing, reductive strain in Thucydides' thought—Thucydides the "value-free" calculator of interest—that appealed to American intellectuals in the forties and fifties. Late fifth-century Greece (or, rather, Thucydides' late fifth-century Greece) served as a model for the postwar world because two superpowers faced one another, in combination dominated the overall shape of events, and based decisions on calculations of their national interests. The political realism seen in Thucydides' history allowed leaders such as Dean Acheson to reconcile the moral principles in which they believed and the ruthless policies they felt compelled to pursue: morality could only be pursued so long as the needs of interest were satisfied. The Athenians on Melos could as well have been Nazis at Munich or the Chinese in North Korea; universal laws, far more significant than the peculiarities of culture, defined their behavior. One can thus read Acheson's account of the Korean War from start to finish without gaining any sense of how (or even if) Korean and Chinese culture and history influenced events.[5] Looking back on his contribution to another American war in Asia, Robert McNamara would bitterly lament American ignorance of Vietnamese culture and cite this as a major cause for America's destructive failure.[6]

The complexity of the world is far clearer now than it was even a decade ago. Regional conflicts in Somalia, Haiti, Rwanda, the former Yugoslavia, and the Kurdish sections of Turkey and Iraq, which have little to do with bipolar power politics, provide us with incentive to attend more closely to the complexity that underlay even the apparently bipolar conflicts of the past. The papers on the Korean War remind readers how much more complex that conflict was than any simple communist/free world contrast. The Thucydides who viewed a world in which two superpowers pursued predominant culturally universal interests has justifiably lost much of its appeal.

But this simplified world is not the only Greece that Thucydides depicts. Even as Thucydides' characters, by the words he puts in their mouths and the motives that he attributes to their actions, provide plenty of material for the "realist" point of view, his history includes evidence that subverts, or at least complicates, this perspective. Thucydides' work continues to draw attention outside of classics in part because it captures the strengths and weaknesses of what would become political realism and shape American actions in Korea and elsewhere.[7]

Much as Athens and Sparta may dominate Thucydides' history, several of the most widely studied episodes—the Mytilenean Debate, the collapse of civil society in the Corcyrean civil war, the Melian Dialogue—center around the smaller powers of the time and the consequences this Greek world war had upon them. My intention is to concentrate on one episode that has attracted surprisingly little attention: the fate of Plataea, a small city-state, perched uneasily between Athens and Thebes. Koreans sometimes refer to their country, situated between Japan and China, as a "shrimp between two whales": a related language binds Koreans more closely to Japan, just as the common Boeotian dialect of Greek linked Plataea with nearby Thebes; in each case, however, fury at previous domination undercuts linguistic solidarity: Plataea (as this chapter will make clear) struggled to maintain its independence from Thebes, while resentment of Japanese colonization is one sentiment on which all Koreans, North and South, can agree.

The first blows of the Peloponnesian War—which, in its length and geographical extent, constituted a world war for the Greeks—were struck in Plataea, a small city-state that belonged, by custom and dialect, to the Greek district Boeotia, but that had been, for almost a century, a client of neighboring Athens. Few, if any, living Thebans in 431 could recall a time when Plataea and Athens were not closely linked, but the Thebans still seem to have viewed Plataea's friendship with Athens as an aberration. When a force of Thebans seized the town in peacetime and under cover of darkness, they sought to reverse that century of history, restoring Plataea to Boeotia and asserting, in their own minds, a stable order of things that was supposed to stretch unbroken, backward and forward in time.

If the Thebans, in attacking Plataea, championed a continuity that defied change and froze history, they championed broad conservative forces that shaped the Archaic World Order as a whole. Their Spartan allies—who ultimately allowed them to wipe Plataea from the map—represent something far more complex. Seen as the guardians of conservative aristocratic values, the Spartans—or at least Thucydides' Spartans—prove to be something very different. Where the Thebans, like a truly conservative force, attempt to fit the future into the past, freezing history by rejecting change or, at least, asserting continuity, the Spartans destroy history. When they finally capture Plataea, they decide the fate of their prisoners by denying the past and restricting history to the brightly lit, narrow bubble of the present. Nothing matters for them that does not directly connect to the work at hand. I will argue that this flattened view of history is present in the most apparently conservative Spartan figure, Archidamos. Strange as these distant figures from the past may be, with their third-world economy, odd rituals, and state-sponsored prophets, they nevertheless anticipate the shallow balance-sheet decision making that bedevils modern leaders. The political realists who shaped American foreign policy during the Cold War would have probably found it difficult to work with the Thebans, but would have immediately understood the Thebans' allies from Sparta.

The Archaic World Order

Before talking about Plataea, however, I need to say a few things as context about the Greek world of the fifth century. The "New World Order" (or Disorder, as the case may be), with its many regional centers of power and its dizzying collection of distinct states, sovereign in name but often satellites to some larger entity, would have been, in many ways, familiar to any member of the educated, strikingly international elite that dominated and, starting in the early eighth century B.C., indeed invented the "Greek" world. Consider, for the moment, several features of the time.

First, the Greeks (in some ways much like contemporary Americans) were newcomers seeking order in a very old world. Greek society may seem ancient to modern eyes, but it had in fact emerged late in the day. If we take c. 2900 B.C. (roughly the time of Gilgamesh, who seems to have been an historical figure) as the beginning of Mesopotamian history, then the halfway point of continuous European history is 447 B.C., or roughly the end of the "first" Peloponnesian War (little known in part because its results were relatively inconclusive as well as because it had no Thucydides to document it). A man such as the Greek historian Herodotus was, temporally, as far removed from the dawn of European history as he is from us. But if Herodotus could not study such early history as closely as we can study the events of the past twenty-five hundred years, that history was nevertheless recorded and its repercussions were felt in Greece in the fifth century. The seventh-century Assyrian king Assurbanipal had, for example, collected a library that preserved two-thousand-year-old traditions—one of our best redactions of the Gilgamesh epic (which dates back to the third millennium) comes in fact from Assurbanipal's library.[8] Furthermore, when Xerxes invaded Greece, disrupting the Archaic Greek "World Order," setting the stage for an Athenian empire, and indeed providing one model for Athenian imperialism,[9] his actions reflected a tradition of international relations that Sumerians, Egyptians, Akkadians, Hittites, Hurrians, Assyrians, Medes, and Lydians (to name only a few) had developed during the course of two thousand years. Pericles' Funeral Oration may foreshadow Lincoln's Gettysburg Address, but it follows a tradition of authoritative political speech that we can trace from the third millennium Stele of the Vultures to the sixth-century Behistun Inscription of Dareios.[10]

Second, the invention of "Hellas" and "Hellene" ("Greece" and "Greek" are Roman terms) as separate, distinct categories was an ongoing process that was still relatively new in the fifth century B.C. The vast majority of Greeks were small farmers with limited direct experience of the world. As observers from the Persian Mardonius (Hdt. 7.9B) to Victor Hanson have pointed out, Greek men regularly hoisted their fifty-odd pounds of armor, massed with their townsmen into dense, clanking scrums, and then spent many summer days throughout their lives shoving one another with shields while attempting to jab spears into their adversaries. It took some time for these men to recognize a common bond with their rivals in the next valley (much less with people who came from a city-state a thousand miles away, dressed oddly, and spoke an incomprehensible dialect of Greek). Geographically, Greece belongs to the Balkans, and the brutal rivalries of the former Yugoslavia would probably not have puzzled any Greek who survived the Peloponnesian War.

Thucydides shows himself to be a great skeptic about Hellenic unity in the opening of his history: where the Trojan War had become in the classical period a great "Panhellenic" expedition in which Greeks defined themselves as a distinct cultural group in opposition to an Asiatic Other, Thucydides points out that even in the Homeric epics, composed long after the fact, the "Greeks" never refer to themselves as such but applied to themselves such labels as Danaans, Argives, or Achaeans (Thuc. 1.3). Not only does the term "Hellene" never appear in the Homeric epics, but we do not even find the generic *barbaros*, from which we derive "barbarian" but more closely designates any "non-Greek" whether civilized or not.[11] Thucydides goes on in his introduction to undermine—with apparent pleasure— the cultural pretensions of his fellow Greeks (Thuc. 1.5–6), and this prejudice, as I will argue, affects the whole of his history.

Nevertheless, if the idea of Hellas was a recent development, many Greeks of the fifth century took to it with great enthusiasm. If the modern Olympic games strive, however unevenly, to be a model of internationalism, their ancient counterparts were self-consciously parochial affairs designed to exclude non-Greeks and indeed to define who was and who was not Greek. Only Greeks were allowed to participate in the Olympic games: when the *Hellanodikai*, the officials who presided over the Olympic games, allowed Alexander of Macedon (the ancestor of Alexander the Great) to participate, Herodotus could use this admission as evidence that the royal family of Macedon was in fact Greek (Hdt. 5.22).

Third, Hellas—to the extent that such a thing existed—was a jumbled, diffuse, chaotic entity by comparison to which the contemporary international scene is a model of order and simplicity. The Greek city-states were frighteningly numerous and dispersed over a geographic space that rivals any ancient empire. There were hundreds of Greek city-states, and the colonizations that began in the "dark ages" and continued into the classical period filled the Mediterranean world with Greek city-states: Hellas stretched from cities such as Olbia in the Crimea to Cyrene in North Africa to Massilia (modern Marseilles) to Gadeira (modern Cadiz) in Spain.

Tensions great and small pulled these Greek city-states in many different directions. I will have more to say about local rivalries when I turn to Plataea, but consider one large-scale problem. On the one hand, the cultural center of the Greek world lay in mainland Greece (more or less the area that modern Greece now occupies). On the other hand, the newer Greek city-states challenged and surpassed the old. The center of Greek intellectual life in the sixth century was Ionia in the western coast of Turkey: even Herodotus, who as Dorian Greek from Halicarnassus openly despised the Ionians (e.g., Hdt. 1.142, 169, 6.12), nevertheless wrote his history in the Ionian dialect rather than his own Doric. In the fifth century, after Lydian and then Persian domination had turned Ionia into a backwater, the most powerful Greek states were in Italy and Sicily. We can measure the cultural aspirations of these new cities by the expense that they lavished on Panhellenic competitions at Olympia and by the many victory odes that Western Greeks commissioned to inscribe their successes in the permanent literary tradition of Greece. We can also glimpse in Herodotus the enormous tension between Western Greek power and the fiercely defended cultural prerogatives of the old Greek elite: even the Athenians would rather risk a Persian conquest than yield to Gelon, master of Syracuse, any of their own prestige (Hdt. 7.157–163).

Fourth, many of these hundreds of Greek city-states aspired to be free and independent. I will turn in the next section to some of the limits and realities of this freedom, but for now I wish to stress that, in this case, the narrowness and bias of our sources is very problematic. Xerxes' invasion of mainland Greece in 480 was a watershed event that changed the Greek world forever and indeed played a crucial role in the refinement and definition of Greek versus barbarian.[12] At the same time, of the hundreds of Greek city-states, only a few dozen opposed Xerxes,[13] and when the allies after the war considered punitive measures against those who had collaborated, they realized that such actions would include several of the most powerful Greek states such as the Thessalians, the Argives, and the Thebans (Plut. *Them.* 3). In part to preserve the balance of power and in part no doubt from expediency, they chose not to pursue collaborators with vigor. Only a small minority of Greeks seem in fact to have participated in the great war. Most were, like their Ionian kinsmen and like smaller states of the Near East for two thousand years, prepared to accept a foreign overlord, but freedom from external restraint and freedom to act remained central to Greek consciousness.[14]

These four factors—the antiquity of those cultures that impinged upon Hellas, the relative newness of their own cultural identity, the complexity of this Greek world, and the general stress on freedom—combined to place Greeks in a difficult situation. On the one hand, they found themselves confronted with a vast and predatory Persian empire, which had at its disposal traditions and cultural mechanisms for imperial expansion that had evolved over millennia. On the other hand, the individual city-states were jealous of their autonomy, and the relatively few major powers of mainland Greece—Athens, Thebes, Corinth, Sparta, Aegina, for example—were quarrelsome at best. Simply making it possible for hundreds of city-states scattered over thousands of miles to communicate with one another and to see in themselves a homogeneous group was a challenging task. The Greeks needed unity, but not too much unity. They wanted cohesion, but not the cohesion that an Assurbanipal or a Dareios could give. They wanted to be Hellenes, but they also wanted to Megarians or Siphnians, Dorians or Ionians, Alcmeonids or Philaids.

Much of the cultural and political history of the archaic and classical periods reflects the anarchic self-structuring of this complex system. Imperial cities are extremely useful: they serve as depositories of wealth and information alike and provide "network hubs" in which peoples from far-flung cities can meet one another. The Greeks developed imperial cities without empires: where the Near East had cities such as Babylon, Sardis, and Sousa, Delphi and Olympia emerged in the eighth century as leading "Panhellenic" centers. The oracle of Apollo at Delphi provided a crucial clearinghouse of information, and any Greek planning to send out a new colony would consult with the god first. The athletic contests at Olympia, which seem at first to have attracted primarily local talent, developed into a major event that drew competitors from all over the Greek world. The games, which took place every fourth year, provided Greeks with, in effect, a convention to which they could come, like modern-day academics, to reaffirm acquaintances and establish new relationships with people from cities scattered around their world.[15] This institution was so successful that, in the early sixth century, more Panhellenic games appeared so that there was a Panhellenic festival to which Greeks could congregate at either Olympia, Delphi, Nemea, or the Isthmos. Greeks (and even, in some cases, non-Greeks such as Croesus) demonstrated their piety (as well as their wealth and good fortune) by making lavish dedications to various sanctuaries, and, when the Peloponnesian War broke out, Olympia and Delphi were seen as, in effect, treasuries that could finance the Peloponnesian side (Thuc. 1.121, 143).

Political control of Delphi was problematic in the fifth century. At one point, the Spartans marched in and placed the people of Delphi in charge, but an Athenian force followed close on their heels and handed the oracle over to the people of Phocis (Thuc. 1.112.4). By the time the Peloponnesian War was about to break out, the situation had reversed itself again: the oracle at Delphi declared that the god would support the Peloponnesians against the Athenians (Thuc. 1.118). Nevertheless, the great Panhellenic sanctuaries were, in theory, supposed to be neutral centers that rose above Greek politics. The Amphictyonic League was an association of Greek states designed, in theory, to preserve the neutrality of Delphi, while Herodotus remarks that Pheidon of Argos (Hdt. 6.127.3) "exhibited the most *hubris* of all the Greeks when he drove out the Olympic officials of Elis and organized the Olympic games himself." Delphi and Olympia could only develop into Panhellenic status because they were not too closely linked to any leading city-state. By contrast, in roughly the same time as the "circuit" (*periodos*) of Panhellenic games took shape in the first half of the sixth century, Athens established its own set of major games, the Greater Panathenaia, which took place once every four years.[16] Unlike the Panhellenic circuit, the Athenian games offered prizes of substantial value. And unlike the Panhellenic circuit, the Athenians games never really achieved the status of a Panhellenic institution. No one seems to have commissioned an ode from Simonides, Pindar, or Bacchylides to celebrate victory at Athens. Even in the sixth century, Athens seems to have been too powerful for the rest of the Greek world to concede it Panhellenic prestige.

The tensions between Panhellenism and local ties, between the need to cooperate and relentless competition are hallmarks of Greek history until Philip and Alexander first began to impose a level of unity on the Greek world. The Periclean building program in the fifth century gave Athens the architectural pretensions of a Panhellenic sanctuary, but Athenian grandeur and Athenian imperialism were closely linked in the popular imagination (although how far is hard to gauge, since our main sources are themselves primarily Athenian). Old patterns continued, as Athens' claim to cultural hegemony grew as Athenian military and political power declined after the end of the Peloponnesian War.

Having briefly sketched some of the dynamics that shaped the Greek world, I will now turn to a particular city-state, the Boeotian town of Plataea, since this city, with its complex relations to the major powers, brings out a number of the forces at play both in the Greek world and in Thucydides' account of the Peloponnesian War.

Plataea

Ernst Badian published a few years ago an essay entitled "Plataea Between Athens and Sparta,"[17] and this title goes well with the Koreans' view of their nation as a "shrimp between two whales," but each of these felicitous phrases is inappropriate to the subject of this book, for in each case the smaller country confronts not two but three competing forces. Strictly speaking, Plataea lies between Athens and Thebes, the leading city of Boeotia, just as Korea finds itself between China and Japan. In each case, however, an external superpower entered the mix. The Spartans played a pivotal role at two junctures in Plataean history, while only energetic American intervention preserved the existence of a separate South Korea. Plataea and South Korea alike confront the complex issues of a triangular diplomacy, balancing the threats of kinsmen against the promises of strangers.

Plataea is a smallish Greek city-state, and even the largest Greek city-states were comparable in size to Rhode Island: Herodotus reports that six hundred Plataeans fought the Persians at the battle named after their city—a bit more than the five hundred from Aegina but less than one-tenth the size of the 8,000-man contingent of Athenian hoplites (Hdt. 9.28.6). A half-century later, Thucydides reports that about four hundred Plataean men stayed behind to defend the city after the women, children, and non-combatants were evacuated (Thuc. 2.78).[18] Clearly, Plataea had a total population of a few thousand at most. Equally important for the Plataeans as their size was their location. Their city lay just over the hills from Attica in the territory of Boeotia, a group of semi-autonomous cities linked by a shared dialect and a common ethnic background, distributed about one of the richest agricultural sections of Greece. Plataea was, however, in an awkward position: roughly twelve kilometers from Thebes, the most powerful city in Boeotia, it was too close to Thebes for comfort and too small to hold its own adequately.

Although Plataea was an ancient city that had presumably coexisted with Thebes for centuries (it shows up, for example, in the "Catalogue of Ships" at Hom. *Il.* 2.504), the situation had evidently become intolerable in the late sixth century. Some time around 520,[19] the Plataeans, "being hard pressed by the Thebans" (Hdt. 6.108.2), tried to form an alliance with the Spartans. The nature of this relationship has attracted a good deal of scholarly attention and deserves close scrutiny, for it is paradigmatic of a client/patron relationship in a society that sets great store by autonomy and freedom. Moreover, the language Herodotus uses is clearly formal, as the identical phrase (i.e., "to give oneself") reasserts itself five times in roughly one page of Greek.

The narrative frame for the background of Plataea is the battle of Marathon in 490. Alone among the Greeks, the Plataeans arrived to help the Athenians in what truly was their darkest hour: a Persian expeditionary force that had just annihilated the city of Eretria and was turning its attentions to the Athens.

> As the Athenians were marshalled in the precinct of Heracles, the Plataeans came to help them in full force. The Plataeans had given themselves (*ededôkesan spheas autous*) to the Athenians, and the Athenians had undergone many labors on their behalf. (Hdt. 6.108.1)[20]

The phrase "giving themselves" clearly indicates both that the Plataeans sought Athenian help and that they accepted a status that was, in some sense, inferior. Badian argues that the phrase "give oneself" was strong and quite technical: he points out that elsewhere in Herodotus this phrase describes people who surrender themselves to, and accept the overlordship of, the Persian king (e.g., Hdt. 1.169.2, 2.19.3, 4.132.1, 5.13.2) or the Pharaoh (e.g., Hdt. 2.162.6, 4.159.4). Badian says of this phrase that Herodotus "invariably uses it of the surrender of an inferior to a superior (normally a king, twice a god), thus into a condition that would be technically described by the Greeks as *douleia* ['slavery']. Whether we believe it or not, there can be no doubt as to how Herodotus meant us to understand Plataea's action."[21]

But there are many flavors of super- and subordination, and hierarchical relationships are often quite complex. Even when people "give themselves" to the great monarchs of the Near East, the element of reciprocity reappears. Thus, when a group of Egyptians became disenchanted with their home, they headed for Ethiopia:

> So they came to Ethiopia, and *gave themselves up* (*didousi spheas autous*) to the king of the country, and he *in return made them a gift* (*antidôreetai*) in the following way: he told them to dispossess certain Ethiopians with whom he was feuding, and occupy their land. The Ethiopians then learned Egyptian customs and have become milder-mannered by intermixture with the Egyptians. (Hdt. 2.30.5, emphasis added.)

In this event, both parties give and both receive: the Egyptians give themselves over as subjects to the Ethiopian king, but the king carefully answers with a "gift" of his own.

If we extend our analysis of this phrase to the other major fifth-century historian, the first attestation reveals a nuanced rather than a technical usage. The Corcyreans arrive at Athens in search of an alliance so that they can defend themselves against an enraged Corinth. They are in a difficult position but they have much to offer Athens, for Corcyra was a rich state with one of the premiere navies in the Greek world. They boast—and rightly—that they could potentially swing the balance of power against Athens if they were forcibly included in the Peloponnesian League (Thuc. 1.33.2, 35–36):

> Yourselves excepted, we are the greatest naval power in Hellas. Moreover, can you conceive a stroke of good fortune more rare in itself, or more dishearten-

ing to your enemies, than that the power whose adhesion you would have valued above much material and moral strength, should present herself self-invited, giving herself (*didousa heautên*) into your hands without danger and without expense, and should lastly put you in the way of gaining a high character in the eyes of the world, the gratitude of those whom you shall assist, and a great accession of strength for yourselves? You may search all history without finding many instances of a people gaining all these advantages at once, or many instances of a power that comes in quest of assistance being in a position to give to the people whose alliance she solicits as much safety and honor as she will receive. (Thuc. 1.32.2)

The pose these Corcyreans strike is proud and stresses the benefits they can confer in return for Athenian help. It may be that, in fact, the Corcyreans would find themselves very much at Athens' mercy, but the language they use to describe their offer clearly balances deference to Athenian superiority against pride in their own status. They may accept a client/patron relationship, but they are certainly not describing themselves as accepting slavery or absolute subjection.[22]

At stake here is not merely terminology but methodology. To see a precise and narrow definition within which to constrain the relationship between Plataeans and Athenians is to misconstrue another culture, as European and American academics sometimes do. Badian is looking for a precise and very narrow definition within which to constrain the relationship between the Plataeans and the Athenians. The Plataeans give themselves to the Athenians, acknowledging that the Athenians occupy a superior position. If the Plataeans and Athenians are not equals, then one must be the master and the other the slave.

Compare Benedict Anderson's analysis of European colonial censuses, where European administrators applied a dizzying array of precise categories to the local populations, even though

> it is extremely unlikely that, in 1911, more than a tiny fraction of those categorized and subcategorized would have recognized themselves under such labels. These 'identities,' imagined by the (confusedly) classifying mind of the colonial state, still awaited a reification which imperial administrative penetration would soon make possible. (Anderson, 1992, 165)

While it is easy to dwell upon a particular interpretation by a particular scholar, we all of us fall victim to the temptation to invent phantom classifications. Anderson goes on to focus upon one particular habit of thought that marked these colonial census takers and still characterizes many of those who analyze history or recent foreign policy:

> One notices, in addition, the census-makers' passion for completeness and unambiguity. Hence their intolerance of multiple, politically 'transvestite,' blurred, or changing identifications. Hence the weird category, under each racial group, of 'Others'—who, nonetheless, are absolutely not to be confused with *other* 'Others.' The fiction of the census is that everyone is in it, and that everyone has one—and only one—extremely clear place. No fractions. (Anderson 1992, 166)

This tendency has grown, if anything, worse: indeed, the models that underlie the databases that structure the great administrative apparatus of our world thrive on one-to-one matches. Even now, most forms ask whether we are white *or* black, Asian *or* Latino.

If we return to the archaic Greek world (and, indeed, to most of the world in which we live), ambiguity is not so much a problem to be solved as it is a space within which actors can maneuver. Everyone knew that the Plataeans were in some sense dependent upon the Athenians, but no one—unless they were, or wished to become, sworn enemies of Plataeans and Athenians alike—would try to squeeze the ambiguity out of their relationship. The Plataeans are paradigmatic for many other smaller Greek states that depended for their safety upon the patronage of a stronger neighbor but that, if they were to maintain their self-respect, needed to keep blurred the distinction in power. At the same time, empires and power politics, ancient and modern, can, as we will see, pour a harsh light upon such shadowy relations and impose a brutal clarity.

The passage that I quoted above (Hdt. 6.108.1) makes the reciprocal nature of the relationship between Plataea and Athens clear from the start: if the Plataeans had given themselves over to the Athenians, the Athenians had, in return, undertaken many toils (*ponoi*) on behalf of their Plataean friends. When the Plataeans devoted the full strength of their polis to the Athenian defense at Marathon, they were repaying the Athenians for their own generous services in the past. Consider now more fully the delicate three-part minuet that the Spartans, Plataeans, and Athenians act out in Herodotus:

> When the Plataeans were pressed by the Thebans, they first tried to *give themselves to* Cleomenes son of Anaxandrides and the Spartans, who happened to be there. But *they did not accept them*, saying, "We live too far away, and our help would be cold comfort to you. You could be enslaved many times over before any of us heard about it. [3] We advise you to *give yourselves to* the Athenians, since they are your neighbors and not bad men at giving help." The Spartans gave this advice not so much out of goodwill toward the Plataeans as wishing to cause trouble for the Athenians with the Boeotians. [4] So the Spartans gave this advice to the Plataeans, who did not disobey it. When the Athenians were making sacrifices to the twelve gods, they sat at the altar as suppliants and gave themselves to the Athenians. When the Thebans heard this, they marched against the Plataeans, but the Athenians came to their aid. (Hdt. 6.108.2–4, emphasis added.)

Note that the offer and acceptance are quite distinct, and that acceptance is not automatic. The Plataeans attempt to "give themselves" to the Spartans, but the Spartans have no obligation to accept the offer: the gap between the offer and acceptance is a blank, but awkward, space upon which Thucydides plays in his account of the Corcyrean and Corinthian offers of friendship in book one.[23] When the Plataeans decide to follow the Spartan advice and seek Athenian protection, they plan an even more spectacular gesture: they appear as suppliants—like something out of the fifth-century Athenian stage—while the Athenians conduct a solemn sacrifice. Ethnographic parallels would suggest that the supplication and acceptance were carefully dramatized, with both sides having worked the result out beforehand: the Plataeans would thus have avoided the humiliation of a rejection, while the Athenians would have escaped the charge of being hard-hearted.

The events that follow are a classic example of archaic Greek international affairs. (Indeed, such multipolar machinations and scuffles spring up even during the Peloponnesian War, both in the Peace of Nicias and when Athenian power reels from the shock of the Sicilian defeat. Complex alliances and backdoor negotiations sprang up as soon as Sparta or Athens appeared weakened, contributing to the confused and messy quality that many readers have found in books five and eight of Thucydides.)

> [5] As they were about to join battle, the Corinthians, who happened to be there, prevented them and brought about a reconciliation. Since both sides desired them to arbitrate, they fixed the boundaries of the country on condition that the Thebans leave alone those Boeotians who were unwilling to be enrolled as Boeotian. After rendering this decision, the Corinthians departed. The Boeotians attacked the Athenians as they were leaving but were defeated in battle. [6] The Athenians went beyond the boundaries the Corinthians had made for the Plataeans, fixing the Asopus River as the boundary for the Thebans in the direction of Plataea and Hysiai. So the Plataeans had *given themselves to* the Athenians in the aforesaid manner, and now came to help at Marathon. (Hdt. 6.108.5–6)

The Spartans, who suggested the alliance between Plataea and Thebes, may or may not have emerged as the preeminent power in Greece, but they have left the Athenians and Plataeans to their own devices and are now nowhere to be seen—there are no equivalents to peripatetic American envoys, bustling to and fro between the sides even as U.S. forces disappear over the horizon. Nevertheless, if the Spartans are absent, the Corinthians conveniently materialize and offer themselves as an intermediary to arrange a settlement by the only international mechanism available: arbitration by a third party. Neither side, however, feels any compunction about seizing an opportunity to gain an advantage: the Boeotians ambush the Athenians and their Plataean allies only to suffer a defeat; the victorious Athenians then cheerfully rearrange the boundary with Thebes to the advantage of their Plataean friends. But if both sides are willing to break, or at least bend, their agreements, the stakes seem relatively modest: a few acres of land exchange hands. There is no sign that the Athenians pressed their advantage to the fullest or sought to annihilate their Boeotian foes. This is precisely the kind of border skirmishes for low stakes that Thucydides deplores as typical of warfare in the introduction to his history (Thuc. 1.15.2).

The battle of Marathon in 490 was a watershed event for the relationship between the Athenians and Plataeans. The Persians had no official complaint with Plataea (technically, a Boeotian city), and few seem to have believed that the Athenians could prevail. The wholehearted Plataean aid came thus at the best possible time and was the strongest possible proof of Plataean loyalty. The Athenians were suitably impressed. In drawing up the line of battle, they placed the Plataeans on the far side of the left wing (Hdt. 6.111.1), a position of honor over which the Tegeans and Athenians would wrangle sixty years later in the battle of Plataea (Hdt. 9.26–28). The Plataeans lived up to this trust, for their position on the far left, along with the Athenian position on the far right, held firm in the battle and made victory possible (Hdt. 6.113.1).

In many ways, Marathon, even more than Salamis and the battles against Xerxes, was the foundation of Athenian pride. Even in the 420s, long after most of them were dead and the few remaining survivors were in their eighties and nineties, the *Marathonomakhoi*, the veterans of Marathon, remained the paradigmatic representatives of Athenian glory in the plays of Aristophanes (Ar. *Ach.* 181, 692–701; *Kn.* 781–785; *Cl.* 986). The Athenians did their best to convert Delphi into a memorial to this victory. Not only did those entering the sacred precinct encounter a set of statues dedicated by the Athenians after Marathon (Paus. 10.10), but roughly at the turn in the sacred way, halfway to the temple, they would encounter an Athenian treasury, also built from the spoils of Marathon (Paus. 10.11.4). When the visitor finally reached Apollo's temple, he would find gold shields hanging from the architrave that the Athenians had dedicated, again after the battle of Marathon (Paus. 10.19.4).[24] The Athenians did not have to share the victory at Marathon with any of their rivals—even the Spartans, who did send help, came too late and could only congratulate the Athenians on a job well done (Hdt. 6.120).

The Plataeans had helped, but they were far from being rivals: the Athenians could lavish their gratitude upon this small city and earn nothing but credit for their *megalophrosunê*, that "great heartedness" on which the elites of the archaic Greek world lavished such admiration.[25] As a sign of particular honor (Thuc. 2.34.5), the Athenians buried the 192 dead at Marathon itself. The Plataeans evidently brought their own dead home with them, but the Athenians dedicated a funeral mound that served as a grave to those liberated slaves who had fought for Athens and as a cenotaph to memorialize the Plataeans (Paus. 1.32.3).[26] From Marathon onward, whenever the Athenians celebrated the Greater Panathenaia, their most important civic festival, the herald prayed during the sacrifices on behalf of the Athenians and Plataeans alike (Hdt. 6.111.2). This gesture was considerable, for the Panathenaic festival—the festival of "All the Athenians"—helped to define who was and was not Athenian. While many non-Athenians competed in the athletic contests and the Athenians welcomed the prestige added by these international visitors, the focus of the Panathenaia was inward to Athens and not outward to Hellas as a whole.

The Plataeans also took great pride in their contribution at Marathon. Pausanias reports that they used their share of the spoils from Marathon to create a sanctuary of Athena Areia (Paus. 9.4). The relationship between the two cities that had begun in c. 520 grew even deeper after the battle of Marathon a generation later.

The Plataea of 490 provides an extreme, and in many ways idealized, example of how a small state establishes a stronger neighbor as its patron, acknowledging hierarchy but maintaining, on both sides, a strong sense of reciprocity and mutual respect. The relationship between Tegea and Sparta represents a case in which very different beginnings led to a similar result: Sparta had sought to conquer Tegea, its neighbor to the north, and presumably to render the Tegeans into helots, as they called the Greeks who worked as their serfs. According to Herodotus, the oracle at Delphi had refused them this conquest, but it had agreed that the Spartans would emerge as *epitarrhothoi*, "champions" or "patrons," of the Tegeans (Hdt. 1.67).[27] It would, however, be wrong to see in this a grudging accommodation to "realist" power politics. The Tegeans were fiercely proud of the honor they received from the Spartans (see Hdt. 9.26–28), and two generations later, when an anti-Spartan coalition was taking shape and the Tegeans had an excellent opportunity to pay the Spartans back for any ills that they had suffered, they "declared that they would not oppose the Spartans in any way" (Thuc. 5.32). When military conflict did come, the Tegeans faithfully came to the aid of their Spartan friends (Thuc. 5.57).

The dysfunctional relationship between Thebes and Plataea pushed the Plataeans toward Athens, but this is not the only instance in which a powerful state finds its expectations frustrated. The scuffle between Corinth and Corcyra that opens Thucydides' history of the Peloponnesian War allows us to see what can happen to a relationship that is supposed to be hierarchical, but that is also based upon mutual respect and affection. The Corinthians had originally founded Corcyra as their colony, and Corinth was thus its *mêtropolis*, "mother city." When the Corcyreans refused, however, to show Corinth the deference and respect that it expected, the Corinthians became implacable enemies of Corcyra. Nor were the Corinthians bashful about their expectations, for they state clearly in their speech at Athens what they believe they should receive:

> The attitude of our colony toward us has always been one of estrangement, and is now one of hostility; for, say they, "We were not sent out to be ill-treated." [2] We respond that we did not found the colony to be insulted by them, but to be their head, and to be regarded with a proper respect. [3] At any rate, our other colonies honor us, and we are very much beloved by our colonists; [4] and clearly, if the majority are satisfied with us, these can have no good reason for a dissatisfaction in which they stand alone, and we are not acting improperly in making war against them, nor are we making war against them without having received signal provocation. [5] Besides, if we were in the wrong, it would be honorable in them to give way to our wishes, and disgraceful for us to trample on their moderation. (Thuc. 1.38)

The Corinthians assert openly that they founded the colony at Corcyra with an eye to the honor and respect that they would receive from their colonists. They expect from the Corcyreans not only the same respect, but even affection that they receive from their other colonists. There is no question in the Corinthian mind about equality. Even if Corinth behaves badly, the proper response of the Corcyreans should be moderation and forbearance. When Athens chooses to intervene and provide limited, but crucial, aid to the Corcyreans, it intrudes upon a social space that the

Corinthians deem private to themselves, and the Corinthians ultimately shift their fury against the Corcyreans to the Athenians instead (e.g., Thuc. 1.68–71).[28]

Plataea after Marathon falls into a recognizable class of smaller states that form mutually beneficial, respectful, and even affective allegiances with more powerful neighbors. A decade later, however, events render their status more complex. The Plataea of the Peloponnesian War has more complex ties to the rest of the Greek world than it did in the decade before Xerxes' invasion. Its subsequent fate at the hands of the Spartans reflects, as I will argue, a hardening rejection of traditional ambiguities that were crucial to the traditional Greek World Order and that neither Athenians nor Spartans are prepared to tolerate. Athenians and Spartans alike began to press for that unnatural clarity—that rejection of "fractions," as it were—which modern historians and colonial census takers imposed as well. The results were brutal, but telling, for they reflected the extent to which international politics changed during the Peloponnesian War.

The emperors of the Near East seem to have spent most of their time, from the third millennium onward, crushing insurrections and retaliating against troublesome barbarians on the borders of their realms. A Persian response to Marathon was inevitable, and in 480 Xerxes personally led a vast invasion force into Greece. After the first year of conflict (and after the Greek allies had dealt a severe blow to the Persian navy at Salamis), Xerxes withdrew along with the bulk of his forces, but he still left behind a massive army of his best troops under Mardonios. The decisive land battle of the war took place in August of 479. Its location, the territory of Plataea, transformed Plataea's status in the Greek world.

Elaborate religious rites preceded each battle, and both the Persians and the Greek allies had at their disposal Greek seers who enjoyed an international reputation within Hellas: Teisamenos of Elis, a member of the famous Iamid clan celebrated in Pindar *Olympian* 6, sacrificed for the Greeks, while the Persians enjoyed the services of Hegesistratos—another Eleian and a bitter enemy of the Spartans who had barely escaped them with his life (Hdt. 9.37). Plutarch preserves an elaborate tradition about the pre-battle sacrifices in which the Plataeans played a crucial role. Both Herodotus (9.36) and Plutarch (Plut. *Arist.* 11.2) report that Teisamenos determined that the Greeks would be successful if they maintained the defensive. Plutarch, however, goes on to report that the Athenian Aristides then sent for an additional oracle from Delphi. The response from Delphi confused the Greeks, for part of it seemed to imply that they should fight in Attica while other aspects pointed toward battle in the territory of Plataea (11.3–5). The Greeks were about to retreat and make their stand in Attica, when Arimnestos, the leader of Plataea (whose territory was about to be abandoned), reported a convenient dream in which Zeus told him that the Greeks were mistaken. Closer scrutiny would, he was told, show that the Delphic oracle directed them to fight in Boeotia, and this proved indeed to be the case (11.5–6). In addition, Arimnestos showed Aristides a good position in Plataean territory for the Greeks, because the terrain limited the Persian advantage in cavalry (11.7).

Nevertheless, Arimnestos did not rely solely upon belief in his dream or faith in the reinterpretation of the oracle. He was determined that the Greeks should fight in his territory and planned a grand gesture to ensure that this would happen, oracle or no oracle:

> [8] And besides, that the oracle might leave no rift in the hope of victory, the Plataeans voted, on motion of Arimnestos, to remove the boundaries of Plataea on the side toward Attica, and to add (*epidounai*) this territory to the Athenians, that so they might contend in defence of Hellas on their own soil, in accordance with the oracle.
> [9] This munificence of the Plataeans became so celebrated that Alexander, many years afterward, when he was now King of Asia, built the walls of Plataea, and had proclamation made by herald at the Olympic games that the King returned (*apodidôsi*) this grace upon the Plataeans in return for their bravery and magnanimity in freely adding (*epedôkan*) their territory to the Hellenes in the Median war, and so showing themselves most zealous of all. (Plut. *Arist.* 11.8–9)

Strictly speaking, Arimnestos "added" (*epidounai* and *epedôkan*, both from *epi-didômi*) the territory of Plataea to Attica, and Plataea ceased to exist as a separate state. Moreover, Plutarch stresses that this gesture was famous in the Greek world, and that one hundred and fifty years later it elicited a complementary gesture from Alexander the Great.

The Greeks, of course, went on to surprise everyone and crush the Persian army. Still, the Greeks remained true to their quarrelsome nature and, according to Plutarch, a dispute arose about which the Greek contingent deserved the *aristeion*:[29] the Athenians refused to concede to the Spartans any formal prize for their performance at Plataea (Plut. *Arist.* 20). Earnest intercession from Aristides convinced the Athenian generals to hand the matter over to general arbitration by the Greeks. After a Megarian prudently argued that, to avoid a rift in the alliance, neither Athens nor Sparta should receive a prize,

> At this point Cleocritus the Corinthian rose to speak. Every one thought he would demand the meed of valor for the Corinthians, since Corinth was held in greatest estimation after Sparta and Athens. But to the astonishment and delight of all, he made a proposition in behalf of the Plataeans, and counselled to take away contention by giving them the meed of valor, since at their honor neither claimant could take offence. [3] To this proposal Aristides was first to agree on behalf of the Athenians, then Pausanias on behalf of the Spartans. (Plut *Arist.* 20.2–3)

This was, in fact, a very traditional Greek solution: just as Delphi and Olympia had developed in large measure because neither was, by itself, powerful, Plataea, because it was small, threatened no one (except for its enemies, the Thebans, who had, in any event, taken the wrong side and fought for the Persians).

> Thus reconciled, they chose out eighty talents of the booty for the Plataeans, with which they rebuilt the sanctuary of Athena, and set up the shrine, and

adorned the temple with frescoes, which continue in perfect condition to the present day; then the Spartans set up a trophy on their own account, and the Athenians also for themselves. (Plut. *Arist.* 20.3)

This solution neatly resolved the tensions of the moment. The sullen Athenians and Spartans could each erect their own monuments, while the Greeks as a whole gave the Plataeans an endowment with which to rebuild and adorn the sanctuary of Athena. Meanwhile, this major financial support allowed the Plataeans to rebuild the sanctuary (reportedly first constructed with the spoils from Marathon) in lavish style (Paus. 9.4). They commissioned an elaborate cult statue of Athena from Pheidias himself and wall paintings from Polygnotos and Onasias, while a portrait of their leader, Arimnestos, was placed at the feet of Athena's statue.

The Plataeans reaped a considerable windfall from their moment in the sun. The Spartan general, who was the supreme commander of the Greek forces, formally "restored" to the Plataeans their land to be inhabited in freedom (Thuc. 2.71.2), thus undoing the formal gesture by which Arimnestos had caused Plataea to be annexed to Attica.[30] Plataea became, in effect, something like a "Panhellenic sanctuary":

> [1] After this, there was a general assembly of the Hellenes, at which Aristides proposed a decree to the effect that deputies and delegates from all Hellas convene at Plataea every year, and that every fourth year festival games of deliverance be celebrated—the Eleutheria; also that a confederate Hellenic force be levied, consisting of ten thousand shield, one thousand horse, and one hundred ships, to prosecute the war against the Barbarian; also that the Plataeans be set apart as inviolable and consecrate, that they might sacrifice to Zeus the Deliverer in behalf of Hellas. [2] These propositions were ratified, and the Plataeans undertook to make funeral offerings annually for the Hellenes who had fallen in battle and lay buried there. And this they do yet unto this day, after the following manner. (Plut. *Arist.* 21.1–2)

The pattern proposed was typical. There would be a ritual of some kind each year, and on every fourth year (as at Olympia or at the Greater Panathenaia) there would be athletic contests as well. It is unnecessary to detail the elaborate rite the Plataeans performed. Not only Plutarch, who was writing five centuries after the battle, but Pausanias, who lived a century later still, specifically mentioned that the Plataean games continued to be celebrated (Paus. 9.2.6).

Both the Spartans and the Athenians have elaborate tombs for their own dead with funerary inscriptions by Simonides, the most famous poet of his day (Hdt. 9.85; Paus. 9.2.5), while the Tegeans, Megarians, and Phliasians also erected monuments in the neighborhood (Hdt. 9.85.2). Other cities erected cenotaphs to memorialize their dead, and Herodotus reports that the Aeginetans were even able to use their connections at Plataea to erect a monument ten years after the battle, despite the fact that they had not participated in the battle (Hdt. 9.85.3). The fact that Herodotus includes such a story—true or not—implies that Plataea continued to exercise a considerable hold upon the Greek imagination for years after the battle.

The yearly rituals and the games every fourth year must have attracted a large number of visitors, and the Plataeans even created a kind of sideshow, displaying curiosities from among the bones of the Persian dead (Hdt. 9.83.2).

Plataea was redesigned to serve as an engaging monument to Greek freedom. The actual monument to the battle was placed about a mile away from the city itself (Paus. 9.2.6), while the graves of the Megarians and Phliasians seem also to have stood in the battlefield, as they commemorated casualties in particular skirmishes with the Persian cavalry. The visitor to Plataea would encounter near the entrance of the city various Athenian and Spartan graves as well as a memorial commemorating the war dead of all the Greeks (Paus. 9.2.5). A little ways beyond the common memorial, in the marketplace of the city (Paus. 9.2.5; Thuc. 2.71.2), stood an altar to Zeus Eleutherios, "the Liberator," where Pausanias had sacrificed to Zeus after the Greek victory. Everything seems to have been calculated to remind the visitor of the sacrifices that the Greeks had made in their heroic struggle for freedom. By the later 430s, when tensions between Athens and Sparta had reached a critical level and while many of the younger veterans were still alive, Greeks must have found the atmosphere at Plataea as somber and grand as Americans find that at the American cemeteries near Normandy or the Vietnam War Memorial in Washington.

But if Plataea came to occupy a special position in the Greek world, its situation incurred dangers as well. Plataea now had two separate, and not wholly compatible, identities. On the one hand, the Plataeans evidently maintained their warm ties with Athens and in this they followed a pattern that had many parallels throughout the Greek world. At the same time, however, they had become a sanctuary that was, if not Panhellenic, at least common to all those states that had stood together against the Persians. As the Athenians gradually began more and more to resemble the Persian overlords whom they had helped expel from Greece, Plataea's condition became more difficult. How could the caretakers of this magnificent memorial to Greek freedom be the closest allies of the Athenians, who maintained an empire and extracted tribute by force?

Plataea might easily have weathered this contradiction—it was no threat to anyone and had little or nothing to offer a conqueror—if it had not been for its unfortunate physical location. Thebes, leading city of Boeotia, inveterate enemy of Plataea and, according to Herodotus, enthusiastic collaborator with the Persians, lay only twelve kilometers away. Every monument and reminder of the great battle kept alive the subsequent humiliation and disgrace which the Thebans had received. Even the wall paintings in the sanctuary of Athena Areia must have grated on Theban sensibilities: Onasias celebrated the original expedition of the seven against Thebes (Paus. 9.4.2).

By the summer of 431 there existed what we might aptly term a "cold war" between Athens and Sparta: all exchanges between the two states, we hear, were conducted by heralds (Thuc. 2.1), and a state of war existed in all but name, as it did between the United States and Revolutionary China in 1950. Nevertheless, as in the

Korean conflict, actual fighting broke out first between two smaller states, each sharing a common language and ethnic identity (Boeotian/Korean) with its rival, but each separated by its ties to enemy alliances. The people of Thebes, like the North Koreans after them, seized the opportunity to strike at their wayward relatives and attempted to unify their homeland, Boeotia, by force. The Korean War brought American troops into battle for the first (and only) time against China, but the conflict, though murderous, luckily did not spiral out of control and precipitate a world war (though not for want of trying on MacArthur's part). The Greeks were not so fortunate, and Thucydides uses the Theban attack on Plataea as the starting point for his account of a war that would last for twenty-seven years and, according to Thucydides, in its destructive power and in the suffering that it inflicted, dwarf all previous conflicts. Still less fortunate were the Plataeans, whose city was wiped from the map, and who either suffered genocide or lived on as refugees.

The basic events are simple in outline. In the spring of 431, a Theban force staged a pre-dawn raid on Plataea and temporarily seized control of the town, but their forces proved inadequate and, when the shock wore off, the Plataeans were able to overpower the Theban invaders. They forestalled a Theban attack upon those Plataeans outside the city walls by promising to return the prisoners, but, when the Thebans withdrew and the Plataeans had collected their people inside the city walls, they slaughtered all one hundred and eighty Thebans in their power (Thuc. 2.5). The Thebans claim that the Plataeans had sworn an oath to this agreement, but, although the Plataeans disputed the oath (Thuc. 2.5.6) and they may thus not have violated their word, their action was ill advised and clashed with the advice that arrived from Athens after the fact (Thuc. 2.6). The massacre of these Theban prisoners would later return to haunt the Plataeans.

The following year, in 430, the Spartan king Archidamus led a Peloponnesian army against Plataea. The Plataeans could not confront this force directly, but, safely ensconced behind their walls, they were, for the moment, in a strong position: the Athenians had removed all the women, children, and non-combatants the year before, leaving provisions and an Athenian garrison. The Spartan king offered to suspend hostilities against the Plataeans if they would simply declare themselves neutral in the war (Thuc. 2.72). The Plataeans refused. Archidamus began a siege. The Athenians, however, never made a serious attempt to relieve Plataea. After three years had passed, the remaining defenders—two hundred Plataeans, twenty-five Athenians, and the women who prepared food for the garrison (originally one hundred and ten: Thuc. 2.78)—surrendered (Thuc. 3.68). To please the Thebans, the Spartans executed all the men—roughly one Plataean for each of the Thebans executed four years before, with the twenty-five Athenians presumably thrown in for good measure. The women were, of course, sold into slavery—only barbarians murder women and throw away the money that they would earn on the slave market (cf. Mykalessos at Thuc. 7.29).

Nothing in this sad little episode would be out of place in the *New York Times* or *Washington Post*. The numbers are, of course, small by some standards, but, if twenty-five American marines were caught in a besieged city during the 1990s, CNN, with graphic video of the trapped marines' families, if not of the marines themselves, would probably have seen to it that the president who had sent them there was impeached long before three months, much less three years, had elapsed. If the Plataeans rejected the Spartan offer of neutrality at least in part because the Athenians implied help that was never forthcoming (Thuc. 2.73.3), many of their brethren among the smaller powers—whether Bosnians trapped in enclaves by the Serbs, Cuban veterans of the Bay of Pigs, Kurds alternately supported and dropped by various powers, or the many unknown pawns dropped by the great powers— could at least lay claim to similar stories.

Four separate vectors of force tugged the Plataeans in different directions. The first three reflect distinct, sometimes conflicting, sometimes complementary, "imagined communities" of the Archaic Greek world: the bonds of subgroups within the Greek world (in this case, Boeotia), the self-consciously forged and maintained ties of friendship (between Athens and Sparta), and the evolving vision of Hellas as a new, reified entity, to which the Persian Wars and the Plataeans had made a substantial contribution. The first two relationships, ethnic identities and manufactured affections, provided much of the fabric out of which the third, Hellenism, constituted. Each of these three shares a conservative vision of history: the Thebans harken back to an idealized time of Boeotian unity, the Plataeans and Athenians seek to preserve and enlarge the ties forged by their fathers, while the Greek-speaking peoples seek to reestablish and perfect that glorious unity of the Trojan War. While all three of these forces may conflict or reinforce one another, the fourth force, blind to the distant past and future alike, substitutes for history an eternal present. It sits like a black hole, empty, terrible, and ready to absorb the other three.

From the beginning of the incident in 431, the Thebans represented their attempt on Plataea as a profoundly conservative act. The Thebans had allies within Plataea and were able to make their way into the city without raising the alarm (Thuc. 2.2.2–3). The Plataean conspirators wanted the Thebans to murder their local enemies, but in this they were disappointed:

> After the soldiers had grounded arms in the marketplace, those who had invited them in wished them to set to work at once and go to their enemies' houses. This, however, the Thebans refused to do, but, determined to make a conciliatory proclamation, and if possible to bring the city into an agreement (*xumbasis*) and friendship (*philia*). Their herald accordingly invited any who wished to serve as allies in accordance with the ancestral usages (*kata ta patria*) of all the Boeotians to ground arms with them, for they thought that in this way the city would readily join them. (Thuc. 2.2.4)

The subsequent fury that the Thebans vent against the Plataeans has attracted a great deal of negative comment, but few have noted the manner in which the Thebans began this adventure. They refused to act as a death squad and to murder their ideological enemies in the dead of night. They were confident that they could create an agreement (*xumbasis*) and render the city-state "friendly" (*philia*). Their confidence in ancestral custom and in persuasion was a mistake: despite the fact that Thebes was much larger than Plataea, the Thebans had not dispatched forces suitable to overawe the Plataeans, who, when they had recovered from the initial shock and determined how few were the invaders, staged a counterattack and took most of them prisoner (Thuc. 2.3–4). Within a few hours, all one hundred and eighty Thebans—including Eurymachos, their leader who had refused to hunt down and assassinate the Plataean leaders when he had the chance—were dead, liquidated in cold blood, presumably under the orders of the Plataean leaders whose lives they had spared (Thuc. 2.5.7). It is hardly surprising that when, four years later, the Peloponnesian allies finally captured Plataea and took prisoner two hundred Plataean soldiers, the Thebans implacably demanded their lives. In their speech, the Thebans reproach the Plataeans for transgressing *ta patria*, ancestral custom (Thuc. 3.61), and bitterly recall the moderation that they had shown (and to which Thucydides attests) in their assault of 431 (Thuc. 3.65.3–66). In their fury of 427 they react against that restraint that they had exercised four years before.

By contrast, however, the Plataeans point to a second force, the value of friendship, and to that invented community distinct from the inherited community that the Thebans strive to restore. Not every state that had repudiated its traditional ties had such creditable moral capital. The Corcyreans, who feuded with their metropolis Corinth, could not cite any close relationship to another state (Thuc. 1.32.4–5): they had, to apply a modern term, no "credit rating" by which a new ally could be confident that good services conferred would be repaid, and the Corinthians pounded them on this score (Thuc. 1.37, 40.4). Confronted with imminent execution, the Plataeans make their friendship with Athens a cornerstone of their defense.

In their remarkable (and remarkably understudied) speech, the Plataeans firmly connect their friendship with Athens to the values of the Greek elite. The friendship, useful as it may be in itself, is more important as a sign that points to something else:

> If we refused to desert the Athenians when you asked us, we did no wrong; they had helped us against the Thebans when you drew back, and we could no longer betray (*prodounai*) them with honor (*kalôs*); especially as we had obtained their alliance and had been admitted to their citizenship (*politeia*) at our own request, and after receiving benefits at their hands (*eu pathôn*); but it was plainly our duty loyally to obey their orders. (Thuc. 3.55.3)

Thucydides' Plataeans may have suggested less exalted motives three years before (2.72.2), but the speech in book three plays upon the importance of character in this period. The fifth century was still largely an oral world: a man's, or a state's, "word" was very much his bond. Moreover, although Pindar, a poet of the Greek elite, speaks constantly of interest and loans, financial mechanisms remained fairly primitive. The best capital—wealth invested where it would bear interest—

was what Pierre Bourdieu called symbolic capital:[31] the rich man did well to lavish gifts upon his friends, for these gifts provided him with insurance and a "diversified portfolio" in times of dire need. And no great power was more dependent upon the intangibles of reputation and remembered obligations than the Spartans, with their ramshackle alliance abroad and their embittered serfs, always ready to revolt at home. The Plataeans, in fact, make a point of reminding the Spartans that they had sent a third of their citizens to help the Spartans put down a serious helot revolt (3.54.5).

The Plataean argument is quite shrewd: you cannot, they tell the Spartans, blame us for living up to our friendship with the Athenians. If you punish men for being loyal to their friends, then you undermine the complex interdependencies that hold the Greek world together:

> We now fear to perish by having . . . chosen to act justly (*dikaiôs*) with Athens sooner than profitably (*kerdaleôs*) with Sparta. [7] Yet in justice the same cases should be decided in the same way, and expediency (*to sumpheron*) should not mean anything else than lasting gratitude (*charis*) for the service of a good ally combined with a proper attention to one's own immediate advantage (*ôfelimon*). (Thuc. 3.56.6–7)

The Spartans may think it expedient, the Plataeans argue, to throw them to their Theban enemies, but true expediency has a healthy regard for justice and for lasting *charis*, that elusive but fundamental word that describes at once the immediate pleasure of a good service and the subsequent gratitude that it wins.

> [2] Our lives may be quickly taken, but it will be a heavy task to wipe away the infamy of the deed; as we are no enemies whom you might justly punish, but friends forced into taking arms against you. [3] To grant us our lives would be, therefore, a righteous judgment; if you consider also that we are prisoners who surrendered of their own accord, stretching out our hands for quarter, whose slaughter Hellenic law forbids, and who besides were always your benefactors. (Thuc. 3.58)

This is very ably put: whatever the Plataeans may or may not have done to the Theban prisoners, they are in Spartan hands and their fate will contribute, for better or worse, to Sparta's prestige and standing in Greece as a whole. Given the importance of reputation, they cannot, so the Plataeans argue—and argue well—afford to miscalculate here. The Odysseus of Sophocles' *Ajax* allows us to glimpse the powerful effect that the magnanimity of an enemy could have.

The brilliance of the Plataean speech—and I think it is brilliant—lies in the connection they forge between the second and third forces. The speech of Diodotus, which precedes it by a few chapters (Thuc. 3.41–48), has attracted vastly more attention and even now retains a strong position in anthologies and histories of political philosophy and international relations. Diodotus has the advantage that his speech, as opposed to that of the doomed Plataeans, was successful. Diodotus, however, spoke to an audience that was eager to be persuaded (Thuc. 3.36.5), while the Spartans had made up their mind before the Plataeans said a word (as the Plataeans rightly fear: Thuc. 3.53.4). More importantly, Diodotus, with his rationalizing argument that justice and expediency coincide, sounds like a professor of international relations or political science. The Plataeans make a similar argument, but overall they sound like fifth-century Greeks, manipulating conventions and assumptions far more foreign to modern sensibilities.

The Plataeans weave back and forth between the importance of loyalty and their contribution to the Persian War because they wish to tie them so inextricably together that the Spartans cannot touch one without damaging the other. The Spartans cannot, they argue, watch the Plataeans be butchered without undermining that imagined community of Hellas upon which Sparta's status itself depends. From the start, the Plataeans locate their defense squarely in the ethics of *charis* and symbolic capital:

We will place before you those just things (*dikaia*) that are at our disposal, not only on the question of the quarrel which the Thebans have against us, but also as addressing you and the rest of the Hellenes; and we will remind you of our good services (*tôn eu dedramenôn*), and endeavor to prevail with you. (Thuc. 3.54.1)

The Plataeans then point to their service for Hellas in the Persian War (3.54.3–4). A brief glance at Plataea's service to Sparta against the helots and its relationship to Sparta in the war then leads back to the Plataean friendship with Athens (3.55.3). They quickly turn back, however, to the contrast between Theban support of Persia and Plataea's steadfastness (3.56.4–5), shifting just as quickly back to its loyalty to the Athenians (3.56.6). Their remark here, partially quoted above, makes the link between past and present, Athens and Hellas, explicit.

To these few we belonged, and highly were we honored for it; and yet we now fear *to perish by having again acted on the same principles*, and chosen to act well with Athens sooner than wisely with Sparta. (Thuc. 3.56.6, emphasis added)

The issue is general not particular: loyalty to Athens and loyalty to Greece are the same quality, differently directed. If the Spartans turn on the Plataeans now to please Thebes, then they will be punishing the Plataeans for being loyal to their friends. In so doing, the Spartans attack loyalty in general, putting at risk the values on which their alliance and indeed the idea of Hellas rest.

At this point, the Plataeans turn once and for all to the special status their city had earned after the Persian Wars. Thucydides is in general famous for his disinterest in religious affairs,[32] and he rarely evinces much sensitivity to the spectacular dedications and sanctuaries that fascinated Herodotus.[33] But to any contemporary observer, the events in and around Plataea must often have seemed surreal. The Thebans begin the affair by occupying the marketplace (Thuc. 2.2.4) where, after the Persian army had been crushed, Pausanias had originally sacrificed at the Altar of Zeus the Deliverer (Thuc. 2.71.2), effectively converting Plataea into a shrine of Greek freedom. The besieging forces who walled in the Plataeans gazed every day upon the graves of those who had fallen for Greece and that the Plataeans had, for half a century, continuously tended. What did defenders and attackers alike say on that day of the year when the besieged Plataeans would normally have proceeded out of their city, in festive procession, a trumpeter at their head, wagons with myrtle wreaths behind them, and young men carrying offerings (Plut. *Arist.* 21.3)? On that day, the chief magistrate of Plataea, dressed in purple (although required on all other days to wear white) and with sword in hand (although he was forbidden during the rest of the year to touch iron), would normally have followed the procession to the graves, where he would have washed the stone monuments with sacred water and anointed them with myrrh (Plut. *Arist.* 21.4–5). Did anyone tend to these graves during the siege? Thucydides does not introduce these factors into his description of the siege itself (although the physical features of Apollo's temple at Delium play a prominent role in his account of the battle there: Thuc. 4.90).

Nevertheless, before the siege begins and after its conclusion, Plataea's status as sacred ground is (for Thucydides, at least) remarkably prominent. The moment a Peloponnesian army enters their territory, the Plataeans waste no time reminding the Spartans of the honor that Pausanias had conferred or the rewards that an earlier generation of Spartans had lavished upon them (Thuc. 2.71.2–4). The Spartan king, Archidamus, treats the Plataeans with care. When he fails to win from them a promise of neutrality, he carefully calls the "gods and heroes of the land" as witnesses that he had done everything he could for the Plataeans (Thuc. 2.74.2). Later, when the captured Plataeans build up to the climax of their speech, they point again and again to the physical signs of Greek triumph, of their own contributions, and of Sparta's role. They imagine the horror that their weapons dedicated as spoils in the Panhellenic sanctuaries will provoke among those who remember Plataea's services to Greece (Thuc. 3.57.1). They recall the Serpent Tripod, which the Spartans had dedicated and on which they had caused to be inscribed the names of the Plataeans, and ask if the Spartans will blot out their city to please the very Thebans who had served with Persia (Thuc. 3.57.2). We do not, of course, know exactly where this debate would have taken place, but Thucydides' speakers play upon the well-known prominence of the Greek tombs there:

[4] Look at the sepulchres of your fathers, slain by the Persians and buried in our country, whom year by year we honored with garments and all other dues, and the first fruits of all that our land produced in their season, as friends from a friendly country and allies to our old companions in arms! Should you not decide aright, your conduct would be the very opposite of ours. Consider only: [5] Pausanias buried them thinking that he was laying them in friendly ground and among men as friendly; but you, if you kill us and make the Plataean territory Theban, will leave your fathers and kinsmen, in a hostile soil and among their murderers, deprived of the honors which they now enjoy. What is more, you will enslave the land in which the freedom of the Hellenes was won, make desolate the temples of the gods to whom they prayed before they over-

came the Persians, and take away your ancestral sacrifices from those who founded and instituted them. (Thuc. 3.58.4–5)

The Plataeans continue in this vein, declaring themselves suppliants at the tombs of the Spartan forefathers and calling upon these dead warriors for help (Thuc. 3.59.2).

Graves and supplications were powerful symbols in the archaic Greek world (cf. Thuc. 1.26.3). The Plataeans, because of their peculiar status as caretakers of a battlefield that meant so much for Sparta and for Greece, can compound the already considerable authority of their gestures: not only are the Plataeans helpless and not only does their care of Spartan graves give them a particular claim upon Spartan benevolence, but they and their city are a living symbol of Hellas and that very freedom the Peloponnesian League claims to be restoring. Whatever the Plataeans may have done to Theban prisoners, they surrendered to the Spartans, and it is the Spartans who decide their fate (Thuc. 3.59.4). When the Spartans relentlessly execute each Plataean prisoner, they turn their backs upon an appeal that few of their fellow Greeks could have matched. Why, then, did the desperate Plataean speech fail?

This brings us to the fourth force, which I will term spreadsheet logic and which finds its expression in a reductive calculation of short-term costs and benefits. The Plataean speech may have had some effect—at least, the Thebans fret that the Spartans will weaken in their resolve and spare the Plataeans (Thuc. 3.60). They launch into a speech that is famous for the hatred and malice that animates it. Central to their argument is the attribution of unambiguous responsibility to the leader: the Thebans advance the argument that following orders relieves the doer of any accountability, good or evil: the Plataeans fought against Persia because of their Athenian masters (Thuc. 3.64), while the Thebans as a whole collaborated with the Persians because a despotic cabal of leaders compelled them (Thuc. 3.62).

This is the logic of Benedict Anderson's colonial census takers: one actor bears all of the responsibility—no fractions. The same oversimplification occurs today when observers assume that, if interest plays a role, then justice or morality must have been irrelevant. But ambiguity was crucial for the Archaic Greek World Order. The Plataeans were subordinate to the Athenians, but they were also free Greeks. They were doubtless expected to come to Athens' help on the field of Marathon, but the safer course might have been to follow the Spartan lead and arrive just too late to participate. The help that they and the Athenians gave to one another was both an obligation and a gift. The Thebans were surely correct in arguing that, had the Athenians gone over to the Persians, the Plataeans would have followed suit, but this had been as obvious in 479 as it was in 427, and neither Pausanias nor the other Greek allies cared about such factors when they pondered the fresh battlefield. The common judgment of the Greek allies lavished praise and admiration upon Plataea.

But the new conditions of the fifth century left little room for the old leeway afforded by multiple identities and criss-crossing loyalties. The act by which the Athenians finally and unalterably alienated Corinth was, by any material measure, trivial: the Poteideians were both tribute-paying allies of the Athenians and colonists who received magistrates from Corinth. As relations with Corinth soured, the Athenians decided that they could not accept this ambiguous situation and ordered the Poteideians no longer to receive these Corinthian magistrates (Thuc. 1.56). Poteideia must be Athenian or Corinthian—no fractions.

If the narrow Theban logic would have been ruinous to the archaic Greek world, unraveling the competing ties that held the complicated structure together, the Spartan position is even more destructive. The Spartans reportedly did not charge the Plataeans with any crime—neither their aid to the Athenians nor their massacre of the Thebans was mentioned (Thuc. 3.52.4). Instead, they pose what the Plataeans would tactfully describe (Thuc. 3.53.2) as a "brief question": what service, they ask, had the Plataeans provided to the Spartans and their allies in *the war currently under way* (*en tôi polemôi tôi kathestôti*)? The long, impassioned speeches of the Plataeans and Thebans follow, but, when the debate is over, the Spartans simply pose again the question with which they had begun (Thuc. 3.68.1): what had the Plataeans done for them in this war? In a brutal parody of legal procedure that even Stalin could have admired, the Spartans ostentatiously refused to judge the Plataeans as a group:

> Having, as they considered, suffered evil at the hands of the Plataeans, they brought them in again one by one and asked each of them the same question, that is to say, whether they had done the Spartans and allies any service in the war; and upon their saying that they had not, took them out and slew them all without exception. (Thuc. 3.68.1)

This personal attention would anticipate the "special treatment" of later ages.

The significance of the Spartan action, brutal as it may have been, did not lie with the execution of the Plataeans. Although the Spartan actions would long be remembered (see Isoc. 12.91–94), why they acted was more important than what they did, for their reasoning brought to the surface, without apologies and for all to examine, the attitude that destabilized their own world. Each time the Spartans questioned the Plataeans, they specifically restricted their horizons to the war at hand. If the Thebans oversimplify and disturb the ambiguities of archaic Greek society, the Spartans truncate time and history. Alliances between states, as between families, evolved over generations. A good service by one generation lived on and could be reclaimed by another. The Plataeans, for example, looked back, for better or worse, to ninety years of close cooperation with Athens. If Sparta, the greatest champion of the conservative values of the Greek elite, could brush aside (Thuc. 3.68) profound attachments established in the memory of men still living and kept alive by solemn rituals year by year, then there was no basis for long-term loyalty to any state.

This is not the first time that actors in history have made such a chronologically shallow decision: after a debate in which the Corcyreans and Corinthians each masterfully manipulate the rhetoric of gift exchange and of ritualized friendship, the Athenians act as if neither had spoken, ignoring the offers of friendship and claims on past good services alike, while choosing a course that would damage both parties in the short-term Athenian interest (Thuc. 1.44.2).[34] But the Athenians were famous for their strange ways and new thinking: indeed, the Corinthians clamored for war precisely because they came to understand how new and different Athenian power was.[35] And later in the war both sides would cynically manipulate friends and foes alike.[36] But if even the Spartans find that the exigencies of the moment dominate all other concerns, then who could be relied upon? Theban aid may have been essential to victory in the war against Athens, but what would victory bring if the Spartans turned their backs on the Plataeans and all that they represented? If history no longer matters, then what hope remains for the future? It is no accident, I think, that the narrative moves immediately from the fate of Plataea to the stasis at Corcyra and the collapse of civil society (Thuc. 3.69–85). We can see already in the reasoning of the Spartans at Plataea that war was indeed a *biaios didaskalos*, a "brutal teacher" (Thuc. 3.82.2).

I would like to conclude with two points, of which one points backward to the time before the execution of the Plataeans while the other briefly links the problems described above with the thought processes of contemporary decision makers. The decline of society under the stress of warfare is a major theme in Thucydides' history, but he tends to stress continuity as well as change. I have already mentioned the solemn exchange by which the Spartan king Archidamus initiates hostilities with Plataea in 430. Archidamus stands out, in some ways, as the chief exemplar of the old-fashioned Spartan in Thucydides—he is, in fact, perhaps the only Spartan whose moral authority Thucydides does not undermine (Thucydides expresses admiration for Brasidas, but for his "Athenian" energy and intelligence and precisely because he does not resemble a Spartan). Archidamus' speech at 1.80–85, and especially the passage at 1.84, presents for us the most positive picture of the Spartan outlook.

In negotiating with the Plataeans, Archidamus makes them a handsome offer:

> You have only to deliver over the city and houses to us Spartans, to point out the boundaries of your land, the number of your fruit-trees, and whatever else can be converted *into a number*, and yourselves to withdraw wherever you like as long as the war shall last. When it is over we will restore to you whatever we received, and in the interim hold it as a trust (*parakatathêkê*) and keep it in cultivation, paying you a sufficient allowance. (Thuc. 2.72.3, emphasis added.)

This is a grand gesture: the Spartans offer to take upon themselves a very onerous and complex responsibility so that the Plataeans can be neutral in this war and reestablish themselves in their memorial city when peace returns. The custom of entrusting to a friend a *parakatathêkê*, a "trust" or "deposit," was, in fact, a well-known device whereby the international Greek elite could cushion themselves against harsh turns of fortune.[37]

Nevertheless, impressive as Archidamus' gesture may be and much as it may strengthen his moral position, it is not clear how realistic it would have been for the Plataeans to accept. Thucydides tells us that Archidamus made this proposal to quiet Plataean concerns, but one of the chief Plataean concerns was the fate of the women, children, and dependents whom they had entrusted to the Athenians. The Plataeans expressed a fear that, if they became neutral and abandoned their support of the Athenians, their families would become, in effect, hostages (Thuc. 2.72.2). Archidamus' offer does not address this concern at all.

Consider, however, another aspect of Archidamus' speech that is perhaps more subtle, but, I think, also suggestive. As in his speech at Thuc. 1.80–85, Archidamus belies the stereotype of the bluff, supposedly unsophisticated Spartan. His offer betrays a clear grasp of the powers of representation: he asks the Plataeans to show him the "boundaries" of their land, to count their trees and, indeed, to create, insofar as they can, a numerical model for their entire property. Ignoring Plataean loyalty to Athens on the one hand and Plataean fear on the other, Archidamus reduces the situation to a balance sheet. When the Plataeans respond that the situation is not that simple and that they cannot accept the offer as stated, Archidamus piously withdraws, prays to the gods, and begins the siege.

Archidamus with his stress on numbers anticipates not only the reductive, yes or no, reasoning of the Thebans and the Spartans who finally capture Plataea, but the realist program of thought. He would have readily understood the academic realisms of Hans Morgenthau and Robert Keohane and the practical realisms of Dean Acheson and George Marshall. Whatever his emotional reactions or professional assessment may have been, he would have had little difficulty seeing how a naive reliance on statistical reasoning would lead to the "body counts" in Vietnam. Numbers can be charted, summarized, mixed in formulas with other numbers to present yet more charts, summaries, and numbers. But numbers (even if they are, unlike the body counts of Vietnam, relatively accurate) can be meaningful and utterly misleading: the Nazis developed a daunting logistical structure to uproot Jews, ship them across Europe to camps, define their rations, extract labor from a selected few, and render their remains into useful commodities, but the mountains of statistics, ledger books, reports, and other bureaucratic instruments lent a veneer of logical analysis to a process that was staggeringly inefficient, drained precious rolling stock from direct support of the war, diverted troops to guard duty, and wasted far more labor than it ever extracted from its brutalized victims.[38]

Numbers can be meaningful and reasonable but not decisive. Thucydides' Melians would, in defending their freedom against Athenian imperialism, defy the numbers and suffer annihilation—the adult males liquidated, the women and children sold into slavery. But the ruthless brutality of the Athenian action points to the weakness of their power politics. They kill the Melians to prove that numbers—in this case, the preponderance of Athenian ships and men—are decisive, that resistance really is futile, that all smaller states should draw up the balance sheet—tribute on one side, survival on the other—and do the "reasonable" thing. The Athenians—like any imperial power—fear nothing more than that those under their control will look past the numbers and pursue a different logic.

Most of this chapter has attempted to tease out the complexity of a single episode in the Peloponnesian War. Koreans from both sides of the DMZ can surely empathize with the suffering of the Plataeans. The destruction of their city was not, however, the end for Plataea. The Athenians accepted the Plataeans into their community afterward, providing them with *isopoliteia*, "an equal share in the commonwealth," until 386, when Sparta restored the Plataeans to their home. Plataean troubles were not, however, at an end: in 373, the Thebans again destroyed Plataea and drove the people from their homes (although this time, the expulsion was bloodless: Paus. 9.1.8). The oldest of these refugees who made their way back to Athens may have been able to recall that night, almost sixty years ago, when the Thebans attempted to seize the city and their troubles began. Those refugees who had been alive in 431 would never see their home again. But the younger refugees of 373 would live to see their home restored a second and final time. Philip of Macedon reestablished Plataea as a check to Thebes after the battle of Chaeronea in 338, almost a century after the Theban assault in 431.

It would be easy to despair and to argue that this ongoing rivalry offers little hope to Koreans or any divided people: violence begets violence in a chain that can extend indefinitely. But the atrocities in this case did not produce an endless chain. Later still, when Macedonians ruled Greece and Cassander, son of Antipater (c. 358–297) rebuilt Thebes, the Thebans sought reconciliation with their ancient enemies so that they could all participate in the common assembly of Boeotians and the shared sacrifices that helped Greeks (like the Boeotians) to define themselves as separate groups (Paus. 9.3.6). In Pausanias' account, written four centuries later, Plataea and Thebes are neighbors, Boeotian cities keeping alive their traditions, celebrating their ancient festivals, tending the graves of the war dead from 479, mindful of their darker history, but at peace.

Notes

1. See his speech delivered at Princeton on 22 February 1947, *Department of State Bulletin* 16, 391.

2. In *Classics of International Relations*, 2nd ed. (Englewood Cliffs, NJ: Prentice Hall, 1990), 3, John Vasquez stressed the influence of Thucydides and Machiavelli on realist thinking in the 1950s and 1960s until Vietnam drove many to focus again upon issues of morality. He includes the Hobbes translation of the Melian Dialogue (16–20) in his anthology of realist authors. Likewise, Paul R. Votti and Mark V. Kauppi, *International Relations Theory: Realism, Pluralism, Globalism* (New York: Macmillan, 1987), 34–36, opens its section on the intellectual precursors and influences on realism with Thucydides. Even Frank W. Wayman and Paul F. Diehl, eds., *Reconstructing Realpolitik* (Ann Arbor: University of Michigan Press, 1994), which concentrates on charts, tables, and heavily quantification rather than historical analysis, begins immediately with Thucydides (5). For Thucydides as a realist, see in general Gregory Crane, *Thucydides and the Ancient Simplicity* (Berkeley: University of California Press, 1997).

3. On Thucydides as a realist, see Michael T. Clark, "Realism Ancient and Modern: Thucydides and International Relations," *PS: Political Science and Politics* 26, 3 (1993): 491–94; James der Derian, ed. "A Reinterpretation of Realism," in *International Theory: Critical Investigations* (New York: Macmillan, 1995), 363–96 at 382–85; Michael Doyle,

"Thucydides and Political Realism," *Review of International Studies* 16, 3 (1990): 223–37; Steven Forde, "Varieties of Realism: Thucydides and Machiavelli," *Journal of Politics* 54 (1992): 372–93; Daniel Garst, "Thucydides and Neorealism," *International Studies Quarterly* 33, 1 (1989): 3–28; Laurie M. Johnson, *Thucydides, Hobbes, and the Interpretation of Realism* (Dekalb: Northern Illinois Press, 1993); and Laurie Johnson-Bagby, "The Use and Abuse of Thucydides," *International Organization* 48 (1994): 131–53.

4. Robert Gilpin, *War and Change in World Politics* (Cambridge: Cambridge University Press, 1981), 226–27.

5. Dean Acheson, *Present at the Creation: My Years in the State Department* (New York: Norton, 1969).

6. The top four "lessons" McNamara draws from the Vietnam War are variations on American ignorance of the "history, culture, and politics of the people in the area": see Robert S. McNamara and Brian VanDeMark, *In Retrospect: The Tragedy and Lessons of Vietnam* (New York: Random House, 1995), 321–22.

7. This is the central theme of Crane, *Thucydides and the Ancient Simplicity*.

8. Jeffrey H. Tigay, *The Evolution of the Gilgamesh Epic* (Pennsylvania: The University of Pennsylvania Press, 1982), 11.

9. See, for example, Margaret Cool Root, "The Parthenon Frieze and the Apadana Reliefs at Persepolis: Reassessing a Programmatic Relationship," *AJA* 89 (1985): 103–20; on Herodotus' complex vision of Hellenism and "barbarism," especially as conditioned by his experience of imperial Athens, see Pericles Georges, *Barbarian Asia and the Greek Experience* (Baltimore: Johns Hopkins University Press, 1994), 115–206.

10. For the classical background of Lincoln's Gettysburg Address, see Garry Wills, *Lincoln at Gettysburg: The Words That Remade America* (New York: Simon and Schuster, 1992); for the Stele of the Vultures, see Jerrold Cooper, *Presargonic Inscriptions*, vol. 1 (New Haven, CT: The American Oriental Society, 1986); for the Behistun inscription, see Roland G. Kent, *Old Persian: Grammar, Texts, Lexicon*, 2nd ed. (New Haven, CT: American Oriental Society, 1953).

11. The one exception—the compound *barbarophônos*, which appears at Hom. *Il.* 2.867—does, however, indicate that term *barbaros* did preexist our *Iliad*.

12. On this, see, for example, Edith Hall, *Inventing the Barbarian: Greek Self-Definition Through Tragedy* (Oxford: Oxford University Press, 1989); and Georges, *Barbarian Asia*, 115–63.

13. The number of city-states inscribed on the Serpent Column, originally erected at Delphi, is thirty-one: see Russell Meiggs and David Lewis, *A Selection of Greek Historical Inscriptions* (Oxford: Oxford University Press, 1988), no. 27; Pausanias only records twenty-seven names from the Greek dedication at Olympia, but the difference may well reflect carelessness on his part (Paus. 5.23). In any event, thirty-one seems to have been a traditional figure for the allies (Plut. *Them.* 20.4) and the real figure, if different, was doubtless very close to this.

14. On the complexities and "archaeology" of Greek freedom, see Kurt Raaflaub, *Die Entdeckung der Freiheit: zur historischen Semantik und Gesellschaftsgeschichte eines politischen Grundbegriffes der Griechen*, vol. 37 (Munich: Beck, 1985); for a fascinating look at Greek freedom from a non-classicist perspective, see the work of the sociologist Orlando Patterson: *Freedom in the Making of Western Culture* (New York: Basic Books, 1991).

15. On the development of Delphi and Olympia, see Catherine Morgan, *Athletes and Oracles: The Transformation of Olympia and Delphi in the Eighth Century B.C.* (Cambridge: Cambridge University Press, 1990).

16. On the Panathenaia, see Jenifer Neils, "The Panathenaia: An Introduction," in Jenifer Neils, ed., *Goddess and Polis: The Panathenaic Festival in Ancient Athens* (Princeton: Princeton University Press, 1992), 13–27.

17. See Ernst Badian, "Plataea Between Athens and Sparta," in Hartmut Beister and John Buckler, eds., *Boiotika* (Munich: Editio Maris, 1989), 95–111, vol. 2 of *Münchener Arbeiten zur alten Geschichte*, ed. Hatto H. Schmitt, which is republished in revised form in Ernst Badian, *From Plataea to Potidaea: Studies in the History and Historiography of the Pentecontaetia* (Baltimore: Johns Hopkins University Press, 1993). Subsequent to, and in part responding to, Badian's 1989 article, see N.G.L. Hammond, "Plataea's Relations to Thebes, Sparta and Athens," *JHS* 112 (1992): 143–50.

18. Thucydides is fairly consistent in this estimate: after two hundred and twelve men managed to break out of the besieged city (Thuc. 3.24.2), two hundred men survived the siege of Plataea and fell into the hands of the Spartans (Thuc. 3.68). When we consider the garrison of eighty Athenians, the defenders totaled four hundred eighty. Thucydides notes that one hundred and ten women accompanied these men as "bakers" (*sitopoioi*), a ratio of one to four(!). He does not specify whether these women were free, slave, or a combination.

19. On the date, see G.S. Shrimpton, "When Did Plataea Join Athens?" *CP* 79 (1984): 295–304.

20. The translations of Herodotus and other authors cited in this paper are based upon those in Gregory Crane, ed., *Perseus 2.0: Interactive Sources and Studies on Ancient Greek Culture* (New Haven: Yale University Press, 1996), but have been freely modified where this has seemed appropriate.

21. Badian, *From Plataea to Potidaea*, 117.

22. Cf. as well Thuc. 2.68.7, where the people of Amphilochian Argos "give themselves" to the Acarnanians: here one set of free Greeks clearly joins themselves to another. There is no question of slavery or dishonor in this exchange.

23. For the dramatic role of the verb "to accept" in the Corcyrean and Corinthian speeches, see Gregory Crane, "Power, Prestige and the Corcyrean Affair in Thucydides 1," *Classical Antiquity* 11 (1992): 1–27.

24. On the Athenian use of space at Delphi and on Thucydides' peculiar treatment of sacred space, see Gregory Crane, *The Blinded Eye* (Lanham, MD: Rowman and Littlefield, 1996), 163–208.

25. For the dynamics of this ethic of generosity in the fifth century, see, for example, Leslie Kurke, *The Traffic in Praise: Pindar and the Poetics of Social Economy*, ed. Gregory Nagy (Ithaca, NY: Cornell University Press, 1991), 163–256.

26. Badian, *From Plataea to Potidaea*, 117–18, relies on Pausanias' statement that one mound served the Athenian slaves and Plataeans alike, but the archaeological evidence and the problems in Pausanias' account argue otherwise: for a summary of the archaeological excavations of both mounds and a plausible analysis of the situation, see Hammond, "Plataea's Relations to Thebes, Sparta and Athens," 147–50 (which responds to the earlier publication of Badian's article, "Plataea Between Athens and Sparta").

27. For the symbiotic relationship implied by the term *epitarrhothos*, see Hom. *Il.* 5.808, 828, 11.366, 12.180, 17.339, 20.453, 21.28, and *Od.* 24.182, where an *epitarrhothos* is the divine helper and patron of a mortal hero. In this relationship, there is no ambiguity about the rank—the god is superior to the man—but the divine help adds to, rather than diminishes, the achievement and glory of the hero.

28. On the relationship between Corinth and Corcyra, see Gregory Crane, "Fear and Pursuit of Risk: Corinth on Athens, Sparta and the Peloponnesians (Thucydides 1.68–71, 120–121)," *TAPA* 122 (1992): 227–56.

29. Hdt. 9.81 claims ignorance as to who received prizes, but he seems to have in mind prizes for individuals rather than any award for a city as a whole. It is quite probable that such a judgment was rendered after the battle: Herodotus reports, for example, that the Aeginetans were judged to have performed best at Salamis, followed by the Athenians (Hdt. 8.93).

30. The meaning of this passage has attracted considerable attention, most recently in

Badian, *From Plataea to Potidaea,* 118–119, and Hammond, "Plataea's Relations," 145 (with Hornblower leaning toward the latter in his commentary). Neither Badian nor Hammond, however, considers Plut. *Arist.* 11.8–9, where the Plataeans formally "give in addition" (*epididômi*) their land to Athens. The verb *apo-didômi* almost always means "to restore" or "give back" a thing and it would thus be the precise term for Pausanias to use if he wished to reverse this action. Hammond argues that the verb simply means "concede" here, but the two parallels that he adduces (Thuc. 1.144.2, 3.36.5) are both cases in which a prior condition or privilege is being restored: Pericles charges the Spartans to allow their allies to be free *again* at 1.144.2, and the Athenians at 3.36.5 wish to consider *again* the Mytilenean decree that they had already passed the previous day. It is problematic that Thucydides does not refer to the Plataeans having "added" their land to Attica, but the whole Plataean episode is allusive, and Plutarch stresses that the Plataean gift of their land was famous (Plut. Arist. 11.9: *periboêtos*). It might be argued that Pausanias did not have the authority to "give back" Attic territory, but it is also perfectly clear that the Athenians had no designs upon Plataea and would in any event only have lost face had they tried to enforce the Plataean gesture after the battle.

31. Bourdieu, *Outline of a Theory of Practice,* 171–83.

32. E.g., Hornblower, "The Religious Dimension"; Badian, *From Plataea to Potidaea,* 112–13, on religion in the Plataean episode.

33. For a comparison of Thucydides' and Herodotus' treatment of religious space, see Crane, *The Blinded Eye,* "The Politics of Religious Space."

34. On this, see Crane, "Power, Prestige and the Corcyrean Affair in Thucydides 1."

35. See their speech at Thuc. 1.68–71, with Crane, "Fear and Pursuit of Risk."

36. See 8.32.3: Astyochos argues that the Spartans should help the Lesbians because they will either have more allies or, at least, harm the Athenians; 8.45.2: Alcibiades advises Tissaphernes to underpay the sailors—keep them weak and dependent; at 8.46, Alcibiades goes on to argue that Tissaphernes should keep the war going so that neither side should win a victory and emerge as a threat. Tissaphernes was a quick study and put this policy into practice (8.56.2, 87.4).

37. See Hdt. 6.86, where the thought of stealing a *parakatathêkê* brings down divine vengeance upon Glaukos of Sparta; even the tyrant Periander—whom Herodotus represents as the paradigm of the odious despot—resorts to extreme measures to make good a *parakatathêkê* deposited with him (Hdt. 5.92G).

38. The wastefulness of Nazi extermination camps constitutes one of the major themes in Hannah Arendt, *The Origins of Totalitarianism,* 2nd ed. (New York: Harcourt Brace, 1973).

Bibliography

Acheson, Dean. *Present at the Creation: My Years in the State Department* (New York: Norton, 1969).

Anderson, Benedict. *Imagined Communities: Reflections on the Origin and Spread of Nationalism* (New York: Verso, 1992).

Arendt, Hannah. *The Origins of Totalitarianism.* 2nd ed. (New York: Harcourt Brace, 1973).

Badian, Ernst. *From Plataea to Potidaea: Studies in the History and Historiography of the Pentecontaetia* (Baltimore: Johns Hopkins University Press, 1993).

_____. "Plataea Between Athens and Sparta." In Hartmut Beister and John Buckler, eds., *Boiotika* (Munich: Editio Maris, 1989), 95–111. Vol. 2 of *Münchener Arbeiten zur alten Geschichte,* ed. Hatto H. Schmitt.

Bourdieu, Pierre. *Outline of a Theory of Practice,* Jack Goody, ed., Richard Nice, trans. (Cambridge: Cambridge University Press, 1977), vol. 16.

Clark, Michael T. "Realism Ancient and Modern: Thucydides and International Relations." *PS: Political Science and Politics* 26, 3 (1993): 491–94.

Cooper, Jerrold. *Presargonic Inscriptions*, vol. 1. (New Haven, CT: The American Oriental Society, 1986).

Crane, Gregory. *Thucydides and the Ancient Simplicity* (Berkeley and Los Angeles: University of California Press, 1997).

_____. *The Blinded Eye* (Lanham, MD: Rowman and Littlefield, 1996a).

_____, ed., *Perseus 2.0: Interactive Sources and Studies on Ancient Greek Culture* (New Haven: Yale University Press, 1996b).

_____. "Fear and Pursuit of Risk: Corinth on Athens, Sparta and the Peloponnesians (Thucydides 1.68–71, 120–121)." *TAPA* 122 (1992a): 227–56.

_____. "Power, Prestige and the Corcyrean Affair in Thucydides 1." *Classical Antiquity* 11 (1992b): 1–27.

der Derian, James, ed. "A Reinterpretation of Realism." In *International Theory: Critical Investigations* (New York: Macmillan, 1995), 363–96.

Doyle, Michael. "Thucydides and Political Realism." *Review of International Studies* 16, 3 (1990): 223–37.

Forde, Steven. "Varieties of Realism: Thucydides and Machiavelli." *Journal of Politics* 54 (1992): 372–93.

Garst, Daniel. "Thucydides and Neorealism." *International Studies* Quarterly 33, 1 (1989): 3–28.

Georges, Pericles. *Barbarian Asia and the Greek Experience* (Baltimore: Johns Hopkins University Press, 1994).

Gilpin, Robert. *War and Change in World Politics* (Cambridge: Cambridge University Press, 1981).

Hall, Edith. *Inventing the Barbarian: Greek Self-Definition Through Tragedy* (Oxford: Oxford University Press, 1989).

Hammond, N.G.L. "Plataea's Relations to Thebes, Sparta and Athens." *JHS* 112 (1992): 143–50.

Hanson, Victor D. *Hoplites: The Classical Greek Battle Experience* (London: Routledge, 1991).

Hornblower, Simon. "The Religious Dimension of the Peloponnesian War," *HSCP* 94 (1992): 169–97.

Johnson, Laurie M. *Thucydides, Hobbes, and the Interpretation of Realism* (Dekalb: Northern Illinois Press, 1993).

Johnson-Bagby, Laurie. "The Use and Abuse of Thucydides." *International Organization* 48 (1994): 131–53.

Kent, Roland G. *Old Persian: Grammar, Texts, Lexicon.* 2nd ed. (New Haven, CT: American Oriental Society, 1953).

Kurke, Leslie. *The Traffic in Praise: Pindar and the Poetics of Social Economy*, Gregory Nagy, ed. (Ithaca, NY: Cornell University Press, 1991).

McNamara, Robert S., and Brian VanDeMark. *In Retrospect: The Tragedy and Lessons of Vietnam* (New York: Random House, 1995).

Meiggs, Russell, and David Lewis. *A Selection of Greek Historical Inscriptions* (Oxford: Oxford University Press, 1988).

Morgan, Catherine. *Athletes and Oracles: The Transformation of Olympia and Delphi in the Eighth Century B.C.* (Cambridge: Cambridge University Press, 1990).

Morgenthau, Hans. *Politics Among Nations* (New York: Knopf, 1948).

Neils, Jenifer. "The Panathenaia: An Introduction." In Jenifer Neils, ed., *Goddess and Polis: The Panathenaic Festival in Ancient Athens* (Princeton: Princeton University Press, 1992), 13–27.

Patterson, Orlando. *Freedom in the Making of Western Culture* (New York: Basic Books, 1991).

Raaflaub, Kurt. *Die Entdeckung der Freiheit: zur historischen Semantik und*

Gesellschaftsgeschichte eines politischen Grundbegriffes der Griechen. Vol. 37 (Munich: Beck, 1985).

Root, Margaret Cool. "The Parthenon Frieze and the Apadana Reliefs at Persepolis: Reassessing a Programmatic Relationship." *AJA* 89 (1985): 103–20.

Tigay, Jeffrey H. *The Evolution of the Gilgamesh Epic* (Pennsylvania: The University of Pennsylvania Press, 1982).

Vasquez, John A. *Classics of International Relations.* 2nd ed. (Englewood Cliffs, NJ: Prentice Hall, 1990).

Votti, Paul R., and Mark V. Kauppi. *International Relations Theory: Realism, Pluralism, Globalism* (New York: Macmillan, 1987).

Wayman, Frank W., and Paul F. Diehl, eds. *Reconstructing Realpolitik* (Ann Arbor: University of Michigan Press, 1994).

Wills, Garry. *Lincoln at Gettysburg: The Words That Remade America* (New York: Simon and Schuster, 1992).

8

The Korean War and North Korean Politics

Dae-Sook Suh

Introduction

Compared with the Peloponnesian War of the fifth century B.C. (431–404), which lasted for twenty-seven years, the Korean War seems to be a short war, ending in three years, 1950–1953. The Korean War, however, ended with only an armistice agreement, not with a peace treaty. The truce on the fighting front ended the killing but without providing a solution to the cause of the war—that is, it did not bring an end to the division of the Korean peninsula. Unlike the Peloponnesian War, there was neither victor nor vanquished in the Korean War. It ended without resolving the most important issue of the Korean people on both sides, reunification of their country. It restored the status quo antebellum, but it solidified the division. There is no passionate description of the Korean War from either side comparable to the account by Thucydides on defeated Athens in the Peloponnesian War. North Korean and South Korean accounts of the war still differ vastly, and they remain at the level of justifying or explaining their own positions. The Korean War will be a subject of controversy as long as Korea remains divided.

Contrary to the study of the Peloponnesian War, many issues of the Korean War are still unresolved, and they are political and legal issues among participants to the war. Nearly half a century after the conclusion of the truce on the fighting front, North Korea is seeking a peace treaty with the United States to legally terminate the Korean War. Scholars of the Peloponnesian War can afford to study the eloquent orations of great leaders, such as the funeral oration of Pericles, and the other splendors and accomplishments of Athens and Sparta. Contemporary issues of the Korean War, however, are far more political, and although there is no fighting at the front, the state of belligerency still exists between North Korea and South Korea. In its effort to reunify the country, the South wanted to continue the war and so did not sign the armistice agreement.

There are important political questions that need to be addressed before a peace treaty can be negotiated. Was the Korean War a civil war between North and South Korea, or was it an international war involving the armed forces of some twenty countries? Was the United Nations a party to the conflict as a belligerent? Is the People's Republic of China, which dispatched Chinese volunteers to the war, responsible for the termination of the war? The Korean War is not legally terminated until a peace treaty is concluded. At present, it is not clear even how the termination of the Korean War should be negotiated and who should be a party to the peace Tteaty.

The classic approach of customary international law to the termination of a war was first to conclude an armistice by military commanders followed by a peace treaty that terminated the war. The Napoleonic Wars were good examples, and similar armistices and peace treaties were signed in this century to terminate the two world wars. In this kind of legal interpretation, the Korean War has yet to be terminated, and in this sense, the duration of the Korean War has already surpassed that of the Peloponnesian War. Legally, the Korean War lingers on, and although everyone recognizes the need to conclude a peace treaty to terminate the war, no one except North Korea seems willing to conclude one. Even North Korea seems to prefer a peace treaty with the United States only, alleging that the Korean War was fought between North Korea and the United States. Such are the issues that scholars of the Peloponnesian War do not have to deal with.

Compared with the Peloponnesian War, the Korean War not only witnessed the improvement in the art of killing, in weapons, and in the techniques of fighting, but also it changed the concept of war by introducing such ideas as "limited war" and "international police action" that mobilized the armed forces of many countries under the flag of an international organization to stave off a blatant act of aggression. Unlike the mourning of the defeat of Athens and chants of victory by Sparta, none of the belligerents in the Korean War admit defeat in the War. At times, North Korea claims victory in the war. Even the question of the responsibility for starting the war is not settled.

In its official account of the war, North Korea accused the United States of launching an armed invasion on North Korea on June 25, 1950.[1] There are equally important official accounts of the war from South Korea and the United States that attest to North Korean aggression.[2] Scholars continue to study the Korean War as more source materials have become declassified and available to the public. Many ambiguous facts of the war have been clarified during the past twenty years. When the captured North Korean documents were declassified from the United States National Record Center, some impressive studies were published using the new archival materials.[3] These studies significantly improved our conventional understanding of the Korean War. North Korean archival materials revealed many facts that would question conventional theories. When the Soviet Union collapsed in 1991, its archival materials became available, and it is anticipated that they will further clarify our understanding of the war. The Soviet Union had played an important role, and there are Russian and American scholars who are beginning to study seriously the new Russian archives.[4]

After the normalization of diplomatic relations between Russia and South Korea, some archival materials relating to the Korean War were given to South Korea by the Russian government in 1994.[5] Not all Russian archival materials are made public and those declassified materials are not yet thoroughly studied, but gradual unveiling of the Russian archival materials on the Korean War has helped clarify many ambiguous facts and improve our understanding of the war.[6]

There are many Chinese source materials that have become public during the past decade, including selected military writings of former Commander Peng Dehuai of the Chinese Volunteer Army and the diary of Wang Dongxing. But the Chinese accounts of the Korean War, while revealing in details of operation, still maintain the cold-war rhetoric and support the North Korean position. The latest two-volume study of the "true record" of the Korean War is a good example.[7] Indeed, it is difficult to assess the true record of the Korean War, because relevant source materials of key participants, such as China and North Korea, have not been made public.

Because so many countries participated in the conflict, the Korean War is often referred to as an international war. It is generally understood as a war that the communist camp, if not Stalin himself, started in order to expand the sphere of communism in Asia, and that the United States responded to in order to contain the expansion. It has now become clear that Stalin was reluctant to give the North Koreans permission to start the war. It is also apparent that the Chinese were extremely cautious about the North Koreans starting another front in the peninsula when the Chinese themselves had not concluded their own civil war. There seem to have been close consultations among the North Koreans, Chinese, and Russians, but it was the North Koreans who initiated the idea and started the war for the purpose of reuniting their divided fatherland.[8] Studies emphasizing the international aspects of the war often neglect the domestic Korean causes of the war. It was fought in the Korean peninsula by the Korean people, who suffered the most casualties, and the peoples of North and South Korea are still suffering from the consequences of the war. While it is important to analyze the international involvement, it is more important to understand the war in terms of Korean domestic politics.

Proper comparison of the Peloponnesian War and the Korean War should examine the impact of these wars on Greece and Korea, but that is beyond the scope of this chapter. This is a modest effort to assess the impact of the Korean War on North Korean domestic politics.

North Korean Domestic Politics

At the close of World War II, Korean revolutionaries who were fighting the Japanese returned to Korea from all over the world.[9] The majority, including the

Korean Communist Party, established their headquarters in Seoul, the capital of Korea located in the South. When the Soviet Union and the United States divided the country along the 38th parallel and occupied the northern and southern parts of Korea, respectively, all Koreans expected the division to be a temporary arrangement until the Japanese were expelled from the peninsula. However, the United States and the Soviet Union were not able to settle their differences on how the two parts of Korea should be reunited, and after three years of unsuccessful negotiation, both sides ended their military occupation by establishing separate governments, each styled after the image of the occupier, in each occupation zone and withdrew their military forces. It was left to the Koreans to reunite their country, and it was the North Koreans, with trained armed forces twice the size of their South Korean counterparts, who launched an attack with this end in mind.

By the time the government of the Democratic People's Republic of Korea (DPRK) was established in Pyongyang on September 9, 1948, four distinct political groups had emerged in North Korea. The first was the partisan guerrilla group, who fought with the Chinese against the Japanese in northeast China (Manchuria) under the leadership of Kim Il Sung. The second was the so-called Yanan group, who fought in mainland China along with the Chinese communists near Yanan. This group was led by Kim Tu-bong, and several officers served in the Chinese army during the Chinese Long March. The third group consisted of ethnic Koreans who lived in the Soviet Union and returned to Korea after Japan was defeated. Some returned to Korea as officers of the Soviet occupation forces, but many returned on their own and helped the Russians to set up a new government. The fourth group consisted of Korean communists who resisted the Japanese within Korea. They are often referred to as the domestic group, which was organized under the leadership of Pak Hŏn-yŏng, chairman of the Korean Communist Party. The domestic group was the largest in number, with a headquarters in the South and branch bureaus in the North. Long before two separate governments with disparate ideologies were established, the Korean Communist Party was outlawed in the South and Pak and his followers fled to the North.[10]

From the time of the liberation of Korea in 1945 to the conclusion of the Korean War in 1953, North Korean domestic politics was dominated by cooperation, competition, factional struggles, and other workings of these four groups. Their struggles lingered on after the Korean War, and it was not until after the Chinese Volunteer Army had left Korea and the Fourth Party Congress was convened in 1961 that the traces of any factional groupings disappeared in North Korea.

Unlike the domestic group, the revolutionary groups that returned to Korea after long years of fighting abroad had little or no roots within Korea. Each had problems unique to its own group. The Soviet-Korean group, for example, consisted of second- or third-generation Koreans who were Soviet citizens, and even their names did not sound Korean. Ivan Pak was changed to Pak Ui-wan and Khegai was changed to Hŏ Ka-i, but the group as a whole exercised significant influence over political and social events because they could speak Russian and North Korea was occupied by the Soviet Union.

The leaders of the Yanan group had returned to Korea after World War II, but their Chinese comrades-in-arms were still fighting the Chinese civil war, and many Korean soldiers who fought alongside the Chinese communists during World War II remained in China and helped the Communist army fight the Nationalists in northeast China. This may have weakened the position of the leaders of the Yanan group in their struggle against other groups, but after the Chinese victory on the mainland, many Koreans were trained in northeast China and returned to Korea to prepare for the Korean War.

At the time of liberation in 1945, the domestic group could have claimed to be the legitimate leader of all leftist groups in Korea, but it was located in the South under the American military occupation. Eventually the leaders of the group had to abandon their supporters in the South and flee to the North and, once there, they had to adjust their positions vis-à-vis other groups. They eventually merged with comrades in the North Korean Branch Bureau of the Korean Communist Party, but in the process, they lost the leadership of the leftist groups in North Korea.[11]

Kim Il Sung's partisan group was small in number, consisting of less than 250 highly trained partisan guerrillas. They were defeated by the Japanese in Manchuria by the beginning of World War II, and spent their war years in the Russian Maritime Province. The Korean and Chinese guerrillas who fled to the Russian Maritime Province were trained at the 88th Rifle Brigade in the village of Vyatskoe, near the Khabarovsk region. The training of the partisans by the Russians was most fortunate, because the Soviet Union later chose their leader, Kim Il Sung, to lead North Korea. Kim Il Sung immediately took over practically all North Korean military and security units. The first factory he built in the North was a munitions factory, and he succeeded in establishing the Korean People's Army even before the North Korean government was officially proclaimed.[12]

Domestic Settings for the War

In the establishment and operation of the party and the government from 1945 to 1950, leaders of the four groups cooperated under the supervision of an able Russian advisor, Colonel Alexandre Mateevich Ignatiev.[13] Kim Il Sung took over the North Korean Branch Bureau of the Korean Communist Party, merged it with the Yanan group, and founded the Workers' Party of North Korea in August 1946. The chairman of the party was Kim Tu-bong of the Yanan group, and Kim Il Sung of the partisan group and Chu Yŏng-ha of the domestic group served as vice chairmen. Hŏ Ka-i of the Soviet-Korean group and Ch'oe Ch'ang-ik of the Yanan group were added to make up the Political Committee. Some changes were made in the leadership two years later at the Second Congress of the party in March 1948, but the core leadership remained the same.[14] After the formation of the party, North Korea called for the founding of the Workers' Party of South Korea, and it was duly organized in November 1946 in Seoul. The two workers' parties of North and South Korea were merged into the Workers' Party of Korea, with Kim Il Sung as its chairman, one year before the Korean War on June 24, 1949, when Kim Il Sung vowed to reunite the country within a year.

Similarly, when the government of the DPRK was launched, leaders of all four groups were represented. There were three partisans, including Kim Il Sung as premier, four members of the Yanan group, and thirteen members, by far the largest number, from the domestic group. It was alleged that Soviet-Koreans, who enjoyed easy access to the Soviet occupation authority, were appointed to the vice ministerial positions to keep the Russians informed as well as to convey instructions from the Soviet Union to North Korea. The Soviet Union seems to have had close contact with the North Korean government by appointing one of the generals of the occupation forces, T.F. Shtykov, as ambassador. In addition, Colonel A.M. Ignatiev, who shaped North Korean politics from liberation to the founding of the republic, remained in North Korea as an advisor to the Soviet embassy in Pyongyang.

After establishing the party and the government, the Soviet Union withdrew its occupation forces by December 1948, and both the Soviet Union and the DPRK repeatedly asked the United States to withdraw its forces from South Korea. More importantly, when they realized that consultation between the United States and the Soviet Union over all kinds of proposals, such as the trusteeship plan and the United Nations plan, would not bring reunification of the Korean people, leaders of North Korea began to prepare to unify their country by themselves and by force, if necessary. While the South Koreans were equally vocal about unification, they were led by a seventy-five-year-old revolutionary who knew nothing about military operations, while North Korea was led by a thirty-eight-year-old leader of partisan guerrillas.

The international political environment surrounding the Korean peninsula after World War II was complex, and many studies have attributed the origins of the Korean War to the intensifying East–West confrontation, while Korean domestic causes for the war are seldom considered.[15] International political and military developments surrounding the Korean peninsula, such as the United States troop withdrawal in 1949, the success of the Chinese communists on the mainland, and the Acheson declaration regarding the U.S. defense perimeter in Asia, all contributed to the decision of the North Korean leaders to launch an attack, but it was neither the Soviet Union, nor the Chinese who conspired with the North Koreans, who started the war. Eventually the Russians supplied munitions and the Chinese supplied the volunteers, but they were extremely cautious in advising the North Koreans to refrain from resorting to military means to reunite the country. Stalin may have acquiesced and Mao may have given permission, but it was North Koreans who launched the attack. They had problems at home. Their efforts to develop the economy in the northern half alone ended in failure, and many groups contended for power after the Russians left, but most importantly, they needed to reunify their country.

Domestic Causes of the War

There are three domestic causes of the war in the context of North Korean poli-
tics.[16] The first cause was the desire of the Korean people to reunify their country.
This feeling was common to political leaders of both North and South Korea.
Syngman Rhee was as vocal as Kim Il Sung in trying to reunify the fatherland.
South Korean slogans were *myŏlkong t'ongil* (Destroy Communism and Unify)
and *pukchin t'ongil* (March North and Unify). The difference was that old Syngman
Rhee was not able to implement his policy and did nothing about it, while young
Kim Il Sung built an army twice the size of that of South Korea's to reunify the
country. Considering his guerrilla fighting in Manchuria, Kim's resolve to bring a
military solution to the Korean division is not surprising. It is clear now that Kim
had appealed repeatedly to Stalin for military action, but that Stalin had restrained
him from attacking the South for fear that the United States might intervene. He is
said to have convinced Kim Il Sung that the Soviet Union would not deploy its
military forces in Korea. Instead, Stalin told the North Korean leaders to make a
proposal to the South for peaceful unification. Indeed, North Koreans formed a
social organization, the Democratic Front for the Fatherland Unification, and proposed a peaceful unification in June 1949, one year before the commencement of
the war. Kim may have agreed to make such a proposal, but he knew it would not
work and continued preparing for war.

Kim Il Sung himself had never advocated peaceful unification prior to the Korean War. On the contrary, Kim advocated military solutions on a number of occasions. He emphasized the strengthening of the Korean People's Army in his New
Year's address of 1950, and delivered a militant speech on January 19, 1950, at the
Third Party Congress of the Ch'ŏndogyo Young Friends Party. He said that the
great task of national unification must be accomplished by "ourselves," and added
that he was not afraid of a war. If there should be a war, he said, North Korea
would be victorious, but he continued that "victory does not come on its own;
victory should be won."[17] When the war was going in his favor in July 1950, Kim
said in an interview with a reporter for the French newspaper *L'Humanité* that he
envisioned a short war, and had it not been for the American intervention the war
would have been over. The difference between the South Korean policy and the
North Korean policy for unification was not in the policy itself, but the method of
implementation or, more accurately, the ability to carry out the policy. The desire
to solve the problems of the Korean division even at the cost of military conflict
was common to all Koreans.

Another cause of the war stemmed from the economic conditions of North
Korea. After launching two one-year economic plans for 1947 and 1948 to improve its economy, it reported that these plans were successful. However, after the
government was established, it launched a modest two-year economic plan for
1949–1950, but this plan ran into considerable difficulty. By February 1950, the
government admitted the failure of the two-year economic plan, pointing to a

certain unforeseen expansion of basic industries.[18] The North Korean leaders may have felt the futility of trying to develop an independent North Korean economy with only half the labor force of the South.

Kim traveled to the Soviet Union in March–April 1949 to ask for economic assistance, and he was able to conclude a treaty for economic and cultural cooperation for ten years, but his economic situation had not improved. Even with the Russian aid, it was difficult to expect any kind of real gains in conventional economic development. Geographically, North Korea was mountainous, with less arable land than the South, and was not self-sufficient in grain production. The Japanese left behind a number of industrial plants when they left Korea, but they were carted away by the Russians during their military occupation. In their place, the Soviet Union did not build new industries for North Korea. What Kim envisioned was the economic development of a unified Korea that included the southern half of the republic.

The immediate domestic cause of the war was the relationship between Kim Il Sung and the domestic group headed by Pak Hŏn-yŏng. For different reasons, the two men seem to have advocated unification of Korea by military means. For Kim, the military conquest of the South at the time was feasible, considering the debilitated security conditions of the South in 1949–1950. The United States had left Korea in 1949, while Kim commanded a battle-trained Korean army twice the size of the South Korean army. There was no greater political ambition for Kim than to reunify the peninsula under his leadership, enabling him to become the supreme leader of a unified Korea, the father of the first Korean republic.

Pak Hŏn-yŏng, on the other hand, was from South Korea, and he thought he and the domestic group would improve their opportunities once the country was united and the capital of unified Korea was moved from Pyongyang to Seoul.[19] Perhaps Pak Hŏn-yŏng would have preferred a popular uprising in the South to a military campaign, but all such disturbances ended in failure. For example, more than 3,000 guerrillas were sent from North to the South during the seven-month period from September 1949 to March 1950, but most of them, including two of the top leaders, Kim San-yong and Yi Chu-ha, were arrested in the South.

There were many reasons for the domestic group to advocate military action. Most of its organizational units and operational bases were in the South, and once a unified government was established there, the domestic group would improve its position over the rival groups. Pak Hŏn-yŏng is alleged to have said to Kim Il Sung that once the Korean People's Army began a military action to liberate the South, more than 200,000 followers of his underground organizations would rise up and overthrow the South Korean regime, and therefore the military campaign in the South would not be a difficult one. Kim Il Sung lamented Pak's remarks long after the conclusion of the war. Reminiscing about the war on the fifteenth anniversary of the North Korean army in February 1963, Kim said that Pak Hŏn-yŏng was a liar, because there had not been 1,000 followers, let alone 200,000, who came to support the North Korean army. Kim said that his army had marched down to the

South as far as the Naktong River, a stone's throw from the city of Taegu, but no one rose up to overthrow the South Korean government.[20]

The Korean War was not simply an international conspiracy to expand the sphere of communism; more importantly, it was an effort on the part of the Korean communists to reunify their country by force. As early as September 1950, when the war was still going in his favor, Kim spoke about the problem of aggression and starting the war. He asked, "How could anyone call the people who rose up to fight for the freedom and independence of their own fatherland aggressors?"[21]

Disputes Over the Party During the War

When the United Nations forces crossed the 38th parallel and occupied Pyongyang, Kim Il Sung panicked. He asked Stalin for Russian military assistance, but Stalin reminded him of what he had said before the war. Kim admitted that American intervention was an unexpected turn of events and that his forces had suffered greatly from sudden retreat. When the Chinese volunteers intervened, they pushed Kim aside and took over management of the war.[22] Therefore, for Kim Il Sung the war had ended as early as December 1950 when the Chinese volunteer army recovered most of the North Korean territory back from the United Nations forces. The two sides began ceasefire negotiations in early 1951, when the military front was stabilized near the 38th parallel. Kim Il Sung convened the Third Joint Plenum of the Central Committee in December 1951 in Kanggye, a town near the border of North Korea and China, and attacked almost every group, including men from his own partisan group. These assaults were directed at North Korean generals and military personnel for their conduct of the war, and those reprimanded were later reinstated, but more a serious confrontation came between Hŏ Ka-i, the leader of the Soviet-Korean group and Kim Il Sung.

Hŏ Ka-i was the undisputed leader of the Soviet-Koreans who moved up in party ranks rapidly by taking over the Organization Committee of the party. By the time of the Korean War, Hŏ ranked third in the party behind Chairman Kim Tu-bong and Vice Chairman Kim Il Sung, and he was also the head of the Inspection Committee. The dispute centered on how to handle the disloyal members during the war and to reorganize the party. Kim said that the war had distinguished the loyal from the disloyal members of the party, and that appropriate measures should be taken to punish the disloyal ones, irrespective of their positions within the party. However, Kim cautioned that each member should be examined individually and that the party should be magnanimous in dealing with its members and should avoid indiscriminate purges. This task was to be carried out by Hŏ Ka-i, the chairman of the Inspection Committee.

Hŏ Ka-i did exactly the opposite of what Kim had instructed. During the one-year period from December 1950, when Kim ordered the investigation, until November 1951, when the Fourth Joint Plenum was convened, Hŏ expelled and punished 450,000 of the party's 600,000 members.[23] The central issue in the confrontation between Hŏ and Kim was whether to build an elite Communist Party in the style of the Soviet Union or a mass party of the kind that Kim wanted. Hŏ argued for an elite Communist Party of fewer than 60,000 members, consisting of industrial workers, while Kim advocated building a mass party in accordance with the conditions peculiar to Korea. Kim emphasized that Hŏ should not be forcing on Korea unsuitable foreign organizational styles. Hŏ often bragged about his knowledge of how the foreign parties were organized and operated, and he was called a "party doctor." Hŏ's challenge was aiming at gaining control of the party in post-war North Korea.

Kim Il Sung ordered most of those expelled to be reinstated, and by the Fifth Plenum in December 1952, 69.2 percent of those expelled were reinstated. Many new members were subsequently recruited, and the members of the party had grown to one million by the Third Party Congress in 1956, approximately 12 percent of the population.[24] Hŏ was purged, and the group most directly affected was the Soviet-Korean group. It was the Chinese and not the Soviet Union that had saved North Korea from certain defeat. Marshal Stalin was dead, Ambassador Shtykov had been replaced, and Colonel Ignatiev had died early in the war in an American air raid. In general, the Soviet-Koreans were not looked upon as favorably as during the Soviet occupation, and most of the Soviet-Koreans started to return to the Soviet Union. Hŏ committed suicide, and Kim condemned his cowardly act.

Challenge of the Domestic Group

Much discussion on the outcome of the war as well as the responsibility for the devastation of the North Korean countryside by the American air force ensued in North Korea. Unlike the Soviet-Korean group, the domestic group directly challenged the leadership of Kim Il Sung by asking who was responsible for the defeat. The challenge was an organized effort to overthrow Kim Il Sung and replace him with their own leader, Pak Hŏn-yŏng, in 1952, while the war was still going on. It was led by Yi Sŭng-yŏp, the Minister of Justice, and twelve loyal supporters of Pak.

When the seesaw battle along the 38th parallel had stabilized, North Koreans once again tried to start underground guerrilla operations in the South, establishing a military training school in Sohŭng County, Hwanghaedo, called Kŭmgang Political Institute, consisting primarily of the members of the domestic group. There were approximately 1,500 cadets in the Institute, and it was headed by Kim Ŭng-bin, a South Korean. The South Korean Liaison Bureau, which coordinates all operations of the party in South Korea, was also headed by South Koreans such as Cho Il-myŏng, and they were all close comrades of Yi Sŭng-yŏp.

According to the trial records, Yi Sŭng-yŏp had prepared for more than a year, from September 1951, for a military coup to unseat Kim Il Sung, and a new cabinet was formed with Pak Hŏn-yŏng as premier. When Kim Il Sung was ousted, North Korea was to institute a collective leadership, and Yi Sŭng-yŏp was to become chairman of the party. All cabinet members were from the domestic group, and it was clear from their preparation that the members of the old Korean Communist Party planned to take over North Korea. The coup was attempted in early 1953, with a rebel force of approximately 4,000 men, but it failed. Details of the actual military confrontation are not known, but all the conspirators were arrested within a short period of time.[25]

Three days after the conclusion of the Korean War, twelve conspirators were indicted in accordance with Article 25 of the criminal law for high treason. The prosecutor, Yi Song-un, who was a partisan guerrilla, charged them with three counts: espionage activities for the United States, indiscriminate destruction of democratic forces in the South during the war, and the attempted overthrow of the government of the republic by military force. None of the defendants contested the charges leveled against them, admitting everything. Ten out of twelve were sentenced to death and one (Yun Sun-dal) was sentenced to fifteen years' and another (Yi Wŏn-jo) was sentenced twelve years' imprisonment. Both died during their imprisonment. Pak Hŏn-yŏng was tried separately two years later, and was sentenced to death.

In retrospect, the chance of the domestic group to unseat Kim Il Sung was slim, because most of the Korean People's Army units were led by Kim's partisans, and the partisans controlled the military and security forces. All the charges leveled against the domestic group may not be substantiated, particularly the charges that the South Korean communist leaders who fled to the North were American spies and that they indiscriminately destroyed progressive forces who supported the North during the brief period when the North Korean forces occupied the South. However, the military units of the domestic group were no match for Kim Il Sung's partisan guerrillas. While there is no evidence of Chinese volunteers interfering in North Korean domestic politics, Kim Il Sung and his partisans had fought with the Chinese for a long time in Manchuria during World War II, and as long as the Chinese forces were stationing in North Korea, a challenge to Kim would have had a difficult time succeeding.

Aftermath of the War

During his guerrilla days in Manchuria, Kim Il Sung often suffered defeat and setbacks from his fighting with the Japanese, but survival was his trade. He was one of the very few leaders of the guerrilla force who was never captured by or submitted to the Japanese. His fight against the Japanese continued in Russia even after he fled Manchuria in defeat, and it lasted to the end of World War II. He survived the Korean War also, and when the ceasefire agreement was signed, he launched serious campaigns of rehabilitation and reconstruction. Kim had successfully destroyed the challenges of both the Soviet-Korean and the domestic groups, but challenges to his leadership did not end there. Sometime after the Korean War, in 1956, the Yanan group challenged the partisans, but it too was crushed by Kim Il Sung and his partisans. The Korean War did bring factional struggles among various groups into open competition in North Korea, but it did not change the political leadership. Kim Il Sung may not have won the war, but he consolidated his position to meet all challenges successfully. By the time the Fourth Party Congress was held in September 1961, there was no group or individual who could challenge Kim and his partisans. One of the consequences of the war is the disappearance in North Korean politics of any factional groupings. There was only one dominant group with one supreme leader, Kim Il Sung.

North Korea was devastated by continuous American bombings during the ceasefire negotiations that lasted more than two years. Chinese volunteers stayed in North Korea five more years after the conclusion of the ceasefire agreement in July 1953, helping to rebuild the country. The idea of devising a plan for economic development throughout North and South Korea had vanished, and the North Koreans had to reconstruct their country from the ashes. Kim received much help from his fraternal socialist countries during and after the war, but none greater than the help received from China. The Chinese, not the Russians, rescued North Korea with arms and soldiers when the United Nations forces reached the Yalu River.

Recently released records reveal that the Russians promised and delivered air cover, but it is widely known that the United States encountered no opposition in their control of the North Korean air space during the War. Even in the post-war reconstruction period, Russians gave loans while the Chinese gave outright grants and forgave any loans they had earlier given to North Korea. One important consequence of the war was the change in Kim Il Sung's attitude toward the Soviet Union. When the Sino-Soviet disputes forced him to take sides, he had no difficulty in choosing China over the Soviet Union. It took the Korean War to liberate Kim and the North Koreans from the Soviet Union. Later, North Korea would become completely independent when Kim Il Sung developed his own idea of self-reliance, *chuch'e*, but this idea had its origin in the Korean War when he first developed an anti-Soviet and pro-Chinese attitude.[26]

Today, North Korea claims victory in the Korean War, but Kim Il Sung had already conceded as early as August 1952 that the truce meant the cessation of military operations by the belligerents on an equal footing, with no victor or vanquished.[27] Regardless of the rival claims, the Korean War has not solved the most important issue for the Korean people—reunion of the Korean people and reunification of the country. This was the cause of the war. In this sense, all Koreans on both sides of the 38th parallel suffered defeat in the Korean War. The fratricidal war solidified the lasting division and reinforced mutual enmity between Koreans. Perhaps the most tragic consequence of the Korean War is not those who died fighting, but rather nearly 70 million Korean people who have suffered the division of their country for the past half century.

Notes

1. The Research Institute of History, Academy of Sciences of the Democratic People's Republic of Korea, *History of the Just Fatherland Liberation War of the Korean People* (Pyongyang: Foreign Languages Publishing House, 1961). For the complete North Korean

history of the Korean War, see *Hyŏngmyong ŭi widaehan suryŏng Kim Il Sung tongji kkesŏ yŏngdo hasin chosŏn inmin ŭi chŏngŭi ŭi haebang chŏnjaengsa* [History of the Just Fatherland Liberation War of the Korean People Led by the Great Supreme Leader of Revolution, Comrade Kim Il Sung], in three volumes (Pyongyang: Sahoe kwahak ch'ulp'ansa, 1973).

2. For the South Korean account, see the Ministry of National Defense, Republic of Korea, *The History of the United Nations Forces in the Korean War*, in seven volumes (Seoul: The Ministry of National Defense, 1972–1979), and *Han'guk chŏllan illyŏnji* [Records of the Korean War, The First Year]; this series has five volumes from the first to the fifth year, 1950–1955, (Seoul: Kukbangbu, 1951–1956). There are many American accounts of the war. All branches of U.S. armed services that participated in the war produced their own accounts, such as Roy E. Appleman, *South to the Naktong, North to the Yalu* (Washington, DC: Office of the Chief of the Military History, Department of the Army, 1961), and James A. Field, Jr., *History of U.S. Naval Operations* (Washington, DC: U.S. Government Printing Office, 1962). From the State Department, *Foreign Relations of the United States, 1950, 1951, 1952–1954*, deals with the Korean War.

3. U.S. National Record Center, United States, Far East Command, Record Group 242, "Captured Enemy Documents," and Record Group 331. General Hq./Supreme Command of Allied Forces. See also United States Military Government in Korea, Record Group 319, G-2, Intelligence Summaries—North Korea. There are many studies using these declassified sources, such as B. Cumings, *The Origins of the Korean War*, vol. I (1945–1947) and vol. II (1947–1950) (Princeton: Princeton University Press, 1981 and 1990).

4. See, for example, an interview with Russian historian Grigoriy Kuzmin reported in *Komsomolskaia pravda*, June 25, 1991. Kathryn Weathersby, "Soviet Aims in Korea and the Origins of the Korean War, 1945–1950: New Evidence from Russian Archives," *Cold War International History Project Working Paper*, No. 8, November 1993.

5. Two groups of documents were given to South Korean president Kim Young Sam by the Russian president Boris Yeltsin. The first group of documents consisted of materials from January 1949 to July 1953, approximately 100 items (279 pages), and the second group of documents consisted of supplementary materials from January 1949 to September 1953, approximately 116 items (269 pages).

6. See the latest study by Haruki Wada, *Chōsen sensō* [The Korean War] (Tokyo: Iwanami shoten, 1995).

7. Jie Lifu, *Chaoxian zhanzheng shilu* [True Records of the Korean War] (Beijing: Shijie zhishi, 1993), in two volumes. For Peng's role, see Peng Dehuai, *Peng Dehuai junshi wenxuan* [Selected Military Documents of Peng Dehuai] (Beijing: Zhongyang wenxian chubanshe, 1988).

8. Controversies are still entertained as to who started the Korean War, but recent studies have uncovered more than ample evidence to support North Korean aggression. Sources are available for military deployments by mid-June and orders to attack on June 25, 1950. See the recent study by a Japanese scholar, Ryo Hagiwara, *Chōsen sensō: Kin Nichisei to Makkasa no imbo* [The Korean War: Conspiracy of Kim Il Sung and MacArthur] (Tokyo: Bungei shunju, 1993).

9. There are many studies on the Korean independence movement. For the nationalist movement, see Chong-Sik Lee, *Politics of Korean Nationalism* (Berkeley: University of California Press, 1965). For the Communist Movement, see Dae-Sook Suh, *The Korean Communist Movement, 1918–1948* (Princeton: Princeton University Press, 1967).

10. It is not the purpose of this paper to recount the details of the Korean division and North and South Korean political systems. For an extensive comparative analysis of the two systems, see Sung Chul Yang, *The North and South Korean Political Systems: A Comparative Analysis* (Boulder: Westview Press, 1994).

11. For a good study of the Workers' Party of South Korea and its fate in North Korea,

see Kim Nam-sik, *Sillok, Namnodang* [True Records of the Workers' Party of South Korea] (Seoul: Sin hyŏnsilsa, 1975).

12. The Korean People's Army was founded on February 8, 1948, and the North Korean government was not organized until September 9, 1948. North Koreans have backdated the founding of the Korean People's Army to April 25, 1932, when it was alleged that Kim Il Sung first organized a band of guerrillas to fight the Japanese.

13. For the liberation of Korea by the Russians, see *Osvobozhdenie Korei* [Korean Liberation] (Moscow: Akademii nauk, 1976). For the Sovietization of North Korea, see Dae-Sook Suh, "A Preconceived Formula for Sovietization: North Korea," in T.T. Hammond, ed., *The Anatomy of Communist Takeover* (New Haven: Yale University Press, 1975).

14. Hŏ Ka-i replaced Chu Yŏng-ha as vice chairman, and two men, one each from the partisan group (Kim Ch'aek) and the Yanan group (Pak Il-u) were added to the Political Committee. For the details of agenda and personnel actions of the party, see Dae-Sook Suh, *Korean Communism, 1945–1980: Reference Guide to the Political System* (Honolulu: University of Hawaii Press, 1980).

15. Many studies attribute the War to the Soviet Union. See David J. Dallin, *Soviet Foreign Policy After Stalin* (Philadelphia: J.B. Lippincott, 1961) and Thomas W. Wolfe, *Soviet Power and Europe, 1945–1970* (Baltimore: Johns Hopkins Press, 1970). For a good survey of the literature of the Korean War, see Hak-chun Kim, *Han'guk chŏnjaeng: Wŏnin, kwajŏng, hyujŏn, yŏnghyang* [The Korean War: Origins, Processes, Ceasefire, Impact] (Seoul: Pagyŏngsa, 1989).

16. I have explained my argument for the three causes in my earlier study. See Dae-Sook Suh, *Kim Il Sung: The North Korean Leader* (New York: Columbia University Press, 1988), pp. 107–36.

17. This speech appeared only in the first edition of Kim's works. See Kim Il Sung, *Kim Il Sung sŏnjip* [Selected Works of Kim Il Sung], First Edition, Vol. II (Pyongyang: Chosŏn nodongdang ch'ulp'ansa, 1953), pp. 485–96.

18. Kim Il Sung, *Kim Il Sung sŏnjip* [Selected Works of Kim Il Sung], Second Edition, Vol. II (Pyongyang: Chosŏn nodongdang ch'ulp'ansa, 1960), pp. 399–408.

19. The 1948 constitution of the DPRK claimed that the capital of the republic was Seoul. This provision was not changed until the new constitution was adopted in December 1972.

20. Kim Il Sung, *Kim Il Sung Selected Works*, Vol. III (Pyongyang: Foreign Languages Publishing House, 1971), pp. 316–25.

21. Kim Il Sung, *Kim Il Sung sŏnjip* [Selected Works of Kim Il Sung], First Edition, Vol. III (Pyongyang: Chosŏn nodongdang ch'ulp'ansa, 1953), pp. 104–15.

22. There are many accounts newly available from China about the conduct of the Korean War. For the joint military operation from the North Korean perspective, see Pak Il-u, *Chosŏn inmingun kwa chungguk inmin chiwŏn'gun kwaŭi kongdong chakchŏn* [Joint Operation of the Korean People's Army and the Chinese Volunteers Army] (Pyongyang: Chosŏn nodongdang ch'ulp'ansa, 1951). For recent Chinese accounts that are more revealing, see among others *Zhongguo renmin zhiyuanjun kangmei yuanchao zhanshi* [History of Resist America and Aid Korea Campaigns of the Chinese Volunteer Army] (Beijing: Junshi kexue chubanshe, 1990).

23. Kim Il Sung, *P'yŏngan-pukto tang tanch'edŭl ŭi kwaŏp* [Tasks of the P'yongan-pukto party organizations] (Pyongyang: Chosŏn nodongdang ch'ulp'ansa, 1956), pp. 2–66.

24. Kim Il Sung, *Kim Il Sung Selected Works*, Second Edition, Volume IV (Pyongyang: Foreign Languages Publishing House, 1971), pp. 433–517.

25. Information is based on the trial records. See *Mijeguk chuŭi koyong kanch'ŏp Pak Hŏn-yŏng, Yi Sŭng-yŏp todang ŏi chosŏn minju chuŭi inmin konghwaguk chŏnggwŏn chŏnbok ŭmmo wa kanch'ŏp sagŏn kongp'an munhŏn* [Trial Records of Pak Hŏn-yŏng, Yi

Sŭng-yŏp and Conspirators, Spies of American Imperialism, Who Attempted to Overthrow the Government of the Democratic People's Republic of Korea] (Pyongyang: Chosŏn nodongdang ch'ulp'ansa, 1952), pp. 153–60. The trial records are also available in *Nodong sinmun*, August 5–8, 1952.

26. The idea of *chuch'e* is the prevailing ideology of North Korea today. It replaced Marxism and Leninism as the official ideology of North Korea in the constitution. It was first proclaimed by Kim Il Sung on December 28, 1955, to curb Soviet influence in North Korea. See Kim Il Sung, *On Eliminating Dogmatism and Formalism and Establishing Jooche in Ideological Work* (Pyongyang: Foreign Languages Publishing House, 1964).

27. Kim Il Sung, *Kim Il Sung sŏnjip* [Selected Works of Kim Il Sung], First Edition, Vol. IV (Pyongyang: Chosŏn nodongdang ch'ulp'ansa, 1954), pp. 230–31.

9

The Korean War and South Korean Politics

Kongdan Oh

The Peloponnesian War and the Korean War differ in several important respects, including their impact on the political institutions of the combatants. The contending Greek city-states enjoyed much greater independence from foreign influence than did the two halves of Korea, which had been separated only five years earlier by the two Cold War superpowers. And whereas the Korean War was fought in the context of a broader Cold War conflict, the Greek city-states, while they might call on other powerful states such as Persia for backing, were much more on their own in terms of marshaling military resources. In regard to the political legacy of the two wars, the greatest difference is that the Greek states, Athens foremost among them, had a history of limited participatory government and norms of debate. Koreans had no experience with even limited democracy, and rather than finding the opportunity to cultivate democracy after liberation from Japanese rule, they were swept into the Cold War and almost destroyed by the Korean War. The pre-war leaders in both Koreas were able to hold on to their positions throughout the war thanks to the backing of their Cold War sponsors. While it now seems clear that the war was launched on the initiative of Kim Il Sung rather than under the direction of Stalin or Mao, the conduct of the war can hardly be said to reflect the personalities, no matter how strong, of either Kim or his counterpart in the South, Rhee. The war was begun as a "Fatherland Liberation War" by Kim Il Sung to reunite the Korean nation under his rule, but after the initial thrust it was fought and resolved on the basis of Cold War calculations that involved much more than Korean personalities or issues.

With only a few exceptions, South Korean politics for thirty-five years after the Korean War was the story of the exercise of monolithic power by a deeply conservative anti-communist government supported by the United States. Political opponents were routinely crushed. Korean politics had not always been so dull. A thousand political flowers had bloomed in the closing years of the nineteenth century, just before the Japanese colonial occupation, and again in the brief interim between the Japanese surrender and the advent of the Korean War. Many of the politicians and political themes from these earlier periods played an important role in the political situation during the Korean War and after. The goal of this chapter is to illuminate the background of conservative post–Korean War politics, in the process highlighting the factors that changed Korean politics from chaotic pluralism to repressive monocracy.

The history of (South) Korea's politics may be divided into five stages. The story begins in the final days of the Yi dynasty, when Korea was torn between maintaining its traditional policy of isolation and adopting an open-door policy. The debate was finally settled by the Japanese, who annexed Korea in 1910, but the next thirty-five years of occupation were an exciting time for Korean underground politics. After smoldering during the Japanese occupation, politics came out into the open with the surrender of the Japanese to allied forces in 1945, only to be quickly smothered by the American and Soviet occupation forces months later. The smothering was more complete in the northern half of the peninsula under Soviet occupation, but what would have developed in the South will never be known, since North Korean troops attacked the South in June 1950, putting South Korea into a life-or-death struggle. When the struggle ended in a stalemate, after the almost complete destruction of South Korea's economy, domestic politics took a back seat to national security, which became the South Korean politics of the Cold War.

The Korean Dynasty at the Turn of the Century

In several respects, Korean politics of the late Yi dynasty provide a foretaste of politics in the middle of the twentieth century. The last Korean dynasty, named after its founder King Yi, lasted from 1392 to 1910, making it the longest lived dynasty in East Asian history. During this period, Korea witnessed the change of power from China's Ming dynasty ruled by Han Chinese to the Ch'ing dynasty ruled by the Manchus. Japan, during this same period, made the transition from the Ashikaga era to the Tokugawa Shogunate, before entering the modern world with the advent of the Meiji Restoration in 1868. The durability of the Yi dynasty conferred legitimacy on Korean monarchs, who were considered to have received the mandate of heaven. But Korean confidence and pride in its monarchy prepared the way for the eventual destruction of the dynasty as Korea resisted the political, economic, social, and military changes that had transformed and strengthened the Western world and brought it to Korea's doorstep.

As inhabitants of a small country squeezed between large countries, Koreans have rarely been free to practice domestic politics unmolested. In the late nineteenth century the Chinese, Japanese, and Russians were especially eager to extend their influence over Korea. The Chinese had traditionally held a suzerain relationship with Korea, but the Chinese themselves had resisted the Western reforms, and had become weak. The Korean peninsula was an attractive land to control: it had year-round warm-water ports coveted by the Russians, a long border shared with the Chinese, and extensive mineral deposits (in the north) and fertile rice fields (in the south), valued by the resource-poor Japanese. Korea's neighbors also recognized the geostrategic value of the peninsula as a bridge connecting the Eurasian landmass to Japan and the Pacific.

Koreans were of two minds about how to respond to foreign approaches. The conservative school, which counted the ruling class among its members, favored maintaining Korea's traditional political relationship with China, which itself was still resisting Western influence. The reform school, consisting of the young educated segment of Korean society as well as some of the more enlightened ruling elite, saw a desperate need for Korea to modernize in order to hold its own in the new world order dominated by the industrial powers. These reformers looked to Japan for a model of modernization, for the Japanese, after a similar debate, had chosen the road to reform in the 1860s, and they were already beginning to act imperially like their Western mentors.

While the Koreans were debating, the surrounding powers were acting. The Korean government was forced to sign treaties opening the Korean economy and society to the Japanese in 1876, the United States in 1882, Britain and Germany in 1883, Russia and Italy in 1884, France in 1886, and Austria-Hungary in 1889. But Korea's strongest ties remained with the Ch'ing government of China.

As the outside world pressed against traditional Korea, the reformers mounted a coup against the Yi court, demanding that the government sever its tributary ties with China and adopt a republican form of government. The coup failed and its leaders were captured and killed, leaving Korea at the mercy of whichever of its neighbors was strong enough to take advantage of its weakness.

The invasion of Korea was actually triggered by the actions of yet a third political faction, the Tonghak school. This group, consisting largely of nationalistic farmers and led by the charismatic peasant leader Chŏn Pong-jun, staged a rebellion against the government in February 1894, seeking to preserve the traditions of Korean society (Tonghak means "Eastern Learning") and resist the encroachment of foreigners. Ironically, these innocent nationalists, who were most opposed to foreign influence, by their actions gave foreign forces an excuse to enter Korea. In June 1894 the weak Korean government appealed to the Chinese to send troops to quell the rebellion. China sent 3,000 soldiers, and Japan, under the excuse of protecting its new economic investments, sent 8,000 soldiers. The Chinese and Japanese were now in direct contention, and when Japanese troops seized the royal palace in Seoul on July 23, they triggered the Sino-Japanese War, which they won.[1]

China's influence in Korea and throughout the region weakened as Japan's grew. Korea became the battleground for another regional conflict when the economic rivalry on the Korean peninsula between Russia and Japan exploded into the year-long Russo-Japanese war, with Japan once again proving how quickly it had modernized its military. The Japanese were now recognized as full-fledged members of the imperialists' club and in the Taft-Katsura Memorandum of 1905, the United States agreed to respect Japan's preeminent interests on the Korean peninsula (in exchange for Japanese recognition of U.S. interests in the Philippines).[2]

The Japanese Colonial Period

Japan's rise and China's fall from power turned the historical order in Asia upside down. Korea was caught in the middle and was annexed by Japan in 1910. There were two sides to the harsh Japanese occupation of Korea. On the positive side, the Japanese modernized Korea's economic infrastructure and educational system, something that the Koreans had been unwilling to do for themselves. On the negative side, the Japanese took complete control of Korea's political system and attempted to replace Korean culture with Japanese culture. Japan's Shinto religion was imposed, the Japanese language was used in schools and government offices, and Koreans were required to transform their names into fabricated Japanese-style names.

As the Japanese took control of all levels of the Korean government, Korean politicians were forced to collaborate, go underground, or flee. The two most popular foreign destinations for expatriates were neighboring China and Russia, where Korean communities had already been established. As early as 1884, the Kando (Chien-tao) region of Southeast Manchuria had a population of 65,000 Koreans, increasing to 109,000 by 1910. From 1910 to 1920, approximately 7,000 Koreans also fled to Hawaii, from which many proceeded to the U.S. mainland. Tens of thousands migrated to the Russian Maritime Province. Among the notable political figures during this period were Yi Si-yŏng and Yi Tong-nyong in the Kando region, Syngman Rhee in the United States, and Yŏ Un-hyŏng and Chang Tŏk-su in Shanghai.[3]

Expatriates enjoyed two kinds of political freedom. On the one hand, they were free to voice their political views (of which there were many) without fear of Japanese punishment. On the other hand, their political ideas were not constrained by the responsibilities of governance since they had no state to govern. The Korean diaspora created a marketplace for political schools of every stripe, from Marxism/Leninism (especially popular among the Chinese and Russian expatriates) to Wilsonian democracy. Proponents of these schools constantly argued among themselves, and during this period it is difficult to organize the many competing views into major schools of political thought, although the labels "right wing/left wing" (up'a/jwap'a) and conservative/progressive were often used. At this time all Koreans could be considered nationalists, although later some of the nationalists would rally to communism. The right- and left-wing groups were constantly forming new parties, merging with like-minded parties and then splitting up again.

In Korea the Japanese violently repressed political activity, but on March 1, 1919, thirty-three "national representatives" signed a Declaration of Independence, triggering nationwide demonstrations against Japanese rule. The Japanese authorities crushed the demonstrators, who are said to have numbered one million, killing several hundred (according to Japanese count) to several thousand (according to Korean estimates). In response to this March First Independence Movement, Koreans underground and expatriate political figures decided to unify their political activities by establishing a Provisional Government of the Republic of Korea (Taehan Min'guk Imsi Chŏngbu; hereafter, KPG) with headquarters in Shanghai. Syngman Rhee was chosen as the KPG's first president; An Ch'ang-ho, Yi Tong-hwi, and other prominent political figures from China, Manchuria, Siberia, and the United States were given cabinet posts.[4]

This spirit of unity was short-lived, as nationalist/conservative figures clashed with socialist/progressive figures. A geographical/functional split also emerged: the Koreans in China and Russia, who were on the frontline of independence fighting against the Japanese, did not support Syngman Rhee from the United States, whose contribution was to raise funds and lobby for American political support. During this period (the 1920s) the communists split from the nationalists. Among Koreans, communism became popular because many of the exiles (as well as many Koreans who stayed behind) had been dispossessed of their land by wealthy Korean and Japanese landlords.

Viewed in terms of the later roles they would play, the two most important political leaders of the colonial period were Syngman Rhee and Kim Ku, both conservative nationalists. Rhee, who would become the first president of the Republic of Korea in 1948, had gone to the United States in 1904, where he received his bachelor's degree from George Washington University, his master's degree from Harvard, and his doctoral degree in education from Princeton, taking courses under Woodrow Wilson (and later using this connection to further his political interests in the Korean community). His plan for Korean independence was to gather support from the United States and the international community and wait for the pressure to mount against Japan. This approach set him against the progressives in China and Russia, who wanted to take the fight directly to the Japanese without relying on foreign powers.

Kim Ku took a more prominent role in the KPG after Rhee was forced out by accusations he had abused his position to strengthen his personal fortunes. Kim's political philosophy was influenced by several sources: Tonghak, Confucianism, and Christianity. In his love of ideas, he was the typical Korean intellectual. He was also a man of action: in 1894 he participated in the Tonghak rebellion; in 1886 he was imprisoned for killing a Japanese military officer; and from 1911 to 1914 he was imprisoned for teaching nationalism. While Rhee's imprint on the KPG was short-lived (his tenure lasted only six years), Kim's influence on the KPG continued until Liberation, sustaining the KPG by trying to hold the different factions together.[5]

The bond that drew all Korean political figures together during the colonial period was the fight for independence from the Japanese. This goal brought factions together after they had split apart. Ironically, despite their fighting and planning, independence was finally gained not by the efforts of Koreans but as a result of the global coalition of forces that defeated the Japanese and forced them to relinquish their colonies. This reality immediately endangered Korea's goals for independence. The Japanese in Korea surrendered on August 15, 1945, Russian troops took control of the northern half of Korea on August 22, and American troops took over administration of Korea below the 38th parallel on September 8. In a sense, Korea was independent for only a few days.

Between Liberation and War

The Korean exiles returned home to find their land occupied by Americans and Russians, who talked not of independence but of international trusteeship, for as Koreans had not had the opportunity to govern themselves for thirty-five years, the major powers doubted their ability to do so now. It soon became apparent that the two occupying powers would not be able to agree on a reunification policy. Kim Ku, as president of the KPG, traveled to Moscow where the United States and the USSR were discussing joint trusteeship, to plead for independence, but the leaders of the emerging Cold War blocs had bigger fish to fry. Keeping Korea out of the hands of the other side was the primary policy goal.

The United States had little knowledge about Korea and the many political groups that sprang up like weeds after Liberation. Weeds, because the garden of politics was tended by the U.S. Military Government in Korea (USAMGIK), the "depository of political authority, which acted to sanction the legitimacy and effectiveness of each political organization."[6] General Hodge, the commander of U.S. forces, adopted the policy of consulting "only with organized political groups," and USAMGIK required that "all groups of three or more persons designed to influence the political life of the country should register with the military government."[7] By March of 1946, 134 groups had so registered.

One of the major political groups was the Preparatory Committee for Founding the Nation (Kŏn'guk Chunbi Wiwŏnhoe, or Kŏnjun), headed by Yŏ Un-hyŏng, who had been involved in the Korean nationalist movement in China. Immediately before the Americans arrived, Kŏnjun formed the People's Republic of Korea (PRK), which included prominent conservative expatriate figures such as Syngman Rhee and Kim Ku as well as communists. USAMGIK refused to recognize this indigenous Korean government, and it melted away. Yŏ then formed a left-leaning political party, the Working People's Party (Kŭllo Inmin Tang, different from the southern branch of the later North Korean Workers Party). The communists in the South, deprived of their political legitimacy, pursued their goals by inciting resistance against the military government.

Meanwhile, conservatives organized the Korean Democratic Party (Han'guk Minju Tang or Hanmindang) a week after the Americans arrived. Hanmindang supported Kim Ku's KPG. Syngman Rhee and his supporters established an organization called the National Association for the Rapid Realization of Korean Independence (NARRKI). As the organizations and parties proliferated and contended against each other, they held one thing in common: the goal of independence rather than trusteeship. When the American-Soviet trusteeship was officially announced, it was roundly denounced by all Korean political factions except the leftists, who saw a silver lining in the likelihood that, whereas an independent Korea would probably choose to remain outside the socialist camp, under Soviet trusteeship at least the northern half of Korea would be socialist.

When Kim Ku and other leaders of the KPG returned to Korea in December of 1945, the USAMGIK refused to recognize their government. The former KPG exiles then formed the Korea Independence Party (Han'guk Tongnip Tang, or Hantoktang), which was conservative in orientation. USAMGIK established a South Korean Interim Government and a Korean Interim Legislative Assembly in December 1946, moving closer to a separate government for the southern half of the peninsula.

Kim Ku's KPG had been viewed by the Americans as a hotbed of radicals since it included progressives as well as conservatives. Looking around for a leader whose anti-communist views were consistent with the emerging Cold War mentality in Washington, the Americans lit on Syngman Rhee, who was a known figure. Rhee was the default choice, not much admired by General Hodge and lacking in supporters within South Korea since he had been living in the United States for thirty-five years. In the search process more legitimate leaders such as Yŏ Un-hyŏng and Kim Ku were passed over by the American king-makers.

The Soviets quickly set up an administration under Kim Il Sung, who had lived in Siberia during the final years of Japanese occupation. Kim's clear goal was to establish in the shortest possible time a monolithic communist regime to serve as a base to unify the entire peninsula under communism.

Even at this late date many Koreans still hoped that a unified independent Korea could be established. The nationalists Kim Ku and Kim Kyu-sik went to Pyongyang in April 1948 to attend a meeting of the North–South Leadership Coalition, whose goal was to reunite Korea. They were warmly welcomed by Kim Il Sung, who was no stranger to united front tactics. Rhee and his American backers decided it was time to forcefully consolidate the southern half of the peninsula for fear it would be torn apart by political dissension and fall into the hands of the communists. Elections were held in the South in May 1948, resulting in the election of Rhee as the first president of the Republic of Korea, which was formally established on August 15 (in the North the Democratic People's Republic of Korea was founded in September).

The new South Korean government had to contend with rebellion, much of it leftist-inspired, from many ordinary Koreans who opposed the U.S.-backed Korean government, which was staffed with many officials who had worked for the Japanese. The epicenters of these rebellions, which spread throughout the South, were on the island of Cheju, off the southern coast, and around Yŏsu, a port city of Chŏlla province. The rebellions were put down by South Korean government forces, resulting in the deaths of thousands of civilians. To forestall future rebellions and demonstrations, the Rhee government promulgated as its first law the National Security Law of December 1, 1948. This law, still on the books in 2000, prohibits any activities or expressions of support for communism or the government of North Korea, although in the late 1990s the interpretation and enforcement of the law was somewhat relaxed. Ironically, the National Security Law was a continuation of the repression that the Japanese had imposed on the Korean people since 1935.[8]

On June 26, 1948, Kim Ku, back in South Korea from his meeting with Kim Il Sung, was assassinated by one of his lieutenants, quite likely at the instigation of Rhee forces. The Rhee government adopted the slogan "march north," but the U.S. government discouraged this adventure, and the last of the U.S. troops were withdrawn from South Korea in December 1949, leaving behind 500 military advisors to strengthen the poorly trained and equipped ROK army. Many disgruntled laborers and farmers, along with communist sympathizers, fled to the North. Kim Il Sung, probing the weakness of the South, and hoping to capitalize on the support of anti-government forces there, periodically staged raids across the DMZ. Then on June 25, 1950, a Sunday, with many ROK soldiers on home leave, North Korean forces launched a full-scale attack across the DMZ.

The Korean War

The origins of the Korean War are complex. Preceding the North Korean attack of June 25, 1950, both sides had sent raiding parties across the border and both governments had threatened to unite the peninsula by force.[9] Contrary to North Korea's assertion that the war was triggered by South Korea, we know that it was Kim Il Sung who, after consulting with Stalin and Mao (neither of whom was enthusiastic), planned and launched a coordinated attack across the 38th parallel.

South Korean domestic politics played some part in encouraging Kim Il Sung to start a war. Pak Hŏn-yŏng, the president of the South Korean Labor Party (Namnodang) had defected to the North and received a cabinet post after the Rhee government crushed the Yŏsu and Sunch'ŏn rebellions and began rounding up communists and communist sympathizers. It seems likely that Pak believed his position and that of other South Koreans who had escaped to the North would be strengthened if the South could be communized, for in the North the Namnodang faction was weak and isolated. In the event, as Kim Il Sung later complained, "Pak Hŏn-yŏng, a spy on the American payroll, bragged that south Korea had 200,000 Party members . . . but in actual fact, this guy, in league with the Yankees, totally destroyed our Party in south Korea. . . . If only a few thousand workers in Pusan had risen to hold a demonstration . . . we would have definitely liberated even Pusan and the American troops would not have been able to land."[10]

Throughout the war, President Rhee remained firmly in control of the political situation in the South. From the birth of the Republic in 1948, Rhee's opponents had favored a parliamentary form of government, which would deprive the president of much of his power. In the early years of the First Republic, coalitions and parties emerged and died, capturing votes in the assembly but never threatening Rhee's rule. Gradually two parties emerged: Rhee's Liberal Party (Chayu Tang) and later the opposition Democratic Party (Minju Tang), but neither of the parties was very cohesive. The Liberal Party has been described as "a motley assortment of opportunists held together by a desire for power and loyalty to Rhee."[11] In the 1950 Assembly election, Rhee's party won only a quarter of the seats. Since the constitution provided for the election of the president by the Assembly, this lack of Assembly support threatened Rhee's reelection in 1952. He proposed an amendment to the constitution providing for popular election of the president, but the proposal was defeated in the Assembly by 143 to 19 votes. On May 25, 1952, Rhee proclaimed martial law, ordered some representatives arrested on minor charges, and forcibly brought the remaining members to the National Assembly building and held them prisoner for two days until they passed Rhee's amendment with a vote of 163 to 0 and one abstention. In 1954, Rhee overturned another important Assembly vote in order to amend the constitution to allow him to run for reelection. Other laws passed by the Assembly were never promulgated by the executive branch. For all intents and purposes, democracy had died in South Korea.[12]

The range of anti-democratic activities of the Rhee government and the grave consequence of these activities have been eloquently summarized by Young-ho Lee:[13]

> The meaning of the electoral process as a democratic means of selecting and holding political leaders accountable to the electorate was largely lost when election irregularities became so frequent and widespread. Interference by the police and government officials (who were by law forbidden from political activities), vote-buying, blackmail, threats, physical violence, and numerous other corrupt practices flourished. All this adversely affected the Korean people's orientation toward the electoral process in particular and toward the representative democracy in general about which they were just beginning to learn.

Over the years, Rhee's presidential opponents met untimely deaths: Kim Ku was assassinated in 1949; Cho Pong-am, a presidential candidate in 1952 and 1956, was hung for treason; Sin Ik-hŭi, a candidate in 1956, died of a heart attack shortly before the election; Cho Pyŏng-ok, a candidate in 1960, also died just before the election.[14]

The military campaigns of the war, after the initial North Korean onslaught, were directed by foreign generals and spearheaded by foreign troops. The United States rescued Syngman Rhee and the Chinese rescued Kim Il Sung. Kim Il Sung stood to lose more in defeat because he had launched the war, but he was forceful and skillful enough to be able to use the war to continue his purge of political opponents, including Pak Hŏn-yŏng, who was executed on a trumped up treason charge. Syngman Rhee continued his dictatorial ways during the war, wielding the weapon of the National Security Law, which effectively banned any anti-Rhee activity. In what would later be seen as a strategic blunder, Rhee refused to sign the armistice agreement ending the fighting, holding out the unlikely option of renewing the attack against North Korean forces. Consequently, North Korea in the 1990s refused to acknowledge that South Korea should be a party to the replacement of the armistice with a peace treaty formally ending the war.

Post-War Politics

The tragedy of the war for South Korean politics was that the terror and destruction suffered by the South Korean people, whose land was overrun by North Korean forces before they were thrown back from the Pusan perimeter, rendered South Koreans docile against the increasingly totalitarian Rhee regime. The variegated pageant of political voices that had briefly contended between Liberation and the Korean War was effectively silenced. After the war Koreans turned to the task of rebuilding their economy and let President Rhee and his authoritarian successors run the country under the cloak of American military support (the ROK-U.S. Mutual Defense Treaty was signed on October 1, 1953). Rhee, true to early indications, single-mindedly pursued the goal of enhancing his personal power, but his dream of becoming South Korea's president for life was thwarted by widespread student demonstrations in 1960. A few months later General Park Chung Hee ended the messy exercise in democracy by seizing power and continuing Rhee's totalitarian tradition. Park's eighteen-year tenure was terminated by the chief of the Korean Central Intelligence Agency, who put a bullet in the president's head. After Park would come Generals Chun Doo Whan and Roh Tae Woo, followed by the civilians Kim Young Sam and Kim Dae Jung, who, thanks to the opening provided by Roh's belated initiative, oversaw South Korea's transition to a respectable democracy.

Legacies of the War

In important respects politics and governance in South Korea after the war were not greatly different from governance during the Yi dynasty or during the Japanese colonial period. Throughout the post-war period both South and North Korea boasted of being democracies, but democracy was only an institutional framework within which strong rulers exercised almost unlimited power. The concept of power in traditional Korean society was hierarchical: a meritocratic class and ruling elite grasped the mandate to rule as they saw fit. Rulers had little tolerance for dissent or complaint: the Yi rulers would not adapt to foreign ways, the Japanese colonial masters would not entertain the idea of coexisting with Koreans, and the Rhee government would not permit political competition. In all three periods, the rulers tried to deflect popular dissatisfaction on to foreigners: the Western barbarians during the last years of the Yi dynasty, the white imperialists during Korea's colonial occupation as a dependent territory of Japan's Greater East Asia Co-Prosperity Sphere, and the North Korean communists during the post-war military regimes. Since the mid-1980s the South Korean people have gradually been given a larger measure of freedom to voice their opinions and participate in governance as political controls have been relaxed and layers of political corruption peeled away from the body politic. The North Koreans have yet to taste political freedom, and may indeed find that their unreformed society is as fragile as Yi dynasty Korea.

The war demonstrated the importance of a strong military for the survival of the state, and this legacy persisted in the form of a quasi-war atmosphere pervading South Korea through the 1970s and continues as strongly as ever in North Korea today. Rhee was a civilian, unlike his most immediate presidential successors, but he initiated the militarization of Korea. The economy was geared for military strength, and military men played an important role in government. Between 1963 and 1986, 14 percent of Assembly seats and 42 percent of committee chairmanships in the Assembly were held by former military men.[15]

The most tragic legacy of the Korean War is that it fostered two kinds of politics of exclusion, both having roots in the pre-war period. In domestic South Korean politics the president and his clique effectively excluded the citizenry and competing politicians from gaining power. In inter-Korean relations, the anti-communist mentality fostered by successive South Korean leaders for the purpose of creating a threatening external environment favorable to the rule of a strongman turned South Koreans against North Koreans. Thus, the greatest damage inflicted by the Korean War was not on economy or politics but on the psychology of the Korean people.

A pervasive fear of North Korean infiltration and invasion undergirds South Korea's domestic politics and foreign policy. The stifling effect of the National Security Law (which until recently prohibited citizens from gaining access to any information about communism or North Korea, and still prohibits them from publicly saying anything deemed beneficial to the North Korean government) has often been used as a political weapon by the ruling party to stifle political opposition.

In the late 1980s what Samuel Huntington called the "third wave" of democratization lapped on South Korea's shores, eroding some of the more restrictive political bonds in place since the Korean War.[16] Multiple causes for this democratization can be adduced, including the emergence of a middle class, a growing confidence founded on greater economic and military power, and the desire of General Roh Tae Woo to secure for himself a place in history as the first South Korean president to voluntarily relinquish power. Indirectly, the Korean War itself was a factor in this delayed democratization in the sense that the continued presence of American troops and the influence of American culture nudged South Korea toward democracy. Some of this influence was intentional, a notable example being President Carter's explicit although largely futile attempts to improve human rights in South Korea by putting pressure on the Park government. Arguably the strongest influence was the constant exposure of South Koreans to American culture in the media coupled with the argument that if the United States was so committed to preserving the freedom of South Korea, why should the South Korean people not enjoy the same political freedoms as Americans?

South Korea's economy has grown far beyond what anyone could have dreamed in the closing days of the war, even considering the economic difficulties accompanying the Asian financial crisis. In fact, the economic crisis may prove to be a blessing by revealing structural weaknesses in South Korea's economic and political systems and by diverting attention from a North Korean threat to South Korea's own problems. The South Korean political system is well on its way to practicing full-fledged democracy with the presidential election of opposition leader Kim Dae Jung, who had been persecuted by earlier South Korean governments. But the bitterness that has sprung up between North and South as a result of the Korean War remains, on a regional level if not on a personal one, and this legacy may rear its ugly head during the tough economic and political period immediately following an eventual reunification. If Northerners and Southerners are unable to work together after unification, democracy in a unified Korea will be difficult to achieve.

Notes

This chapter does not necessarily represent the opinions of the Institute for Defense Analyses or its clients.

1. Song Kwang-song, *Migun chŏmnyŏng 4-nyŏnsa* [The History of the 4-Year U.S. Military Occupation] (Seoul: Hanul Pub. Ltd., 1995), 33–35.

2. Song Hwang-yong, *Ilbonŭi Taehan chŏngch'aek* [Japan's Korea Policy] (Seoul: Myongji Pub. Ltd., 1981), 70–76.

3. Carter J. Eckert, Ki-baik Lee, Young Ick Lew, Michael Robinson, and Edward W. Wagner, *Korea Old and New: A History* (Seoul: Ilchokak Publishers, distributed by Harvard University Press, 1990), 273–77.

4. Ibid., 278–80.

5. Kim Hŭi-gon, Han Sang-su, Han Si-jun, Yu Pyŏng-yong, *Taehanminguk imsi chŏngbu ŭi chwau hayja undong* [Merge Movements Between Left-Wing and Right-Wing Groups of the Provisional Government of the Republic of Korea] (Seoul: Hanul Pub. Ltd., 1980); pages 24–225 are devoted to a description of Kim Ku's vivid career.

6. C.I. Eugene Kim and Young Whan Kihl, "Development of Party Politics (1945–1961): An Overview," in C.I. Eugene Kim and Young Whan Kihl, eds., *Party Politics and Elections in Korea* (Silver Spring, MD: The Research Institute on Korean Affairs, 1976), 8.

7. Ibid., 9.

8. Kim Hyung-tai, "Yangsimsu wa kukka poanbŏp" [Prisoners of Conscience and National Security Law], *WIN* (October 1997): 100–3.

9. The immediate and more distant events leading up to the war are described by Bruce Cumings in *The Origins of the Korean War*, Vol. I. (1981) and Vol. II. (1990) (Princeton: Princeton University Press).

10. Kim Il Sung, *Works, Vol. 17, January–December 1963* (Pyongyang: Foreign Language Publishing House, 1984), 112–13.

11. Eckert et al., *Korea Old and New*, 350.

12. Young-ho Lee, "The Politics of Democratic Experiment: 1948–1974," in Edward Reynolds Wright, ed., *Korean Politics in Transition* (Seattle: University of Washington Press, 1975), 21–23.

13. Ibid., p. 23.

14. C.I. Eugene Kim and Young Whan Kihl, "Development of Party Politics," 15.

15. Chae-Jin Lee, "The Effects of the War on South Korea," in Chae-Jin Lee, ed., *The Korean War: 40-Year Perspective* (Claremont, CA: Claremont McKenna College, 1991), 122.

16. Samuel Huntington, *The Third Wave: Democratization in the Late Twentieth Century* (Norman: University of Oklahoma Press, 1991).

Part IV

Demagogues? or Domestic Politics in Democracies at War

10

McCarthyism and the Korean War

Ellen Schrecker

No one burned his draft card during the Korean War. Though the war was contro-versial, unpopular, and, in many respects, unsuccessful, there was little organized opposition to it—no teach-ins or mass demonstrations in Washington, no civil disobedience or student protests.[1] Unlike the Vietnam conflict, no serious ques-tions about the legitimacy of American intervention in Korea reached the political mainstream during the course of the conflict. The reason seems obvious. The do-mestic political repression that we now call McCarthyism was at its peak during the Korean War and it effectively silenced or marginalized most domestic criti-cism of the war.

But the relationship between the Korean War and McCarthyism ran in both directions, for the conflict broke out at a moment when it might have been pos-sible to have reversed or at least mitigated some of the more extreme manifesta-tions of the anti-communist furor. Instead, the repression intensified, legitimized by the heightened concern for national security that the actual fighting provoked. Take Joe McCarthy, for example. His career as a demagogue might well have been aborted had the North Koreans not marched south a few days before the release of a special Senate committee's report calling his charges of treason in high places "a hoax and a fraud."[2] A similar quirk of timing may have helped the federal judi-ciary overlook the McCarthy era's incursions against the First Amendment. The key case, the criminal prosecution of the leaders of the Communist Party, was actually being argued before an influential appeals court when the war broke out.[3] Would the judges have reached a different decision had there been no conflict in Korea? We will never know.

Such counterfactual speculations may well be beside the point. McCarthyism and the Korean War occurred at the same moment in time and played off against each other in a mutually reinforcing manner. The "moral chauvinism" that Victor Hanson considers the inevitable accompaniment of warfare in a democracy surely heightened McCarthyist political repression after June 1950.[4] So, too, did the Korean War's seeming legitimation of the partisan scenarios about the "loss" of China that were such an important element of the Wisconsin senator's crusade. But, as we shall see, there was more to McCarthyism than McCarthy. And the anti-communist crusade that so effectively stifled opposition to the Korean War had been well under way before Kim Il Sung's troops crossed the 38th parallel.

When we look more closely at the phenomenon that has been misleadingly identified with the bizarre antics of the senator who gave it a name, we encounter what may well have been the most widespread wave of political repression in American history. In a frenzied effort to stamp out the perceived menace of domestic communism, the nation's politicians, bureaucrats, businessmen, unionists, educators, journalists, and religious and civic leaders mounted a massive investigation into the political beliefs and activities of millions of men and women. By the time the Korean War began, this inquisition had already done most of its damage, demonizing dissent and marginalizing all serious criticism of the status quo.[5]

Obviously, compared to the state-sponsored terrorism of Stalin's Russia or Syngman Rhee's Korea, what happened during the McCarthy era was rather mild. Most of the men and women who ran afoul of the anti-communist crusade lost their livelihoods, not their lives. Two people (Julius and Ethel Rosenberg) were killed and a few hundred others went to prison or were deported. The main sanctions were economic. Though figures are hard to come by, it is likely that about ten thousand men and women lost their jobs for political reasons during the 1940s and 1950s.[6]

The relative leniency of the witch hunt should not, however, blind us to its efficacy. As I will argue in the rest of the chapter, McCarthyism was devastatingly effective in eliminating meaningful dissent within the American polity. As the historians of American slavery have noted, it did not take more than a few whippings to ensure the docility of a plantation's workforce in the antebellum South, and in a similar way the limited economic sanctions of the McCarthy era sufficed to keep the resistance in line. The prospect of unemployment and social ostracism prevented all but a tiny handful of activists from openly challenging the status quo or opposing the cold war policies of the federal government. More important, the institutional structure through which it might have been possible to offer a political and cultural alternative to the American mainstream was largely destroyed.

The process that achieved these results was a complicated one; its very complexity and the multiplicity of its constituent elements was, in fact, central to its efficacy. Like a piece of rope, McCarthyism got its strength from the intertwining of dozens, perhaps hundreds, of separate strands. There were congressional committee hearings, criminal prosecutions, passport denials, loyalty-security programs, blacklists, deportations, dishonorable discharges, disbarments, and dozens of other public or private actions that imposed sanctions of one kind or another on the groups and individuals tainted by communism. The interrelations between these sometimes competing, sometimes collaborating operations were crucial to the success of the project.

Actually, we may be oversimplifying matters to speak of McCarthyism as a single entity, for there was not one, but several McCarthyisms, each with its own agenda and modus operandi. Though they all shared a commitment to eliminating the threat of domestic communism, they often differed considerably in how they defined that threat. The McCarthyism peddled by the traditional patriots of the American Legion, with its contention that "pinkos" were as dangerous as party members, could not have been more different ideologically from that of the anti-Stalinist left that viewed communists as traitors to the socialist ideal.[7] Liberals had their own type of McCarthyism (though, of course, they would never have called it that); they supported sanctions against communists, but not against non-communists.[8] There was also a partisan brand of McCarthyism, purveyed by ambitious politicians who wanted to further their own careers and boost the Republican Party. Business groups, conservative trade unionists, Catholics, ex-communists, segregationists, and J. Edgar Hoover's FBI—all had their own reasons for signing onto the anti-communist crusade. Their agendas overlapped and there was considerable collaboration. Many of the most active anti-communists worked together, forming a surprisingly self-conscious network of professionals, which one of them not so facetiously referred to as "Red-Baiters, Inc."[9]

This network played a pivotal role in disseminating the most widely accepted interpretation of the communist threat and designing the machinery that would be used to combat it. Most of its members were full-time ideologues and activists who had been fighting the Communist Party (CP) since the 1930s. When the cold war legitimized their cause and brought the apparatus of the American state into the anti-communist crusade, these people were already on the scene with a fully developed scenario for dealing with the newly threatening CP. Not only did they have long-standing connections with those elements of the federal government that were to become the main instruments of repression, but their alleged expertise in identifying party members made them central to the two-stage procedure that characterized the purges of the McCarthy era.[10]

The first stage, that of identification, was usually handled by the professional anti-communists of an official agency like the FBI or a congressional committee; the second, the imposition of sanctions, by an employer.[11] In a typical case, a college teacher might take the Fifth Amendment at a congressional hearing and then be fired by his university's president or board of trustees on the grounds that, as Rutgers University stated with regard to the classicist M.I. Finley, his refusal to answer the committee's questions "impairs confidence in his fitness to teach."[12] This diffusion of responsibility enhanced the efficacy of the process by making it possible for many of the participants to distance themselves from those aspects they found distasteful and claim that they opposed rather than supported McCarthyism.

The anti-communist operatives did not select their targets at random. During the 1950s, it was commonplace within liberal and left-wing circles to insist that almost anybody could be victimized by the false accusations of anonymous informers. But, in fact, there were only a few such "innocent liberals," people who had never been involved with communism and who lost their jobs or were otherwise punished because of an ex-daughter-in-law's political activities or because their names were found on the mailing list of a defunct left-wing organization. The cases of these people often got considerable publicity; they provided an easy and politically acceptable way for anti-communist liberals to dramatize the evils of McCarthyism.[13]

The victims of the purges of the 1940s and 1950s also contributed to the myth-making process. Many of them were or had been communists; but because they were trying to conceal their political affiliations, they often lied or refused to talk about those affiliations, describing themselves as ordinary "progressive" citizens, victimized by an organized assault on anyone who advocated peace and freedom. Such disingenuous behavior on the part of these people and their defenders complicates our assessment of the impact of McCarthyism, mainly because it makes it difficult to figure out exactly how far the party actually reached. But, perhaps we do not need such a distinction, for, even if we cannot identify the political affiliation of any one specific individual, it is, nonetheless, clear that most of the people who appeared before anti-communist investigating committees or lost their jobs for political reasons during the 1940s and 1950s were either communists or people who had been involved in the broader left-wing movement that had flourished within the party's orbit.

The Party

It is also clear that American communism played a considerable role in shaping the political repression that it encountered during the McCarthy years. The people who created the mechanisms of that repression, the FBI officials and anti-communist professionals who ran the congressional investigations and devised the political tests for employment, followed the contours of the party and structured their operations in accordance with its policies and practices. Their knowledge about how the CP functioned and whom it appealed to helped these anti-communists decide what groups and individuals to target. This is not to say that the party caused McCarthyism or that it could have escaped some kind of repression during the cold war, merely that it had some responsibility for the *form* of that repression. Or, in other words, to understand anti-communism, we must understand communism.

Though an insignificant political sect by the late 1950s, the Communist Party had been much more influential during the 1930s and 1940s when it was at the heart of the nation's most dynamic movement for social and economic change. It was never a major player (as an organization it was always marginal), but it did supply key personnel to that movement and offer an alternative collectivist and internationalist vision of the world that had some impact on the political discourse of the period. In part, this influence derived from the historical accident that the party had adopted the reformist strategy of the Popular Front and was espousing causes, like anti-fascism and unionization, that were acceptable to mainstream liberals, but it was also due to the CP's unique nature and the extraordinary commitment of the people who joined it.[14]

The party was not just another political party. It was a highly disciplined organization that required its members to devote much of their time and energy to party work. Thus, even though the CP never had more than 75,000 card-carrying members at the height of its influence, the dedication of these men and women enabled it to have a much larger impact on American society than its size might suggest. Its members were the most devoted radicals of their generation, long-term political activists who joined the CP because it seemed *at the time* to be the most effective vehicle for them to accomplish their own goals. Many of them became full-time cadres, professional political organizers who worked for the party or staffed the dozens of left-wing groups and labor unions that operated within its sphere.[15]

Their activities addressed a wide range of political, social, and cultural issues. Unlike today's political scene in which hundreds of single interest groups vie for the allegiance of individual activists, the CP offered its members one-stop shopping. It was at the center of a larger movement, a constellation of groups and causes that enabled it to service its adherents and mobilize them in support of specific projects. The labor unions in which thousands of American communists became active during the 1930s and early 1940s were by far the most important of these organizations; by the end of World War II communist-led unions controlled nearly 20 percent of the membership of the CIO (Congress of Industrial Organizations). But there were also dozens of other outfits, the so-called front groups, through which the party tried to extend its influence into other areas of American life. Besides the communist-led labor unions, there were party-linked peace groups, summer camps, civil rights organizations, youth groups, literary magazines, lawyers' groups, choral societies, fraternal orders, adult education centers. There was even a communist insurance company, the International Workers' Order, a party-run fraternal order that sold cheap life insurance and cemetery plots.[16]

Together, the Communist Party and the organizations within its periphery created the institutional framework for the development of a unique Popular Front political culture. It embraced both capital letter "C" Communists who submitted to the CP's discipline and lower case ones, the so-called fellow travelers and left-wing activists, who supported the same causes, worked in the same organizations, and sent their kids to the same summer camps. It was not a monolithic movement (its members signed on for different reasons and participated in different ways), but it was a cohesive one, with a shared sense of belonging to a larger common enterprise. Thus, for example, front groups and left-wing unions took stands on issues that were not directly related to their specific missions and participated as organizations in other movement activities. Much of that involvement took the form of what one ex-communist called "the resolution bit"—public statements or editorials in a union paper that championed the party's position on foreign policy or other matters. But it could also involve more direct action: recruiting people for mass meetings, demonstrations, and May Day parades; encouraging members to circulate petitions or take classes at CP-run labor schools; or raising money for one or another of the movement's many causes.[17]

Racial justice was one such cause, and by the 1940s a particularly important one. Not only did the front groups and left-led unions oppose the Jim Crow practices that were common throughout American society at that time, but they were also among the few organizations not specifically devoted to civil rights that promoted racial equality within their own ranks.[18] Left-wing unions, for example, pressed for the elimination of job discrimination within their industries and encouraged minority-group members to take leadership positions. Communist summer camps recruited African-American children; communist schools taught black history. In the Southwest, the movement worked with Mexican Americans.[19]

Though party members did not advertise their affiliation, it was easy for anti-communist investigators to identify them by the pattern of their political activities and their premature multiculturalism. These investigators were looking for communists and they made few distinctions between the upper case and lower case varieties. Accordingly, they targeted the groups and causes that had been associated with the Popular Front left. They relied on the well-known duck test: "If it looks like a duck, waddles like a duck, and quacks like a duck . . ." And they assumed—not always incorrectly—that any white who was active in a lot of front groups or socialized with Negroes was probably some kind of duck.[20]

There were, however, more critical problems with the Communist Party than the fact that any contact with it exposed people to McCarthyism. In large part because of its dynamism, internationalism, and cohesion, the CP virtually preempted all other left-wing organizations during the 1930s and 1940s and attracted an entire generation of idealists and activists into its orbit. But the left's association with the Communist Party proved self-destructive, for the party's allegiance to the Soviet Union and its lack of internal democracy made it an unsuitable vehicle for an American political movement. As a result, for all its advantages as a source of organizational strength and solidarity, the CP ultimately undermined the political activities of all the individuals and organizations that were associated with it.

The party was, after all, an ostensibly revolutionary organization that saw itself as the official American representative of the Third International and its Russian directors. Because of the importance of its Soviet connection to the CP's overall identity and sense of mission, the party's leaders struggled to ensure that they retained their franchise by coordinating their policies and activities with what they assumed Moscow wanted. In short, they followed the Soviet line. This did not mean that they were mindless robots whose every action had been dictated by Stalin, but rather that on major issues the party tried to have its policies conform to the interests of the Soviet Union and to the worldwide movement that it headed. This posture caused considerable embarrassment at those moments, like the signing of the Nazi-Soviet pact in August 1939, when the party had to scramble to keep up with the Kremlin's twists and turns. The process was also far from democratic and often encouraged the most rigid sort of sectarianism. And, of course, the

CP's rationalization of the Soviet purges and recruitment of espionage agents deeply compromised the entire movement.[21]

Though directives from the Kremlin did not govern the daily activities of the party's rank-and-file members, the perception that they did was widespread. As a result, when the cold war transformed the Soviet Union from an ally to an enemy, the CP was converted from an unpopular but tolerated political organization to a threat to national security. American policymakers, with J. Edgar Hoover in the lead, claimed to believe that all party members were under Moscow's orders and available for whatever spying or sabotage the Kremlin might demand.[22] The widely publicized espionage cases of the early cold war bolstered that conviction and gave added impetus to the drive to eliminate communists and their associates from positions where they might endanger the nation's security.[23] Exaggerated notions about the red peril received widespread currency within the most respectable of venues and were routinely cited as justifications for specific programs or investigations.[24] That these dangers never materialized had little impact on the widespread perception that communists imperiled the nation.

Almost as devastating to the communist movement as its Soviet connection was the party's penchant for secrecy. The CP had begun its existence as an illegal underground organization and, even at the height of its influence, it retained a conspiratorial aura. As we have already seen, many individual communists kept their affiliation hidden. To a certain extent, this policy made sense. The party had long been a pariah and an open admission by its members that they belonged would have cost them their jobs or credibility as political organizers and labor leaders. Nonetheless, in the long run the CP's secrecy was self-defeating. By refusing to let its members reveal themselves, the party contributed to its own demonization. It made it impossible for the American public to discover what communists really did and it enabled its enemies to characterize the entire movement as a conspiracy. The CP's secrecy also contributed to making the identification of individual communists central to the machinery of anti-communism and rendered the party's members, former members, and the men and women who traveled within its orbit uniquely vulnerable to exposure.[25]

The Purges

That exposure took an enormous personal toll. Thousands of the men and women who were called before congressional committees or otherwise identified as politically undesirable lost their jobs or suffered in other ways. Much of what they endured, however, was not designed to punish them as individuals, but to reach the organizations in which they had been active, curb those groups' activities, and force them out of business. The campaign was successful; by the end of the 1950s, few of the organizations that been connected to the communist movement would still exist. In the long run, the institutional destruction caused by McCarthyism was to be its main legacy. Most of the men and women who had been affected by the criminal prosecutions and blacklists of the 1940s and 1950s eventually put their lives back together. But the network of causes and organizations to which these people had devoted themselves disappeared forever. When political radicalism revived in the 1960s, it did so in a more ephemeral way. Groups formed on an ad hoc basis to oppose segregation in the South or the war in Vietnam, but there was no unifying institutional structure behind them.

The left-wing unions were the chief organizational casualties of the McCarthy era. Only the Communist Party was harassed as much. These unions had, after all, been the CP's main link to the only mass constituency it ever reached; their destruction would seriously undermine the communist movement. These unions and their leaders had been under attack for years. Hostile employers, rival unions, internal dissidents, the Catholic church: left labor's enemies were many. Until the cold war, however, most of these unions had managed to hold their own and retain the loyalty of their overwhelmingly non-communist members. But once anti-communism became a national priority, the federal government joined the campaign, bringing both legitimacy and all the resources of the national security state to the multipronged assault on the left-wing unions. Washington's intervention proved decisive; within a few years, most of the unions that had been such a vital part of the communist movement had disappeared or dwindled into insignificance.[26]

Some of the first initiatives came from the Immigration and Naturalization Service, which targeted almost every foreign-born left-wing labor leader for deportation.[27] Congressional committees were also active; by early 1947, several investigations of the left-led unions were already under way.[28] The FBI and military intelligence agencies were keeping tabs on these unions and Catholic activists had begun serious campaigns to oust the left-wing leaders.[29]

The most damaging blow was the passage of the Taft–Hartley Act in 1947. The culmination of a long-term campaign by the business community to reverse the gains made by the entire labor movement since the 1930s, the measure also contained a provision requiring all union officials to sign a non-communist affidavit. Unions that did not comply with the law would forfeit the services of the National Labor Relations Board (NLRB). Though there was considerable hostility to the new legislation within the labor movement, the left-led unions were to be its main victims. Hostile employers wrapped themselves in the flag and refused to bargain with the allegedly subversive organizations; rival unions stepped up their raids. Deprived of the legitimacy and practical help conferred by NLRB recognition, the party-linked unions found it increasingly difficult to fend off these attacks. Survival dictated compliance, and by the middle of 1949, all the left-wing unions had capitulated and ordered their leaders to sign the affidavits.[30]

A few months later, these unions were expelled from the CIO, an action that codified their marginality and rendered them even more vulnerable to raids and secession movements. The expulsion was not unexpected, for the CIO's moderate leaders had been moving toward a close alignment with the Democratic administration ever since President Truman's unsuccessful veto of the Taft–Hartley Act.

The left-wing unions' criticism of Truman and his foreign policy and their support for the third-party candidacy of Henry Wallace in the 1948 presidential election precipitated the break. At its 1949 convention, the CIO expelled two left-wing unions and voted to bring charges of communist domination against ten others. The most interesting aspect of the trials that followed was not their outcomes—the expulsion of the left-led unions was a foregone conclusion—but the fact that they employed exactly the same witnesses and evidence as congressional committees and federal prosecutors.[31]

Almost as devastating to the labor left as the Taft–Hartley Act and the CIO expulsions were the various public and private loyalty-security programs. Though these programs ostensibly affected individuals rather than organizations, they targeted an inordinate number of the left-wing unions' most active members. National security, not union-busting, was the rationale, but the outcome was the same. The Truman administration's 1947 loyalty program was the prototype. Because most of the contemporary criticism directed against the program centered on its unfair procedures, the fact that it essentially destroyed the left-wing United Public Workers union got little attention. The military ran a similar screening program in defense plants and other "vital facilities" that had the same impact on the left-led unions in the electronics industry.[32]

The Korean War intensified all the pressures the left-led unions faced. Employers and rival unions took advantage of the heightened sense of crisis to push for measures that would weaken the left-wing unions. The most drastic was the little-known port security program that was instituted during the summer of 1950. Ostensibly designed to protect the nation's ships and harbors from sabotage, the program actually targeted the left-wing maritime unions of the West Coast. Before the federal judiciary stopped the Coast Guard from applying political tests to the nation's waterfront workers, thousands of longshoremen and sailors had lost their jobs and the small West Coast Marine Cooks and Stewards Union had been completely destroyed.[33]

The other left-led unions were similarly affected by the tightened loyalty-security procedures that accompanied the Korean War. By the mid-fifties they were under attack from so many different quarters that they could hardly perform their ordinary economic functions. Their most active members had lost their jobs. Their national leaders had been repeatedly summoned before grand juries, congressional committees, and immigration panels. Many of these leaders also faced criminal charges for contempt of Congress or else, in a few cases, for falsifying their Taft–Hartley non-communist affidavits. Though the unions often won these cases, the struggle took years and involved such an enormous drain on their resources that the organizations rarely survived. By the mid-1960s all but two of the left-led unions had disappeared or merged with their rivals.[34]

Most of the other organizations within the communist movement went the way of the left-led unions. They too faced an unrelenting barrage of attacks from all manner of sources. Their tax exemptions were withdrawn; they were placed on the attorney general's list and ordered to register with the Subversive Activities Control Board (SACB). Their leaders were hauled before congressional committees and grand juries, prosecuted, deported, and spied upon; their ordinary members were threatened with unemployment and blacklisting. The harassment occurred at every level. New York State's superintendent of insurance drove the largest front group out of existence when he ordered the liquidation of the International Workers' Order (IWO) as a "political hazard."[35]

Given the onslaught against these organizations as well as the fact that simply belonging to them could cost someone a job, it was no wonder that most of them had either folded by the mid-fifties or else had transformed themselves into defense groups that serviced the embattled remnants of a once vibrant movement. Even more than with the unions, the Korean War accelerated the front groups' demise. In the anti-communist frenzy that gripped the nation's politicians in the summer of 1950, Congress overrode a presidential veto to pass the so-called McCarran Act. This measure, whose passage the administration had previously been able to prevent, sought to put the front organizations out of business by forcing them to register with the Subversive Activities Control Board. And it succeeded.[36]

Though the Supreme Court ultimately declared the McCarran Act's registration provisions unconstitutional, the SACB's attempt to implement them finished off most of the front groups. In most cases, these already debilitated organizations had neither the human nor the financial resources to defend themselves against a registration order and they usually went out of business instead.[37] The destruction of these groups ended whatever influence communism had within the United States. The party survived, but as a tiny, deracinated fringe group. The broader communist movement, with its coherent central vision of social and political change, had ceased to exist. In the rest of this chapter, I want to look at the main consequences of that loss.

The Impact: Lost Voices, Lost Opportunities

If nothing else, McCarthyism repressed freedom of speech. Accordingly, traditional assessments of the period tend to focus mainly on its impact on civil liberties. They describe the chilling effects of the anti-communist crusade, what Justice William O. Douglas called "The Black Silence of Fear," and the ways in which it inhibited people from speaking out on controversial issues or joining organizations that promoted unpopular causes.[38] Less dramatic and, therefore, less obvious at the time were the structural consequences of McCarthyism, the ways in which it narrowed political options and shaped the long-term contours of American political discourse.[39]

As we assess this aspect of the anti-communist crusade, we encounter a world of "might-have-beens." There is a tendency among some scholars to view the immediate post-war period, the years between 1945 and 1948, as a political watershed, "that brief hopeful note," a moment of lost opportunity when it might have been possible to build upon the reforms of the New Deal, reconstruct the American government along social democratic lines, find an equitable solution to the nation's racial problems, and reach some kind of an accommodation with the Soviet Union.[40] Instead, in large part because of McCarthyism, that promising moment vanished. This notion, that the United States was on the verge of a revivified New Deal in the late 1940s and could have avoided the cold war, underestimates the power of American conservatism, but it does, nonetheless, focus our attention on a real turning point. Whether or not the anti-communist crusade aborted a promising reform movement or merely intensified an already ongoing drift to the right, it did destroy the most important left-wing movement of the period and it diverted more moderate groups and individuals from pressing for social and political change.[41]

Destroyed as well was the mind-set that had characterized so much of the broader communist movement. Marxist with a little "m" was a set of preconceptions about how the world worked that enabled the men and women who subscribed to it to view their daily activities as part of a larger whole and to link those activities to a worldwide struggle. It was that sense of connectedness that gave the movement so much of its force. Expressed in the institutions that were destroyed, it was also expressed in a set of ideas, a kind of Popular Front sensibility that created conceptual linkages between race, class, and international affairs. McCarthyism silenced the voices that articulated these connections, as it marginalized the ideas these voices propounded. When the left rebounded in the late 1950s, it threw itself into many of the same causes as the communist movement, but it did so in a fragmented way that did not offer the same comprehensive approach to the issues of the day.

One of the most important themes to disappear was that of a kind of progressive internationalism. Because of the problematic nature of the CP's ties to the Soviet Union, there is a tendency to overlook the other elements of the communist movement's international program. It is, after all, easy to caricature this aspect of the old left's political culture, to lampoon the impulse that makes a union local adopt resolutions about the situation in China, and to portray such actions as meaningless gestures primarily indicative of Russian control. But such an interpretation, though partially correct, misses the internationalist sensibility that characterized American communism. The movement's Marxist orientation gave its adherents the intellectual tools for situating overseas events within a broader context; its international ties brought those events to their consciousness. Communists and the men and women who worked with them involved themselves in the rest of the world because Moscow told them to, but also because they realized the interconnectedness of all human affairs. And at least during the Popular Front years of the 1930s, many people gravitated into the communist movement precisely because of its stand on international issues.[42]

That internationalism continued into the post-war years. Much of it involved an often mindless pro-Soviet stance, but it also touched on issues that certainly warranted attention. Imperialism was one such issue. It was by no means certain that the main Western powers would regain their colonies after World War II and the fate of the subordinated populations was very much an open question. Communists and their allies had long championed the independence struggles of colonial peoples throughout the Third World. In Korea, Indochina, and elsewhere, communist parties often led the movements for national liberation.

Within the United States, the CP's identification with anti-imperialism had particular resonance within the African-American community. For black leaders within the party's orbit, like W.E.B. DuBois and Paul Robeson, support for Pan-Africanism was a natural outgrowth of their commitment to the struggle for racial equality. Their views caught on; by the early forties, the essentially Marxist brand of anti-colonialism that they espoused had become commonplace within the African-American community. Mainstream black leaders like the NAACP's Walter White were denouncing imperialism in essentially the same language as Robeson and DuBois.[43]

Within a few years, however, White abandoned his anti-imperialist stance and simply stopped talking about Africa. White's defection was emblematic of the changes that the hardening of the cold war had produced and of the narrowing of debate on international issues. The NAACP, like the labor movement, had cut an implicit deal with the Democratic Party. In an attempt to gain the Truman administration's support in the domestic sphere, it would back the U.S. position in the cold war. Such compromises isolated left-wing critics of American foreign policy. Rejected by their former liberal allies and harassed by the forces of McCarthyism, the organizations and individuals that had advocated an alternative version of internationalism lost their audience. Not only were their voices silenced, but the ideas that they had espoused, the interconnections that they had stressed, simply disappeared from mainstream discourse. Black Americans abandoned the Pan-African movement, and when they finally reconnected to the diaspora, they did so in cultural or nationalist terms rather than in the language of economic exploitation that people like DuBois had used.[44]

By the time of the Korean War, there were no alternatives left. Both the administration's drive to mobilize public support for its increasingly militarized view of the cold war and the GOP's partisan manipulation of issues like the so-called loss of China made the discussion of foreign policy peculiarly sensitive. Dissenting interpretations of world events, especially ones that paralleled those advanced by the Communist Party, came to be treated as signs of disloyalty. Peace was a dangerous issue. So, too, was disarmament. There was no serious debate about nuclear weapons during the early 1950s; communist-linked peace groups had raised the issue and been prosecuted for the attempt. Non-communist organizations got the message, distanced themselves from the CP, and kept a low profile.[45]

The elimination of these alternative viewpoints did not occur automatically. It was the result of a concerted campaign on the part of all the different agents of McCarthyism. The most vigorous advocates of the internationalist and anti-imperialist world view like Robeson and DuBois experienced the same massive persecution that had been unleashed against the leaders of the left-wing unions. They faced criminal prosecutions, passport denials, deportations, and congressional investigations. The organizations they led lost their tax exemptions and were put on the attorney general's list. Often, thanks to the undercover activities of the FBI, they could not even find a place to hold a public meeting.[46]

Here again, the Korean War intensified an already ongoing campaign against dissent. It allowed critics of the Truman administration to utilize McCarthyism in what was really, as Bruce Cumings and others have demonstrated, an ongoing struggle between different sectors of the American polity that played itself out in the realm of foreign affairs. In a successful move to restructure the debate over the nation's East Asian policy, right-wing nationalists took advantage of the opportunities the Korean War created to transform what had originally been a matter of foreign policy into one of loyalty. They employed the machinery of McCarthyism to disseminate their charges that traitors within the State Department had engineered the so-called loss of China to the communists before a congressional investigation.

By the time Senator Pat McCarran had finished his Internal Security Subcommittee's year-long hearings into the alleged influence of the communists in the Institute of Pacific Relations and forced the Justice Department to indict the liberal, but certainly not communist, China scholar Owen Lattimore for perjury, it was no longer possible for public officials to advocate a realistic policy with regard to East Asia. Korea was pivotal, for without the legitimization that the Korean War conveyed to the right-wing agenda, it would not have been possible for McCarran, McCarthy, and their allies to have so thoroughly anathematized previously mainstream views about the Chinese Revolution.[47]

Debates about domestic issues were also stunted. Organized feminism, for example, may well have been considerably diminished by the disappearance of the Congress of American Women, a little-known organization that had created a connection between the traditional women's movement of the earlier part of the century and the broader social and international concerns of the communist world.[48] A similar narrowing of debate occurred in other areas of social concern where the destruction of the organizations that comprised the communist world kept the policy alternatives that these groups espoused from getting a hearing. From housing to health care, projects like the IWO's incipient network of clinics and pre-paid medical plans went down the tubes with the organizations that sponsored them, ensuring that in the truncated post-war debate about a national health insurance policy, there would be no left-wing presence.[49]

Had there been such a presence, it might have come not only from an organization like the IWO, but also from the left wing of the labor movement. The destruction of the communist-influenced unions may well have been the single most important outcome of the anti-communist crusade of the 1940s and 1950s. These organizations had, after all, constituted the most dependable advocates for the extension of the welfare state. Though I do not want to overemphasize their uniqueness (other unions also subscribed to a broad program of social reforms) or their political power, the disappearance of these unions had ramifications that extended far beyond the hundreds of thousands of workers who belonged to them.

But those workers also lost. Within the economic sphere, the left-led unions had been quite effective. They had won wages and working conditions that were as good as or better than other unions had obtained. Their efficacy was widely recognized; at no point during the elaborate proceedings that led to the expulsion of the communist-led unions from the CIO did any of their critics attack them on economic grounds. By the late 1940s, however, the cumulative impact of all the raiding and repression these unions faced made it increasingly more difficult for them to win strong contracts and service their members as well as they once had done.[50]

The demise of the labor left coincided with a transformation in the character of the labor movement as a whole. Though certain segments of it continued to support broader programs of social and economic change, the movement lost its militant edge. Gone was the adversarial stance toward the nation's corporations that had characterized the CIO, for example, during its early organizing struggles of the late 1930s. Most labor unions now began to see themselves as partners in a pluralistic capitalist system that had replaced divisive issues of redistribution with an emphasis on sharing in the benefits of increased productivity. They entered the establishment and concentrated on wage hikes, pensions, and health benefits for their members. The confrontational view of industrial relations associated with the left-led unions simply disappeared from the agenda of the mainstream labor movement.[51]

So too did the language in which that position was articulated. One of the most striking transformations that accompanied the destruction of the labor left was the loss of class consciousness. Central to the communist worldview was the notion that the labor movement was the most important organized expression of the working class. The left-led unions shared this vision and when they became marginalized so too did their conception of international working-class solidarity. Even the language of class seems to have disappeared. This conclusion is a tentative one, based mainly on the impressions I got from my own exposure to the sources of the period. Class-laden terminology that was common in the mainstream media during the 1940s was gone by the 1950s. References to "working stiffs" disappeared and the word "boss" lost its negative connotations. A recent study of the topics cited in the *Readers' Guide to Periodical Literature* notes a similar linguistic shift.[52]

The shift, of course, was more than rhetorical. Just as the labor movement was losing its militancy, it was also entering the earliest stages of its present decline. The economic boom of the 1950s and 1960s concealed labor's underlying problems, its heavy reliance on the skilled trades and traditional mass production industries and its failure to expand into other areas of the economy. To a certain extent, the adherence of the labor movement to the anti-communist crusade diverted it from addressing these problems. Instead of trying to organize unorganized workers, mainstream union leaders raided the left-wing unions. They also ignored large segments of the workforce and, after an abortive attempt to organize Southern workers, abandoned their efforts to reach out beyond their Northeastern and Midwestern base.[53]

We will never know to what extent the destruction of the left-led unions contributed to the domestication and later decline of the labor movement. These unions did have a greater commitment to organizing and had, in fact, been active in just those sectors of the economy that needed unionization: low-paid women and minority service workers and white-collar and public sector employees. We should not, however, delude ourselves about the autonomy the labor movement actually enjoyed. The corporate sector had inaugurated an aggressive rollback campaign against all of organized labor at the end of World War II. Anti-communism was useful to that campaign, allowing the anti-union businessmen and their political allies who led that campaign to portray their reassertion of managerial prerogatives as a defense of national security. But all unions suffered from the corporate sector offensive and it may well be the case that, even if the cold war had not legitimized the destruction of the left-led unions, the overall position of labor within the American polity would still have declined.[54]

One can, perhaps, make a similar assessment of the impact of McCarthyism on the civil rights movement, though, again, we must be careful not to overestimate the influence of the communist movement or underestimate that of white racism. Still, a case can be made that the anti-communist crusade may have deflected the civil rights movement from making economic as well as legal and political demands. In that moment of opportunity in the late 1940s, when a left-liberal coalition still seemed possible, there was the promising kernel of a Southern black working-class movement rooted in some of the CP-led unions that seemed to offer leadership and community support for a broad-scale attack on Jim Crow.[55]

The moment passed quickly as segregationists, employers, rival unions, and anti-communist politicians moved in for the kill. Red-baiting had long been a weapon in the battle to preserve white supremacy and, since it offered a way to attach the segregationist struggle to something more legitimate than racism, it blossomed during the McCarthy era.[56] When the civil rights movement rebounded, as it slowly did in the years after *Brown v. Board of Education*, it did so under middle-class leadership and without the emphasis on economic issues that had characterized it only a few years before.[57] Again, all of these speculations need qualifications. It is possible that there had never been a moment of opportunity in American race relations. Southern racism was much stronger than southern liberalism; and there is little evidence to suggest that white workers, North or South, would ever have been open to the kind of racially integrated working-class movement that the communist movement promoted. Still, the option had not been offered.[58]

In almost every area of American life, as we examine the impact of the McCarthy years, we confront lost opportunities, missing options, unheard voices. It is hard to evaluate something that never happened or put a price tag on something that never went on sale. The speculative nature of the project makes estimates of the impact of McCarthyism so unsatisfying. And yet, we *know* the anti-communist crusade affected more than the people who were directly caught up in it. There was a cultural and political shift. Class consciousness and the emphasis on collective action that had been so important to the worldview of American communism did disappear from the nation's discourse. Hollywood stopped making films about social issues; the infant television industry never even began. Academicians embraced consensus history and celebrated the end of ideology. Organized religion abandoned its earlier concern with peace and social issues.[59]

In short, American culture and politics experienced a rollback. Movements and ideas that had once been widely acceptable were now beyond the pale. Though communists and their allies were the direct victims, mainstream liberals within the Truman administration and elsewhere were the indirect ones. Adhering to the anticommunist crusade did not protect them from being attacked for "losing" China or supporting desegregation in the South. Because McCarthyism had destroyed the left, when liberals came under attack they had to defend themselves from a more politically exposed position than they would otherwise have occupied. This may seem obvious, but it is a point that needs to be stressed. The disappearance of the communist movement weakened liberalism and strengthened the right. Liberals had less political space within which to maneuver. The defense of the social and economic status quo to which they were committed had become a far more tenuous operation once American politics shifted to the right and they found themselves on the left of the political spectrum instead of at its center. At the same time, conservative and even reactionary groups and ideas that once seemed extreme entered the mainstream, while previously acceptable ones became marginalized by being associated with communism.[60]

The Korean War was a genuine watershed in this process. During the summer of 1950 the abrupt infusion of hyperpatriotism that the war inspired allowed for a considerable racheting up of the level of political repression. Congress passed the McCarran Act; loyalty-security programs were tightened up; Joe McCarthy gained power. Perhaps the transformation of the political climate that occurred during the summer of 1950 would have taken place anyhow, though in a more gradual manner. After all, the destruction of the communist-linked left was already far along by the time the North Korean army moved south. Nonetheless, it is by no means certain that the onslaught against the Democratic administration would have been as successful as it was without the war. Liberals had long been under attack, but it was not until the Korean War seemingly vindicated the most extreme anti-communist scenarios that the defense of the New Deal and its successors became precarious.

Ultimately, the chill receded, McCarthyism disappeared, and moderation triumphed.[61] Or did it? Though the most extreme manifestations of the anti-communist furor dwindled away, there were no barriers to a reoccurrence. The willingness of the nation's elites to use repressive measures to limit political debate in the name of national security set unfortunate precedents. So, too, did the undemocratic practices that were employed. It is hard to calculate just how seriously the national polity was damaged by the unfairness and dishonesty that characterized so much of the anti-communist crusade. We do not know to this day, for example, whether the police state apparatus that J. Edgar Hoover and his allies tried to put in place during the 1950s and 1960s has been entirely dismantled. Nor can we assume that the callous disregard for constitutional limitations and illegal behavior that characterized Richard Nixon's Watergate and Ronald Reagan's Iran-Contra have disappeared with the cold war. All three men, after all, built their careers on McCarthyism, and the sleaziness that it introduced to American politics may well be its main legacy.[62]

Notes

This chapter is distilled from the larger study of McCarthyism that was subsequently published as *Many Are the Crimes: McCarthyism in America*. Though I have relied heavily on archival materials and FBI (Federal Bureau of Investigation) files in that larger work, the following notes refer mainly to secondary sources. I am grateful for the advice and comments of Marvin Gettleman, Barry Strauss, and Luise White.

1. On the unpopularity of the Korean War, see the survey data collected in John E. Mueller, *War, Presidents and Public Opinion* (New York: John Wiley, 1973).

2. On the political career of Senator McCarthy, see David M. Oshinsky, *A Conspiracy So Immense: The World of Joe McCarthy* (New York: Free Press, 1983); Thomas Reeves, *The Life and Times of Joe McCarthy* (New York: Stein and Day, 1983); and Robert Griffith, *The Politics of Fear: Joseph R. McCarthy and the Senate*, 2d ed. (Amherst: University of Massachusetts Press, 1987).

3. Michal Belknap, *Cold War Political Justice: The Smith Act, the Communist Party, and American Civil Liberties* (Westport, CT: Greenwood Press, 1977), 123–51.

4. Comments by Victor Hanson at "War and Democracy" conference, Woodrow Wilson International Center for Scholars, Washington, DC, May 31–June 2, 1995.

5. The literature on McCarthyism is large, but not particularly satisfying. For the most recent study, see Ellen Schrecker, *Many Are the Crimes: McCarthyism in America* (New York: Little, Brown, 1998).

The most influential early interpretations of McCarthyism were the essays collected in Daniel Bell, ed., *The New American Right* (New York: Criterion, 1950). They viewed the phenomenon as a kind of political aberration, the product of a strand of right-wing populism that history was on the verge of passing by. The main works of early revisionism were Richard Freeland, *The Truman Doctrine and the Origins of McCarthyism: Foreign Policy, Domestic Politics, and Internal Security, 1946–1948* (New York: Knopf, 1971); Athan G. Theoharis, *Seeds of Repression: Harry S. Truman and the Origins of McCarthyism* (Chicago: Quadrangle, 1971); and Michael Paul Rogin, *The Intellectuals and McCarthy: The Radical Specter* (Cambridge: MIT Press, 1967). Rogin's later work is in *Ronald Reagan: The Movie and Other Episodes in Political Demonology* (Berkeley: University of California Press, 1987). The most useful treatments of the FBI can be found in Theoharis and John Stuart Cox, *The Boss: J. Edgar Hoover and the Great American Inquisition* (Philadelphia: Temple University Press, 1988); Richard Gid Powers, *Secrecy and Power: The Life of J. Edgar Hoover* (New York: Free Press, 1987); and Kenneth O'Reilly, *Hoover and the Un-Americans: The FBI, HUAC, and the Red Menace* (Philadelphia: Temple University Press, 1983).

For fuller discussions of the historiography of McCarthyism, see Ellen Schrecker, *The Age of McCarthyism: A Brief History with Documents* (Boston: Bedford Books, St. Martin's Press, 1994), 253–62; and Robert Griffith, *The Politics of Fear: Joseph R. McCarthy and the Senate*, 2d ed. (Amherst: University of Massachusetts Press, 1987), ix–xxv.

6. The most thoughtful attempt to assess the quantitative impact of McCarthyism is in Ralph S. Brown, Jr., *Loyalty and Security: Employment Tests in the United States* (New Haven: Yale University Press, 1958), 164–82, 487–88. He comes up with a figure of 11,500, based largely on official reports, but admits that this may be a considerable understatement,

since it does not, and cannot, factor in the hidden dismissals of people who were fired under other pretexts or resigned under pressure.

7. There is surprisingly little scholarship on the far right and little of it that goes beyond the 1930s. The most thoughtful work by far is Leo Ribuffo, *The Old Christian Right: The Protestant Far Right from the Great Depression to the Cold War* (Philadelphia: Temple University Press, 1983). See also William Pencak, *For God and Country: The American Legion, 1919–1941* (Boston: Northeastern University Press, 1989); and Rodney G. Minott, *Peerless Patriots: Organized Veterans and the Spirit of Americanism* (Washington, DC: Public Affairs Press, 1962).

There is, however, a lot of work on the left-wing anti-communists. Three good recent studies are Alexander Bloom, *Prodigal Sons: The New York Intellectuals and Their World* (New York: Oxford, 1986); Terry A. Cooney, *The Rise of the New York Intellectuals: Partisan Review and Its Circle* (Madison: University of Wisconsin Press, 1986); and Alan Wald, *The New York Intellectuals: The Rise and Decline of the Anti-Stalinist Left from the 1930s to the 1980s* (Chapel Hill: University of North Carolina Press, 1987).

8. A good introduction to liberal anti-communism is Mary Sperling McAuliffe, *Crisis on the Left: Cold War Politics and American Liberals* (Amherst: University of Massachusetts Press, 1978). See also Stephen M. Gillon, *Politics and Vision: The ADA and American Liberalism, 1947–1985* (New York: Oxford, 1987). The classic liberal anti-communist statement is Arthur M. Schlesinger, *The Vital Center: The Politics of Freedom* (Boston: 1949).

9. Eugene Lyons to J.B. Matthews, November 18, 1948, J.B. Matthews papers, Box 672, Special Collections Department, William R. Perkins Library, Duke University, Durham, NC. One professional witness even tried to organize a group called "The Federation of Former Communists, Inc." See Paul Crouch, "Certificate of Incorporation of the Federation of Former Communists, Inc.," typescript, n.d., Crouch papers, Box 2, The Hoover Institution on War, Revolution and Peace, Stanford, CA.

10. There is exhaustive evidence for the existence of this network in the personal papers of its leading members. J.B. Matthews, a former fellow traveler turned anti-communist expert, worked for HUAC (House Un-American Activities Committee) and then Hearst. He helped to operate the Hollywood blacklist, and put together a lucrative consulting business checking out people's political backgrounds for corporations and right-wing philanthropists. His papers are full of correspondence with other red-baiters, as are the papers of Alfred Kohlberg, Paul Crouch, Karl E. Mundt, Karl Baarslag, and the Senate Internal Security Subcommittee, to name a few of the collections where I have encountered evidence for the operations of this network and the interconnections between its members.

The FBI was, of course, the most important single agency involved here. Files released under the Freedom of Information Act reveal something of the scope of its operations. For a look at the work of the bureau and its all-powerful leader, see Richard Gid Powers, *Secrecy and Power: The Life of J. Edgar Hoover* (New York: Free Press, 1987); Kenneth O'Reilly, *Hoover and the Un-Americans: The FBI, HUAC, and the Red Menace* (Philadelphia: Temple University Press, 1983); Theoharis and Cox, *The Boss*; Sigmund Diamond, *Compromised Campus: The Collaboration of Universities with the Intelligence Community, 1945–1955* (New York, Oxford, 1992); and Schrecker, *Many Are the Crimes*, 203–39.

11. I have developed the discussion of this two-stage process more fully in my book about the operations of McCarthyism in the academic community, *No Ivory Tower: McCarthyism and the Universities* (New York: Oxford, 1986).

12. Ibid., 178.

13. For examples of some of the more egregious abuses of the federal government's loyalty-security system, see Adam Yarmolinsky, ed., *Case Studies in Personnel Security* (Washington, DC: The Bureau of National Affairs, 1955). The case of the guy with the former daughter-in-law is in George Brenner to Corliss Lamont, January 6, 1955, Box 1, Bill of Rights Fund papers; Rare Book and Manuscript Library, Columbia University. The

best book about the operations of the loyalty-security program is Eleanor Bontecue, *The Federal Loyalty-Security Program* (Ithaca: Cornell University Press, 1953). See also, Brown, *Loyalty and Security*, and David Caute, *The Great Fear: The Anti-Communist Purge Under Truman and Eisenhower* (New York: Simon and Schuster, 1978).

In 1953, Edward R. Murrow brought considerable attention to the case of Milo Radulovich, an Air Force reservist who was being forced to resign because of the political activities of his father and sister. Fred W. Friendly, *Due to Circumstances Beyond Our Control...* (New York: Random House, 1967), 4–17.

14. The literature on the American Communist Party is quite extensive. For a thoughtful short introduction, see Mark Naison, "Remaking America: Communists and Liberals in the Popular Front," in Michael E. Brown, Randy Martin, Frank Rosengarten, and George Snedeker, *New Studies in the Politics and Culture of U.S. Communism* (New York: Monthly Review Press, 1993), 45–73. Other useful works (and this is by no means an exhaustive list) include Theodore Draper, *The Roots of American Communism* (New York: Viking, 1957); Draper, *American Communism and Soviet Russia: The Formative Period* (New York: Viking, 1960); Harvey Klehr, *The Heyday of American Communism: The Depression Decade* (New York: Basic Books, 1984); Fraser M. Ottanelli, *The Communist Party of the United States: From the Great Depression to World War II* (New Brunswick: Rutgers University Press, 1991); Maurice Isserman, *Which Side Were You On? The American Communist Party During the Second World War* (Middletown, CT: Wesleyan University Press, 1982); and Joseph Starobin, *American Communism in Crisis, 1953–1957* (Cambridge: Harvard University Press, 1972).

15. There is a growing memoir literature by ex-communists. Some of the most useful examples are Peggy Dennis, *The Autobiography of An American Communist: A Personal View of a Political Life* (Berkeley: Lawrence Hill and Creative Arts, 1977); Dorothy Healey and Maurice Isserman, *Dorothy Healey Remembers: A Life in the American Communist Party* (New York: Oxford, 1990); Steve Nelson, James Barrett, and Rob Ruck, *Steve Nelson: American Radical* (Pittsburgh: University of Pittsburgh Press, 1981); Al Richmond, *A Long View from the Left: Memoirs of an American Revolutionary* (Boston: Houghton Mifflin, 1973); Junius Irving Scales and Richard Nickson, *Cause at Heart: A Former Communist Remembers* (Athens: University of Georgia Press, 1987).

Most of these memoirs are by party leaders, not rank-and-file members. For the experiences of the latter see Paul Lyons, *Philadelphia Communists, 1936–1956* (Philadelphia: Temple University Press, 1982) and Vivian Gornick's somewhat fictionalized *The Romance of American Communism* (New York: Basic Books, 1977). Oral history collections at depositories like New York University's Tamiment Library, Columbia University's Oral History Project, and the Bancroft Library at Berkeley contain useful tapes and transcripts of interviews with former communists.

16. For a general overview of communist activities in the labor movement, see Harvey A. Levenstein, *Communism, Anticommunism, and the CIO* (Westport, CT: Greenwood, 1981); Bert Cochran, *Labor and Communism: The Conflict That Shaped American Unions* (Princeton: Princeton University Press); and Steve Rosswurm, ed., *The CIO's Left-Led Unions* (New Brunswick: Rutgers University Press, 1992).

There is no overall study of the front groups. Klehr's *Heyday* treats the main ones from the 1930s. See also, Robert Cohen, *When the Old Left Was Young: Student Radicals and America's First Mass Student Movement, 1929–1941* (New York: Oxford, 1993); Lowell K. Dyson, *Red Harvest: The Communist Party and American Farmers* (Lincoln: University of Nebraska Press, 1982); Gerald Horne, *Communist Front? The Civil Rights Congress, 1946–1956* (Rutherford, NJ: Fairleigh Dickinson University Press, 1988); Roger Keeran, "The International Workers Order and the Origins of the CIO," *Labor History* 30 (Summer 1989): 385–408; Robbie Lieberman, *"My Song Is My Weapon": People's Songs, American Communism and the Politics of Culture 1930–1950* (Urbana: University of Illinois Press,

1989); Thomas Joseph Edward Walker, "The International Workers Order," (Ph.D. diss., University of Chicago, 1982).

17. The memoir literature and oral histories cited in note 11 above discuss the interconnectedness of the party and the movement. They also give numerous examples of the kinds of activities ordinary members of the communist movement took part in. As for the editorials and resolutions, all the newspapers of the left-led unions contain examples, many of them collected and collated in preparation for the CIO expulsion trials. The quote is from Lyons, *Philadelphia Communists*, 126.

18. There are two outstanding books on the party's work with African Americans: Mark Naison, *Communists in Harlem During the Depression* (Urbana: University of Illinois Press, 1983); and Robin D.G. Kelley, *Hammer and Hoe: Alabama Communists During the Great Depression* (Chapel Hill: University of North Carolina Press, 1990).

19. On communist unions and minorities, see Jack Cargill, "Empire and Opposition: The 'Salt of the Earth' Strike," in Robert Kern, ed., *Labor in New Mexico: Unions, Strikes, and Social History Since 1881* (Albuquerque: University of New Mexico Press, 1983); Joshua B. Freeman, *In Transit: The Transport Workers Union in New York City, 1933–1966* (New York: Oxford, 1989), 151–56, 254–57; Mario T. Garcia, "Border Proletarians: Mexican-Americans and the International Union of Mine, Mill, and Smelter Workers, 1939–1946," in Robert Asher and Charles Stephenson, eds., *Labor Divided: Race and Ethnicity in United States Labor Struggles, 1935–1960* (Albany: State University of New York Press, 1990), 85–103; Michael K. Honey, *Southern Labor and Black Civil Rights: Organizing Memphis Workers* (Urbana: University of Illinois Press, 1993); Horace Huntley, "Iron-Ore Miners and Mine Mill in Alabama: 1933–1952" (Ph.D. diss., University of Pittsburgh, 1977); Howard Kimeldorf, *Reds or Rackets? The Making of Radical and Conservative Unions on the Waterfront* (Berkeley: University of California Press, 1988), 165–68; Robert Korstadt and Nelson Lichtenstein, "Opportunities Found and Lost: Labor, Radicals, and the Early Civil Rights Movement," *Journal of American History* 75 (1988): 786–811; August Meier and Elliott Rudwick, "Communist Unions and the Black Community: The Case of the Transport Workers Union, 1934–1944," *Labor History* 23 (Spring 1982): 165–97; Bruce Nelson, *Workers on the Waterfront: Seamen, Longshoremen, and Unionism in the 1930s* (Urbana: University of Illinois Press, 1988), 259–61; and Nancy Quam-Wickham, "Who Controls the Hiring Hall? The Struggle for Job Control in the ILWU During World War II"; Bruce Nelson, "Class and Race in the Crescent City: The ILWU, from San Francisco to New Orleans"; and Karl Korstad, "Black and White Together: Organizing in the South with the Food, Tobacco, Agricultural & Allied Workers Union (FTA-CIO), 1946–1952," all in Rosswurm, ed., *The CIO's Left-Led Unions*.

On other topics, see Marvin E. Gettleman, "The New York Workers School, 1923–1944: Communist Education in American Society," in Brown et al., eds, *New Studies in the Politics and Culture of U.S. Communism*, 266–67; and Paul Mishler, "Communism and Youth in the Country: Summer Camps and Communist Education," paper delivered at the annual meeting of the Organization of American Historians, March 1995, Washington, DC.

20. The NAACP (National Association for the Advancement of Colored People) was especially concerned about the racist implications of federal security officials asking people about their civil rights activities. See Walter White to President Truman, November 26, 1948, Papers of the NAACP, Part 13, Series C, (Bethesda: University Publications of America, n.d.), reel 5.

For examples of the kinds of questions security officials asked, see Yarmolinsky, *Case Studies in Personnel Security*, passim; and Roland Watts, *The Draftee and Internal Security* (New York: Workers Defense League, 1955).

21. In addition to the works quoted in notes 10 and 11 above, see Fernando Claudin, *The Communist Movement from Comintern to Cominform* (London: Penguin, 1975). Recently released materials from the Soviet archives may help settle the continuing debate among

the historians of American communism about the extent to which the Soviet Union determined the CPUSA's (Communist Party of the United States of America) policies. See Harvey Klehr, John Earl Haynes, and Fridrikh Igorevich Firsov, *The Secret World of American Communism* (New Haven: Yale University Press, 1995); and Harvey Klehr, John Earl Haynes, and Kyrill M. Anderson, *The Soviet World of American Communism* (New Haven: Yale University Press, 1998).

22. The papers of President Truman and his aides are full of memos from Hoover describing the CP threat. See, for example, FBI Report, "Soviet Activities in the United States," July 25, 1946, Clark M. Clifford papers, Box 15, folder (Russia—folder 5) in Harry S. Truman Library, Independence, MO.

23. It is clear that there was some spying; how much and what kind still remains an open question. The intercepted wartime KGB cables recently released as the VENONA documents do indicate that perhaps a hundred party members were involved. The most useful books on these cases are Allan Weinstein, *Perjury: The Hiss-Chambers Case,* 2d ed. (New York: Knopf, 1997); Ronald Radosh and Joyce Milton, *The Rosenberg File: A Search for the Truth* (New York: Holt, 1983); Robert Chadwell Williams, *Klaus Fuchs, Atom Spy* (Cambridge; Harvard University Press); Walter and Miriam Schneir, *Invitation to an Inquest,* rev. ed. (New York: Pantheon, 1983); Merrily Weisbord, *The Strangest Dream: Canadian Communists, the Spy Trials, and the Cold War* (Toronto: Lester and Orpen Dennys, 1983); and Gary May, *Un-American Activities: The Trials of William Remington* (New York: Oxford, 1994).

24. For a few examples of the kinds of exaggerations that were current about the dangers of communism, see *Red Channels* (New York: American Business Consultants, 1955), 5; report of the Broyles Commission, cited in E. Houston Harsha, "Illinois," in Walter Gellhorn, ed., *The States and Subversion* (Ithaca: Cornell University Press, 1952), 83–84; Bruno Stein, "Loyalty and Security Cases in Arbitration," *Industrial and Labor Relations Review* 17, no. 1 (October 1963): 107.

25. There was considerable debate within the party about its secrecy. See, for example, Starobin, *American Communism in Crisis,* 97–98; Schrecker, *Many Are the Crimes,* 23–26; Schrecker, *No Ivory Tower,* 39–43.

26. For a survey of the overall impact of McCarthyism on the left-wing labor unions, see Ellen W. Schrecker, "McCarthyism and the Labor Movement: The Role of the State," in Rosswurm, ed., *The CIO's Left-Led Unions,* 139–57. A useful anthology of short pieces about the left-led unions during the McCarthy years that contains information about most of the left-wing unions is Ann Fagin Ginger and David Christiano, eds., *The Cold War Against Labor* (Berkeley: Meiklejohn Civil Liberties Institute, 1987). Besides the books already cited, see Charles P. Larrowe, *Harry Bridges* (New York: Lawrence Hill, 1972); Louis Goldblatt, oral history, Bancroft Library, University of California (I saw a copy at the Columbia Oral History Project).

27. There is very little scholarly literature on the use of immigration procedures during the McCarthy period. Milton R. Konvitz, *Civil Rights in Immigration* (Ithaca: Cornell University Press, 1953); Ellen Schrecker, "Immigration and Internal Security: Political Deportations During the McCarthy Era," *Science and Society* 60, no. 4 (Winter 1996–1997). Louise Pettibone Smith, *Torch of Liberty* (New York: Dwight-King, 1959) is an in-house history of the American Committee for Protection of Foreign Born, the party-sponsored immigration defense group.

28. Though HUAC was actively investigating the left-wing unions, the most important early set of hearings were by the House Education and Labor Subcommittee in the early part of 1947: "Hearings, Amendments to the National Labor Relations Act," House of Representatives, Committee on Education and Labor, 80th Cong., 1st sess., February–March 1947.

29. Both Cochran, *Labor and Communism,* and Levenstein, *Communism, Anticommunism and the CIO,* give solid accounts of the various campaigns against the labor left. See

also Ronald L. Filippelli and Mark McColloch, *Cold War in the Working Class: The Rise and Decline of the United Electrical Workers* (Albany: State University of New York Press, 1995); Vernon H. Jensen, *Nonferrous Metals Industry Unionism, 1932–1954* (Ithaca: Cornell University Press, 1954); and Ronald W. Schatz, *The Electrical Workers: A History of Labor and General Electric and Westinghouse, 1923–1960* (Urbana: University of Illinois Press, 1983). On the Catholic opposition, see Joshua B. Freeman and Steve Rosswurm, "The Education of an Anti-Communist: Father John Cronin and the Baltimore Labor Movement," *Labor History* 33, no. 2 (Spring 1992): 217–47; Rosswurm, "The Catholic Church and the Left-Led Unions: Labor Priests, Labor Schools, and the ACTU," in Rosswurm, ed., *The CIO's Left-Led Unions*, 119–38; and Douglas P. Seaton, *Catholics and Radicals: The Association of Catholic Trade Unionists and the American Labor Movement from Depression to Cold War* (Lewisburg: Bucknell University Press, 1981).

30. Both Cochran, *Labor and Communism,* and Levenstein, *Communism, Anticommunism, and the CIO,* discuss the impact of the Taft–Hartley Act on the left-led unions. The best single survey of the Act is Harry A. Millis and Emily Clark Brown, *From the Wagner Act to Taft-Hartley* (Chicago: University of Chicago Press, 1950). See also James A. Gross, *The Reshaping of the National Labor Relations Board: National Labor Policy in Transition* (Albany: State University of New York Press, 1981); and George Lipsitz, *Class and Culture in Cold War America* (South Hadley, MA: J.F. Bergin, 1982).

31. A good analysis of the CIO's alliance with the Democratic Party can be found in David Brody, *Workers in Industrial America,* 2d ed. (New York: Oxford, 1993), 199–241. In addition to the treatment of the expulsions in Cochran, *Labor and Communism,* and Levenstein, *Communism, Anticommunism, and the CIO,* see McAuliffe, *Crisis on the Left,* passim.

32. On the federal loyalty-security program, see Bontecue, *The Federal Loyalty-Security Program*; Brown, *Loyalty and Security*; and Carl Bernstein, *Loyalties: A Son's Memoir* (New York: Simon and Schuster, 1989). In 1951, the FBI inaugurated a secret "Responsibilities Program" to give public officials information about the alleged subversives in their employ. Sigmund Diamond, *Compromised Campus: The Collaboration of the Universities with the Intelligence Community, 1945–1955* (New York: Oxford, 1992). Even in Hollywood, the main casualties of the congressional investigations and subsequent blacklists were people who had been active in the film industry's unions. See Larry Ceplair and Steven Englund, *The Inquisition in Hollywood: Politics in the Film Community, 1930–1960* (Garden City, NY: Anchor/Doubleday, 1980).

33. There is almost no scholarship about the Port Security program. The most important study is a 1953 law review article that, for obvious reasons, only covers the first few years of the program. Ralph S. Brown, Jr., and John D. Fassett, "Security Tests for Maritime Workers: Due Process Under the Port Security Program," *Yale Law Journal* 62, no. 8 (July 1953): 1163–208. See also Caute, *The Great Fear,* 392–400. The papers of the labor law firm, Gladstein, Leonard, Patsey, and Anderson, in the Manuscript Collections of the Bancroft Library, University of California, Berkeley, contain voluminous material on the Marine Cooks and Stewards' fight against the port security program.

34. Among the decisions that overruled the criminal convictions of labor leaders were: *Emspak v. U.S.,* 349 U.S. 190 (1955); *Quinn v. U.S.,* 349 U.S. 155 (1955); *Jencks v. U.S.,* 353 U.S. 657 (1957). The Court of Appeals for the Ninth Circuit ruled that the Coast Guard's proceedings were a violation of due process. *Parker v. Lester,* 227 F, 2d 708 (9th Cir. 1955).

For a survey of the fate of the left-led unions, see F.S. O'Brien, "The 'Communist-dominated' Unions in the United States Since 1950" *Labor History* 9, no. 2 (Spring 1968): 184–205.

35. Bontecou, *The Federal Loyalty-Security Program,* 168–202. Horne, *Communist Front?* and Arthur J. Sabin, *Red Scare in Court: New York Versus the International Workers Order* (Philadelphia: University of Pennsylvania Press, 1993), describe the impact of McCarthyism on two important front groups.

36. William R. Tanner and Robert Griffith, "Legislative Politics and 'McCarthyism': The Internal Security Act of 1950," in Robert Griffith and Athan Theoharis, eds., *The Specter* (New York: New Viewpoints, 1974), 176–89.

37. There has not been very much scholarship on the Subversive Activities Control Board. For a preliminary treatment of its work, see Ellen W. Schrecker, "Introduction," to *Records of the Subversive Activities Control Board, 1950–1972* (Frederick, MD: University Publications of America, 1989), v–xvi.

38. Justice Douglas's classic statement, "The Black Silence of Fear," appeared originally in the *New York Times Magazine*, January 13, 1952. For other important contemporary statements about McCarthyism's damage to civil liberties, see Alan Barth, *The Loyalty of Free Men* (New York: Viking, 1951); Francis Biddle, *The Fear of Freedom* (Garden City, NY: Doubleday, 1951); and Elmer Davis, *But We Were Born Free* (Indianapolis: Bobbs Merrill, 1954).

39. A useful and balanced assessment of the overall impact of McCarthyism is in Robert J. Goldstein, *Political Repression in Modern America* (Cambridge: Schenkman, 1978), 548–59; see also Schrecker, *Many Are the Crimes*, 359–415.

40. See, for example, Nelson Lichtenstein, "From Corporatism to Collective Bargaining: Organized Labor and the Eclipse of Social Democracy in the Postwar Era," and Ira Katznelson, "Was the Great Society a Lost Opportunity?" in Steve Fraser and Gary Gerstle, eds., *The Rise and Fall of the New Deal Order, 1930–1980* (Princeton: Princeton University Press, 1989).

Studs Terkel, a participant observer if ever there was one, notes the sense of anticipation that followed the immediate end of the war. Terkel, "The End and the Beginning," *The Nation* (May 15, 1995), 670–72. For a discussion of the possibilities for peace and nuclear disarmament, see Lawrence S. Wittner, *The Struggle Against the Bomb: Vol. I, One World or None: A History of the Nuclear Disarmament Movement Through 1953* (Stanford: Stanford University Press, 1993).

41. On the influence of the right, see James T. Patterson, *Congressional Conservatism and the New Deal: The Growth of the Conservative Coalition in Congress, 1933–1939* (Lexington: University of Kentucky Press, 1967); and Elizabeth Fones-Wolf, *Selling Free Enterprise: The Business Assault on Labor and Liberalism, 1945–1960* (Chicago: University of Illinois Press, 1994).

In his recent study of New Deal liberalism during the late 1930s and 1940s, Alan Brinkley makes a strong case that liberalism changed and became much less concerned with economic redistribution matters during this period. He does not indicate whether and to what extent the New Deal liberals he studied had any contact with groups or ideas to their left. See Brinkley, *The End of Reform: New Deal Liberalism in Recession and War* (New York: Knopf, 1995).

42. Naison, "Remaking America," 55–58.

43. Much of this discussion of the rise and decline of pan-Africanism within the American black community comes from Penny M. Von Eschen, *Race Against Empire: Black Americans and Anticolonialism, 1937–1957* (Ithaca: Cornell University Press, 1996); and Brenda Gayle Plummer, *Rising Wind: Black Americans and U.S. Foreign Affairs, 1935–1960* (Chapel Hill: University of North Carolina Press, 1996). See also Gerald Horne, *Red and Black: W.E.B. DuBois and the Afro-American Response to the Cold War, 1944–1963* (Albany: State University Press of New York, 1986); and Martin Bauml Duberman, *Paul Robeson* (New York: Knopf, 1988).

44. Von Eschen, *Race Against Empire*, passim.

45. In 1950 the Justice Department tried to prosecute W.E.B. DuBois and the other leaders of the Peace Information Center for refusing to register their organization as the agent of a foreign power. The case was so far-fetched that it was thrown out of court. W.E.B. DuBois, *In Battle for Peace* (Millwood, NY: Kraus-Thomson Organization, Ltd.,

1976, reprint of 1951 ed.); Horne, *Black and Red*, 151–82. For the impact of McCarthyism on the non-communist peace movement, see Wittner, *One World or None*, 319–23.

46. Duberman, *Paul Robeson*, is particularly chilling on the repression one international-ist spokesman experienced. See also Horne, *Red and Black*, passim. There was also con-siderable repression within the Chinese-American community of the main supporters of the People's Republic of China; see Renqiu Yu, *To Save China, to Save Ourselves: The Chinese Hand Laundry Alliance of New York* (Philadelphia: Temple University Press, 1992), 179–91.

47. Bruce Cumings, *The Origins of the Korean War: Volume II, The Roaring of the Cataract, 1947–1950* (Princeton: Princeton University Press, 1990), 3–32; Benjamin Fordham, "Building the Cold War Consensus: The Political Economy of National Security Policy, 1949–1951" (Ph.D. diss., University of North Carolina, Chapel Hill, 1994).
There is a large literature on the impact of McCarthyism on America's China experts. See, for example, Robert P. Newman, *Owen Lattimore and the "Loss" of China* (Berkeley: University of California Press, 1992); Harvey Klehr and Ronald Radosh, *The Amerasia Spy Case: Prelude to McCarthyism* (Chapel Hill: University of North Carolina Press, 1996); Stanley Kutler, *The American Inquisition: Justice and Injustice in the Cold War* (New York: Hill and Wang, 1982), 181–214; Gary May, *China Scapegoat: The Diplomatic Ordeal of John Carter Vincent* (Washington, DC: New Republic Books, 1979); John N. Thomas, *The Institute of Pacific Relations* (Seattle: University of Washington Press, 1974); E.J. Kahn, Jr., *The China Hands* (New York: Viking, 1975); Ross Y. Koen, *The China Lobby in Ameri-can Politics* (New York: Macmillan, 1960).

48. Amy Swerdlow, "The Congress of American Women," in Linda K. Kerber, Alice Kessler-Harris, and Katherine Kish Sklar, eds., *U.S. History as Women's History: New Femi-nist Essays* (Chapel Hill: University of North Carolina Press, 1995).

49. Because there has been so little attention paid to the work of the CP's front groups, it is hard to find much scholarship on their activities. For some examples of the movement's activities in areas like housing and health care, see Henry Kraus, "In the City Was a Gar-den," and Stephen Fritchman, "A Community Medical Center for Workers," in Ginger and Christiano, *The Cold War Against Labor*; Walter J. Lear, "A Sampler of the Past," *Health and Medicine* 4, no. 1 (Spring 1987): 4; Fitzhugh Mullan, *Plagues and Politics: The Story of the United States Public Health Service* (New York: Basic Books, 1989), 131; Beryl Gilman, interview with the author, October 6, 1993.

50. There is a growing body of scholarship on the economic practices of the left-led unions. See, for example, Mark McCulloch, "The Shop-Floor Dimension of Union Ri-valry: The Case of Westinghouse in the 1950s," in Rosswurm, ed., *The CIO's Left-Led Unions*, 183–200; Judith Stepan-Norris and Maurice Zeitlin, "'Red' Unions and 'Bour-geois' Contracts? The Effects of Political Leadership on the 'Political Regime of Produc-tion,'" *American Journal of Sociology* 96, no. 5 (March 1991): 1151–200.

51. Brody, *Workers in Industrial America*, 155–95; Steven Rosswurm, "Introduction: An Overview and Preliminary Assessment of the CIO's Expelled Unions," in Rosswurm, ed., *The CIO's Left-Led Unions*.

52. Margo Anderson, "The Language of Class in Twentieth-Century America," *Social Science History* 12 (December 1988): 349–75.

53. Rosswurm, "Introduction: An Overview and Preliminary Assessment of the CIO's Expelled Unions," 1–17; Barbara Griffith, *The Crisis of American Labor: Operation Dixie and the Defeat of the CIO* (Philadelphia: Temple University Press, 1988); Lichtenstein, "From Corporatism to Collective Bargaining"; Honey, *Southern Labor and Black Civil Rights*, 215–77; Michael Goldfield, *The Decline of Organized Labor in the United States* (Chicago: University of Chicago Press, 1987).

54. For insight into the business community's attempt to roll back union gains, see Howell J. Harris, *The Right to Manage: Industrial Relations Policies of American Business*

in the 1940s (Madison: University of Wisconsin Press, 1982); Kim McQuaid, *Big Business and Presidential Power from FDR to Reagan* (New York: Morrow, 1982), 122–49; and Fones-Wolf, *Selling Free Enterprise*, passim.

55. Korstadt and Lichtenstein, "Opportunities Found and Lost," which describes the civil rights activities of a communist-led tobacco workers' local in North Carolina, offers the most detailed case study of such a working-class civil rights movement destroyed by anti-communism. See also Honey, *Southern Labor and Civil Rights*, passim. On the potential for a left-liberal political realignment in the South, see Patricia Sullivan, *Days of Hope: Race and Democracy in the New Deal Era* (Chapel Hill: University of North Carolina Press, 1996).

56. For an introduction to some of the ways in which Southern white supremacists enlisted anti-communism in their struggle against integration, see, for example, John Egerton, *Speak Now Against the Day: The Generation Before the Civil Rights Movement in the South* (New York: Knopf, 1995); Adam Fairclough, *Race and Democracy: The Civil Rights Struggle in Louisiana, 1915–1972* (Athens: University of Georgia Press, 1995), 135–47; John M. Glen, *Highlander, No Ordinary School, 1932–1962* (Lexington: University Press of Kentucky, 1988); Anne Braden, *The Wall Between* (New York: Monthly Review Press, 1958); Numan V. Bartley, *The Rise of Massive Resistance: Race and Politics in the South During the 1950's* (Baton Rouge: Louisiana State University Press, 1969).

57. Aldon D. Morris, *The Origins of the Civil Rights Movement: Black Communities Organizing for Change* (New York: Free Press, 1984).

58. Robert J. Norrell, "Caste in Steel: Jim Crow Careers in Birmingham, Alabama," *The Journal of American History* 73 (December 1986): 669–94; Rick Halpern, "Interracial Unionism in the Southwest: Fort Worth's Packinghouse Workers, 1937–1954," 158–82, and Norrell, "Labor Trouble: George Wallace and Union Politics in Alabama," 250–72, in Robert H. Zieger, ed., *Organized Labor in the Twentieth-Century South* (Knoxville: University of Tennessee Press, 1991).

59. John Cogley, *Report on Blacklisting*, 2 vol. (New York: Fund for the Republic, 1956); Erik Barnouw, *Tube of Plenty: The Evolution of American Television* (New York: Oxford University Press, 1975); J. Fred MacDonald, *Television and the Red Menace: The Video Road to Vietnam* (New York: Praeger, 1985); Ceplair and Englund, *The Inquisition in Hollywood*, 299–324; Fones-Wolf, *Selling Free Enterprise*, 219–54; Peter Novick, *That Noble Dream: The "Objectivity Question" and the American Historical Profession* (Cambridge: Cambridge University Press, 1988).

60. Bruce Cumings has a good discussion of how this process operated with regard to foreign policy. His argument can be extended to other areas as well. Cumings, *The Origins of the Korean War: Volume II*, 79–121. For an overview of the impact of McCarthyism on mainstream liberals, see Gillon, *Politics and Vision*, 57–103.

61. There is a tendency to celebrate the decline of Joseph McCarthy's power and the legal victories of the late 1950s as an indication that McCarthyism was a trivial and fleeting episode in American history. One of the best statements of this triumph-of-the-law thesis can be found in Kutler, *American Inquisition*, passim.

62. For information about the expansion of the FBI's illegal activities, see William W. Keller, Jr., *J. Edgar Hoover and the Liberals: Rise and Fall of a Domestic Intelligence State* (Princeton: Princeton University Press, 1989); Athan Theoharis, *Spying on American: Political Surveillance from Hoover to the Huston Plan* (Philadelphia: Temple University Press, 1978); and Frank Donner, *The Age of Surveillance: The Aims and Methods of America's Political Intelligence System* (New York: Knopf, 1981).

11

Korea, the Cold War, and American Democracy

Stephen J. Whitfield

What Walt Whitman predicted would befall "the seething hell and the black infernal background" of the Civil War is even more applicable to Korea: "The real war will never get in the books."[1] In the early 1950s the real war on a distant and obscure peninsula spawned only a tiny literature, inspired perhaps only a dozen minor movies, and thereafter generated little imaginative literature or memoirs of combat. No novelist of stature drew upon his experiences fighting in Korea, where Dwight D. Eisenhower promised to go if elected in 1952. But his pledge was symbolic and non-committal, and the glow of his reputation was emitted from the crusade in Europe (and on the prudence of avoiding military involvement in Indochina in 1954) far more than from the truce in Panmunjom. Usually treated as an interlude, the Korean War has tended to remain a remote and unmemorialized conflict. It is remembered—if at all—because it is slotted between the Second World War and the Vietnam War.

Perhaps it is hard to get a fix on a conflict that ended in a draw, like the War of 1812, rather than as a clear-cut victory or defeat. Nor can the Korean War be associated with an ideological justification or slogan; there was no equivalent or echo of the Gettysburg Address, the Fourteen Points, or the Four Freedoms. There was no Alamo to remember, no *Maine* to avenge. Even a massive recent anthology like *Cultures of United States Imperialism* (1994), edited by Amy Kaplan and Donald E. Pease, includes four essays on Operation Desert Storm, but fails even to mention the Korean War. In any event the historical niche is oblique. For example, *Catch-22* (1961), according to its author, "wasn't really *about* World War Two. It was about American society during the cold war, during the Korean War, and about the possibility of a Vietnam," a debacle which continues to loom over and dwarf the trauma of the Korean War; few readers would share Joseph Heller's own sense of his true subject. Though a film like *M*A*S*H* (1970), opens with two quotations from the Korean War (General Douglas MacArthur's Farewell Address to the U.S. Congress and Ike's promise to go to Korea), the film might not exist had it not been for the ordeal of Vietnam. Director Robert Altman denied that his comedy drew upon war as anything other than a backdrop: "We hear the firing, but the only gun we actually see in the entire picture is that used by the timekeeper at the football game to mark the end of each half."[2] The real war in Korea did not get into the movies either.

Or consider the only novel written by one particular veteran of the D-Day landings at Normandy: *The Catcher in the Rye* (1951). An early story about Holden Caulfield has him reported missing in action in the Pacific theatre of the Second World War. By 1951, however, J.D. Salinger makes Holden just about young enough to be drafted for Korea. Better the firing squad, because "it'd drive me crazy if I had to be in the Army and be with a bunch of guys like Ackley and Stradlater and old Maurice all the time, marching with them and all. . . . I'm sort of glad they got the atomic bomb invented. If there's ever another war, I'm going to sit right the hell on top of it. I'll volunteer for it, I swear to God I will." (Such a posture would be akin to the climax of *Dr. Strangelove* [1964], in which Major Kong descends toward apocalypse astride a missile.) Holden's older brother D.B. had served in the Second World War, though not in combat, and had "hated the war" so much that he gave Holden a copy of *A Farewell to Arms*. D.B. once remarked to Holden that, if given a rifle, "he wouldn't've known which direction to shoot in. He said the Army was practically as full of bastards as the Nazis were"; this anti-authoritarian near-nihilism was a stance that Heller's novel would amplify a decade later.[3] Korea is as far away and dimly understood in Peter Bogdanovich's film, *The Last Picture Show* (1971), which hints at how the war offers an escape from the dreary isolation of Amarene, Texas, where the frontier as well as the only movie house in town have closed. For Duane Jackson (Jeff Bridges) to ship off to Korea is merely the next inexorable stage of manhood, a form of maturity to be attained naturally—that is, thoughtlessly.

Though about 37,000 Americans died in the conflict (as did over a million Koreans),[4] this unwritten war had repercussions, which can be traced, however tentatively, within the larger geopolitical conflict as well as its domestic effects. The demonstration of superpower resolve coincided with a weakening of democratic values at home. This chapter is intended to explore how the era of the Korean War poses historiographical challenges that are more formidable than merely rectifying its omission from the narrative of the cold war. For Americanists who study the 1950s must figure out how to join what is usually left asunder. The prosperity of the era induced a complacency that made dissidence hardly welcome; political and social criticism was like a stink bomb thrown into a suburban barbecue party. Nay-sayers seemed vaguely unpatriotic; in the civic culture of an era of boosterism, the ayes always had it. Anti-communist hysteria put liberalism on the defensive and sanctioned an intolerance of radicalism, which could be stigmatized as merely fortifying the propaganda of the Soviet Union and its sympathizers. The effect was to reduce the political options, to stifle intellectual independence, and to promote stability at the expense of change, order at the expense of reform, and authority at the expense of freedom.

Yet consider a paradox that this chapter proposes to sharpen and illuminate, a paradox that a new phase of historiographical synthesis might resolve. Political historians who look at the 1950s tend to find the muffling of conflict in the name of consensus: the cold war freezes the two major parties in place during the Ike Age. Political historians tend to highlight the suppression of dissent, the enfeeblement of the Bill of Rights, and the short-circuiting of struggles for racial and sexual equality and for the rights of labor and other progressive causes. The emotion that is commonly depicted is fear. But social historians do not look through the 1950s darkly. They find not paranoia but pride, not anxiety but an affluence-bred assurance, not the mindless zealotry that McCarthyism fostered but a mindless privatism associated with withdrawal from public concerns. Such historians discern what the Athenians labeled "idiocy," and limn an innocence that is barely distinguishable from ignorance and smugness, a radiant faith (however exaggerated or misplaced) in "the best years of our lives." This is the version of the 1950s that induces sentimentality, a slumber party to which the likes of Roy Cohn and Karl Mundt and J. Edgar Hoover and John Foster Dulles have pointedly not been invited. One embodiment of this paradox was Ronald Reagan, who ended the decade speaking for Democrats for Nixon. Reagan bounced through the period jauntily promoting General Electric whiggishness ("progress is our most important product"), and exuded sunny-side-up optimism. Yet he also felt so beleaguered, so concerned that communists in Los Angeles were out to get him, that he carried a gun.[5] When the president of the Screen Actors' Guild packs heat, vigilance has become manic.

The efforts of both political and social historians of cold war America have not been persuasively synthesized, nor have their perspectives been seen in tandem. The only significant—and invaluable—exception is Michael S. Sherry's *In the Shadow of War* (1995). But otherwise the political historian generally associates the 1950s with terror and repression, in the shadow of the dreadful either/or of being either Red or dead, or at best choosing between a cold war and Armageddon. The social historian associates the 1950s with the soporific comfort of "happy days," when men were (still) men and women were housewives, and the conflict between them seemed muted. In that era, even the rebels were without causes. They had only a smoldering, barely articulated non-conformity that elevated James Dean, Marlon Brando, and Elvis Presley into youthful apostles of grudging submission and spasmodic resistance who (along with Marilyn Monroe) have not been superseded as international icons of American culture. The political historian on the left looks back with a shudder, on the right with bemusement; the social historian looks back in anger (if a feminist) or with nostalgia (if a centrist). Neither sort of historian has much interest in the Korean War itself. Both, however, should consider efforts at synthesis, so that separate visions might become bifocal.

What, for instance, does the Korean War have to do with the texture of social life in the era that political historians refer to as McCarthyism? What if anything did that war have to do with the career of the guttersnipe whose name summarized a democratic crisis? Yet too little has ever been made of periodicity: his influence was almost exactly synchronous with the duration of the bloody conflict itself. McCarthy began to show up on page one only four months before North Korean troops crossed the 38th parallel. Roughly a year after the armistice accords were concluded, he could be found next to the shipping news. Perhaps that was coincidence, a caprice of Clio. For example, the social scientists who contributed to Daniel Bell's *The New American Right* (1955) did not highlight the immediate stresses of the cold war in accounting for the impact of McCarthy. Only one contributor, Talcott Parsons, focused on foreign policy, and only in a 1962 postscript even mentioned the Korean War as precipitating McCarthyism, as "a 'last straw' [which] frustrated the expectations of relaxation that many Americans held after the end of the big war." The only historian among the contributors, Richard Hofstadter, acknowledged the backdrop of the Korean War only later.[6] Even Richard H. Rovere's *Senator Joe McCarthy* (1959), the best interpretation of McCarthy's personality and modus operandi, mentions the war only once, in passing.

But historians can now speculate that the Korean War both amplified McCarthy's charges and yet limited his appeal, made domestic anti-communism plausible and yet revealed how unserious he was. In the spring of 1950, the Senate commissioned a panel headed by Millard Tydings (D-Md.) to investigate McCarthy's first major round of charges—that communism had knowingly been permitted to fester in the Department of State. Yet the Tydings Committee failed to crush McCarthy at the outset because its business was finished soon after the Korean War had begun. The patriotism that it stirred made the Bill of Rights look like legal niceties, and in such an atmosphere McCarthy could flourish. In the fall the Internal Security Act was passed (and President Truman's courageous veto easily overridden), though McCarthy himself was indifferent to the mechanics of legislation even as communists would be systematically stripped of their rights. The stalemate in Korea had undoubtedly made forbearance and tolerance more difficult to achieve at home, since no unalloyed victory could be promised; and the unconditional surrender of the enemy on V-J Day was a memory that was still fresh. In the early 1950s, a frustrating war juiced up the paranoia that McCarthyism promoted. No wonder then that columnist George Sokolsky could inquire: "If our Far Eastern policy was not betrayed, why are we fighting in Korea?" Senator William E. Jenner (R-Ind.) asked: "How can we get the Reds out of Korea if we cannot get them out of Washington?"[7]

Yet the war could also serve as a rationale to oppose the tactics of deception and fraudulence that were McCarthy's specialty. According to the majority report of the Tydings Committee, "the totalitarian technique of the 'big lie' employed on a sustained basis" was splitting the country when unity was most needed. While G.I.s were dying in Asia, McCarthy's demagoguery had "confuse[d] and divide[d] the American people . . . to a degree far beyond the hopes of the Communists themselves." Liberal columnist Elmer Davis tried to put McCarthy's conspiracy-mongering in perspective: his "campaign against imaginary communists looks sillier than ever when the very people he has attacked most bitterly are trying to stop the real Communists." The *Nation* grasped the point too: "McCarthyism will have a hollow sound when applied to the government that stood up to the Russians."[8] By standing up to Kim Il Sung—the putative pawn of the Kremlin—the White House could not be discredited as being soft on communism.

McCarthyism nevertheless imposed restraints on the exercise of statecraft. Two administrations had to demur, without quite saying so, from MacArthur's aversion to any substitute for victory. Two administrations had to acknowledge, however tacitly, the limitations on American arms and diplomacy. To be sure, the Soviet Union lacked the capacity to hit the United States directly with nuclear weapons (though Manhattan's Sherry-Netherland Hotel took the precaution of placing Geiger counters in its fanciest suites).[9] But the Kremlin could credibly threaten Western Europe, and NATO allies were therefore adamant in opposing a wider war that risked atomic devastation. By cracking the Anglo-American nuclear monopoly, the Soviet Union had ensured a balance of terror that undercut the yearning for total victory that many Americans took to be almost their birthright. Limited war, Senator Robert A. Taft (R-Ohio) remarked, resembled "a football game in which our team, when it reaches the 50-yard line, is always instructed to kick. Our team can never score." McCarthy himself blamed Secretary of State Dean Acheson for a Korean strategy that was "like advising a man whose . . . family is being killed . . . not to take hasty action for fear he might alienate the affection of the murderers." Such slashing partisanship made it impossible for a Democratic president to sign the truce accords that his successor accepted. "I would have been crucified for that armistice," Truman surmised. Indeed the point was not lost on the more primitive Republicans. With Korea still divided roughly as it had been in 1950, with Eisenhower leaving aggression unpunished and settling a deal that Truman could have cut too, the cease-fire was, Jenner fumed, "the last tribute to appeasement."[10] But in offering no alternative but Armageddon, McCarthyism betrayed its own political weakness. The junior senator from Wisconsin exhibited a talent for mendacity, mischief, and bamboozlement—but not for thoughtful criticism. Lacking even a sincere abhorrence of Stalinism, he was one of the biggest frauds since Piltdown Man.

The complicated question of McCarthy's constituency and appeal lingers, and cannot be confined to the era of the Korean War—or even of the cold war itself. Because his followers claimed that "Nobody's for Joe but the People," the tradition he may have re-activated and represented is still a live political issue. According to the *New American Right*, the political resentment he managed to tap mobilized a populism turned sour but still potent. Michael Paul Rogin disagreed, and tried to separate McCarthyism from the populist heritage by insisting that the rule of educated elites need not be valorized at the expense of popular will. Though Rogin claimed that the Korean War heightened McCarthy's appeal and broadened—or at least deepened—his traditionally Republican constituency, the evidence to sustain such a claim is thin. When asked who can be trusted to handle communism, only 8 percent of those polled replied McCarthy. Three times as many citizens preferred Eisenhower, the candidate who promised not victory in Korea but peace. (The director of the Federal Bureau of Investigation got a slightly higher percentage than Ike.)[11] This dispute pivots in part on how "populism" is defined. It can denote almost anything, making room even for a billionaire like H. Ross Perot, or for the most recent president willing to accept the label. Jimmy Carter described himself as "a Populist in the tradition of Richard Russell,"[12] which makes as much sense as to be called "a Southerner in the tradition of Pierre Trudeau," or "a Democrat in the tradition of William McKinley." But populism also signifies a certain cultural homogeneity, since the villages and the farms, where differences stood out, instigated the populist revolt in the 1890s; and in this sense McCarthyism was the last gasp of the conformism that foreign visitors as early as Alexis de Tocqueville abhorred. Soon the ideal of homogeneity (*the* American Way of Life) would gradually but dramatically yield to a celebration of diversity and the enhancement of dissent and variety ("do your own thing"). The 1950s were pre-multiculturalist— the last decade of which that could be said. Afterward "the tyranny of the majority" would cease to be an axial principle of democracy.

The conformist ethos that so characterized the 1950s thus looks like a clumsy imbalance in the perennial tension between freedom and order, as unity tipped over into uniformity. Especially while the G.I. (*Time*'s "Man of the Year" in 1951) was facing communist troops in Korea, the case for protecting the rights of communist sympathizers in America could not be forthrightly advanced—to say nothing of the hypocrisy of formal economic discrimination against Koreans in the United States, where their right to become citizens, to collect old-age pensions, to own land, and to marry Caucasians was widely infringed upon.[13] So extensive was the tarnishing of democratic ideals during the Red Scare that scholarly efforts to document its scope have hardly been exhausted. Though *No Ivory Tower* (1986) is tinctured by pretty-in-pink sympathies that obscure the beliefs that Stalinists presumably harbored, Ellen Schrecker's book makes an important contribution to the history of academic freedom, a right that was first incorporated into the Constitution in the decade when *Lehrfreiheit* was being traduced in practice. So much has now been published on the role of the legislative investigating committees (especially the House Committee on Un-American Activities) and on the FBI that the fate of an annual bibliography should be recalled. In 1972 the huffing-and-puffing editors of *American Literary Scholarship* were already pleading for a moratorium on further articles about Faulkner's "A Rose for Emily."[14] The writ of HUAC and the FBI—these official but furtive monitors of the nation's intellectual and cultural life—was stretched to include such threatening figures as scenarists and directors, artists and literati usually deemed incapable of inflicting harm beyond their immediate families. To the music lover, for example, Leonard Bernstein was an adornment of American culture. But to the FBI, he was worth 666 pages of reports, in a dossier that has recently been released, even though (need it be said?)

he was no communist. With the likely approval of Chief Justice Fred Vinson, even the Supreme Court's First Amendment "absolutists," Hugo L. Black and William O. Douglas, were subjected to FBI surveillance.[15]

Political historians have generally echoed the criticism of liberals of the 1950s that the domestic cold war exacted a penalty upon the very notion of citizenship, which has tended toward a minimal definition. The cold war demanded that, for the sake of democracy, Americans take sides, and stand up in a certain way. To be a good citizen did not mean championing civil rights or guaranteed medical care or liberal values in education, but rather resisting such an agenda in the name of a higher loyalty and national security. Citizenship was simply patriotism, which left virtually no institution unaffected. Even the Girl Scouts of America were found wanting, after the Illinois chapter of the American Legion discovered that the 1953 edition of the moppets' *Handbook* was "collectivist." After all, the index omitted reference to the U.S. Constitution; and yet the chapter on "One World" encouraged the girls to support the United Nations and its Charter—a document that "the arch-traitor, Alger Hiss," had allegedly crafted. Reeling under the American Legion attack, the Girl Scouts altered its "One World" badge to "My World," while the hope that "you are preparing yourself for world citizenship" became "you are preparing yourself to be a friend to all."[16] Civic awareness did not include a careful exploration of what the FBI or the Central Intelligence Agency or other agencies were doing in the name of democratic values, nor did the pressures of citizenship permit too close or too skeptical a look at the military.

Especially while the brutal, limited war was being fought in Korea, men in uniform enjoyed tremendous prestige. The military was largely unchallenged in the 1950s—the only decade in the twentieth century when a general was elected president, which did not stop his young successor from complaining during the 1960 campaign that the defense budget was too *small*. The only politician bold enough to take on the Army was McCarthy, who reviled a World War II hero, General Ralph W. Zwicker, for having failed to block the automatic promotion from captain to major of a leftist Army dentist named Irving Peress: "You are a disgrace to the uniform. You're shielding Communist conspirators. You're not fit to be an officer. . . ."[17] Such viciousness was so disturbing, however, that soon thereafter the overreacher destroyed himself. Small pacifist organizations like the War Resisters League and the Fellowship of Reconciliation got even smaller; and ministers of the Gospel told activist A.J. Muste that peace could not be introduced into sermons, because the subject was considered communistic. In 1954 the most popular of the new Revell hobby kits was a plastic model of a battleship, the U.S.S. *Missouri*.[18] Major General Claire L. Chennault, the retired former chief of the American Air Force in China who headed the CIA's Civil Air Transport Company on Taiwan, was so well known that he endorsed Camel cigarettes in magazine advertisements. The military stress upon authority, leadership, and strength reinforced other patriotic implications—as the first line of defense against communism and atheism too, and as a bulwark of order against subversion and chaos.

In the spring of 1951, when Truman announced that General Douglas MacArthur had been dismissed for insubordination, the press conference was sneaked in after midnight. Shock and anger were nevertheless widespread. The Michigan state legislature passed a resolution that began, *"Whereas*, at 1 a.m., of this day, World Communism achieved its greatest victory of the decade in the dismissal of General MacArthur. . . ."* By the time the startling news reached Los Angeles, the city councilors were so upset that they adjourned for the day. Old Glory was flown at half-mast. According to Jenner, the dismissal proved that "agents of the Soviet Union" were actually running the Democratic administration: "Our only choice is to impeach President Truman to find out who is the secret invisible government." Others were more temperate than Jenner, merely demanding the resignation of the commander-in-chief. Over two out of every three Americans told pollsters of their opposition to the firing of the five-star general, whose dismissal the Reverend Billy Graham put in an other-worldly perspective: "Christianity has suffered another major blow."[19] After all, MacArthur's "masterstroke"—the landing at Inchon in September 1950—would become his compatriots' last famous triumph until Operation Desert Storm.[20] When New York City gave MacArthur a far bigger ticker-tape parade than Charles A. Lindbergh had endured in 1927, members of the crowd were seen crossing themselves as the limousine passed. Eighteen New Yorkers—a race renowned for sangfroid—were hospitalized for hysteria. The general himself rumbled ominously about "Communist influences" in the executive branch (until his wife finally muzzled him), and continued to insist that the art of war allowed "no substitute for victory."[21] Many citizens agreed. In November 1951, 51 percent of the public told Gallup pollsters that they favored dropping of nuclear weapons on "military targets" in Korea.[22] But Truman's policy of no wider conflagration was sustained, and, true to the general's own prediction at a joint session of Congress, MacArthur soon faded away.

In that same year, the popularity of a novel by Herman Wouk demonstrated the esteem that military authority enjoyed during the Korean War in particular. The winner of the Pulitzer Prize in 1952, *The Caine Mutiny* quickly sold about 3 million copies and was translated into seventeen foreign languages. Wouk's Broadway version of the forensic climax, *The Caine Mutiny Court-Martial*, sold out the house for two seasons (1954–56). The author had served in World War II as a lieutenant in the Naval Reserve, and had become an executive officer (second in command) of a destroyer-minesweeper. "No topic is more popular in officers' clubs throughout the wide Pacific than 'Captain Bligh' stories of this war," Wouk noted in a memorandum. Yet a prefatory note in his novel insists upon an ideological point: American naval annals disclosed no "court-martial resulting from the relief of a captain at sea" during the previous three decades. The invocation of Regulations 184, 185, and 186 in *The Caine Mutiny* was therefore completely fictional, and was not intended to reflect badly on the Navy. The author added that both captains under whom he served were "decorated for valor."[23] But what if a commander is not brave at all, and scarcely capable of exercising leadership? What if the skipper is like Philip Queeg—mentally unbalanced, exhausted, tending toward paranoia, so cowardly that the crew nicknames him "Old Yellowstain"? Is mutiny, under conditions of peril to the entire ship and crew, justified?

That is the moral dilemma the novel poses. Though attorney Barney Greenwald wins acquittal for the mutineers by making Captain Queeg break down, he later berates his clients, turning Queeg into the hero, Lieutenant Steve Maryk into a dupe, and Lieutenant Tom Keefer into the villain. The twist may not give the plot its punch;[24] and the huge popularity of *The Caine Mutiny* was due to Wouk's silken narrative abilities, not to the ending. But Greenwald's rearrangement of moral categories was not peripheral to his creator's purposes; it was essential. For Wouk wrote in his journal that "the crux of the tale will come in the realization of Maryk and [Willie] Keith that the mutiny was a mistake even though Maryk was acquitted." Thus, as Wouk told an interviewer in 1972, what began as a panoramic war novel became "a novel about authority versus responsibility."[25] This counterpoint is puzzling. For Wouk seemed to take the side of authority *against* responsibility, leaving unexplained why—in defense of democracy—blind obedience warranted a higher priority than individual conscience and independent judgment during a crisis. For *The Caine Mutiny* implies that losing the ship in a typhoon is better than challenging a skipper whose powers of command have failed. Or, as the U.S.S. *Alabama*'s Captain Frank Ramsey (Gene Hackman) insists in Hollywood Pictures' *Crimson Tide* (1995): "We're here to preserve democracy, not practice it." This was a message that millions of readers did not seem to reject or resist, a message at variance with the anti-military or pacifist novels that became famous in the aftermath of World War I, like *A Farewell to Arms*, Dos Passos's *Three Soldiers* and Remarque's *All Quiet on the Western Front*. Other post–World War II best-sellers, like Norman Mailer's *The Naked and the Dead* (1948) and James Jones's *From Here to Eternity* (1951), echoed earlier laments for the crushing of the spirit of individuality. But what might be inferred from the runaway success of *The Caine Mutiny* is how smoothly popular taste during the Korean War could accommodate an authoritarian ideology, justifying submission to a demented superior.

The joke, however, was soon on Wouk, who had composed *The Caine Mutiny* as a valentine to the Navy, which at first adamantly refused to cooperate with any movie studio seeking to transfer his sizzling property to the screen. Again the Navy insisted that it had never suffered a mutiny, and two studios that tried to adapt Wouk's novel simply gave up. Warner Brothers, for example, was informed that *The Caine Mutiny* was "extreme" and "derogatory." It took independent producer Stanley Kramer a year and a half to secure an agreement, going all the way up to the secretary of the Navy. In 1954, two years after the chief of naval operations gave his approval, Columbia Pictures released the film, which explicitly assures audiences (again) that no mutiny had ever marred naval annals. The edgy skipper (Humphrey Bogart) still turns out to be the hero after all—a misunderstood, unappreciated professional. Though the Navy had cooperated, permitting the filming of actual ships, its help could not be directly acknowledged in the lapidary final credits: "The dedication of this film is simple: To the United States Navy." *The Caine Mutiny* was also an early directing assignment for an emeritus of the Hollywood 10, Edward Dmytryk, who emerged from prison to recant his communist past and testify voluntarily before HUAC.[26]

Another movie released in 1954 was also dedicated to the Navy: Mark Robson's *The Bridges at Toko-Ri*, based on James Michener's best-selling novel of the previous year. Though Lieutenant Harry Brubaker (William Holden) dies in a ditch in Korea, his sacrifice is a rebuke to the comfort and complacency of the home front, which is ever in need of reminders that vigilance and martyrdom are the price of freedom. Though Mike Forney (Mickey Rooney) remarks, "I hate those Commies," the political or ideological sources of such animosity are unexamined, and seem to derive from personal volatility and stone Irish bellicosity rather than from principled hostility to tyranny. *The Bridges at Toko-Ri* shows military resistance to communism to be more a test of national and individual character, a moral responsibility to be borne, than a calculation based on collective security considerations. Initially embittered that a home, a family, and a law practice in Denver must be given up, World War II veteran Brubaker eventually realizes the justice of the Korean War: "You fight simply because you're here." The geopolitical value of "here" is scarcely considered; stalwart and unreflective heroism seems to be a sufficient rationale *pro patria mori*. In the question posed by Rear Admiral George Tarrant (Fredric March), the film wonders: "Where do we get such men [as Lieutenant Brubaker]? . . . Where do we get such men?" One sign of the continuities of cold war political culture is that Ronald Reagan later asked the same question, attributing it to an actual admiral, and later quoted the line as though he had come up with it himself, obliterating the distinction between the real military and the "reel" one.[27]

In the 1950s Soviet culture was notoriously straight-jacketed by official dogma and completely constrained by censorship. But perhaps it was a little too easy for Hollywood to feel superior to the State Cinema Institute in Moscow, since the producers of American popular art willingly subjugated themselves to political controls as well. The champions of free enterprise saw nothing contradictory in submitting to such governmental interference. For the American motion picture industry had long relied on the military to provide technical expertise and personnel, as well as equipment and locales, to convince audiences of the verisimilitude of its stories. It was virtually unthinkable in the 1950s to proceed without the cooperation of the Pentagon, which exercised script approval—and therefore censorship. The aesthetic analyses of the Department of Defense were crude, and its reaction to criticism was oversensitive; and of course it was under no obligation to assist the studios. But what is striking is how eager the producers were to be coopted into the military system, to relinquish the independence that they sometimes praised on the screen. Though Kramer ran his own production company and cultivated a reputation for maverick liberalism, he did not try to adapt *The Caine Mutiny* without the Navy's cooperation. Such a movie would have been not only intolerable but also virtually inconceivable in an era when most men—even the very rich, like

G. David Schine and Elvis Presley—were expected to take one step forward and slip into uniform.

Of course not everyone submitted to the patriotic pressures of the decade, even while the cold war had turned hot in Korea. Independent radicalism was vigorous enough to sustain at least the possibility of the 1960s. In the name of public accountability, C. Wright Mills condemned the anti-democratic insularity of "the power elite," the resonant title of his 1956 book that listed five decisions that the sociologist claimed ignored popular sovereignty. All involved foreign policy and warfare, and all involved Asia: Pearl Harbor; Hiroshima and Nagasaki; the resistance to Pyongyang; Dienbienphu; and Quemoy and Matsu. Interest in his political and social criticism would grow, especially posthumously. "Progressives" (often a euphemism for communists) survived too. In 1952 I.F. Stone even published an iconoclastic book that raised doubts about the North Korean aggression of June 1950; and in 1956, when the gallant gadfly flew to the 20th Party Congress in Moscow and described the USSR in unflattering terms, about four hundred readers canceled their subscriptions to the tiny *I.F. Stone's Weekly*.[28] Stone himself did not impugn the claims of citizens who *were* communists to be serving humanity. That task was often left to sullen patriots who imposed loyalty tests that stifled the vigor of democratic debate.

While the Korean War raged, such impositions were especially intense; and thus the standard veered, according to British philosopher Isaiah Berlin's 1958 dichotomy, from negative liberty toward positive liberty. The former protected autonomy from the intrusions of others; the latter facilitated the realization of one's best interests. Berlin, who developed the distinction from Benjamin Constant's essay *De la liberté des anciens comparée à celle des modernes*, insisted upon the validity of both conceptions of freedom. But Berlin clearly preferred the ideal of freedom *from* coercion and of maximal non-interference,[29] because negative liberty could be less easily twisted into something evil. "Deeply influenced by the monstrous misuse of the word liberty in totalitarian countries," he noted that "totalitarian regimes declared that they were the homes of true liberty. That seems to be a cruel caricature of the word." The civic fulfillment that positive liberty represented, he argued, could be twisted into a redemptive fanaticism that would deny privacy. The right to be left alone, the right to gratify merely personal interests, was unknown to the ancients but needed to be asserted against the pressures of official orthodoxy. That is why the Athenians could be admired but not emulated. Not even in the Funeral Oration of Pericles could Berlin detect a legitimation of the apolitical; the sanctity of the individual personality would not be envisaged until well over a millennium later.[30] (Paradoxically, the American legal thinker who first enunciated the civilized value of privacy was Louis D. Brandeis, whose favorite book, Alfred Zimmern's *The Greek Commonwealth*, situated the Funeral Oration as the pinnacle of democracy.)[31]

Nor could the modern need to articulate a theory of civil liberties against the rigidities of homogeneity draw upon Athenian wisdom. The acceptance of truth as multiple, the recognition that "there may be two or more incompatible answers, any one of which could by accepted by honest, rational men—that is a very recent notion," Berlin observed. "Some think that Pericles said something of this kind in his famous Funeral Speech. He comes close to it but does not reach it. If Athenian democracy is good then Sparta or Persia cannot be accepted. The merit of a free society is that it allows of a great variety of conflicting opinions without the need for suppression."[32] Anyone alert to the dangers of unity at the expense of individuality and to demagogic versions of mass democracy might therefore have proclaimed: *Ich bin ein Berliner*, though negative liberty is less glamorous, less noble than the republican "civic virtue" of republican Athens. Negative liberty could can also have melancholy repercussions. Consider, for example, the controversy that erupted after the Korean War over the behavior of American prisoners. Many were reported to have abandoned their sick and wounded comrades, to have fraternized and collaborated with the enemy, and to have taken food from weaker prisoners. "Most of the repatriates," according to an influential article in the *New Yorker*, "came home thinking of themselves not as part of a group, bound by common loyalties, but as isolated individuals." Though such reports of misconduct proved wrong or misconceived,[33] they could be imagined as stemming from too pliable a notion of citizenship, too feeble a sense of political community.

Because the status of civil liberties was so fragile during the cold war, pity for the victims has sometimes included sympathy for their political affiliations as well. But by now the historiographical axis has shifted decisively. Claiming to begin with the leftist assumption that the president of the Carnegie Endowment for International Peace had somehow been framed, Allen Weinstein concluded that "the jurors in the second trial made no mistake in finding Alger Hiss guilty as charged." Beginning with the leftist assumption that the most notorious American atomic spies had somehow been framed, Professor Ronald Radosh realized, along with his co-author Joyce Milton, that "Julius Rosenberg . . . [had] become the coordinator of an extensive espionage operation . . . Ethel Rosenberg probably knew of and supported her husband's endeavors." Indeed "the Rosenberg spy ring was surprisingly productive."[34] The historic implications still reverberate. For if the verdict in the Hiss case suggested that anyone could be a communist, the verdict in the Rosenberg case permitted the inference that any communist could be an atomic spy. Recent scholarship does not completely falsify the fears of superpatriots during the era of the Korean War.

Nor were the Hollywood 10 merely "liberals" whose commitment to social justice somehow ran afoul of the yahoos serving as HUAC's grand inquisitors. "Red-baiting" implies that such charges are inaccurate, that the targets of investigation and innuendo are *not* communists. But that is precisely what scenarists Dalton Trumbo and Ring Lardner, Jr. were. By 1960, incidentally, Trumbo was back at work under his own name; and nearly two decades after leaving Danbury penitentiary, Lardner finally got screen credit under his own name for adapting *M*A*S*H*.[35] Whether the Rosenbergs should have been executed (if they were indeed guilty of espionage), or Hiss jailed (if he had indeed transmitted classified documents to the Soviets), is separable from the empirical evidence of what they did. That evidence is now so strong that the "progressive" idea of martyrdom, according to which the victims of an American fascism had not committed the crimes attributed to them, is in tatters. The status of American democracy when the cold war turned hot in Korea was insecure, and constitutional protections did not inspire confidence. But the abuses associated with the most famous political trials were not quite as capricious as once believed or charged. Such are the uses of disenchantment.

For scholars still resistant to the weight of evidence, the most promising fall-back position can no longer be overt pro-communism but rather post-modernism, which ignores empirical procedures entirely in order to "explore the narrative structures and political operations of language as variously deriving from, discontinuous with, and productive, reproductive, or critical of historical circumstances." No reputable historian would concur with Judge Irving R. Kaufman, who sentenced the Rosenbergs to death in April 1951 for a crime he called "worse than murder," since they "caused, in my opinion, the communist aggression in Korea, with the resultant casualties exceeding 50,000 and who knows but what that millions more innocent people may pay the price of your treason." Yet one recent book on the Rosenberg case insists that the "story" of the couple is "undecidable," then proceeds to take for granted a frame-up (without bothering to show its likelihood), and further fudges the issue of guilt around which the case pivots with prose so dense that it is as though Casey Stengel had tried to parody the late style of Henry James. But the method of "materialist-feminist criticism," if not the ideal of a detached quest for truth, has been served.[36] Or as Abraham Lincoln once remarked, after listening to a paper on the occult: "Well, for those who like that sort of thing, I should think it is just about the sort of thing they would like."[37]

The excesses of positive freedom and of "populist" anti-communism are deservedly central to any assessment of democracy during the cold war and the Korean War. But such centrality poses an historical problem. If "the tyranny of the majority" proved so intractable a feature of American self-government, then how could a conformist public culture have liberated itself so strikingly from McCarthyism by the 1960s? Indeed, how soon after the Korean War *did* the domestic cold war come to an end? For historians of foreign policy, the alphabet makes the task of periodization easier. Their chronology spans barely more than four decades, a trajectory from X to Z—from George F. Kennan's identification of "The Sources of Soviet Conduct" to Martin Malia's account of a free fall so complete that one reporter picked as "the ultimate post–cold war headline: 'Ukraine Recognizes Independence of Croatia and Slovenia.'"[38] But when did domestic passions get banked? For anyone interested in how democratic tolerance could be reinvigorated, this is an intriguing and largely unexplored issue.

The dynamic of change was surely underneath the surface, waiting to erupt in the 1960s, a firebreak of a decade that no seer or social scientist could have foreseen. During the Big Sleep, no one could have envisaged how Great the Awakening would become. Indeed, at the very start of the 1960s, the year the Student Nonviolent Coordinating Committee *and* the National Liberation Front in Vietnam were formed, the year student riots helped topple the regime of Syngman Rhee, Seymour Martin Lipset foresaw only incremental change. Conflict, which he imagined only in class terms, would be barren of ideological passion. Thus even an eminent political sociologist could not have predicted the mass movements, the bombings, the urban riots—such convulsive violence that in 1968 a president of the United States found it imprudent if not dangerous to attend the nominating convention of his own party. Undergraduates had been so apolitical in the 1950s that their passivity was lamented, their silence deplored. Yet by October 1968, students had shifted so far to the left that, in the presidential contest among Richard Nixon, Hubert Humphrey, and George Wallace, a plurality of those polled would have cast their ballots for Ernesto "Che" Guevara (notwithstanding his death a year earlier). Had the race been confined to college campuses, the fiery revolutionary would at least have made it to the Electoral College.[39]

The year that Guevara helped Fidel Castro take power in Havana can suggest how the 1960s were already emerging from the womb of the previous decade. In 1959 ex–Secretary of State Dulles died of cancer, an event that prompted Acheson, during a dinner party, to praise a righteous deity: "Thank God Foster is underground."[40] (The personal is political.) Until then the beneficence, if not the wisdom, of the foreign policy they had both directed had been ritualistically praised, and almost never examined from a radical perspective. But that year William Appleman Williams pushed past the envelope with *The Tragedy of American Diplomacy*, which so stunned a contributing editor of the *National Review* that M. Stanton Evans raised doubts about Professor Williams's very sanity. How could anyone *conceive* of American diplomacy as governed by the need for overseas markets, Evans wondered; and he added: "America is now confronted with a danger . . . that its reasoning class—the segment of the population that deals professionally in ideas—has given over the orderly employment of reason." The reviewer generalized that the "greatest terror, to a sane man, must always be the possibility of losing his sanity." That is apparently what has happened to the Madison historian, whose invocation of an "open door for revolutions" punctuated one of the most terrifying books that Evans had ever read. The reviewer concluded: "This kind of analysis, offered as sober counsel on foreign relations by an American professor, sends more chills through me than any vision of atomic holocaust, or the lurking menace of Soviet power. Other dangers may promise death in the future; this is death here and now." (Incidentally, Williams would later express admiration for Herbert Hoover.) Investigated both by the Internal Revenue Service and by HUAC, Williams may have been a radical; but he was no communist. He had, after all, graduated from the Naval Academy, in an era when even a burnt-out case like Captain Queeg could be transformed into a hero.[41]

Nineteen fifty-nine was also the year novelist Richard Condon published *The Manchurian Candidate*, which United Artists adapted three years later into what has been called the last as well as "the most sophisticated film of the cold war." No other artifact of the era so closely connects the Korean War, the domestic cold war, and the larger geopolitical struggle. This "demonologically explicit" thriller, Professor Michael Paul Rogin has argued, is "a brilliant, self-knowing film. But far from mocking the mentality it displays, it aims to reawaken a lethargic nation to the Communist menace."[42] This interpretation is dubious. From the opening scene of libidinous G.I.s carousing with Korean bar-girls beneath the photo of a stern General MacArthur, cold war conventions are shaken. At the center of the bizarre plot is Sergeant Raymond Shaw (Laurence Harvey), who is brainwashed during captivity in the Korean War. Programmed to kill when told to "pass the time by playing a little solitaire," Shaw is released and repatriated to the United States, where he is turned over to a top Soviet agent—his own mother (Angela Lansbury). She intends to use her son the *somnambule* to assassinate the presidential nominee from a booth high above Madison Square Garden. A clear path to the White House would thus be blazed for her husband, Johnny Iselin (James Gregory), a nutty right-winger who is the party's vice-presidential candidate. This calamity is avoided only when Major Ben Marco (Frank Sinatra), who had been captured in Shaw's unit in Korea but has freed himself from Red brainwashing control, picks open the lock of Shaw's mind as well. Marco races to the booth just as the trigger is about to be squeezed, but Shaw assassinates his mother and stepfather on the podium before killing himself. (Here the film deviates from the novel, in which Marco orders the mother and stepfather to be executed and kills Shaw himself.)[43]

In the form of a weird espionage nightmare, *The Manchurian Candidate* shows the diabolical cunning of the Eastern bloc. But what must have made the film so disorienting is its satire of McCarthyism as well. Shaw's stepfather is obviously modeled on the Wisconsin demagogue, with whom Iselin shares a spurious war record. In first running for the Senate, "Tail Gunner Joe" had himself photographed in the cockpit of a plane that he had never flown in combat, while Iselin likewise became a "one-man battleship" by testing some Navy guns on harmless targets. Senator Iselin accuses the secretary of defense of harboring fifty-seven communists—a figure derived from spotting a bottle of catsup. A drunken ignoramus who cannot even remember the numbers he invents, Iselin is also a blackmailer who dominates the fanatical and paranoid ambience of anti-communist politics. The movie version of *The Manchurian Candidate* omits much of the novel's spoof of right-wingers bristling with "opinions they rented that week from Mr. Sokolsky, Mr. [David] Lawrence, Mr. [Westbrook] Pegler, and that fascinating younger fellow who had written about men and God at Yale," William F. Buckley.[44] But the film is unsparing in showing that, while communism is fiendish and still dangerous, the far right is hypocritical and foolish.

Director John Frankenheimer recalled that "at one stage we were going to be picketed by both the American Legion and the Communist Party at the same time, which we tried to encourage of course; after all, the whole point of the film was the absurdity of any type of extremism." When cadres felt free to picket again, when a movie director could welcome such controversy, and when the American Legion was considered as "extremist" as the Stalinists, the historian can surmise that the 1950s were going into remission. But then something terrible happened. A year after *The Manchurian Candidate* was released, a young ex-military man, who had returned to the United States from a communist nation, carried a rifle outfitted with telescopic sights up several flights of steps, intending to shoot through the window on a presidential target surrounded by a crowd below. But this time no savior rushed up to stop the assassin, a loner whose name—Lee Harvey Oswald—even bore an eerie resemblance to actor Laurence Harvey's, and who fit D.H. Lawrence's definition of the archetypal American soul as "hard, isolate, stoic and a killer." For the next quarter of a century, Sinatra blocked the commercial rerelease of the film.[45]

In 1959 MGM also released Alfred Hitchcock's *North by Northwest*. Its communists—or at least those who spy for "the other side"—are certainly villainous. The dapper Philip Vandamm (James Mason) is very dangerous, perfectly capable of orchestrating cold-blooded murder; and his secretary Leonard (Martin Landau) is either epicene because he is sinister or sinister because he honors his "woman's intuition." But *North by Northwest* breaks with the tradition of films exemplified by *Casablanca* (1943), in which Rick and Ilsa realize that romantic attachments must be subordinated to the larger cause of humanity. A guy who sticks his neck out for nobody must relearn the value of political engagement. That is the lesson that wives and other women must learn too. But Roger Thornhill (Cary Grant) falls so hard for secret agent Eve Kendall (Eva Marie Saint) that he briefly becomes a conscientious objector to the cold war, and seeks her exemption from its demands. "I don't like the games you play, Professor," he tells the spymaster (Leo G. Carroll) who is running Kendall, the unmarried blonde whose assignment is to defect to "the other side." The CIA operative replies: "War is hell, Mr. Thornhill, even when it's a cold one." "If you fellows can't lick the Vandamms of this world" except with morally dubious counter-espionage methods, Thornhill rebuts, "perhaps you ought to start learning how to lose a few cold wars." In this debate the professor gets the last word ("I'm afraid we're already doing that"). But at least a mild dissent is registered against the "long twilight struggle" invoked in the Inaugural Address of President Kennedy (who had hoped Cary Grant would play Lieutenant Kennedy in *PT-109*).[46]

Despite the square-jawed severity of the cold war rhetoric that resonated in that 1961 speech and thereafter, Kennedy's election marked a certain emancipation from the dogmas of the 1950s; and it must remain conjectural how long such rigidities would have lingered had that close election produced a GOP victory instead. Perhaps there were deeper forces at work in the republic that were pushing toward liberation from so cramped and repressive a consensus (and therefore encouraging a more flexible foreign policy). Perhaps some subterranean receptivity to liberalism had lifted the Democratic candidate into the White House with so narrow a mandate that, with the departure of a White House aide whom the press had hailed as "coruscatingly" brilliant, Kennedy's envoi was self-deprecating: "Fifty thousand votes the other way, and we'd all be coruscatingly stupid."[47]

But (to paraphrase Bob Dylan) it was not necessary to be a meteorologist to know which way the wind was blowing. The atmospherics yielded to increasingly searing social criticism and to fierce political conflict. Cocky young radicals would soon *seek* the media exposure that congressional investigating committees provided, and handing out a HUAC subpoena would no longer be a guarantee of how to make an American wilt. Martin Luther King's "Letter from a Birmingham Jail" (1963) vanquished the spirit of moderation and compromise that had been so valorized as essential to democracy in the previous decade. One year earlier, *The Other America* activated the war on poverty. Written by a socialist (though Michael Harrington was coy about his exact political affiliations), the book implicitly challenged John Kenneth Galbraith's *Affluent Society* (1958), which had bestowed a name on a seemingly robust economy only four years earlier. In terms of future impact on the very texture of American life, *The Feminine Mystique* (1963) would prove to be the most explosive book of all. The problems that middle-class women faced were cultural and psychological, what Marxists called the superstructure; and Betty Friedan's lack of interest in economics marked a return to the ancient Greek meaning of that word: household management. Contrast the fight for racial equality, which sometimes used the cold war for its own aims. President Kennedy once asked: "Why can a Communist eat at a lunch counter in Selma, Alabama, while a black American veteran cannot?" In 1960 Diane Nash, a leader of the Nashville student protests, demanded equal educational opportunities in the South so that "maybe some day a Negro will invent one of our missiles."[48] But Friedan advanced no case for social reform by tabulating the advantages that communism might gain so long as women are excluded from the capitalist system. That makes *The Feminine Mystique* a book of the 1960s indeed.

Even without counting the productivity of housewives, the economic gap between the two most powerful victors in the Great Patriotic War would not be closed—though the gap between political and social histories of the 1950s can thus be narrowed. The USSR stretched across eleven time zones, but its government's persistent fears of capitalist encirclement did prove justified. Indeed, for all the anxieties that were generated in Washington because of Soviet advances in heavy industry and in military and space technology, for all the control that the *nomenklatura* exerted over staggering natural resources (producing oil from reserves exceeding the Middle East's), inefficiency permeated the command economy of the Kremlin. Had it ever gained control of the Sahara, Western analysts liked to quip, there would soon have been a shortage of sand. Backwardness had already been obvious, of course, before the Second World War, when John Ford's *The Grapes of Wrath* (1940) played in Soviet cinemas to heighten awareness of the desperate misery of the "Okies." The film had to be withdrawn after six weeks because audiences were extracting the wrong lesson: even the most dispossessed of California's rural proletariat were shown driving automobiles.[49] Better to treat Soviet citizens to Volume 39 of the *Great Soviet Encyclopedia*, which emphasized, in its treatment of the postwar American economy, extensive industrial unemployment and frequent strikes, as well as the sharp decline of farm income. Prosperity was completely ignored, since the correlation of forces was supposed to be aligned against the anachronistic system of capitalism.[50] Yet American opulence was genuine, as its beneficiaries partook of what a frustrated General George S. Patton, after V-J Day, called "the horrors of peace."[51]

The cornucopia pouring out of postwar factories would mean Satisfaction Guaranteed, generating such distractions and pleasures that the real Korean War would never get in the books, and also keeping customers from detecting any advantages to an economic system founded on different "laws." By the end of the 1960s, General Motors vice president John DeLorean would warn that the chief threats to the auto industry were safety, environmentalism, and consumerism.[52] But in the 1950s, the enemy was the poverty that breeds communism; and, in River Rouge if not on the Yalu River, the foe was getting licked. When *Time* made the president of G.M. its "Man of the Year" in 1955, Harlow Curtice's company had become the first in history to earn a billion dollars in a single year and had consolidated the primacy of Detroit in the hierarchy of American capitalism. About three decades later, Soviet per capita income was still only one-tenth of the American standard, and roughly equal to the per capita income of Mexico. Soviet standards were also *declining*.[53]

No Armistice Day or formal signing aboard a battleship marks the end of this most awesome and terrifying of geopolitical conflicts. But the self-destruction of the Soviet imperium was decisive. Indeed, since the late 1980s, many an ardent cold warrior in the United States has sought to take credit for assisting in so dramatic a suicide, bucking for the chance to play Dr. Jack Kevorkian. The dilemma of reconciling communism and market reform proved too difficult even for Mikhail Gorbachev to accomplish, and the Marxist-Leninist program could not be carried out except on a stretcher. Defense spending was crucial in accounting for the duration *and* the demise of the cold war. With the Pentagon running the second largest planned economy on the globe, the United States has had an enormous military-industrial complex. But the USSR *was* a military-industrial complex, in which the means of destruction flourished more than the means of production. The Soviet failure to match American and Western abundance is in retrospect the hidden term that should link how political and social historians fathom both the anxieties and the complacencies of cold war America.

The Korean War caused Pentagon spending to balloon to four times its previous size, and the defense budget never returned to its earlier dimensions. That the cold war was the health of the state is a generalization American social historians have not yet fully pondered. But consider the caesura, the historic break with previous patterns that militarization after 1950 represented. In his famous report, *The American Commonwealth* (1888), Lord Bryce did not bother to devote even one of his 119 chapters to the military, which was then overshadowed by, say, the armies of France and Germany—forces that were ten times larger and consisted mostly of conscripts. Between 1820 and 1900, Britain, France, and Germany had annually averaged $3–$6 per capita for military expenditures. The United States was spending an average of 70 cents to $1 per capita. Or consider the fury of President Wilson in the summer of 1915, when he summoned the acting secretary of war and pointed to a story in the *Baltimore Sun* describing contingency plans for war with Germany. Were the story true, Wilson heatedly announced, the officers on the General Staff were to be demoted and kicked out of Washington. (Wilson had not foreseen that war with Germany would be only a year and a half away.) As late as the Great Depression, the Navy turned down the job application of a chemist because it already *had* one.[54]

Before the Second World War, the U.S. Army ranked as the nineteenth largest in the world, slightly exceeding Bulgaria's. In 1940 the biggest peacetime maneuvers in American history were conducted in upstate New York. But broomsticks had to serve as machine guns, stovepipe had to be used as anti-tank guns, combat planes were not available, and the trucks driven in the mock battles bore placards on which the word "TANK" had been written.[55] The Second World War and then the remobilization during and after the Korean War shattered the traditional reluctance to keep one's powder dry. NSC-68, the top secret blueprint for the militarization of American statecraft and infrastructure, had been, according to one diplomatic historian, "a policy in search of an opportunity. That opportunity arrived on June 25, 1950, when, as Acheson and his aides later agreed, 'Korea came along and saved us.'" The shift in power has been so emphatic and persistent that the story of the militarization of the economy must be woven into the political and the social narratives of the cold war. So permanent had a war economy become during the half-century after V-J Day, so closely connected had defense spending and national prosperity become during and after the Korean War, that Joseph Heller quipped: "Peace on earth would mean the end of civilization as we know it."[56]

A mail-order catalogue that I recently received illustrates the triumph of the capitalist encirclement that ended the cold war, and offers the historian a tongue-in-cheek reminder of the consumer spending that, even during the Red Scare, even during the Korean War, Americans made integral to their Way of Life: $34.95 now enables customers to buy ("Your Satisfaction Is Unconditionally Guaranteed") KGB spy binoculars—a bargain because of their "restricted issue." Purportedly "designed by a leading producer of optics for the Soviet military," this item is enticingly advertised as "small enough to tuck into a shirt pocket or purse. Ideal for concerts, the theatre, and covert operations."

Notes

1. Quoted in Edmund Wilson, *Patriotic Gore: Studies in the Literature of the American Civil War* (New York: Oxford University Press, 1962), p. 481.

2. Quoted in Sam Merrill, "Playboy Interview: Joseph Heller," *Playboy* 20 (June 1975): 68; and in Lawrence H. Suid, *Guts and Glory: Great American War Movies* (Reading, MA: Addison-Wesley, 1978), p. 268.

3. J.D. Salinger, *The Catcher in the Rye* (New York: Signet, 1953), pp. 127–28; John Seelye, "Holden in the Museum," in *New Essays on The Catcher in the Rye*, ed. Jack Salzman (Cambridge: Cambridge University Press, 1991), pp. 29–30.

4. Max Hastings, *The Korean War* (New York: Simon and Schuster, 1987), p. 9; Michael S. Sherry, *In the Shadow of War: The United States Since the 1930s* (New Haven: Yale University Press, 1995), p. 181; see Introduction, n. 5.

5. Michael Paul Rogin, *Ronald Reagan, the Movie and Other Episodes in Political Demonology* (Berkeley: University of California Press, 1987), p. 30.

6. Talcott Parsons, "Social Strains in America: A Postscript (1962)," and Richard Hofstadter, "Pseudo-Conservatism Revisited: A Postscript (1962)," in *The Radical Right*, ed. Daniel Bell (Garden City, NY: Doubleday Anchor, 1964), pp. 97, 98, 231.

7. Quoted in David M. Oshinsky, *A Conspiracy So Immense: The World of Joe McCarthy* (New York: Free Press, 1983), p. 169; Sherry, *Shadow of War*, p. 185.

8. Quoted in Richard M. Fried, *Nightmare in Red: The McCarthy Era in Perspective* (New York: Oxford University Press, 1990), pp. 128, 130, and in Oshinsky, *Conspiracy So Immense*, p. 166.

9. Fried, *Nightmare in Red*, pp. 128–31; Oshinsky, *Conspiracy So Immense*, p. 172.

10. Quoted in Oshinsky, *Conspiracy So Immense*, pp. 193, 346, and in Richard H. Rovere, *Senator Joe McCarthy* (Cleveland: Meridian, 1960), p. 15; Sherry, *Shadow of War*, pp. 180–82.

11. Michael Paul Rogin, *The Intellectuals and McCarthy: The Radical Specter* (Cambridge, MA: MIT Press, 1967), pp. 224–25, 242–43, 247–48, 259; Michael Kazin, *The Populist Persuasion: An American History* (New York: Basic Books, 1995), pp. 183–93.

12. Quoted in Walter LaFeber, *America, Russia and the Cold War, 1945–1992*, 7th ed. (New York: McGraw-Hill, 1993), p. 287.

13. "The Shape of Things," *The Nation* 171 (August 26, 1950), 179.

14. Werner Sollors, *Beyond Ethnicity: Consent and Descent in American Culture* (New York: Oxford University Press, 1986), p. 11.

15. Howard Temperley and Malcolm Bradbury, "War and Cold War," in *Introduction to American Studies*, 2nd ed., ed. Bradbury and Temperley (London: Longman, 1989), p. 306; Ralph Blumenthal, "FBI's Endless Bid to Peg Bernstein as 'Red,'" *International Herald Tribune*, July 30–31, 1994, p. 18; Roger K. Newman, *Hugo Black: A Biography* (New York: Pantheon, 1994), pp. 422–23.

16. "Girl Scouts Scored by Legion in Illinois," *New York Times*, August 7, 1954, p. 4; "Girl Scouts Deny Charges by Legion," *New York Times*, August 8, 1954, p. 39; Ben H. Bagdikian, "What Happened to the Girl Scouts," *Atlantic Monthly* 195 (May 1955): 63–64; Jocelyn Wilk, "'To Do My Duty to My Country': The Girl Scouts and American Society" (senior honors thesis, Department of History, Brandeis University, 1995), pp. 63–71.

17. Quoted in Richard H. Rovere, *Senator Joe McCarthy* (Cleveland: Meridian, 1960), p. 30.

18. Lawrence S. Wittner, *Rebels Against War: The American Peace Movement, 1941–1960* (New York: Columbia University Press, 1969), p. 220; Karal Ann Marling, *As Seen on TV: The Visual Culture of Everyday Life in the 1950s* (Cambridge, MA: Harvard University Press, 1994), pp. 58–59.

19. Richard H. Rovere and Arthur M. Schlesinger, Jr., *The MacArthur Controversy and American Foreign Policy* (New York: Noonday Press, 1965), pp. 3–16; Eric F. Goldman, *The Crucial Decade—and After: America, 1945–1960* (New York: Vintage, 1960), pp. 202–8; William Manchester, *American Caesar: Douglas MacArthur, 1880–1964* (Boston: Little, Brown, 1978), pp. 647–51.

20. Hastings, *Korean War*, pp. 99, 109, 114; William Stueck, *The Korean War: An International History* (Princeton: Princeton University Press, 1995), pp. 85–88, 91–92, 98; David Rees, *Korea: The Limited War* (New York: St. Martin's, 1964), pp. 96–97; Clay Blair, *The Forgotten War: America in Korea, 1950–1953* (New York: Times Books, 1987), pp. 269–78.

21. Geoffrey Perrett, *A Dream of Greatness: The American People, 1945–1963* (New York: Coward, McCann and Geohegan, 1979), pp. 165–67.

22. Tom Engelhardt, *The End of Victory Culture: Cold War America and the Disillusioning of a Generation* (New York: Basic Books, 1995), p. 64.

23. Herman Wouk, *The Caine Mutiny* (New York: Pocket Books, 1973), pp. vi, xii–xiv.

24. Ibid., p. 599; William H. Whyte, Jr., *The Organization Man* (New York: Simon and Schuster, 1956), pp. 243–48.

25. Arnold Beichman, *Herman Wouk: The Novelist as Social Historian* (New Brunswick, NJ: Transaction Books, 1984), p. 46.

26. Suid, *Guts and Glory*, pp. 129–39; Beichman, *Herman Wouk: The Novelist*, pp. 43–50; Harvey Swados, "Popular Taste and *The Caine Mutiny*," *Partisan Review* 20 (March–April, 1953): 248–56; Stephen J. Whitfield, *The Culture of the Cold War* (Baltimore: Johns Hopkins University Press, 1991), pp. 60–62.

27. Christian G. Appy, "'We'll Follow the Old Man': Sentimental Militarism and Cold War Cinema of the 1950s" (1994), unpublished paper in author's possession, pp. 26–30; Rogin, *Ronald Reagan, the Movie*, p. 7.

28. C. Wright Mills, *The Power Elite* (New York: Oxford University Press, 1956), p. 22; Todd Gitlin, *The Sixties: Years of Hope, Days of Rage* (New York: Simon and Schuster, 1987), pp. 31, 34, 77, 174.

29. Isaiah Berlin, "Two Concepts of Liberty," in *Four Essays on Liberty* (New York: Oxford University Press, 1969), pp. 122–44, 167–72.

30. Quoted in Ramin Jahanbegloo, *Conversations with Isaiah Berlin: Recollections of an Historian of Ideas* (London: Phoenix, 1993), pp. 41–42, 147; Shlomo Avineri, "The Dilemmas of Freedom According to Isaiah Berlin," in *On the Thought of Isaiah Berlin* (Jerusalem: Israel Academy of Sciences and Humanities, 1990), p. 30; Berlin, *Four Essays*, pp. xl–xli.

31. Anthony Lewis, *Make No Law: The Sullivan Case and the First Amendment* (New York: Vintage, 1992), p. 87.

32. Quoted in *Conversations with Isaiah Berlin*, pp. 42–43.

33. Dwight Macdonald, *Discriminations: Essays and Afterthoughts, 1938–1974* (New York: Grossman, 1974), pp. 54–56; H.H. Wubben, "American Prisoners of War in Korea: A Second Look at the 'Something New in History' Theme," *American Quarterly* 22 (Spring 1970): 3–19.

34. Allen Weinstein, *Perjury: The Hiss–Chambers Case* (New York: Alfred A. Knopf, 1978), p. 565; Victor Navasky, "Weinstein, Hiss, and the Transformation of Historical Ambiguity into Cold War Verity," in *Beyond the Hiss Case: The FBI, Congress, and the Cold War*, ed. Athan G. Theoharis (Philadelphia: Temple University Press, 1982), p. 221; Ronald Radosh and Joyce Milton, *The Rosenberg File: A Search for the Truth* (New York: Vintage, 1984), pp. xix–xxv, 450–51.

35. Ring Lardner, Jr., *The Lardners: My Family Remembered* (New York: Harper and Row, 1976), pp. 325–26.

36. Quoted in Radosh and Milton, *Rosenberg File*, p. 284; Virginia Carmichael, *Framing History: The Rosenberg Story and the Cold War* (Minneapolis: University of Minnesota Press, 1993), pp. xii–xv, 270.

37. Quoted in Sydney E. Ahlstrom, *A Religious History of the American People* (New Haven: Yale University Press, 1972), p. 489.

38. Thomas L. Friedman, "Cold War Without End," *New York Times Magazine* (August 30, 1993): 30.

39. Seymour Martin Lipset, *Political Man: The Social Bases of Politics* (Garden City, NY: Doubleday Anchor, 1963), pp. 445, 454–56; Gitlin, *The Sixties*, p. 344.

40. Quoted in Douglas Brinkley, *Dean Acheson: The Cold War Years, 1953–71* (New Haven: Yale University Press, 1992), p. 67.

41. M. Stanton Evans, "Some Voices from the Grave," *National Review* 7 (April 25, 1959): 23–25; David Green, *The Language of Politics in America: Shaping Political Consciousness from McKinley to Reagan* (Ithaca: Cornell University Press, 1992), pp. 245–46; "Interview with William Appleman Williams," in *Visions of History*, ed. Henry Abelove, Betsy Blackmar, Peter Dimock, and Jonathan Schneer (New York: Pantheon, 1983), pp. 125–35; William A. Williams, "My Life in Madison," in *History and the New Left: Madison, Wisconsin, 1950–1970*, ed. Paul Buhle (Philadelphia: Temple University Press, 1990), pp. 264–71.

42. Rogin, *Ronald Reagan, the Movie*, p. 252.

43. Richard Condon, *The Manchurian Candidate* (New York: Jove Books, 1988), pp. 307–8.

44. Ibid., p. 252.

45. Quoted in *The Celluloid Muse: Hollywood Directors Speak*, ed. Charles Higham and Joel Greenberg (Chicago: Henry Regnery, 1969), p. 81; Rogin, *Ronald Reagan, the Movie*, pp. 252–54; D.H. Lawrence, "Fenimore Cooper's Leatherstocking Novels," in *Selected Literary Criticism* (London: Mercury Books, 1961), p. 329; Richard Corliss, "From Failure to Cult Classic," *Time* 131 (March 21, 1988): 84.

46. Robert J. Corber, *In the Name of National Security: Hitchcock, Homophobia, and the Political Construction of Gender in Postwar America* (Durham: Duke University Press, 1993), pp. 193–202; Pauline Kael, *When the Lights Go Down* (New York: Holt, Rinehart and Winston, 1980), p. 5.

47. Benjamin Bradlee, in *A Tribute to John F. Kennedy*, ed. Pierre Salinger and Sander Vanocur (Chicago: Encyclopedia Britannica, 1964), p. 75.

48. Quoted in Friedman, "Cold War Without End," p. 45, and in Clayborne Carson, *In Struggle: SNCC and the Black Awakening of the 1960s* (Cambridge, MA: Harvard University Press, 1981), p. 13.

49. Kenneth T. Jackson, *Crabgrass Frontier: The Suburbanization of the United States* (New York: Oxford University Press, 1985), pp. 187–88.

50. *A Soviet View of the American Past: An Annotated Translation of the Section on American History in the Great Soviet Encyclopedia*, ed. O. Lawrence Burnette, Jr., and William Converse Haygood, trans. Ann E. Yanko and Peter A. Kersten (Madison: State Historical Society of Wisconsin, 1960), pp. 56–60.

51. Quoted in Martin Blumenson, *Patton: The Man Behind the Legend, 1885–1945* (New York: William Morrow, 1985), p. 280.

52. Emma Rothschild, *Paradise Lost: The Decline of the Auto-Industrial Age* (New York: Vintage, 1974), p. 8.

53. LaFeber, *America, Russia and the Cold War*, p. 343.

54. Sherry, *Shadow of War*, pp. 182–83; Temperley and Bradbury, "War and Cold War," in *Introduction to American Studies*, p. 290; Richard J. Barnet, *Roots of War* (Baltimore: Penguin, 1973), pp. 23, 31.

55. William E. Leuchtenburg, *Franklin D. Roosevelt and the New Deal, 1932–1940* (New York: Harper and Row, 1963), pp. 306–7; Don Graham, *No Name on the Bullet: A Biography of Audie Murphy* (New York: Viking, 1989), p. 22; Geoffrey Perrett, *Days of Sadness, Years of Triumph: The American People, 1939–1945* (Baltimore: Penguin, 1974), p. 29.

56. LaFeber, *America, Russia and the Cold War*, pp. 96–98; Sherry, *Shadow of War*, p. 179; Joseph Heller, *Picture This* (New York: G.P. Putnam's Sons, 1988), p. 100.

12

Warfare, Democracy, and the Cult of Personality

Jennifer T. Roberts

The publication not long ago of a new biography of Douglas MacArthur is a salutary reminder of the spell that charismatic figures can cast even in democratic societies. The cover design of Geoffrey Perret's *Old Soldiers Never Die: The Life of Douglas MacArthur* (1996) features a full photograph of MacArthur's famous corncob pipe . . . and just the corner of MacArthur's face, recognizable only because of the sunglasses and the pipe itself. The symbolism is easy enough to decode: the myth conceals the man and the image runs the risk of overshadowing the reality.

As those who aspire to power and prestige have always known, myths and images are important. It is a contradiction in terms for a politician not to care how he or she appears to others. Glamor and romance frequently attach to important public figures, creating larger-than-life heroes who must nonetheless discipline themselves to attend to the mundane minutiae of political life. Inevitably, being larger than life creates special problems in a democratic society; it is something a democratic people does and does not want in its leaders. The subject matter before us in this project, the Peloponnesian and Korean wars, affords a rich opportunity to explore how two societies dealt with the paradoxes that inhere in democratic leadership. The problems are especially acute in states that are prone to warfare, where the aura that attends on leadership is particularly charged. What I propose to do in these pages, therefore, is to probe some of the issues surrounding leadership by exploring Douglas MacArthur's removal in 1951 in the context of the cases of generals who were relieved of their commands in fifth-century Athens. Because the Athenians made liberal use of accountability procedures of many kinds, the instances of leaders who found themselves in trouble with the government were many; the most revealing for my purposes, I think, is the case of Alcibiades, a flashy aristocrat who had much in common with MacArthur despite patent differences in personality, and it is the comparison with Alcibiades that is the core of my essay. In her popular 1985 book *The Reign of the Phallus: Sexual Politics in Ancient Athens*, Eva Keuls argued that far from representing a stage in social development so early as to provide a contrast with modern times, in fact classical Athens offers "a kind of concave mirror in which we can see our own foibles and institutions magnified and distorted."[1] I hope to explore in these pages in what ways this is and is not true in the area of leadership, accountability, and the cult of personality. Are moderns dramatically different from ancients, or can American democracy in some ways be construed as a sanitized version of the Athenian prototype?

Powerful Men and the Need to Control Them: Tyranny and Democracy in Greece

Though legend told of leaders like Odysseus and Achilles, the first historical Greek leaders whose names are known to us are the men known as *turannoi*, tyrants. The word "tyrant" serves as a reminder that the Greek world was in some ways different from our own (though American revolutionaries used the word in speaking of King George III, and certainly both Americans and Athenians have been known to contrast their liberty with the autocracy that it dethroned). The men the Greeks called *turannoi* played a special role in the evolution of the Greek city-state. Despite the negative connotations it later developed with the birth of democracy, a *turannos* originally indicated only a man who had come to power through extra-constitutional means, generally a military coup; Sophocles' most famous play, usually rendered in English as *Oedipus Rex* or *Oedipus the King*, was titled by its author *Oedipus Turannos*. Strongmen who had overthrown the narrow oligarchies of birth that ruled Greek communities in the seventh and sixth centuries B.C., tyrants were at once champions of popular rule and its worst enemies. Ambivalent about the merits of strong leadership, Greeks were fascinated by tyranny and often attached myths to tyrants that jumbled up strongly positive characteristics (e.g., tyrants as heroic benefactors) with patently negative ones (e.g., tyrants as murderers). The tales that congregated about tyrants frequently involved sex, often of an unusual nature: Peisistratus was known for practicing contraception by having intercourse with his wife *ou kata nomon*, "not in the accepted way," and it was believed that Periander of Corinth had sex with his wife Melissa after she was dead.[2]

Though Plato and Aristotle saw Athenian democracy as a tyranny of the majority, Athenian democratic rhetoric identified tyranny as the opposite of democracy. The charter myth of Athenian democracy involved the expulsion of Peisistratus' sons, and if we may believe Thucydides, Alcibiades, trying to endear himself to the Spartans by showing that he wasn't really a democrat, claimed that his family had been identified as democratic simply because they always opposed tyrants (6.89.4). When the oligarchy was overthrown in 410 B.C. and democracy restored, each male citizen took an oath swearing to kill with his own hand anyone who sought to establish a tyranny at Athens or collaborated with a tyrant.[3]

The Athenians' often-noted apprehension about individual eminence probably had its roots in the experience of tyranny. In the 480s the Athenians first made use of the procedure of ostracism, an inverse popularity contest that involved sending the people's choice into a non-punitive exile for ten years: every spring the ostracism law permitted the Athenians to hold a vote to determine whether there was in the city some citizen who seemed so dangerous that his removal could be viewed as essential to the preservation of democratic institutions against the threats of civil war or tyranny; since the last tyrant of the Peisistratid family had been expelled as late as 510, the danger of tyranny was more than a fiction, and the first person known to be ostracized bore a Peisistratid name.[4]

Growing up in the shadow of tyranny, Athenian politics thrived on the tension between two contradictory and seemingly mutually exclusive principles: the suspicious jealousy of eminent individuals and the total dependence of political groups on charismatic leaders. The Athenians had no word for political party; they referred simply to "*hoi peri tou Perikleous*" ("those around Pericles") or around Cleon or around Alcibiades. Not that such groups had no agenda beyond the promotion of a particular man—far from it. But with no formal party organization to keep the ambitious politician in check (either by the demands of his own party's platform or by the balance provided by an institutionalized opposition), individuals took on powerful significance.

Where there is power, however, there is anxiety, and the nervous Athenians structured their government to dilute both as much as possible. The Athenian knew no president, no premier, no prime minister—no chief executive as such. Supreme power in fifth-century Athens lay with the assembly; among the many officials of the state elected or chosen by lot, the greatest authority attached to the ten *stratêgoi* or generals, elected annually to one-year renewable terms. Because the ten *stratêgoi* were the highest officials in the state, outranked by no individual (as they would be by a president in the United States), military men were the object of intense scrutiny, and the demos meeting in the assembly was sleepless in its vigilance over the conduct of war. Minimally, our sources identify the following impeachments during the Peloponnesian War:

- Pericles, in 430, when the Athenians were frustrated that they could neither make peace nor meet the Spartans in battle (Thuc. 2.65.3; Plato *Gorg.* 516a; Plut. *Per.* 35.3); he was fined and removed from office, but the Athenians reelected him the following year;
- Xenophon, Hestiodorus, and Phanomachus, shortly afterward, when they had made terms with the residents of rebellious ally Potidaea after its capitulation to Athens (Thuc. 2.70.4) without consulting the assembly; the impeachment was aborted;
- Paches, who was reported to have committed suicide in the courtroom when

accused of war crimes surrounding the surrender of rebellious Mytilene (Plut. *Nic.* 6.2; Plut. *Aristides* 26.5; Anth. *Pal.* 7.614);

- Sophocles, Pythodorus, and Eurymedon, who had failed to dissuade the Sicilians from making peace among themselves (Thuc. 4.65.3); two were exiled, one was fined;
- Thucydides himself after his failure to save Amphipolis from the charismatic Spartan general Brasidas (Thuc. 4.102–108, 5.26.5); he did not stay to stand his trial and spent the rest of his life in exile;
- Alcibiades (among others) in the religious scandals of 415 (Thuc. 6.61.7; Plut. *Alc.* 18–22; Thuc. 6.60 61); Alcibiades was sentenced to death in absentia and forced to remain in exile until he engineered his recall several years later;
- Phrynichus and Scironides, on the grounds that they had betrayed Iasus to the Peloponnesians probably in fact because of their opposition to the recall of Alcibiades (Thuc. 8.54.3); they were deposed from the generalship but not, as far as we know, tried on criminal charges;
- Anytus, the future prosecutor of Socrates, after his failure to relieve the Messenian garrison at Pylos (Diod. 13.64.3, *Ath. Pol.* 27.5); he was acquitted and suspected of bribing the jury;
- Alcibiades, after he left a subordinate in charge in Ionia who disobeyed orders and engaged the Spartans (Xen. *Hell.* 1.5.11–18; Diod. 13.71; Plut. *Alc.* 36 and *Lys.* 5; Lysias 21.7; Nepos. *Alc.* 7); he was removed from his generalship and presumably exiled;
- the eight generals who failed to retrieve soldiers from the water after the victory off the Arginusae islands in 406 (Xen. *Hell.* 1.7 and Diod. 13.101–2); the six who returned to stand trial were executed.[5]

One of these episodes arose from a fascinating, though non-military, series of events. Alcibiades and many others got into difficulty in 415 in the scandals that erupted surrounding the alleged parodying of the so-called mystery religion at Eleusis near Athens and a sensational episode of vandalism in the city itself: the large-scale nocturnal mutilation of the images of Hermes that guarded Athenian homes sparked panic in the citizenry and persuaded many Athenians that a plot to undermine the democracy was afoot. It was as a consequence of these accusations that Alcibiades defected to Sparta, on whose side he remained until suspicions that he had seduced (and impregnated) the king's wife forced him to flee and seek a reconciliation with the Athenians. Most charges against generals were less colorful and arose in the wake of failed operations or when *stratêgoi* ceased to share the policy views of their constituents; still, efforts were frequently made to obscure the real causes of an impeachment. As H.D. Westlake points out in discussing the trial of Paches, "it must be remembered that the formal charge was not necessarily identical with the real cause of complaint"—hence the frequent accusations of treason and the taking of bribes.[6] We know of no charge on the Athenian books that would have enabled the assembly to remove a general from office simply for doing a bad job, or for having policy differences with the assembly or its leaders.[7]

In refuting accusations of "leveling" against the Athenian democracy, A.H. Greenidge, the author of a handbook of Athenian constitutional history published around the turn of the century, contended that in reality, "Few states have ever been more completely under the sway of great personalities." It is not democracy but rather oligarchy, Greenidge maintained, that is the true leveler, whereas "it is one of the oldest lessons in history that . . . democracy brings a hero-worship generally of an extravagant kind"[8] The truth is, however, that American high school students or even college graduates called upon to name great heroes of the ancient world are not likely to select a single Athenian leader for mention. Why should this be? Was there something about the Athenian ethos that was really engaged in a powerful struggle with the cult of personality? Is democracy a safe environment for great personalities? Few of my students recall ever hearing the name of Pericles before taking my classes; but they have all heard of Alexander and Caesar.

MacArthur and Truman

Since MacArthur's mystique involved comparisons with both Caesar (after whom William Manchester named his biography of the general *American Caesar*) and Alexander (to whom a member of his staff apparently compared him, to Alexander's disadvantage), it may be useful to examine his career in the light of the sorts of issues raised by the many impeachments of the Peloponnesian War. Around him hovered precisely the sort of aura that poses problems in a democracy. MacArthur enjoyed circulating a postcard he had received from India that was addressed

> To His Most Gracious Majesty
> The Old Friend
> The Most Honorable
> General MacArthur, Sahib, Bahadur,
> Military Governor and Crowned King of Japan.[9]

Others added divinity to his attributes. A Japanese observer labeled the general "a second Jesus Christ," Herbert Hoover identified him as "the reincarnation of Saint Paul," and when MacArthur addressed Congress after his removal in April 1951, Representative Dewey Short proclaimed "We heard God speak here today, God in the flesh, the voice of God."[10] And Dewey Short was from Missouri.

So was the man who had sacked MacArthur—a man who had indeed pointedly written to his cousin Nellie Noland of his impending visit to Wake Island "Have to talk to God's righthand man."[11] Unlike cashiered Athenian generals, MacArthur was removed not by a large assembly calling him to account from below but by an identifiable individual disciplining him from above.

Truman's decision to recall MacArthur was overdetermined; policy differences and cavalier insubordination both played a role. The Truman administration could scarcely be termed dovish on the war in Korea; the great bulk of the offensives taken by MacArthur had Truman's full support, and both Omar Bradley and the president himself had raised the possibility, however tenuous, of using atomic weapons to bring the war to a conclusion. But MacArthur's yearning to flex his muscles in a face-off with the Chinese was at odds with Truman's wish to contain the war as much as possible; the general's insistence on his own autonomy, moreover, was provocative and, ultimately, intolerable.[12] Either problem alone might have been sufficient to make MacArthur's position untenable; together they posed a hopeless obstacle to his continuing in his post. MacArthur's trip to Formosa on July 31, 1950, and his independent negotiations there with Chiang Kai-shek were conducted in such a way as to highlight rather than obscure his unwillingness to work cooperatively with the White House. Diplomat William Sebald recalled that when he related Dean Acheson's request for information,

> MacArthur made it clear that he had no intention of providing details. He replied, in effect, that the talks were purely military in nature, that they were limited to matters of military coordination between the Chinese government and himself as theater commander, and hence what was said and done was his sole responsibility and not that of the State Department. But, I insisted, military agreements of this type could eventually have a direct bearing upon the formulation of foreign policy. The General irritably replied: "Bill, I don't know what you are talking about."

During this interview, Sebald reported, he sensed "a growing rift between the American authorities in Tokyo and Washington which, if not corrected, could only lead to disaster."[13]

Truman flew to Wake Island in mid-October to confer with MacArthur, whom he had never met. There MacArthur assured Truman that the Chinese would not enter the war, and on the basis of this assurance Truman authorized him to proceed north to the Yalu. The first UN troops arrived at the Yalu on October 26. On November 24, the Chinese attacked in force, sending the UN forces reeling back over the 38th parallel.

Many observers blamed the setback in Korea on the recklessness of MacArthur, who had promised to have American boys home by Christmas, and relations between MacArthur and Truman deteriorated further when MacArthur continued to release statements to the press that expressed his policy disagreements with Truman. Truman gave serious consideration at this time to the possibility of relieving MacArthur. "Every second lieutenant," Truman remarked, "knows best what his platoon commander ought to do. He thinks the higher-ups are just blind when they don't see things his way. But General MacArthur—and rightly too—would have court-martialed any second lieutenant who gave press interviews to express his disagreement." In the end Truman decided to keep MacArthur—because, he wrote, he knew a general "couldn't be a winner every day of the week," and because he didn't want to appear to be relieving MacArthur simply because his offensive had failed.[14] Truman felt the need, however, to put a stop to MacArthur's public criticisms of his policy, and in December he instructed Dean Acheson and George Marshall to issue directives ordering that no speeches, press releases or other kinds of statements should be distributed before the Department of State or of Defense had cleared them to "insure that the information made public" was "accurate and fully in accord with the policies of the United States government."[15]

The president's gag order did not have the desired effect. MacArthur continued to issue public statements rejecting the concept of a limited war and advocating not only a blockade of the Chinese coast but even direct naval and air attacks on China's industrial capacity to wage war. In March MacArthur issued an inflammatory statement to the communist military commanders that thoroughly undermined Truman's peacemaking efforts, exaggerating UN and U.S. successes and minimizing the threat posed by China.

At the same time another crisis was evolving in Washington. On February 12, House Minority Speaker Joseph Martin, an Asia-First advocate, had given a speech in Brooklyn urging that the forces of Chiang Kai-shek be deployed with American support to open a "second Asiatic front" on mainland China.[16] Foreseeing MacArthur's approbation, Martin sent the general a copy of the speech, to which he responded on March 20. This letter, endorsing Martin's views, contained MacArthur's famous dictum, "There is no substitute for victory." On April 5 Martin released copies of MacArthur's letter to the press and read it aloud on the House floor. MacArthur professed astonishment at the brouhaha occasioned by what he dismissed as a routine communiqué, but his refusal to comply with Truman's orders to clear his policy statements with Washington was patent. Truman had already decided to relieve MacArthur, but the general's public statements of March had given the president substantial ammunition in his bid to make his dismissal seem prudent and, indeed, unavoidable.[17] When the Prime Minister of Iran was assassinated early in March, a crowd of thousands, suspecting American involvement, was reported to have demonstrated outside the American embassy, registering their complaints against the United States by shouts not of "Down with Truman" but rather "Death to MacArthur."[18] Cartoonists in the American papers did not fail to make capital of the problematic two-headed government. One chief would have to go, and it would not be Truman.[19]

In Athens ostracism could have eliminated the weaker rival, but the American system is different, and the carefully calibrated rhetoric that surrounded MacArthur's dismissal revealed much about American attitudes. While seeking to undermine Truman in a variety of ways, MacArthur nonetheless was careful to lay stress on his acceptance of Truman's right to remove him; Truman, for his part, who had frequently commented on MacArthur's arrogance, plainly was bending over backward reiterating how much he admired the general and how averse he was to hurting his feelings. Several months previously, it seems, after MacArthur's comforting predictions about the Chinese had proven false, he "gave serious thought to relieving

General MacArthur as our military field commander in the Far East and replacing him with General Bradley," but he decided against this since "It would have been difficult to avoid the appearance of a demotion, and I had no desire to hurt General MacArthur personally."[20] Truman was careful to include high praise of MacArthur in the public statement announcing his dismissal, citing the gratitude Americans owed him "for the distinguished, and exceptional service which he has rendered his country in posts of great responsibility." For this reason, he concluded, "I repeat my regret at the necessity for the action I feel compelled to take in his case." (Ten year afterward, talking with Merle Miller, Truman expressed himself more candidly, recalling that when Bradley warned him that MacArthur might resign if he knew he was about to be fired, he had replied, "The son of a bitch isn't going to resign on me, *I want him fired.*")[21]

Statements in his *Memoirs* make plain that Truman also felt the need to go on record about the absence of a connection between MacArthur's dismissal and any of the military setbacks in Korea.[22] These pious disclaimers, however, did not prevent Truman from introducing criticism of MacArthur into statements that were cast in a mode of graciousness and generosity. "Now, no one is blaming General MacArthur," he wrote,

> and certainly I never did, for the failure of the November offensive. He is no more to be blamed for the fact that he was outnumbered than General Eisenhower could be charged with the heavy losses of the Battle of the Bulge. But—and herein lies the difference between the Eisenhower of 1944 and the MacArthur of 1950—I do blame General MacArthur for the manner in which he tried to excuse his failure. . . . [W]ithin a matter of four days he found time to publicize in four different ways his view that the only reason for his troubles was the order from Washington to limit the hostilities to Korea.[23]

Similarly, MacArthur, in testifying before the Armed Services and Foreign Relations committees of the Senate, insisted that he did not in the slightest question the authority of the president, saying, "The authority of the president to assign officers or to reassign them is complete and absolute. He does not have to give any reasons therefor or anything else. That is inherent in our system."[24] He went on to draw attention to problems in the mode of his relief and claim that it jeopardized security.[25] Upon MacArthur's death in 1964 Roger Baldwin of the American Civil Liberties Union observed that a fellow general had once remarked to him that MacArthur "had about him a little of Dr. Jekyll and Mr. Hyde."[26] While Dr. Jekyll was solemnly affirming Truman's absolute right to do what he did, Mr. Hyde was informing Matthew Ridgway that the scuttlebutt on the medical grapevine was that Truman "was suffering from malignant hypertension," an affliction "characterized by bewilderment and confusion." The president, MacArthur told Ridgway, would not live six months.[27] As so often in politics, a yawning gulf separated private feelings from public professions.[28]

The dissenting voices that were heard—and they were loud and many—skirted the issue of Truman's right to his decision in the abstract and focused instead on the specific instance at hand, enumerating MacArthur's virtues and suggesting a communist conspiracy in the White House. (Predictably, Joe McCarthy was among the most vituperative, but he was far from alone.) Ironically, one journalist writing in the *U.S. News and World Report* cast Truman's right to discipline MacArthur not as a democratic one but as very much the opposite: "Yes," wrote conservative columnist David Lawrence of Truman,

> as Commander in Chief, he can, as in a totalitarian state, "purge" high military officers. . . . And, yes, he can with one blow impose the sentence of death on the military record of a great General and detach him from his command so quickly that he could not have that final privilege so dear to a military man—to say a few words of farewell to his troops. . . .

But Truman, Lawrence claims, lost sight of the American sense of fairness.[29] In affirming Truman's right to do what he did, Lawrence casts him as a Soviet dictator; but underneath the calculated irony, that right is upheld.

Unlike most dismissed Athenian generals, Douglas MacArthur lived out his life as a military hero as well as a martyr. Savaging Truman by billing him as a totalitarian leader engaged in a "purge," David Lawrence could portray the president as imposing "the sentence of death on the military record of a great General"; the Los Angeles City Council could and did respond to MacArthur's dismissal by adjourning "in sorrowful contemplation of the political assassination" of MacArthur. But all this melodramatic rhetoric of killing was just that. Douglas MacArthur did not die at the hands of Truman in 1951; he remained on active duty without assignment until his death in 1964. On April 11 of that year, thirteen years to the day after he had been relieved of his command, MacArthur was buried with full military honors after his body had lain in state in Washington during a period of national mourning declared by President Lyndon Johnson. The fates of dismissed Athenian generals were usually quite different—frequently exile or execution. The metaphorical "death sentence" imposed on MacArthur in Lawrence's rhetoric was imposed at Athens in all its physicality; the six generals who had failed to retrieve sailors from the waters off the Arginusae islands were immediately led off to execution and were dead by sundown. MacArthur's inability to say farewell to his troops pales by comparison.

Other aspects of the Athenians' wartime conduct make plain their failure to share Americans' squeamishness about death. In his *Memoirs* Truman reported that he had to ask himself whether, as MacArthur suggested, the threat of enslavement to communism was so great that "we had to move to the destruction of cities and the killing of women and children."[30] Viewed in a Greek context, it is ironic that Truman pairs the notion of enslavement with the destruction of cities and the mistreatment of women and children, for of course during the Peloponnesian War the Athenians did destroy cities, not killing the women and children but rather making slaves of them. Truman's rhetorical question about the communist threat was designed to show readers that he was a good fellow—that he was of course against the destruction of cities, the killing of women and children. The Athenians too—at Mende, at Scione—opposed the killing of women and children, but they did not oppose slavery; rather they imposed it on those very women and children of whom Truman speaks, the inhabitants of destroyed cities.

Unquestionably MacArthur's mystique at home was enhanced by his absence. It was a remarkable decision for an American patriot who was the darling of McCarthyite conservatism not to set foot on American soil for fourteen years. Both MacArthur's lack of respect for the opinions of military advisors and his need for adulation were evident to the men who served under him, and they made him the object of considerable ribbing, sometimes affectionate but often bitter; at the same time, however, his cult was flourishing stateside among civilians. After being relieved by Truman he returned to San Francisco to the first of an astonishing series of parades, including one in New York during which the litter thrown in his honor was tabulated by the Department of Sanitation as weighing over 2,852 tons; 40,000 union longshoremen struck for the afternoon, and eighteen spectators became so hysterical that they had to be removed from the demonstration. The Washington office of Western Union was overwhelmed by the unprecedented volume of telegrams calling for the impeachment of Truman, billing him as a "pig," an "imbecile," a "red herring," and "the Judas in the White House who sold us down the river to left wingers and the UN."[31]

MacArthur and Athenian Leadership: Parallels with
Miltiades and Alcibiades

MacArthur's career brings to mind two Athenian generals who, like him, spent a great deal of time away from home. Before proceeding to the case of Alcibiades, it may be instructive briefly to explore another parallel—the one that links MacArthur with another man who earned his reputation in an exotic locale. Miltiades, revered as the architect of the pivotal Greek victory against the Persians at Marathon in 490, ended his life ignominiously not long afterward; having promised the Athenians in 489—when his credit stood very high—that he would make them rich if they voted him ships to sail against an unspecified enemy, he was wounded violating a sacred precinct of the goddess Demeter on the island of Paros and returned a failure. Impeached in the Athenian assembly, he was heavily fined, and there was discussion of putting him to death; in the end he died of his wound, which had begun to gangrene. (Alcibiades paid less promptly for the intimations of immortality he offered the Athenians, hinting first in 416 at the world conquest that might follow on their expedition against wealthy Sicily and then later in 411 that his recall would win Athens Persian support against Sparta.) MacArthur was also in the habit of offering false reassurance—boys home by Christmas, Chinese not going to invade—and of greatly exaggerating his own centrality to events and his power to shape the future.[32] Americans do not impeach or execute you for being wrong the way the Athenians did, but they do fire you.

Several generations before the war, Miltiades, a member of the powerful Philaid family and the father of Pericles' rival Cimon, had been sent by the tyrant Hippias to succeed his uncle in maintaining Athenian power in the Thracian Chersonese; after ruling as a king over the natives, he returned to Athens thirty years later to face charges of "tyranny" in Thrace—plainly a politically motivated accusation, since that was precisely what he had been sent to Thrace to do. Surviving the impeachment, he went on to become the hero of the Battle of Marathon. His predilection for making autonomous decisions may have contributed to his downfall the following year, when he was impeached once more after the failure of his secret mission against the island of Paros.

Like MacArthur, Miltiades hailed from a family with a distinguished record of public service—and with a grudge. Miltiades' uncle, Miltiades the elder, had been a rival of the Athenian tyrant Peisistratus and had been effectively exiled to the Black Sea on the pretext of consolidating Athenian power there; Miltiades the younger—our Miltiades—was posted to the same locale, and it seemed to be the family fate to be kept from the centers of power. Similarly, MacArthur—whose aristocratic family was dotted with Arthurs and Douglases just as Miltiades' family was populated by men named Cimon and Miltiades—was the son of a distinguished general (Arthur MacArthur) who also served extensively in the Pacific and who saw his career as truncated by the opposition of the home government represented by William Howard Taft, then the secretary of war. Like Miltiades, moreover, Douglas MacArthur was exalted in his foreign domain in a way that would have been impossible in a democratic country: "As the years went by," Richard Rovere and Arthur Schlesinger Jr. have suggested, MacArthur "almost seemed to take on the divinity renounced on January 1, 1946, by Hirohito."[33] MacArthur became inured to undemocratic ways of thinking, and his autocratic mind-set got him into difficulty with authorities at home. Certainly Truman saw problems with MacArthur's long absence from the United States and was troubled to think that MacArthur held the power he did without ever having met the president of the United States. In explaining why he decided to go to Wake Island to meet with the general, Truman observed:

> I thought that [MacArthur] ought to know his Commander in Chief and that I ought to know the senior Field Commander in the Far East. I have always regretted that General MacArthur declined the invitations that were extended to him to return to the US, even if only for a short visit, during his years in Japan. He should have come back to familiarize himself with the situation at home. This is something I have always advocated for our foreign service personnel—that they should spend one year in every four in their own country. Then they would understand what the home folks were thinking.[34]

While the parallel with Miltiades illustrates the force of jealousy in political life, the suspicions that attach to long absences from home, the danger that attend on easy promises, and the sensitivity of aristocrats who believe their families have been slighted, analogies with Alcibiades may be more fruitful if only because there is vastly more evidence about Alcibiades, most of it in the writings of Thucydides and Plutarch. At first blush no two men may seem more different than MacArthur and Alcibiades—MacArthur disciplined, Alcibiades self-indulgent; MacArthur austere, Alcibiades sensual; MacArthur a father figure, Alcibiades a perpetual adolescent; MacArthur an accomplished military man, Alcibiades a rake. Alcibiades, we can safely conclude, was not indebted to his mother for getting his career off the ground. Alcibiades was arrested and nearly lost his life in connection with his irreverent attitude to religion; MacArthur was ecstatic at the opportunity to Christianize Japan during the occupation and boasted to George Kennan about his role in the process. (In conversation with James Forrestal he compared his own service in Tokyo to Christ's agony on the cross.)[35] Alcibiades' speeches, as far as one can judge, were free of sermonizing, while a sanctimonious tone of moral superiority marked MacArthur's.

At bottom, however, each man was very much out for himself and had few, if any, guiding principles beyond that of his own glory. Alcibiades' self-serving temperament is evident from his every recorded utterance; MacArthur enjoyed nothing more than striking pious poses, but astute observers often commented on the moral hollowness of his worldview. British liaison officer Gerald Wilkinson wrote in 1943:

> He is shrewd, selfish, proud, remote, highly strung and vastly vain. He has imagination, self-confidence, physical courage and charm, but no humor about himself, no regard for truth, and is unaware of these defects. He mistakes his emotions and ambitions for principles. With moral depth he would be a great man; as it is he is a near miss which may be worse than a mile.[36]

Alcibiades rarely if ever mistook his emotions or ambitions for principles; still, Wilkinson's portrait sustains the parallel at many points. Both men cultivated an elusive complexity and were difficult to assess; where soldiers talking about MacArthur with Roger Baldwin spoke of Dr. Jekyll and Mr. Hyde, Plutarch commented on "how difficult public opinion found it to judge Alcibiades, because of the extreme inconsistency of his character" (*Alc.* 16.6). Both Alcibiades and MacArthur were outstanding orators skilled in manipulating occasions and audiences; both were highly competitive and craved fame. Both men were the object of considerable romantic attention, though MacArthur was not known for amours with partners of both sexes and Alcibiades was never dogged by the same unflattering gossip about sexual inadequacy that surrounded MacArthur (fomented by his first wife, Louise Brooks, after their brief unsuccessful marriage).[37] Both were vain and cared deeply about how they were perceived by their countrymen. Plutarch tells a striking tale about Alcibiades' dog, whose fine tail its owner cut off: "His friends," Plutarch wrote, "scolded him and told him that everyone was angry for the dog's sake," but "Alcibiades only laughed and retorted, 'that is exactly what I wanted. I am quite content for the whole of Athens to chatter about this; it will stop them saying anything worse about me'" (*Alcibiades* 9).

Offbeat grooming habits—from garments to accessories—also link the two men. Both were strikingly handsome and wore their clothes well. Curiously, both manifested an ambiguous gender identification in their clothing. Alcibiades, Plutarch maintains (16.1), was effeminate in his dress and walked through the *agora* trailing incongruous purple robes; and MacArthur, whom his doting mother had dressed in skirts (and groomed in curls) until he reached the age of eight, in adult life enjoyed sporting a muffler that belonged to her. Plutarch reports that on the night of his death Alcibiades dreamed that he was covered in his girlfriend's clothing and that she was cradling his head in her arms while she put makeup on his face as if he were a woman; after his death she buried him in her clothes (*Alcibiades* 39.2). Plutarch reports that Alcibiades' shield was distinguished by a personal logo, and MacArthur while in France in World War I gradually adapted his uniform to his own idiosyncratic preferences, discarding the metal support inside his cap and insisting on carrying a riding crop that served no apparent practical purpose.[38] To complete his outfit, he generally wore a brightly colored turtleneck sweater and, of course, his mother's muffler. Increasingly defensive of his eccentric habits of dress, he apparently threatened to shoot an emissary from Chaumont unless discussion of his outfits ceased at once.[39]

Both men spent a good deal of time abroad, and both found their reputations enhanced by their absence. Both men were ultimately alienated from the democratic states into which they had been born. Both were suspected of plotting to overthrow the duly constituted government. Both laced their speeches with pleas for sympathy based on the troubles they—and their ancestors—had suffered at the hands of "enemies" at home. The cult of personality enabled both to return from their long absences to heroes' welcomes despite what might in the case of more ordinary men have been inauspicious circumstances—MacArthur had gravely miscalculated the intentions of the Chinese with fatal consequences for many American soldiers, and Alcibiades had defected to the enemy. Yet their returns were triumphal. Athenians did not heap litter on the heads of their heroes, but Plutarch reports garlands—and a crown:

> When [Alcibiades] landed, people scarcely seemed to have eyes for the other generals they met, but they ran and crowded round Alcibiades, crying out and embracing him. As they escorted him on his way, those who could press near crowned him with garlands and the rest gazed at him from a distance, the old men pointing him out to the young. (*Alcibiades* 32)

Soon afterward, Plutarch maintains, Alcibiades addressed the assembled Athenians on the hill of the Pnyx, speaking so effectively that the audience responded by placing a gold crown on his head (*Alcibiades* 33). Years later, Alcibiades, like MacArthur, played the role of the unheeded warner. His advice to move the fleet to Sestos was ignored by the generals in 405 at Aegospotami, the battle that lost the war for Athens; and he did not outlive the war by much. MacArthur advised Lyndon Johnson to stay out of Vietnam in 1964; his warning was also ignored, and he died shortly afterward.

Both Alcibiades and MacArthur sparked extraordinary emotions in individuals and groups. The language in which Plutarch seeks to convey Alcibiades' popularity among the masses is strikingly evocative of the popular enthusiasm of American civilians for MacArthur:

> The sway which he held over the humbler and poorer classes was so potent that they were filled with an extraordinary passion for him to rule them as *tyrannos*: some of them proposed this and actually visited him to press the idea. They wanted him to be placed in a position that was out of the reach of envy. Then he could sweep away decrees and laws as he thought fit and rid them of those loud-mouthed wind-bags who were the bane of the city, and he would be free to handle affairs and act without fear of informers. (*Alcibiades* 34)

"We could do with a military government for a change," a fan had written to MacArthur; and as for loud-mouthed windbags, compare the dismissal of Truman as a "little ward politician" by another MacArthur champion. Not for nothing had F.D.R. in 1932 identified MacArthur as one of the two most dangerous men in America. (The other was Huey Long.)

Democracy and the Military

Though analogies with classical Greece may be a useful tool for examining some of the issues surrounding leadership, personality, and accountability, MacArthur's focus when he drew on antiquity for models was chiefly Roman (as was Truman's, though Truman was also intrigued by Alexander). On the wall of his office in the Dai-Ichi building MacArthur had framed a long quotation from Livy purporting to give the views of Lucius Aemilius Paulus:

> In every circle, and truly, at every table there are people who lead armies into Macedonia; who know where the camp ought to be placed; what posts ought to be occupied by troops; when and through what pass that territory should be entered; where magazines should be formed; how provisions should be conveyed by land and sea; and when it is proper to engage the enemy, when to lie quiet. . . . These are great impediments to those who have the management of affairs. . . . I am not one of those who think that commanders ought at no time to receive advice; on the contrary, I should deem that man more proud than wise, who regulated every proceeding by the standards of his own single judgment. What then is my opinion? That Commanders should be counselled chiefly by persons of known talent . . . who are present at the scene of action, who see the country, who see the enemy . . . and who, like people embarked on the same ship, are sharers of the danger. If, therefore, anyone thinks himself qualified to give advice respecting the war which I am to conduct . . . let him not refuse his assistance to the state but let him come with me into Macedonia. He shall be furnished with a ship, a horse, a tent; even his traveling charges shall be defrayed. But if he thinks this too much trouble, and prefers the repose of a city life . . . let him not . . . assume the office of a pilot. The city in itself furnishes abundance of topics for conversation; let it

> confine its passion for talking within its own precincts and rest assured that
> we shall pay no attention to any councils but such as shall be framed within
> our camp.[40]

There is much merit in this thinking. As Rovere and Schlesinger have observed in their discussion of this passage, "there is something to be said for the insights—political as well as military—of the professional soldier. To insist on the principle of civilian control is nothing less than to insist on the principle of a free society; to insist on too sharp a distinction between civilian and military functions is, however, to misunderstand the problems of a free society."[41] These observations remind us of what developments in Nazi Germany had so pointedly taught a generation before—that facile divisions between the political and military spheres are ultimately impossible. Insofar as it is about means, politics—a good Greek word—can mean underhanded machinations, manipulative rhetoric, and the stuffing of ballot boxes; but it is also about ends, and some of the ends are of a profoundly moral character. How, then, could generals be devoid of politics? Except for people with unusually mellow temperaments—a quality rare in generals—to be devoid of politics is to be devoid of beliefs about right and wrong, and who would want this in a person whose business it is to deal out death?

Although MacArthur was eager to bill himself as both innocent of politics—whatever that meant in context—and astounded by his removal, evidence suggests otherwise. Apparently the day before MacArthur was shorn of command General Almond visited him in Tokyo and found him markedly disconsolate.

"I may not see you any more," MacArthur reportedly remarked, "so goodbye, Ned." When Ned Almond asked why this should be so, MacArthur replied "I have become politically involved" and "I may be relieved by the president."[42] His only politics, MacArthur proclaimed upon his return to the country he had not seen for fourteen years, were contained in the simple phrase "God Bless America." This cloying contention, meant to be profoundly endearing, is intrinsically improbable, but MacArthur's need to dissociate himself from politics reflects a fundamental American belief in the separation of politics from the military. George Kenney recalled a conversation with MacArthur in which he announced that when he returned home he would settle down in Milwaukee—a far cry from the Waldorf Suite to which he and his wife in fact retired—and, he continued, "on the way to the house I'm going to stop in at a furniture store and buy the biggest red rocker in the shop. I'll set it up on the porch and alongside it put a good-sized pile of stones. Then I'll rock." Taking the bait, Kenney asked what the stones were for. "They are to throw at anyone who comes around talking politics," MacArthur said.[43]

Athenians were not under the same pressures to separate military and political careers. During the fourth century the Athenian *stratēgos* Phocion, who recurred constantly to the theme of Athens' decline, lamented the passing of the era in which the same men served the state as generals and politicians.[44] His regrets have often been echoed by modern historians of Athens, though it is puzzling that men and women raised in an age that takes such a division to be axiomatic should find in it a sign of a civilization's decadence. In fact, Phocion was profoundly involved in politics and was eventually put to death for Macedonian sympathies. Athenians did not demand that their generals hold aloof from political life; given to pithy self-serving aphorisms, Phocion never claimed that his politics were subsumed in the phrase, "May the Gods preserve Athens." Applauding MacArthur's removal, the *New Republic* rejoiced that the general would be constrained to "carry on his political activity in the open without benefit of the military disguise" and predicted that "The common sense of the American people will doubtless reject the undemocratic philosophy of militarism for which he stands when it is nakedly presented to them."[45] That militarism and democracy are at odds might have struck the Athenians as distinctly peculiar. As in so many areas, they were franker about the complexities of political life than Americans—and a good deal less squeamish. Americans are far more anxious and timid with regard to the disciplining of high officials than were the Athenians. Like Harry Truman, the Athenian assembly was hesitant to be perceived as relieving a commander for incompetence. But Truman's carefully articulated apprehension about hurting MacArthur's feelings seems eminently un-Athenian; the Athenian solution was to level accusations of treason or the taking of bribes against a general who had lost the people's confidence. Now, being accused of betraying your country for money generally wounds a person. . . .

American and Athenian approaches to military and political accountability were distinctly different. From the founding of the nation, Americans have hesitated to humiliate or to impeach, preferring to wait for time to take its course; Athenians impeached routinely. When MacArthur was removed, Truman took great care to appear as respectful as possible; dismissed Athenian generals were often put to death. Because of his high position, Truman could afford to be amiable; voters in the Athenian assembly, fearful precisely of the kind of power people like Truman could come to wield, were determined to be tough and uncompromising. Even the Athenian Diodotus, arguing in the assembly against executing all the adult males in Mytilene after an unsuccessful rebellion from the Athenian empire, takes care to couch his argument in the toughest of terms, explicitly disavowing any motives of compassion and putting self-interest forward as the only basis of judgment (e.g., 3.44). But to be uncompromising is not to be unbending; when it suited their interests to change their minds—as in fact the Athenians did in short order over Mytilene and in a rather longer time frame about Alcibiades—they were quite comfortable doing so. Acting on principles of naked self-interest, the gods were meteoric and changeable; why not people too?

These capricious gods, moreover, were the gods of the city. Like other ancient societies, Athens espoused no principles separating church and state. In impeaching generals and politicians, Athenians could be both more and less open than Americans. It was not legal to remove generals from their positions simply for doing a bad job—hence the frequent trumped-up accusations of treason and the taking of bribes. But it was perfectly legal to arrest people for religious transgressions, as the case of Socrates would soon make plain. And so the proceedings against Alcibiades make a quaint contrast to those against MacArthur. The indictment preserved in Plutarch reads:

> Thessalus, the son of Cimon [a later Cimon, not the son of Miltiades], of the deme of Lacia, accuses Alcibiades, the son of Cleinias, of the deme of the Scambonidae, of committing sacrilege against the goddesses of Eleusis, Demeter, and Kore, in that he made a mockery of the Mysteries; that he enacted them in his own house, wearing a robe such as the High Priest wears, when he displays the rites to the initiates; that he styled himself High Priest, Polytion Torch-Bearer, and Theodorus, of the deme of Phegaea, Herald; that he addressed the rest of his companions as Initiates and Novices, contrary to the laws and ceremonies established by the Eumolpidae, Heralds, and Priests of Eleusis. (*Alcibiades*, 22)

Truman's announcement read:

> With deep regret I have concluded that General of the Army Douglas MacArthur is unable to give his wholehearted support to the policies of the United States Government and of the United Nations in matters pertaining to his official duties. In view of the specific responsibilities imposed upon me by the Constitution of the United States and the added responsibility which has been entrusted to me by the United Nations, I have decided that I must make a change of command in the Far East. I have, therefore, relieved General MacArthur of his command and have designated Lieutenant General Matthew B. Ridgway as his successor.
> Full and vigorous debate on matters of national policy is a vital element in the constitutional system of our free democracy. It is fundamental, however, that military commanders must be governed by the policies and directives issued to them in the manner provided by our laws and Constitution. In time of crisis, this consideration is particularly compelling.
> General MacArthur's place in history as one of our greatest commanders is fully established. The nation owes him a debt of gratitude for the distinguished and exceptional service which he has rendered his country in posts of great responsibility. For that reason I repeat my regret at the necessity for the action I feel compelled to take in his case.
> Harry S. Truman.

There is no percentage in endlessly rehearsing the need for civilian authority to affirm its ascendancy over the military. It is in civilian authority that the moral principle asserts itself. It is in the civilian sphere that the intellectual underpinnings of the state are developed and refined, and in which the conscience of a nation is vested. Inevitably, the principle of civilian control over the military would articulate itself in one way in America, where the government is headed by a civilian executive who is the titular commander-in-chief of the armed forces, and in another in Athens, which lacked (by choice) a chief executive. Truman could afford to be patient and (outwardly, at least) gracious; social as well as political realities dictated this behavior. Voters in the Athenian assembly, lacking not only a chief executive but also telephones, cameras, radios, and airplanes, were less disposed to compromise, to deliberate, to bide their time. The harder it was to know, the quicker they were to judge.

In fact, Truman had little choice about his bearing in the confrontation with MacArthur. For his hesitation to remove the Far Eastern commander had been grounded in a realistic awareness of the danger that MacArthur's cult might pose. The intensity of the initial response to Truman's dismissal of MacArthur was unprecedented and not a little frightening. For every letter and telegram received by the White House commending Truman's decision, twenty arrived in opposition. A Gallup poll showed that 69 percent of American voters supported MacArthur. A veteran in Atlanta returned his Bronze Star to Washington to protest Truman's decision. In Los Angeles a husband and wife were both jailed when their disagreement over the Truman–MacArthur controversy culminated in a fistfight. A Seattle logger tried to drown a friend in a bucket of beer for siding with Truman. More alarmingly, a senator who refused to permit his name to be mentioned revealed his concern that MacArthur's supporters might storm the White House.

Truman had reason to be afraid; what he feared was not so much Douglas MacArthur as the American people. For this reason, he told Merle Miller, he waited to relieve MacArthur until the general had given him a clear example of insubordination "that everybody would recognize for exactly what it was."[46] But the impeachers of Miltiades, of Cimon, of Pericles, of Alcibiades, of the generals who were put to death after the Battle of Arginusae—these were the people. Having nobody to fear, they had no reason to be gracious or amiable, hesitant or timid. The likelihood that MacArthur's supporters would stage a coup that would bring American democracy to an end was minimal, though the likelihood that the combined forces of MacArthurism and McCarthyism would deliver the White House to the Republicans in 1952 was considerable.[47] The danger of a coup in Athens was much greater; though history showed that oligarchy, not tyranny, was the form anti-democratic coups would take, this may not have been obvious even to the most prescient of Athenians. Although no tyrant seized Athens at any time after the expulsion of the Peisistratids in 510, still the fear of tyranny was real and powerful. Miltiades had been exposed to non-democratic ways during his years abroad, and Alcibiades was plainly a wild card. Thucydides was well aware of the connection between tyranny and the mass hysteria that accompanied the mutilation of the herms in 415, identifying fear of tyranny as the cause of the panic that erupted. It is hard to tell why exactly one would imagine, waking up to find the phalluses knocked off statues of Hermes, that a plot was afoot to establish a tyrant, but according to Thucydides, it was the fear and suspicion engendered by the memory of tyranny that provoked the outburst. Instead of investigating the character of their informers, Thucydides writes, the Athenians

> had regarded everything they were told as grounds for suspicion, and on the evidence of complete rogues had arrested and imprisoned some of the best citizens, thinking it better to get to the bottom of things in this way rather than to let any accused person, however good his reputation might be, escape interrogation because of the bad character of the informer. (Thuc. 6.53.2)

The climate Thucydides describes bears a chilling resemblance to the McCarthyism that infected America at the very time of the Truman–MacArthur conflict, and McCarthy's popularity compounded that of MacArthur in impressing on Truman the force of emotionalism and anti-communist hysteria in shaping American thought and behavior.

In many ways what is most noteworthy in Truman's conflict with MacArthur is how long it took for the president to brace himself for dismissing the runaway general. In fifth-century Athens, MacArthur would not have lasted anywhere near as long as in twentieth-century America; but if the career of Alcibiades is any indication, he would have stood a better chance of a festive reinstatement once the temper of the assembly had turned. (Unfortunately for historians, MacArthur's age—he was over seventy when Truman relieved him—prevented history from revealing what might have happened to the general in the presidency, say, of his champion Richard Nixon, which he did not live to see.) Alcibiades' recall is a stern reminder of the Athenians' ambivalence about principle. The Athenians of the war era—at least those who speak in the pages of Thucydides—evince a frank concern for practical realities that contrasts markedly with American respect for conventional morality. At one level the chilling dialogue at Melos is just that; at another, taken with the Athenians' blunt speech at Sparta in 432 and with Diodotus and Cleon's paired speeches on Mytilene in 428, it makes an almost refreshing contrast with the evangelical pieties of American cold war rhetoric. The Athenians never claimed they were destroying Melos in order to save it. After his defection to Sparta, Thucydides claims, Alcibiades curried favor by speaking against the deepest of Athenian principles, dismissing democracy as an "acknowledged folly" (6.89.6). MacArthur, speaking with Chief of Staff Dick Sutherland during World War II, spoke quite differently about the principles of American government. General George Kenney recalled that Sutherland "thought a democracy was a weak type of government in a national emergency. He thought that we might even do without elections in time of war and that there was too much debating by Congress on issues that the President could settle if he had a little more power." This, one should note, was pretty much exactly what Athenian oligarchs (and supporters of Alcibiades) thought they could and should do in 411.

Kenney continues:

> General MacArthur said, "No, Dick, you are wrong. Democracy as we have it in the United States is the best form of government that man has ever evolved."
> He went on to explain that as long as people are allowed freedom of thought and speech they will keeps their minds flexible and progressive. In the dictator state, true freedom disappears, and the mind becomes rigid and regimented. Regimented minds are not flexible enough to meet changed situations quickly in times of war. Eventually something goes wrong in the dictator's plan and some flexible-minded, free-thinking democratic leader takes advantage of the dictator's troubles and defeats him.[48]

The contrast between MacArthur's paean to democracy and Alcibiades' cavalier dismissal of it is dramatic, revealing not only Alcibiades' chameleon-like nature but also his Athenian pragmatism. But for all this, the two men shared a flexible brand of patriotism or pseudo-patriotism. Alcibiades was famous for redefining patriotism to suit his ends; in a famous passage in Thucydides, he tells the Spartans that his defection does not really indicate a lack of civic loyalty. Reframing his apparently treasonous conduct, he explains that

> The Athens I love is not the one which is wronging me now, but that one in which I used to have secure enjoyment of my rights as a citizen. The country that I am attacking does not seem to be mine any longer; it is rather that I am trying to recover a country that has ceased to be mine. And the man who really loves his country is not the one who refuses to attack it when he has been unjustly driven from it, but the man whose desire for it is so strong that he will shrink from nothing in his efforts to get back there again. (Thuc. 6.102.4)[49]

MacArthur also sought to recast patriotism to embrace his disregard for his commander-in-chief. In a speech delivered in the Massachusetts Legislature three months after his dismissal, he decried "laxity" and "corruption" in the Truman administration and "the threat of reprisal if the truth be expressed in criticism of those in higher public authority," declaring:

> I find in existence a new and heretofore unknown and dangerous concept, that the members of our armed forces own primary allegiance or loyalty to those who temporarily exercise the authority of the executive branch of the government rather than to the country and its Constitution which they are sworn to defend.[50]

The argument is somewhat different, but the special pleading is similar.

Examining MacArthur against the backdrop of Alcibiades makes plain that in both Athens and America, verbal skills, aristocratic heritage (real or imagined), charm, good looks, complaints of persecution, and the mystique that is enhanced by absence from home can work together to cover a multitude of sins not excluding flirtation with treason. Differences, however, stand out as well. The more developed and elaborately articulated democracy of the United States, with its checks and balances (as well as a tightly organized two-party system), is ultimately a less fluid system, with less tolerance accorded to erratic behavior and clearer definitions of what is and is not acceptable. Though Truman and MacArthur each embodied something important in American democracy, consensus about the power of the presidency ultimately prevailed, whereas Athenians, who had no chief executive and no written constitution, were amenable to embracing rogue aristocrats who deprecated democracy when it suited their whims. There is nothing like the specter of a sober bicameral legislature to discourage caprice. Memories of tyranny, moreover, worked in more than one way in Athens. On the one hand, the lack of a chief executive reflected a positive paranoia about according too much power to any one man. At the same time, however, because tyrants were somewhat ambivalently regarded, the notion of government in the hands of a charismatic autocrat held appeal for some and might under the strains of war seem to present a realistic alternative to democracy. Few Americans really wished to see a military government under MacArthur replace the traditional civilian government.

Erôs, Heroes, Leaders, and Followers

The relationship between civilian and military authority will always be problematic in wartime, and no solution is likely to be found to the problems that beset it. Where there is war there are heroes, and heroes are trouble. Indispensable for leadership and inspiration, the heroic temperament plays a problematic role in a democratic society. By definition, heroes are not like other people. Viewed in historical perspective, they are male; struggling against the influence of women (Hector is a conspicuous example), they renounce the comforts of family to risk life—and to take it. For the Greek hero—and the Greeks gave us the word—heroism was inescapably tied up with suffering and killing. Heroes do not play by the same rules as others, and are not judged by the same standards. They do not always act in their own best interests; MacArthur's biographer D. Clayton James entitled the chapter dealing with the events surrounding the general's dismissal "The Furies of Self-Destruction."[51]

Like tyrants, heroes are the object of erotic and romantic fascination at many levels—something that has not been suggested of Harry Truman. At the crudest level, both MacArthur and Alcibiades were much sought after as romantic partners, despite MacArthur's extreme difficulty in establishing a relationship until the death of his mother, when he was well into his fifties. Japanese women proclaimed their eagerness for him to father their children, and Alcibiades appears to have made reality from fantasy and impregnated the wife of the Spartan king during his stay in the Peloponnese. (The fathering of a child in exotic and potentially enemy territory seems to be part of the mystique of both the hero and the general; one thinks of Cleopatra's son Caesarion.)[52] But this is only the most superficial manifestation of the phenomenon. *Erôs,* the Greeks well knew, could only be directed at what one does not possess; absence, in the Greek view, makes the heart grow very much fonder indeed. By definition, *erôs* must be of limited duration, unless one has the good fortune never to attain the object of one's desires. Politicians know well the dangers of overexposure; Pericles knew it best of all, withdrawing from social life and cultivating aloofness. Still waters are imagined to run deep. But absent generals come home; when Alcibiades' fans got his wish and the longed-for hero returned, his popularity dropped, and he was exiled again, this time for good. And the hysterical outcry that had greeted MacArthur's dismissal died down promptly as the pendulum swung back to Truman. Both Alcibiades and MacArthur played better abroad than at home; significantly, MacArthur's support came from civilians long distant from the theater of war, while many of the men who served under him bitterly resented his grandstanding. (Not for nothing was the arena of death known as a "theater." In his letters to his wife, General Robert Eichelberger, who served under MacArthur in World War II, rarely mentioned MacArthur by name, referring to him rather by the code name Sara Bernhardt.

Noting this revealing choice, Robert Smith argues that MacArthur's compulsive self-advertisements led many colleagues and subordinates "to despise him with a fury not even his death could extinguish.")[53]

And the sexiness of equality has always been dubious; just as *erôs* appears to be enhanced by scenarios of dominance and submission, of struggle and division, of difference and distinction, so it runs the risk of being dampened by cooperation and familiarity. Even in classical Athens, where a good case could be made that men enjoyed having sex with one another in part because it was more of an accomplishment to have sex with another voter than with a disfranchised woman, sexual relationships among democrats were anything but equal; an unbalanced relationship between an older man and a younger one was normative while sex between age-mates was considered deviant. Ultimately Pericles' exhortation to his countrymen to fix their eyes on the polis and fall in love with her was probably designed to divert their *erôs* not only from individual romantic relationships but also from its natural outlet in competition and war. When Diodotus, speaking in the pages of Thucydides, brings up *erôs*, it is the *erôs* that drives men to undertake plans against others—to murderous, combative enterprises like rebelling against the Athenian empire. Arguing that executing all the rebellious Mytileneans who have surrendered is not in Athens' interest, he dismisses the argument for deterrence by suggesting that the motives for such undertakings as rebellions are not rational but emotional: *erôs* and *elpis* (hope) are everywhere, he says—"invisible factors, but more powerful than the terrors that are obvious to our eyes" (Thuc. 3.45.5).

Erôs, however, works in both directions; the people long for the hero, and the hero longs for combat. Aristophanes in *The Frogs* (1.1425) commented on the people's yearning for Alcibiades: "They long for him; they hate him; and then again they want him." Alcibiades, meanwhile—or so the story went—delighting in the attention he was attracting, departed from custom by having a golden shield made for him and emblazoned with an image of Erôs armed with a thunderbolt: make love *and* war. (Respectable people, Plutarch says, were horrified and found his style *turannika*, characteristic of a tyrant [16.2].) MacArthur, similarly, drew attention by his idiosyncratic gear, and as he was longed for, so he longed. Looking back on the First World War while addressing veterans in 1935, he expressed a yearning for that war that was profoundly romantic, as he spoke of "those days of old that have vanished tone and tint" and "gone glimmering through the dreams of things that were":

> Their memory is a land where flowers of wondrous beauty and varied colors spring, watered by tears and coaxed and caressed into fuller bloom by the smiles of yesterday. Refrains no longer rise and fall from that land of used-to-be. We listen vainly but with thirsty ear for the witching melodies of days that are gone. . : .[54]

If this yearning for war is not romance, not *erôs*, then what is it? The spectacle of a man listening "vainly but with thirsty ear" for the "witching melodies" of a horrific war is eerie and compelling, but not uncommon or surprising.[55]

Erôs, then, tinges and tightens not only the reciprocal bond between leaders and followers but between warriors and war. War needs warriors, but warriors also need war to be complete; and leaders require followers not only definitionally but also for the adulation they provide. Men like Alcibiades and MacArthur need to be needed, want to be wanted. Despite their drive to uniqueness—a preoccupation both men revealed in their dress and their battle gear—they could not exist in a vacuum. It is this reciprocal dynamic to which Plutarch is referring when he writes, "Now at least Alcibiades was overcome with the longing to see his native country again, and still more to let his countrymen see him crowned with the honours of all his victories over their enemies" (*Alc.* 32.1). As the classical scholar Jean-Pierre Vernant has put it in his essay "One . . . Two . . . Three: *EROS*,"

> The sexual dichotomy or duality of roles in an erotic relationship forces each partner to experience his own incompleteness in the impulse toward the other. That relationship is evidence of the individual's inability to remain within his limits, to be satisfied with what he is, to accept himself in his uniqueness, without seeking to duplicate himself in and through the other, the object of his amorous desire. This is what I have called the reflexive relationship.[56]

Not surprisingly, Vernant soon finds himself talking about Narcissus.

* * *

The narcissist Douglas MacArthur would have seen nothing peculiar in the endeavor in which we are engaged in this volume, exploring possible parallels between modern and ancient civilizations; neither would Truman. An avid reader who greatly enjoyed his Plutarch, the president spent one April building a replica of a Roman bridge described in Caesar's *Gallic War* with a couple of friends. MacArthur also took considerable pride in his familiarity with history in general and with the classical world in particular. In his *Reminiscences* he recalled how "dull Latin and Greek seemed a gateway to the moving words of the leaders of the past" and "laborious historical data led to the nerve-tingling battlefields of the great captains."[57] Contemplating his handiwork in Japan, the general found the miracle of the occupation a source of wonder and took pleasure in comparing his land reforms to those of the Gracchi in Rome; he even suggested that his work represented an advance from the days in which Plato wrote his *Republic*.[58] MacArthur's eager admirers at *Life* magazine did not fail to pick up on MacArthur's erudition, rhapsodizing in the spring of 1951: "Greece, Rome, the Middle Ages, the Renaissance, the age of Britain's greatness—all the splendid and tragic meanings of the drama of these centuries are the constant prompters of his mind and spirit." MacArthur no doubt delighted in these words. He would probably also have been pleased by the dignity with which his biographer William Manchester imbued him—until he reached the end of the story. "In the Attic tragedies of Aeschylus, Euripides, and Sophocles," Manchester wrote in a paragraph that brings together all the customary buzzwords that mark modern discussions of Greek tragedy, the hero is a figure of massive integrity and powerful will, a paradox of outer poise and inner passion who recognizes the inevitability of evil, despair, suffering, and loss. Choosing a perilous course of action despite the counsel of the Greek chorus, he struggles nobly but vainly against fate, enduring cruelty and, ultimately, defeat, his downfall being revealed as the consequence of a fatal defect in his character which, deepened by tumultuous events, eventually shatters him.

So it was with Douglas MacArthur. Brave, brilliant, and majestic, he was a colossus bestriding Korea until the nemesis of his hubris overtook him.[59]

It was the unfortunate end of such heroes that underlay Lawton Collins's construction of MacArthur as marching "like a Greek hero of old to an unkind and inexorable fate."[60]

Like Greek heroes, old soldiers do die; they do not simply fade away. Miltiades died of a wound in his thigh, sustained in the Paros campaign that was his downfall; because of the gangrene in his leg, he had been forced to attend his trial on a stretcher (Herod. 6.136). The Roman historian Cornelius Nepos contends that he in fact died in prison; according to Plato, the Athenians considered executing him by throwing him into the pit known as the *barathron*, a distinctly unpleasant way to die (Nepos, *Miltiades* 8; Plato, *Gorgias* 516A). Alcibiades was murdered by assassins in Asia Minor for an uncertain purpose—under orders from Sparta, or at the behest of the oligarch Critias, or because he had seduced his assassins' sister, or perhaps for other reasons entirely (Plut. *Alcibiades* 39). His assailants set fire to his house during the night and shot him down with javelins and arrows as he sought to escape the flames; Plutarch said that his girlfriend Timandra buried him with as much brilliance and honor as she could—in her clothes (*Alcibiades* 39.4). Douglas MacArthur died in Walter Reed Hospital at the age of eighty-four after a series of operations and was buried with full military honors.

Caesar, Aemilius Paulus, Alexander, a Greek hero marching to his fate— MacArthur, it seems, could identify or be identified with anyone from the Greco-Roman world but a historical Athenian. But though the eminently accountable world of classical Athens was not MacArthur's milieu, in many ways it provides a context in which the issues raised by his career and its close can usefully be explored—issues of accountability, of the relationship between civilian and military authority, and of the cult of personality in its fragile relationship with democracy.[61] As an emergent political form, classical Athens tolerated a kind of openness not common in the United States today or half a century ago—candor about imperialism, flagrant sexuality in public men (including open bisexuality, a rare behavior pattern in distinguished American generals), and a frank willingness to humiliate and, indeed, kill. Cold war America manifested a more controlled style, substituting civilities for overt attacks of both a verbal and a physical nature. "Executions" are metaphorical; the modern democracy is more prone to sublimating emotion and violence, attenuating and transforming them. For all these vital distinctions, many of the problems that surround issues of leadership and heroism in a democratic society remain constant. In both cold war America and fifth-century

Athens, democracy was held together by a creative tension between the stability of ordinary folk and the infusions of energy and romance (not to mention turmoil) that came from extraordinary individuals. Unquestionably, men like Alcibiades and Miltiades were disciplined from below whereas MacArthur was disciplined from above; but a case could be made that in his studied down-home folksiness Harry Truman, the quintessential ordinary man, set himself up as the American counterpart to the Athenian demos. If Athenian generals were called to account by all men, MacArthur finally met his match in Everyman.

Notes

1. Eva Keuls, *The Reign of the Phallus: Sexual Politics in Ancient Athens* (New York: Harper and Row, 1985), 12.
2. The extraordinary ambivalence Greek political mythology manifested toward tyrants and the legends that surrounded them is explored in James McGlew, *Tyranny and Political Culture in Ancient Greece* (Ithaca: Cornell University Press, 1993).
3. The fourth-century orator Demosthenes went so far as to suggest that anyone who tries to change a single law of the Athenian democracy was obviously trying to establish a tyranny (Dem. 24, passim). Anti-democratic thinkers saw things very differently. Fourth-century political theorists like Plato and Aristotle would draw close connections between tyranny and democracy. In the "cycle of constitutions" in Book 8 of the *Republic* Plato saw democracy as leading to tyranny, and Aristotle in the *Politics* portrayed radical democracy as a form of tyranny, complaining that democratic government "becomes analogous to the tyrannical form of single-person government. Both show a similar temper; both behave like despots to the better class of citizens; the decrees of the one are like the edicts of another" (1292a). I cite the *Politics* in the translation of Ernest Barker, *The Politics of Aristotle* (Oxford: Oxford University Press, 1946.)
4. On ostracism, see J.L. Carcopino, *L'ostracisme athénien* (Paris: F. Alcan, 1935); A. Raubitschek, "The Origin of Ostracism," *American Journal of Archaeology* 55 (1951): 221–29; Charles Hignett, *A History of the Athenian Constitution to the End of the Fifth Century B.C.* (Oxford: Clarendon Press, 1952); C.A. Robinson, "Cleisthenes and Ostracism," *American Journal of Archaeology* 56 (1956): 23–26; Donald Kagan, "The Origin and Purposes of Ostracism," *Hesperia* 30 (1961): 393–401; R. Thomsen, "The Origin of Ostracism. A Synthesis," *Humanitas* 4 (Copenhagen, 1972); Jennifer Roberts, *Accountability in Athenian Government* (Madison: University of Wisconsin Press, 1982), 142–60; Thomas Figueira, "Residential Restrictions on the Athenian Ostracized," *Greek, Roman, and Byzantine Studies* 28 (1987): 281–305; Lindsay G.H. Hall, "Remarks on the Law of Ostracism," *Tyche* 4 (1989): 91–100; Peter Siewert, "Accuse contro i candidati all'ostracismo per la loro condotta politica e morale," in Marta Sordi, ed., *L'immagine dell'uomo politico: vita pubblica e morale nell'antichità*, Pubblicazione della Università Catolica del Sacro Cuore storiche No. 17 (Milano: Vita e Pensiero, 1991): 3–14; Mogens H. Hansen, *The Athenian Democracy in the Age of Demosthenes: Structure, Principles and Ideology* (Oxford: Blackwell, 1991); and Matthew Christ, "Ostracism, Sycophancy, and Deception of the Demos: [Arist.] *Ath. Pol.* 43.5," *Classical Quarterly* 42 (1992): 336–46.
5. Accountability trials in Athens are discussed in M.H. Hansen, *Eisangelia: The Sovereignty of the People's Court in Athens in the Fourth Century B.C. and the Impeachment of Generals and Politicians* (Odense: Odense University Classical Studies 6, 1975); P.J. Rhodes, "EISANGELIA in Athens," *Journal of Hellenic Studies* 99 (1979): 103–14; Roberts, *Accountability*; Edwin Carawan, "Apophasis and Eisangelia: The Role of the Areopagus in Athenian Political Trials," *Greek, Roman, and Byzantine Studies* 26 (1985): 115–40, and

"Eisangelia and Euthyna: The Trials of Miltiades, Themistocles, and Cimon," *Greek, Roman, and Byzantine Studies* 28 (1987): 167–208; and Richard Bauman, *Political Trials in Ancient Greece* (London: Routledge, 1990). Hansen assembles all the ancient sources on the sensational impiety trials of 415; see also Douglas MacDowell's edition of Andocides' speech relating to the affair (*Andokides on the Mysteries* [Oxford: Clarendon, 1962]) and the discussion in Donald Kagan, *The Peace of Nicias and the Sicilian Expedition* (Ithaca: Cornell University Press, 1981), 192–209. The other celebrated case is the Arginusae trial of 406, in which eight generals were sentenced to death for failing to retrieve sailors from the water after a stunning naval victory; the sources disagree as to whether these sailors were already dead or not. Because of the religious importance of proper burial, in Greece even dead sailors had a considerable claim on their generals' attention. The most elaborate discussion of the episode appears in Donald Kagan, *The Fall of the Athenian Empire* (Ithaca: Cornell University Press, 1987), 325–75; it is also discussed in Paul Cloché, "L'affaire des Arginuses," *Revue historique* 130 (1919): 3–68; Antony Andrewes, "The Arginousai Trial," *Phoenix* 28 (1974): 112–22; Roberts, *Accountability*, 64–69; Graham Wylie, "The Battle of the Arginusae: A Reappraisal," *Civiltà classica e cristiana* (Genoa) 11 (1990): 234–49; Mabel Lang, "Illegal Execution in Ancient Athens," *Proceedings of the American Philosophical Society* 134 (1990): 2429, and "Theramenes and Arginousai," *Hermes* 120 (1992): 267–69.

6. Westlake's remarks appear in "Paches," *Phoenix* 29 (1975): 109. It is difficult to place much credence in the accusations of bribery frequently leveled against Athenian generals—against Xenophon, Hestiodorus, and Phanomachus, for example, who accepted the surrender of Potidaea in 430 without checking first with the assembly, or of Sophocles, Pythodorus, and Eurymedon, who had withdrawn from Sicily in 424 without preventing the Sicilians from coming to terms among themselves. The fact that Pericles began his political career in the 460s by accusing the wealthy and patriotic Cimon of taking bribes in exchange for not attacking Macedonia is a clear index that an accusation of *dôrodokia*—the taking of bribes—was a standard way of undermining a politician and did not always suggest that anyone really believed such a thing had happened.

Athenians did not share the purist American view of what constitutes bribery; accepting gifts from foreign powers was acceptable as long as it did not result in actions contrary to the interest of the state (see, e.g., Hyperides 5.24–25). For Greeks, the giving and receiving of gifts was central to the ways in which aristocrats affirmed and enhanced their status. This phenomenon is apparent to many societies; see for example M. Sahlins, *Tribesmen* (Englewood Cliffs, NJ: Prentice Hall, 1968). In Greece it is particularly conspicuous in the Homeric poems, on which see W. Donlan, "Reciprocities in Homer," *Classical World* 75 (1982); on the ethos and practice in classical Athens, see J.T. Roberts, "Aristocratic Democracy: The Perseverance of Timocratic Principles in Athenian Government," *Athenaeum* n.s. 64 (1986): 355–69. The trials of the period immediately after the Peloponnesian War shed light on the Athenian ethos as a whole; see Barry Strauss, "The Cultural Significance of Bribery and Embezzlement in Athenian Politics: The Evidence of the Period 403–386 B.C.," *Ancient World* 11 (1985): 67–74. The issues are also discussed in F.D. Harvey, "Dona Ferentes: Some Aspects of Bribery in Greek Politics," in Paul Cartledge, ed., *Crux: Essays in Greek History Presented to G.E.M. de Ste. Croix on His 75th Birthday* (London: Duckworth, 1985), 76–113.

One must be similarly skeptical of the frequent charges of treason, such as the one made in 409 against Anytus, the future prosecutor of Socrates, when he failed to round Cape Malea in order to relieve Pylos, under attack by the Spartans.

7. In Athens as in America, there was no law against being stupid, but it could get you into a lot of trouble. Compare the observations of Harry Truman about his removal of Douglas MacArthur: "I didn't fire him because he was a dumb son of a bitch, although he was, but that's not against the law for generals" (spoken in conversation with Merle Miller, in *Plain Speaking: An Oral Biography of Harry S. Truman* [New York: Berkley, 1974], 287).

8. *Handbook of Greek Constitutional History* (New York: Macmillan, 1902), 179.

9. Cited in John Gunther, *The Riddle of MacArthur: Japan, Korea and the Far East* (New York: Harper and Row, 1951), 23–24.

10. Richard Rovere and Arthur M. Schlesinger, Jr. *The General and the President, and the Future of American Foreign Policy* (New York: Farrar, Straus and Young, 1951), 91; D. Clayton James, *The Years of MacArthur. Volume III: Triumph and Disaster, 1945–1964* (Boston: Houghton Mifflin, 1985), 616; Laurence S. Wittner, ed., MacArthur (Englewood Cliffs: Prentice Hall, 1971), 3.

11. Cited in Robert J. Donovan, *The Tumultuous Years: The Presidency of Harry S. Truman 1949–1953* (New York: W.W. Norton, 1982).

12. On the White House policy, particularly as formulated in NSC-68, see John Lewis Gaddis, *Strategies of Containment: A Critical Appraisal of Postwar American National Security Policy* (New York: Oxford University Press, 1982), chapter 4.

13. From William J. Sebald and Russell Brines, *With MacArthur in Japan: A Personal History of the Occupation* (New York: W.W. Norton, 1965), 123.

14. Harry S. Truman, *Memoirs, Volume Two: Years of Trial and Hope* (New York: Doubleday, 1956), 381.

15. Ibid., 382–84; see also other sources cited in William Manchester, *American Caesar: Douglas MacArthur 1880–1964* (Boston: Little, Brown, 1978), 741.

16. Truman comments on the force of the "China lobby" at this time in Miller, *Plain Speaking*, 287–89.

17. The text of MacArthur's statement of March 24 and his letter to Martin, along with their context, can be found in James, *The Years of MacArthur, Vol. III*, 585–90.

18. *London Times*, March 10, 1951.

19. Useful discussions of the circumstances leading up to MacArthur's dismissal include those in Rovere and Schlesinger, *The General and the President*; Omar Bradley and Clay Blair, *A General's Life: An Autobiography* (New York: Simon and Schuster, 1983), 624–36; Rosemary Foot, *The Wrong War: American Policy and the Dimensions of the Korean Conflict, 1950–1953* (Ithaca: Cornell University Press, 1985), 131–48; Sebald, *With MacArthur in Japan*, 211–30; Michael Schaller, *Douglas MacArthur: The Far Eastern General* (New York: Oxford University Press, 1989), 228–40; Manchester, *American Caesar*, 621–47; Ronald Caridi, *The Korean War and American Politics: The Republican Party as a Case Study* (Philadelphia: University of Pennsylvania Press, 1968), 141–75; and James, *The Years of MacArthur*, Vol. III, 584–621.

20. Truman, *Memoirs*, 355.

21. Cited in Miller, *Plain Speaking*, 305.

22. Truman, *Memoirs*, 384.

23. Ibid., 381.

24. Allen Guttmann, ed., *Korea and the Theory of Limited War* (Boston: D.C. Heath, 1967), 17.

25. Concerns about security were also voiced by Richard Nixon, who would have his own career brought to an end by the machinery of accountability: a resolution introduced by Nixon in the wake of MacArthur's dismissal claimed that it had "tragically weakened" the "strength and morale of the Armed Forces of the U.S. now engaged in the defense of the Nation against our enemies" (cited in Wittner, *MacArthur*, 105.)

26. "General Douglas MacArthur," *Nation*, 198 (April 20, 1964): 385.

27. Cited from Ridgway's diary entry for April 12, 1951, in Donovan, *The Tumultuous Years*, 360.

28. Reminiscing with Merle Miller about his trip to Wake Island, Truman said that MacArthur was "just like a little puppy at that meeting. I don't know "which was worse," he said, "the way he acted in public or the way he kissed my ass at that meeting." Miller was intrigued by the incompatibility of this assessment with Truman's observation in his *Mem-*

oirs to the effect that "The general seemed genuinely pleased at this opportunity to talk with me, and I found him a most stimulating and interesting person. Our conversation was very friendly—I might say much more so than I had expected." Miller wanted to ask Truman about the discrepancy but decided against it "since I wanted our conversations to continue" (*Plain Speaking*, 295–96).

29. Lawrence's remarks appear in "A Salute to Courage," *U.S. News and World Report*, 30 (April 27, 1951): 76.

30. Truman, *Memoirs*, 382.

31. Distressed commentators on the relief of MacArthur often focused on his superiority to the man who had dismissed him. One telegram read: "When an ex–National Guard Captain fires a five-star general impeachment of the National Guard Captain is in order" (Rovere and Schlesinger, 8). Compare Xenophon's depiction of Alcibiades' return to Athens: "Everyone wanted to see and to wonder at the sight of the great Alcibiades. He, it was said, was the best citizen they had got and he alone had been banished not because he deserved it but because of the intrigues of people who were inferior to him in power, who lacked his abilities to speak and whose only political principle was their own self-interest" (Hell. 1.4.13, cited in Rex Warner's translation, *A History of My Times [Hellenica]* [Harmondsworth: Penguin, 1979].)

32. Bradley and Ridgway both complained of MacArthur's erroneous reportage, and the general's habit of exaggerating his own role in military successes was well known; see Bradley and Blair, *A General's Life*, 624–36 and passim.

Devastating indictments of MacArthur's veracity and his unwillingness to share credit with others appear in Rovere and Schlesinger, *The General and the President*, 71–72 and 75–76; Hamby, *Man of the People: A Life of Harry S. Truman* (New York: Oxford University Press, 1994), 558; and Schaller, *Douglas MacArthur*, 72–74. See also Matthew Ridgway's frustration with MacArthur's self-promotion expressed, e.g., in *The Korean War: How We Met the Challenge: How All-Out Asian War Was Averted; Why MacArthur Was Dismissed; Why Today's War Objectives Must Be Limited* (Garden City, NY: Doubleday, 1967), 108–9; Ridgway's remarks are discussed in James, *The Years of MacArthur: Vol. III*, 576.

33. In Rovere and Schlesinger, *The General and the President*, 90–91.

34. Truman, *Memoirs*, 362–63.

35. Both conversations are cited in Schaller, *Douglas MacArthur*, 126–27.

36. Cited in Schaller, *Douglas MacArthur*, 74. Truman observed that MacArthur struck him "as a man there wasn't anything real about" (cited in Miller, *Plain Speaking*, 316).

37. Schaller (passim) comments on Louise Brooks's denigration of MacArthur in the sexual arena and on MacArthur's sexual life generally; see also Carol Petillo, *Douglas MacArthur: The Philippine Years* (Bloomington: Indiana University Press, 1981). Perret provides a revisionist view (*Old Soldiers Never Die*, 127).

38. Barry Strauss has pointed out to me that Plutarch is suspected of having taken this tale from comedy; see D.A. Russell, "Plutarch: Alcibiades 1–16," *Proceedings of the Cambridge Philological Society* 12 (1966): 37–47; but even if this is so, it is significant that the story should have become part of the literature that surrounded Alcibiades. See Strauss' excellent discussion of Alcibiades' persona in *Fathers and Sons in Athens: Ideology and Society in the Era of the Peloponnesian War* (Princeton: Princeton University Press, 1993), 148–53.

39. Plutarch tells how Alcibiades, biting a wrestling opponent to save himself from a fall, prompted the remark that he bit the way women do. He replied that, no, he bit the way lions do (Alcibiades 2.2). MacArthur's clothing is discussed in Petillo, *Douglas MacArthur: The Philippine Years*, 121.

40. Cited in Rovere and Schlesinger, *The General and the President*, 120–21.

41. Ibid., 121.

42. Cited in James, *The Years of MacArthur: Vol. III*, 599.

43. George C. Kenney, *The MacArthur I Know* (New York: Duell, Sloan and Pearce, 1951), 248–49.

44. On Phocion and Athenian politics, see Lawrence Tritle, "Leosthenes and Plutarch's View of the Strategia," *Ancient History Bulletin* 1 (1987): 6–9; and my (confirming) response, "Paradigm Lost: Tritle, Plutarch and Athenian Politics in the Fourth Century," *Ancient History Bulletin* 1 (1987): 34–35, as well as Tritle's *Phocion the Good* (London and New York: Croom Helm, 1989).

45. "MacArthur's War Party," *New Republic*, 124 (April 23, 1951): 5. Curiously, Truman ascribed MacArthur's failure to ride the wave of his popularity into public office to his lack of political sense. If the general "had had any background in politics at all instead of being completely military," Truman later observed, "he could have made good use of the situation, but he didn't do it" (Miller, *Plain Speaking*, 311).

46. Miller, *Plain Speaking*, 302. Seeking precedents for the situation in which he found himself, Truman found himself contemplating Lincoln's decision to relieve General George B. McClellan. It did not escape his notice that McClellan allied himself with Lincoln's opposition in Congress and ultimately ran against him for president.

47. As Truman foresaw, the uproar over MacArthur lasted only a few weeks. Yet the troubled Korean War, it should be noted, cost the democrats the election in 1952 and sent a general to the White House for the first time during the twentieth century.

48. Kenney, *The MacArthur I Know*, 113.

49. The discussion by Nathan Pusey of "Alcibiades and to philopoli" in *Harvard Studies in Classical Philology* 51 (1950): 215–23 remains of interest here.

50. MacArthur made these remarks on July 25, 1951; the lines that follow immediately contain a reference to the Roman empire in which MacArthur speaks less kindly of Rome than was his custom: "No proposition could be more dangerous. None could cast greater doubt upon the integrity of the armed services. For its application would at once convert them from their traditional and constitutional role as the instrument for the defense of the Republic into something partaking of the nature of a pretorian guard, owing sole allegiance to the political master of the hour."

51. James, *The Years of MacArthur*, 560, 604.

52. Alcibiades was also said (Plut. *Alc.* 16.4–5) to have fathered and raised a child by a Melian woman after the Athenian execution of all the Melian men.

53. Robert Smith, *MacArthur in Korea* (New York: Simon and Schuster, 1982), 11.

54. Cited in Rovere and Schlesinger, *The General and the President*, 37.

55. The aural element in this romance also appears in MacArthur's famous speech at West Point in May of 1962, in which he concluded with the promise that "when I cross the river, my last conscious thoughts will be of the corps, and the corps, and the corps." Much of that speech was pirated from earlier pieces; once more the speaker is listening "vainly, but with thirsty ear," and in his dreams, he reports, "I hear again the crash of guns, the rattle of musketry, the strange, mournful mutter of the battlefield. But in the evening of my memory I come back to West Point. Always there echoes and re-echoes: Duty, honor, country." War for MacArthur had great auditory appeal; he identified the sound of bugles as his first memory (see, e.g., Perret, *Old Soldiers Never Die*, 3). Writing in his Reminiscences of his father's eagerness to serve in the Civil War, he tells how seventeen-year-old Arthur MacArthur joined the 24th Wisconsin Volunteer Infantry despite Lincoln's efforts to put him off. "But young Arthur did not wait," his son writes: "Perhaps he could hear the roll of the war drums and the old clan war cry 'Listen! O Listen!'" (*Reminiscences* [New York: McGraw Hill, 1964], 6). Again and again, MacArthur conceives the lure of war in the guise of a call—a vocation, but also a voice. After the war, he reports, his father studied law for a year but "the call of the West was in his blood" and he soon was "engaged in the onerous task of pushing Indians into the arid recesses of the Southwest." Lamenting the disappearance of "the vital vestiges of that blazing era," he comments that "The West of old is gone, quite beyond re-*call*" (*Reminiscences*, 11–12, italics mine). MacArthur is

far from alone in conceiving the attraction of war in these terms; after a preamble entitled "Reveille," Manchester's prologue bears the title "First Call." The last chapter, of course, is "Taps."

56. Trans. Deborah Lyons, reprinted in David Halperin, John Winkler, and Froma Zeitlin, ed., *Before Sexuality: The Construction of Erotic Experience in the Ancient Greek World* (Princeton: Princeton University Press, 1990), 465–78; these remarks appear on p. 469.

57. MacArthur, *Reminiscences*, 17.

58. Ibid., x.

59. Manchester, *American Caesar*, 616–17.

60. J. Lawton Collins, *War in Peacetime: The History and Lessons of Korea* (Boston: Houghton Mifflin, 1969), 142.

61. I consider it a great privilege to have participated in this project and in the conference connected with it. We Greek historians, for whom war is something we read about in books and have begun to see on color television, learned more than we can say from the experience of talking about war and killing with the Korean panelists. Their courage and heroism in the face of devastating loss and unspeakable brutality inspires us to renew our attempts, however doomed, to recapture the experience of the far-away and long-ago Greeks whose wars speak to us largely from the printed page, devoid of blood and tears.

The debts I have incurred in preparing this essay are many. Peter Euben's comments on the paper delivered in May of 1995 were thoughtful and incisive. Paul Cartledge was kind enough to offer numerous valuable comments of his own. Ellen Schrecker took time to suggest additional (and better) books for me to read. Robert Lejeune and Thomas Knock read drafts of the paper carefully and made numerous useful suggestions. Most of all, Barry Strauss read numerous versions of this paper and offered much appreciated guidance at every turn. His patience has been great, and he is in no way responsible for the faults of my nervous venture into comparative history. I am greatly indebted to him, to David McCann, to James Morris, and to Susan Nugent for their hard work in making this project a reality.

Bibliography

Adcock, F.E., and D.J. Mosley. *Diplomacy in Ancient Greece*. London: Thames and Hudson, 1975.

Andrewes, Antony. "The Arginousai Trial," *Phoenix* 28 (1974): 112–22.

Arnheim, M.T.W. *Aristocracy in Greek Society*. London: Thames and Hudson, 1977.

Aurenche, O. *Les groupes d'Alcibiade, de Léogoras et de Teucros. Remarques sur la vie politique athénienne en 415 avant J.C.* Paris: Belles Lettres, 1974.

Bauman, Richard. *Political Trials in Ancient Greece*. London: Routledge, 1990.

Berve, H. *Miltiades: Studien zur Geschichte des Mannes und seiner Zeit. Hermes* Suppl. 2, 1937.

Black, C. *Impeachment: A Handbook*. New Haven: Yale University Press, 1974.

Bloedow, E. *Alcibiades Re-examined. Historia* Suppl. 21. Wiesbaden, 1973.

Bradley, Omar, and Clay Blair. *A General's Life: An Autobiography*. New York: Simon and Schuster, 1983.

Carawan, Edwin. "*Apophasis* and *Eisangelia*: The Role of the Areopagus in Athenian Political Trials." *Greek, Roman, and Byzantine Studies* 26 (1985): 115–40.

———. "*Eisangelia* and *Euthyna*: The Trials of Miltiades, Themistocles, and Cimon." *Greek, Roman, and Byzantine Studies* 28 (1987): 167–208.

Cloché, Paul. "Les hommes politiques et la justice populaire dans l'Athènes du IVe siècle." *Historia* 9 (1960): 80–95.

Collins, J. Lawton. *War in Peacetime: The History and Lessons of Korea*. Boston: Houghton Mifflin, 1969.

Connor, W. Robert. *The New Politicians of Fifth Century Athens*. Princeton: Princeton University Press, 1971.

Cumings, Bruce. *The Origins of the Korean War: Liberation and the Emergence of Separate Regimes 1945–47, Vol. 1.* Princeton: Princeton University Press, 1981.

———. *Child of Conflict: the Korean-American Relationship, 1943–1953.* Seattle: University of Washington Press, 1983.

Donlan, Walter. *The Aristocratic Ideal in Ancient Greece: Attitudes of Superiority from Homer to the End of the Fifth Century B.C.* Lawrence, Kansas: Coronado Press, 1980.

Donovan, Robert J. *The Tumultuous Years: The Presidency of Harry S. Truman 1949–1953.* New York: W.W. Norton, 1982.

Finley, Moses. "The Athenian Demagogues." *Past and Present* 21 (1962): 3–24.

———. *Politics in the Ancient World.* Cambridge: Cambridge University Press, 1983.

Foot, Rosemary. *The Wrong War: American Policy and the Dimensions of the Korean Conflict, 1950–1953.* Ithaca: Cornell University Press, 1985.

Forde, Steven. *The Ambition to Rule: Alcibiades and the Politics of Imperialism in Thucydides.* Ithaca: Cornell University Press, 1989.

Gaddis, John Lewis. *Strategies of Containment: A Critical Appraisal of Postwar American National Security Policy.* Oxford and New York: Oxford University Press, 1982.

Greenidge, A.H.J. *Handbook of Greek Constitutional History.* London and New York: Macmillan, 1902.

Guttmann, Allen, ed. *Korea and the Theory of Limited War.* Boston: D.C. Heath, 1967.

Hamby, Alonzo. *Man of the People: A Life of Harry S. Truman.* Oxford and New York: Oxford University Press, 1994.

Hansen, Mogens Herman. *Eisangelia: The Sovereignty of the People's Court in Athens in the Fourth Century B.C. and the Impeachment of Generals and Politicians.* Odense: Odense University Classical Studies 6, 1975.

———. "*Rhetores* and *Strategoi* in Fourth-Century Athens." *Greek, Roman and Byzantine Studies* 24 (1983): 151–80.

———. *The Athenian Democracy in the Age of Demosthenes: Structure, Principles, and Ideology.* Oxford: B. Blackwell, 1991.

Hanson, Victor. *The Western Way of War: Infantry Battle in Classical Greece.* New York: Knopf, 1989.

Harrison, A.R.W. *The Law of Athens: Procedure.* Oxford: Oxford University Press, 1971.

Harvey, F.D. "*Dona Ferentes*: Some Aspects of Bribery in Greek Politics." In Paul Cartledge, ed., *Crux: Essays in Greek History Presented to G.E.M. de Ste. Croix on His 75th Birthday* (London: Duckworth, 1985), 76–113.

Hegel, G.W.F. *The Philosophy of History*, trans. J. Sibree. Reprinted New York: Dover, 1956.

Herodotus. *The History*, trans. David Grene. Chicago: University of Chicago Press, 1987.

James, D. Clayton. *The Years of MacArthur. Volume III: Triumph and Disaster, 1945–1964.* Boston: Houghton Mifflin, 1985.

Kagan, Donald. "The Origin and Purposes of Ostracism." *Hesperia* 30 (1961): 393–401.

Kenney, George C. *The MacArthur I Know.* New York: Duell, Sloan and Pearce, 1951.

Kinzl, Konrad. "Miltiades Parosexpedition in der Geschichtsschreibung," *Hermes*, 104 (1976): 280–307.

Lipset, Seymour. *Political Man*, updated edition. Baltimore: Johns Hopkins University Press, 1981.

MacArthur, Douglas. *Reminiscences.* New York: McGraw-Hill, 1964.

McCullough, David. *Truman.* New York: Simon and Schuster, 1992.

MacDowell, Douglas. *The Law in Classical Athens.* Ithaca: Cornell University Press, 1978.

McGlew, James. *Tyranny and Political Culture in Ancient Greece.* Ithaca: Cornell University Press, 1993.

Manchester, William. *American Caesar: Douglas MacArthur 1880–1964.* Boston: Little, Brown, 1978.

Miller, Merle. *Plain Speaking: An Oral Biography of Harry S. Truman.* New York: Berkley, 1974.

Perlman, Shalom. "The Politicians in the Athenian Democracy of the Fourth Century B.C." *Athenaeum* 41 (1963): 327–55.

————. "Political Leadership in Athens in the Fourth Century B.C." *Parola del Passato* 22 (1967): 161–76.

Perret, Geoffrey. *Old Soldiers Never Die: The Life of Douglas MacArthur.* New York: Random House, 1996.

Plutarch. *The Rise and Fall of Athens,* trans. Ian Scott-Kilvert. Harmondsworth: Penguin, 1960.

Pritchett, W. Kendrick. *The Greek State at War.* Berkeley and Los Angeles: University of California Press, 5 vols., 1974–1991.

Raubitschek. A.E. "Athenian Ostracism," *Classical Journal* 48 (1952–53): 113–22.

Reverdin, Olivier. "Remarques sur la vie politique d'Athènes au Ve siècle." *Museum Helveticum* 2 (1945): 201–12.

Ridgway, Matthew. *Soldier: The Memoirs of Matthew B. Ridgway.* New York: Harper, 1956.

————. *The Korean War: How We Met the Challenge: How All-Out Asian War Was Averted; Why MacArthur Was Dismissed; Why Today's War Objectives Must Be Limited.* Garden City, New York: Doubleday, 1967.

Roberts, Jennifer T. *Accountability in Athenian Government.* Madison: University of Wisconsin Press, 1982.

————. "Athens' So-Called Unofficial Politicians," *Hermes* 110 (1982): 354–62.

————. *Athens on Trial: The Antidemocratic Tradition in Western Thought.* Princeton: Princeton University Press, 1994.

Rovere, Richard H., and Arthur M. Schlesinger, Jr. *The General and the President and the Future of American Foreign Policy.* New York: Farrar, Straus and Young, 1951.

Schaller, Michael. *Douglas MacArthur: The Far Eastern General.* Oxford and New York: Oxford University Press, 1989.

Sealey, Raphael. *The Athenian Republic: Democracy or Rule of Law?* University Park: Pennsylvania State University Press, 1987.

Sebald, William J., and Russell Brines. *With MacArthur in Japan: A Personal History of the Occupation.* New York: W.W. Norton, 1965.

Spanier, John W. *The Truman–MacArthur Controversy and the Korean War.* Cambridge: Harvard University Press, 1959.

Strauss, Barry S. "The Cultural Significance of Bribery and Embezzlement in Athenian Politics: The Evidence of the Period 403–386 B.C." *Ancient World* 11 (1985): 67–74.

————. *Athens After the Peloponnesian War: Class, Faction and Policy 403–386 B.C.* Ithaca: Cornell University Press, 1986.

Thucydides. *The Peloponnesian War,* trans. Rex Warner. Harmondsworth: Penguin, 1954.

Todd, Stephen C. *The Shape of Athenian Law.* Oxford: Clarendon, 1993.

Tritle, Lawrence A. *Phocion the Good.* London and New York: Croom Helm, 1989.

Truman, Harry S. *Memoirs, Volume Two: Years of Trial and Hope.* New York: Doubleday, 1956.

Walcot, Peter. *Envy and the Greeks: A Study of Human Behaviour.* Warminster: Aris and Phillips, 1978.

Westlake, H.D. "Paches." *Phoenix* 29 (1975): 107–16.

Wittner, Lawrence S., ed. *MacArthur.* Englewood Cliffs, NJ: Prentice Hall, 1971.

Part V

Realism, Militarism, and the Culture of Democracies at War

13

Thucydides Theoretikos/Thucydides Histor: Realist Theory and the Challenge of History

Josiah Ober

Thucydides intended his account of the Athenian-Peloponnesian War to be a "possession for all time" (*ktêma es aiei*: 1.22.4); not just a record of past events, but an education in the realities of power and human relations that would help a future reader to understand his own situation. The success of Thucydides' ambitious undertaking can be measured by the eagerness with which subsequent generations of readers have gone about the project of assimilating their diverse theoretical agendas to his history—and vice versa. For several decades after the Second World War, Thucydides was read by many American opinion-makers (and by those academics who taught them) as a prototypical cold war policy analyst. Jennifer Roberts discusses this cold war reading of Thucydides and offers a particularly vivid Korean War era example:

> In the 1950s *Life* magazine ran a series of articles cautioning Americans about the disasters that might attend on ignoring the lessons of the Greek past. Robert Campbell's piece "How a Democracy Died" was designed for high drama, beginning with an account of deadly powers facing one another across the 38th parallel, only to reveal a bit later on that the author is describing fifth-century Greece and not the endangered universe of his own era.[1]

The ancient historian's depiction of the operations of raw state power within a polarized international environment seemingly justified a foreign policy predicated on the brutal and simplistic logic of hegemonic rivalry—a logic that left little room for liberal scruples about morality or interstate justice. This cold war reading was implicitly predicated on the authority of a univocal authorial voice.

The job of the reader of Thucydides was to grasp the author's lesson, and the combination of wise author and astute reader assured that the "right" lesson was transmitted. Thucydides' wisdom was generally seen as unimpeachable—he was a grand exemplar of Western political thought. Although it would certainly be foolish to suppose that the Korean War, or any modern conflict, was the product of a particular approach to reading Thucydides, he was then—and remains—a living presence in American thinking about international relations.[2]

Thucydides, however, offers his readers much more than an abstract theory of state power. His text presents us with a noteworthy example of a shrewd and subjective historical intelligence attempting to make narrative sense of extraordinarily complex events on the basis of fragmentary documentation and the ideologically slanted accounts of participants (1.22.1–2). Moreover, he inaugurates the tradition of presenting his narrative as a rigorously scientific, objective record of events and their causes. Thucydides defines not only a starting point for the Western tradition of political thought, but also for the tradition of writing history that is, ab initio, didactic, argumentative, self-consciously skeptical of its own sources, and yet confident of its own authority and veracity.[3] Thucydides' account of the Peloponnesian War may help us to make sense of the events of the Korean War by offering a better understanding of the complex interrelationship between domestic policy, concerns about national security, and the clash of great hegemonies through minor-power proxies. Moreover, reading Thucydides' ostensibly "objective" narrative as policy analysis *and* as critical history may offer the outsider some perspective on the issues at stake in the highly contested field of Korean War studies, where the problems of fragmentary sources, biased accounts by participants, ideologically loaded interpretations, and attempts to monopolize the high ground of objective historical truth by silencing hostile critics appear to be so prominent and so potentially problematic.

That Thucydides remains a significant presence in the construction (and defense) of theories of international relations is clear enough from a spate of recent work.[4] Michael Doyle summarizes much of this scholarship and argues with great cogency that Thucydides can fairly be adopted by the contemporary Realist school of international relations theory as an intellectual ancestor, a Realist *avant la lettre*. Realist models take states as quasi-individuals, as primary actors in the international arena that tend to mimic the behavior of rationally self-interested, profit-maximizing, risk-managing individuals in the marketplace—that is, in crude terms, as individuals as they are understood by modern market-centered economic theories. Thucydides obviously had no access to the psychological or behavioral theories employed by modern economists. But that does not preclude his thinking in similar terms. Doyle succeeds in showing that a reading of Thucydides as a "minimalist" Realist—one who focuses on states as primary actors in an environment of international anarchy in which no general restraint is sufficient to eliminate conflicts or to guarantee their nonviolent resolution—need not be tendentious or two-dimensional. Moreover, the text seems, at first, to provide the basis for developing a considerably stronger Realist theory of interstate behavior, a theory that will focus on the rationally self-interested behavior of state-actors in seeking to accumulate the resources that will allow them to successfully deploy power in pursuit of hegemonic ends and self-defense in a world composed of similarly rational, self-interested, and power-hungry state-actors.[5]

Granted that Thucydides' account of states and power supports the development of some sort of Realist international relations theory, the question I would like to pose is whether reading Thucydides as a strong Realist will do justice to the complex text that is Thucydides' history of the Peloponnesian War. My eventual conclusion, like Doyle's, is no. I will suggest that Thucydides the "strong Realist" theorist is indeed a centrally important presence in the text. But the careful reader eventually finds Thucydides the theorist of state power (hereafter Thucydides Theoretikos) challenged and even confuted by another of the text's central authorial presences: Thucydides the historian (hereafter: Thucydides Histor).[6] The initial theoretical premise of the war as a conflict between unitary and rationally self-interested state-actors pursuing hegemonic ends—a premise that seems to be securely established in book 1 and the first part of book 2—is demonstrated by the narrative of events in the rest of the (incomplete) history to be a dangerous oversimplification. Theory proves incapable, in and of itself, of predicting or explaining the actual behavior of states, groups, or individuals under the pressure of a protracted and violent power struggle. This does not mean that we should regard Thucydides' work to be anti-theoretical. Rather, Thucydides reminds us that any proper theory of power must be grounded in a close analysis of human behavior in actual circumstances. Theory, by its nature, seeks to be transhistorical. Thucydides' text, however, suggests to the system-builder that the best theorizing will be informed and chastened by the same attention to the complexities and contingency that characterizes the best historical narratives.[7]

Thucydides Theoretikos

Book 1 and the first part of book 2 of Thucydides' history certainly encourage the reader to develop something very much like a strong Realist theory of international relations. Thucydides makes it clear that the primary actors in his story will be poleis. Whether he refers to a polis by its given name ("Athens") or by the name of its citizenry ("the Athenians"), he seems to regard the polis as a "quasi-person"— an entity capable of making decisions and acting in an environment inhabited by similar "quasi-person" entities.[8] Moreover, he regards the key aspects of the environment in which these entities operate to be what we would, nowadays, not hesitate to describe as economic factors: the human and material resources sought and fought over by states. The antithetical structure of the Greek language allows Thucydides to contrast these sorts of "real" factors and the willful efforts of poleis to acquire them with superstructural, epiphenomenal, or merely ideological factors that sometimes disguise or mask the motivations of state-actors.[9]

This material/ideological contrast is most often expressed by Thucydides' antithetical use of the terms *ergon* (fact, deed, action) and *logos* (speech, account, story): when the author claims that in *logos* (according to what people said) factor A was in play, but in *ergon* (according to the underlying facts of the matter) it was factor B, he makes it clear to his readers that B was the true issue and more worthy of their serious attention.[10] By extrapolation, state policy that was predicated on an understanding of the relevant *erga* is likely to be rational; policy that is based primarily on *logos* is likely to be irrational. Thus Thucydides Theoretikos, as a good (proto-)Realist, establishes the ideal norm of rational state-actors making policy on the basis of a more or less thorough understanding of the relevant data. Of course, in practice states may be inadequately informed about certain important, but complex or cryptic, realia or they may be misinformed about them. Under these conditions, the states in question will be likely to make policy errors and their mistakes may have a profound effect on the international order. But the likelihood that bad policy will sometimes be made as a result of poor information does not negate the general model of rational state-actors that consciously seek the best available information and are in general agreement with other state-actors about the categories of information necessary to make realistic policy.

Furthermore, Thucydides Theoretikos is clearly and explicitly concerned with the operations of state power. He is interested in power defined as the material and efficient forces whereby a state seeks to gain its policy ends (the general term he uses for an aggregate of such forces is *dunamis*). He is also interested in the supranational regime that is established as an effect of successfully deployed power—the hegemonic structure or empire (*arkhê*) through which a strong state-actor controls the actions of weak ones, and whereby resources are extracted by the powerful hegemon from its subjects. When various impediments to efficient deployment of power have been removed, and the necessary prerequisite resources have been acquired, a state becomes strong. A strong state will employ *dunamis* in an attempt to establish and extend its *arkhê*. In the so-called "Archaeology" (1.2–19—an account of Greek prehistory), Thucydides succinctly establishes for his readers the nature of the impediments to the growth of strong hegemonic states and the material prerequisites of *dunamis*. He then demonstrates the means by which *dunamis* was employed to achieve the end of constructing an *arkhê* by those states that succeeded in removing the impediments and in gaining the prerequisites of power.

In the Archaeology, the most serious impediment to the development of state power is internal division—that is, the failure of the society in question to coalesce into an integrally coordinated state-actor.[11] This lack of coherence is attributed variously to extreme economic differentiation among the inhabitants of naturally fertile districts (1.2) and to the selfish individualism typical of dynastic tyrants (1.17). Thus a measure of social unity adequate to the maintenance of political unity is established as a necessary precondition to the growth of state power. The polis of Sparta provides Thucydides with his prime example of the advantages of unity: due to their early adoption of a set of social norms that emphasized simplicity and moderation (and thus deemphasized distinctions between social classes), the Spartans were able to become united and strong, and to extend their control over other states (1.6 with 1.18). But Spartan power, is, the reader soon realizes, limited because it does not accommodate the major material factors that make possible the unconstrained growth of *dunamis* and *arkhê*. Primary among these material factors are strong city walls, a "modern" navy of large oared warships (triremes), and significant state treasure (*khrêmata*).[12] After the social unity issue had been (at least provisionally) settled, it is the triad of (1) walls for security,[13] (2) ships as a mobile striking force, and (3) the capital reserves that allow advance planning and the survival of reverses that undergirds Thucydides' analysis of the growth of state power.

The developmental scheme laid out in Thucydides' Archaeology subtly intertwines the three elements of the "material triad." In the earliest period, we are told, before Greek towns were walled, the populations of the peninsula simply wandered from one area to the other, pushed here and there by unstable coalitions of strongmen. There were no capital reserves, no way of protecting accumulated capital, and no organized military presence on the seas (1.2). Settlements without walls were especially vulnerable to the surprise attacks of sea pirates and thus early towns tended to be built inland (1.5, 7–8). We come to realize that the material impoverishment of this early period was based on a vicious circle: walls were expensive to build (1.8), and given their lack of access to the sea, the early settlements were incapable of engaging in lucrative overseas commerce. The reduction of piracy by force was thus a precondition to the growth of wealth, which was in turn a precondition to the building of walls. Enter King Minos of Crete. According to Thucydides (1.4, 8) Minos was the first to build a great navy and "it is reasonable to suppose" that he attempted to suppress piracy in order to secure his own revenues.[14] Because of Minos' suppression of the pirates, maritime commerce flourished and capital reserves were soon accumulated by various parties. This meant that walled settlements could be built on the coasts (1.7–8) and once walled, these places were secure from piracy—presumably even after the end of Minos' thalassocracy. Communities that invested their capital in walls (and, at least in the case of Minos, in ships) became strong. The remaining, weaker, communities in turn recognized that their own material interests would be furthered by submission and thus they willingly accepted the hegemony of the stronger (1.8). Both strong and weak communities evidently made their decisions on rational grounds, as state-actors, and thus an "international" regime came into being. This was, we are told, the condition of Greece before the Trojan War.

Agamemnon's Mycenae (despite its inland location) is clearly intended by Thucydides as an early Greek example of a strong, hegemonic state. Thucydides spends some time in demonstrating that Agamemnon had the most powerful navy of his day and that he used it to control a considerable *arkhê* (1.9). Considerable, that is, by the relatively paltry standards of the time. As it turns out, in comparison to the standards of Thucydides' day, the force against Troy was unimpressive in total size and the Greek commanders at Troy were incapable of concentrating their forces. The Hellenic expedition's failure to use power effectively to achieve its policy goals (we might say, its low *"dunamis* rating") was, according to Thucydides' analysis, the effect of insufficient capital funds. Without the cash reserves to victual and pay the soldiers, too much effort was dispersed in supply-gathering raids; thus the war dragged on despite the overwhelming manpower superiority of the invaders (1.11–12).

The Thucydidean power equation is clear: walls + ships − capital = limited *dunamis*. Thus each of the Greek hegemonic confederations (*arkhai*) in the "premodern" period (that is, in the period before the rise of the Athenian empire) was similarly limited in scope and effectiveness.[15] But these imperfect early examples nonetheless demonstrated the fundamental importance of the "material triad" of walls/ships/treasure in the formulation of the comparable "conceptual triad" of defensive security/*dunamis*/*arkhê* (see Figure 1). And Thucydides' account of early Greek political development allows, even invites, the presumption that the power equation was capable of being perfected. A big, powerful navy would always be a source of considerable strength (1.15.1). That naval force could be used to defend against the depredations of pirates, to encourage trade, and thus to gain capital resources. Those resources could then be employed in constructing a system of walls capable of ensuring adequate defensive security against the relatively meager military forces that could be brought overland by one state against another (1.15.2).

With its walls built and thus its defensive security ensured, the successful Thucydidean city-state could use its naval power more aggressively in consolidating and extending its *arkhê*. As that *arkhê* expanded, weaker states would recognize that their own interests lay in voluntary submission. As these subject, tributary states contributed to the treasure of the hegemon, the hegemon's treasury would grow accordingly. With deeper reserves came a corresponding ability to increase and focus its growing *dunamis* upon appropriate objects and a concomitant potential for engaging in the sorts of ambitious, high-stakes imperial enterprises that would be avoided as too risky by less secure states with shallower capital reserves. Because wealth and naval power encouraged enterprise, while secure walls limited risk, the upward cycle of power seemed, on the face of it, potentially unlimited. The only significant external obstacle would be the presence of another comparable or superior hegemonic power that would see continued growth of a rival power as a threat to its own security or ambitions (e.g., 1.16: example of Ionian power stymied by the rise of imperial Persia).

The Archaeology, with its explicit focus on the development of the Greek *poleis*, reveals that Thucydides' model of international relations is intended to apply specifically to Greece. This is significant in that before the mid- to late fifth century, the tendency among the Greeks had been to associate large-scale, hegemonic imperial structures with the East—first with Lydia and then with Persia. Athenian naval forces had been a key factor in the Greek victory over Persia in 480–478 B.C.

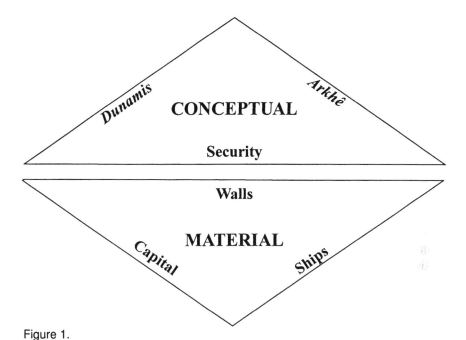

Figure 1.

and it was Athens, not Sparta, that was determined to pursue the anti-Persian cru-sade after the defeat of the Eastern invaders. In the decades after the Persian Wars, some Greek writers, notably Aeschylus and Herodotus, had begun hinting that the dramatic growth of Athenian power might have a darker side; perhaps Athens risked self-corruption as a result of having assimilated the dangerous "Eastern" passion for creating overseas empires.[16] Thucydides' Archaeology, by contrast, gives hegemony a firmly Greek prehistory.

The post–Persian War activities of the polis of Athens provided Thucydides Histor with his real-world test of the model of imperial power developed by Thucydides Theoretikos in book 1. But by the same token the *longue durée* survey of the Archaeology demonstrates that the Athenian empire of the mid-fifth century B.C. was not merely a contingent result of unique and irreproducible circumstances; it was not to be attributed to Athenian exceptionalism, not the result of a distinc-tive and innate Athenian national character (*êthos*).[17] The impulse of strong states to self-aggrandizement, their tendency to seek power and to deploy *dunamis* in pursuit of *arkhê*, is shown in the Archaeology to be a constant and predictable factor in Greek interstate relations. Rather than focusing on Athens' peculiar na-tional character, Thucydides' implication is that Athens, under the fortuitously prescient leadership first of Themistocles and then of Pericles, was the polis that perfected (or seemed to have perfected) the symmetrical arrangement of the same triad of material factors that had been operative since the beginning of settled, civilized Greek life. Those factors had led to the power of Agamemnon's Mycenae but their imperfect alignment had placed limits on Mycenaean power. It remained to be seen whether the Athenian *arkhê* had actually solved the geometric equation of interlocking the material and conceptual triads (Figure 1) and had thereby achieved the strong Realist's grail: a secure hegemony with a potentially limitless capacity to extend its power.

Another set of questions remained: If Athens' behavior in building its *arkhê* was, in hindsight, predictable in light of Thucydides' "proto-Realist" model, could the behavior and performance of Athens as a polis in the course of the Peloponnesian War also be successfully predicted and understood in terms of that model's assumption that poleis could be treated as rationally self-interested state-actors? Or would Greek behavior under the stress of war actually prove to diverge from that of the ideal-type "state-actor"? If there were significant deviations from the predicted patterns of behavior, how are those deviations from rationality to be explained? At what level of deviation would the strong Realist model have to be significantly modified or even abandoned? What sort of additional theories might be developed to explain major deviations from the Realist scheme? If Athens' rise to power could be explained without recourse to Athens' distinctive "national *êthos*," might appeal to that *êthos* help explain its subsequent failure to act rationally? Or might that failure be explained in terms of the distinctive structural properties of Athens' internal political regime—the *dêmokratia*?

To anticipate the argument: Thucydides' narrative encourages his reader to suppose that after the death of Pericles Athens did indeed deviate from the patterns of behavior appropriate to the strongly Realist state-actor; Athenian policy lost its coherence and became at least sporadically "irrational." The reader is led to suppose that the breakdown of orderly and rational policymaking is the result of the failure of the domestic Athenian political system to resolve conflicts among rival politicians and among politically and sociologically defined groups within the Athenian citizen body. In spite of Pericles' optimistic claims in the Funeral Oration, the democratic government is depicted as being unable to negotiate between the demands of freedom and equality; of state, group, and individual interests; of public and private orientations; of *logos* and *ergon*. The reader is taught that the practice of democracy at Athens ultimately provided a fertile ground for the growth of parochial individual and factional self-interest, for destructive competitions among would-be leaders for popular favor, for the willful dissemination and reception of tendentious misinformation, and for the instantiation of highly dubious models of human behavior as the foundation for policy formation.[18] These domestic issues eventually had a profound effect on the conduct of international policy. The ultimate outcome is Athens' failure in the Peloponnesian War—a failure that belies the confident assessment of Athens' resources and chances in books 1 and 2.1–64.

The expectations initially generated by the reader's close attention to Thucydides Theoretikos are confounded by Thucydides Histor's detailed narrative of events at home and abroad. History as *erga* fatally complicates theory as *logos*, and the reader is left sadder but wiser.[19] By the time she arrives at the abrupt end of the narrative, Thucydides' reader has been taught to be less arrogant about the human ability to understand group behavior on the basis of generalizations and abstractions. She is taught that under the pressure of events, even the best and most sophisticated of political systems can fail in its rational policy goals and come to betray its moral and cultural ideals. But she has also come to understand the cryptic comment of 1.22.4: Thucydides' history is a "possession for all time" because of its simultaneous attention to the explanatory abstractions by which humans try to organize the world about them and to the much more complex reality of the lived experience of individuals, groups, and states. In the ashes of theory and idealism, the text itself remains. And thus each weary, chastened generation of readers crowns its author, who stands victorious and without rival in the games his text inaugurated. Or so, I suppose, Thucydides imagined it. Thus, far from lending unconditional support to a strong Realist view of the world, reading Thucydides' history should invite opinion-makers and policymakers to think twice—to take into account history, sociology, politics, and the diversity of ideological commitments—before making the move from abstract Realist regime theory to foreign policy.

Pericles and Athens

Perusing Thucydides' account of the Pentecontaetia—the half-century between the Persian and Peloponnesian Wars (ca. 478–431 B.C.: 1.89–117)—the reader quickly realizes that in fifth-century Athens the three material factors essential to the development of national strength and hegemonic authority were abundantly present. Moreover, she learns that the importance of walls, ships, and treasure, their relationship to one another, and their relationship to the conceptual triad of security/*dunamis*/*arkhê* were well understood by Athens' outstanding post–Persian War political leaders, Themistocles and Pericles. In close conformity to the behavior of weaker states after the establishment of Minos' thalassocracy, the various Greek states of the Aegean assume that their material interests will be furthered by submission to the authority of the hegemon—and thus, looking to their immediate material advantage rather than concerning themselves with ideological notions like "autonomy," they willingly pay Athens to undertake the burden of ongoing naval operations against Persia (1.97–99, esp. 1.99.3). Thucydides also presents his reader with a detailed account of the building programs that resulted in the great fortification walls of mid-fifth-century Athens—the Piraeus circuit (begun in the archonship of Themistocles before the Persian Wars and completed after the Persian Wars by his encouragement), the city circuit (built immediately after the Persian Wars at Themistocles' urging, despite the opposition of the Spartans), and the Long Walls that connected the city with Piraeus (the product of the leadership of Pericles).[20] Various narrative passages discuss the size and efficiency of the Athenian navy and the extent of Athenian financial reserves.[21] But it is in considering the two Assembly speeches delivered by Pericles to the Athenian Assembly, one just before the outbreak of the war and the other in the war's second year, that Thucydides' reader is most fully and explicitly apprised of the tight fit between the model of state power developed in the Archaeology and imperial Athens under the leadership of Pericles. These speeches are particularly revealing when read in the context of other speeches, including the Funeral Oration.[22]

In his first Assembly speech (1.140–44), Pericles begins by asserting that his policy recommendations have been consistent over time and that he would now simply reiterate the advice he had previously given his fellows citizens. After explaining that accommodation with a hostile Sparta was truly dangerous to Athens and that the main Spartan demand (rescinding the so-called Megarian Decree) was therefore unacceptable, Pericles lays out the resources available to the two opposing sides. The Peloponnesians, he bluntly explains, have no financial reserves and no significant naval forces. Thus they will fight the war at a severe disadvantage. They were used to fighting only short, simple wars among themselves—the sort of contests typical of impoverished people (1.141.1–3; cf. Archaeology: 1.15). This lack of the essential resources that (as we have seen) undergirded *dunamis* is exacerbated by the diffuse Peloponnesian alliance structure; the alliance was incapable of acting as a single entity and thus, divided in their councils, the Peloponnesians fritter away their opportunities (1.141.6). Pericles assures the Athenians that they need not be concerned for their security in the face of Peloponnesian invasions of Attica: the invaders will not be able to construct fortifications capable of threatening Athens' own fortified city-harbor complex (1.142.2–3).

The bulk of the speech expands on these central themes. Pericles argues that, having begun the war with inadequate reserves, the Peloponnesians will not easily be able to acquire capital surplus nor will they easily develop a credible sea power. He explains that the Athenians will be completely secure as long as they preserve intact their citizen manpower and the empire that is protected by the deployment of that manpower through the instrumentalities of Athenian treasure and sea power. This entailed withdrawing Athens' population from rural Attica (Athens' home territory) behind Athens' impregnable walls and refusing to meet the Peloponnesian invaders in open battle since a major land battle in Attica would put Athenian manpower resources at risk (esp. 1.143.5).

In the ensuing narrative, the reader learns that despite the psychological and material suffering that this austere policy inflicted on those citizens (a majority of the total) with holdings outside the city walls, the Athenians went along with Pericles' recommendations. They refused the Spartan ultimatum and declined to meet the Peloponnesians in battle when the latter invaded Attica in the summer of 431.[23] The invaders left after a couple of weeks, having accomplished nothing of substance other than demonstrating their incompetence at siegecraft by a failure to capture the fortified town of Oinoe.[24] After the first, ineffectual, Peloponnesian invasion, the war seemed to be going according to Pericles' strategic plan and Pericles was chosen by his fellow citizens to deliver the traditional oration over the war dead.

The Funeral Oration (1.34–46), in which Pericles praises the polis of Athens and the unequaled *dunamis* that Athenian activity abroad had brought about, originally appears to be a monument to his vision of Athens at this optimistic high point of successful, strong Realist policymaking.[25] But on closer inspection, when read in the context of the rest of the history, the Funeral Oration also reveals the fragile social and political bases of Athens' standing as a unitary state-actor. The contrast between *logos* and *ergon* is reiterated throughout the oration.[26] Two passages, which concern the substantiality of Athenian *dunamis*, are particularly noteworthy.

> Furthermore, the power (*dunamis*) of the polis itself, [a power] established by those [Athens'] very qualities, demonstrates (*sêmainei*) that this [Pericles' statement regarding Athenian excellence] is the truth (*alêtheia*)—and not a product of words (*logoi*) produced for the present occasion rather than [a product] of fact (*erga*). (2.41.2)

Here, Athenian *dunamis* is presented as a fact, an *ergon* that is capable of demonstrating the truth of words, *logoi*, that might otherwise be suspect given their location in a public speech of praise. By appealing to Athenian *dunamis* as an exterior reality, Pericles calls the rest of his speech into question: he tacitly admits that since his speech was (by definition) merely a construct of words, and was prepared for an honorific occasion, it might not be true. But, he claims, the self-evident power of the city, a "fact" rather than a product of words, will establish the truth of Athens' greatness. A similar sentiment, reprising the terminology of power, proof, truth, word, and deed, is expressed a few lines later:

> Our power (*dunamis*) is not without the witness of great proofs (*megalôn sêmeiôn*) and we will be the source of wonder for those yet to come, as we are for our contemporaries. Furthermore, we have no need of a Homer to sing our praises, or of any suchlike whose fine verses please only for the moment, since the truth (*alêtheia*) will show that in comparison with the facts (*erga*) [the verbal depiction] is an underestimate. (2.41.4)

Again, the contrast is between the potentially misleading verbal praise and the trustworthy evidence of facts. It is through the witness of great proofs, *megala sêmeia*, that future generations will be amazed at Athens. What are these proofs that will survive the current generation to convince "those yet to come"? In light of the rejection of Homer, the proofs in question can hardly be in the form of words; we must imagine seemingly permanent, material monuments of some sort. The audience is, I would suggest, being put in mind not only of victory monuments established in enemy lands (implied at 2.41.4) but also of the city's new public buildings, and perhaps especially of the great fortification walls of the city. The public cemetery in which the Funeral Oration was delivered stood just outside the city walls and the mighty walls themselves would have provided the speaker's backdrop.

We modern readers know that Pericles was right to suggest that future generations will be amazed at the great fifth-century architectural monuments to Athenian power. But does our amazement reflect an accurate assessment of Athens' actual power? Not if we are to judge by Thucydides' earlier comment, in the Archaeology (1.10.1–2). There he specifically states that to criticize Homer's *logos* on the Trojan War on the basis of the proof (*sêmeion*) of the small size of the existing town of Mycenae (the imperfect, proto-Athens of the Archaeology), would be improper method. In this same passage Thucydides points out that if future generations were to judge by the physical magnificence of the city alone, the *dunamis* of Athens would appear twice as great as it was in fact. Read in light of this earlier, "archaeological" passage, Pericles' rejection of Athens' need for a Homer because the great proofs, the *megala sêmeia*, will impress future generations with the truth about Athenian *dunamis*, seems an empty boast.

In a key passage of the Funeral Oration Pericles describes the structure of Athens' internal government:

> and it [our *politeia*] is called by the name (*onoma*) *dêmokratia* because government (*to oikein*) is not oriented toward the few (*es oligous*) but toward the majority (*es pleionas*). However, in regard to access to the law for resolving disputes all are equal. Yet again in regard to acknowledged worth, it is a matter of individual reputation; the nature of a man's public contribution is not decided in advance on the basis of class (*ouk apo merous*), but rather on the basis of excellence. And if someone is worthy and can do something worthwhile for the polis, he is not excluded by poverty, nor because of his obscurity [of birth]. (2.37.1)

In the first clause of this passage we learn that *dêmokratia* is the name used for the "the *politeia* [internal regime] of Athens" because government (*to oikein*) is oriented toward the majority (*es pleionas*) rather than toward the few (*es oligous*). This is a somewhat ambiguous statement, when viewed from the perspective of power and self-interest. Pericles does not go so far as to say that *dêmokratia* is the rule of the many in their own self-interest *over* and *against* the interests of the few, but neither does he suggest that it is the unified policy apparatus of a rational state-actor. Pericles hints that Athens is divided into two interest groups: the few and the many, and the *politeia* is called a democracy because it tilts toward one group rather than the other.[27]

The next two clauses of the passage, which should explain and clarify the *politeia*, are spectacularly antithetical. The clause "but in regard to access to the law for resolving disputes all are equal," does not explain the term *dêmokratia*, but contrasts to it (*onoma men . . . metesti de*). Thus, the equality in regard to the law is grammatically opposed to the government favorable to the majority, and therefore terms for "many" cannot stand for "all citizens" (as they did in democratic ideology, see 6.39.1). The third clause, "but in regard to acknowledged worth, it is a matter of individual reputation," contrasts the individual citizen to the grouping of the citizenry into "few" and "majority" in the first clause, as well as contrasting the citizen's individual worth to the generalized equality of "all" in the second clause. What happens if there is a conflict between the perceived interests of the groups, or between the equality of all and the merit of individuals? Can a political balance based on such a complex set of contrasts hold up under the stressful circumstances of a long, hard war?

One other key passage in the Funeral Oration deals explicitly with the functioning of the Athenian polity:

> and we [Athenians] ourselves can [collectively] judge rightly regarding affairs, even if [each of us] does not [individually] originate the arguments; we do not consider words (*logoi*) to be an impediment to actions (*erga*), but rather [regard it as] essential to be previously instructed (*prodidachthênai*) by speech (*logôi*) before embarking on necessary actions (*ergôi*). We are peculiar also in that we hold that we are simultaneously persons who are daring and who debate what they will put their hands to. Among other men ignorance (*amathia*) leads to rashness, while reasoned debate (*logismos*) just bogs them down. (2.40.2–3)

This passage helps to clarify the contrast between group-interest oriented government and individual merit, alluded to in the earlier passage. The Athenians recognize that not everyone is equally capable of coming up with plans (this will be the job of the individual political leader), but the many can and do participate in making the decision (as assemblymen). This division of labor is presented as nonproblematic, as is the way Athenians move from speech to action. Here, however, Pericles must appeal to Athenian uniqueness. "Other men" either act hastily out of ignorance or find that debate renders them incapable of efficient action. What, the reader may ask, renders the Athenians different from other people? What is the secret of their ability to avoid the common problems involved in moving from deliberation about policy to its implementation? The requirement that the Athenians be substantively different from other people in this important regard seemingly threatens the value of the general theory of power established in the Archaeology, which was predicated on a generally applicable understanding of human (or Greek) nature. The Archaeology had, we remember, demonstrated the need for a state to transcend internal conflict arising from social differentiation if it were to achieve (and maintain) the status of unitary state-actor. The Funeral Oration underlines the complexity and diversity of the government and the society that constituted the Athenian state-actor. Is this complex form of government—which necessarily draws members of diverse social groups, with their diverse interests and talents, into the process of policy formation—capable of weathering the stresses of war? Thucydides Histor sets out to answer that question.

After Pericles' Funeral Oration, in the second year of the war, came the second Spartan invasion, a devastating plague (described in vivid and clinical detail by Thucydides, who tells us that he himself survived an attack of the disease), a crisis in Athenian confidence, and the second (and last) of Pericles' Assembly speeches recorded in Thucydides' history (2.60–64). In this speech, Pericles focuses less than before on the material triad and emphasizes instead the conceptual triad of defensive security/*dunamis*/*arkhê*.

Pericles begins the speech by asserting that he had accurately predicated the downturn in the Athenian mood—the reader is led to suppose that very pronounced shifts in the national climate of opinion were factored into Pericles' planning and thus that mood swings within the electorate could be accommodated by Pericles' political calculus. He boldly reasserts the priority of the unified public interests of the state over the diverse private interests of each individual Athenian (2.60.2–4). In the Funeral Oration we became aware of the tension between public and private, between the self-identity of the Athenian as a citizen and as a private person or a member of a social subgroup. Now Pericles bluntly underlines the essential precondition of the Realist's focus on the unitary state-as-actor: the individual's identification with the state must take priority over all other loyalties or allegiances. Pericles then claims that he will let his audience in on a secret that lay at the heart of Athenian policy: the special character of Athenian sea power. He describes sea power in quasi-mystical terms as the control of one of two earthly spheres and as a power completely unlike anything produced on land: as it stands, because of their command of the seas, there is no power on earth, not even the king of Persia, who can stop the Athenians from going where they might choose (2.62.2–3; contrast Archaeology 1.16: Ionians stymied by Persia).

This passage in Pericles' second Assembly speech stands as a particularly evocative attempt to define the mysterious essence of Athenian *dunamis*: sea power is a latent strength: the ships can lie motionless in their sheds until needed. So too were Athenian reserve capital and manpower latent sources of strength: Capital lay dormant, most strikingly in the form of the cult statue of Athena Polias, with its removable golden drapery. Athenian men remained passive behind their walls while the enemy invaded Attica. But at the moment a need arose, capital could be conjoined with men and ships to produce extraordinary levels of deployed power and virtually limitless freedom of action. The Athenians could go where they wished, do as they liked because Athens' latent strength could be almost instantaneously materialized as the mobilized fleet cutting through the waves of the Aegean to effect Athens' will upon any object identified by the policymaker. Pericles describes this as a secret, because when regarded from the set of traditional assumptions about the sort of power produced and delivered by land armies, sea power was nearly incomprehensible. The rapidity and precision with which power could be deployed by the state commanding superior naval forces collapsed conventional assumptions about the relationship between power, time, and space—just as, in the United States of the 1950s, new technology (notably atomic weapons and jet power) conjoined with the dynamism of postwar capitalism to encourage a sense of limitless opportunities.[28] Athenian sea power narrowed to the disappearing point the gap between desire and fact, between the wish of the state-actor that something should occur and the accomplishment of that wish in the material world. In 430 B.C. the sum of Athens' accomplished wishes was the Athenian empire.

Much of the rest of the speech focuses on the nature of the *arkhê* itself, which Pericles reminds his listeners is "now like a tyranny" (2.63.2). Pericles is at pains to explain to the Athenians that the possession of the empire entailed grave security risks—but only if they were foolish enough to suppose that their freedom of action was a freedom from imperial responsibilities. The entirely unrealistic and pseudo-altruistic policy advocated by certain apathetic and useless Athenians who wanted to give up the empire would, according to Pericles, put Athens in grave danger (2.63). Pericles does not need to spell out the equation in detail; its outlines are clear enough from what had gone before: without the empire Athens would lose the revenues that had provided for the fortified security of the city and that maintained the navy. This was especially problematic in light of the anticipated reaction of Athens' subjects to a condition in which Athens had stripped itself of imperial possessions. The weight of Athenian tribute and the rigor of Athenian punishment of recalcitrant subjects had eventually brought home the real meaning of lost autonomy to the states of the empire. Tribute and punishment had made their original, self-serving decision to pay Athens to maintain Aegean security appear shortsighted. Resentment at their own past folly led to hatred of Athens (2.63.1, 2.64.5), and this meant that other state-actors would take the Athenian attempt to drop imperial responsibilities as an opportunity to seek revenge against the one-time hegemon. Deprived of the resources by which the conjoined material and conceptual triads were supported, Athens would become weak in fact and would suffer the consequences of the rationally self-serving policies that had created and that maintained her current strength. The realistic policy that led Athens to a position of hegemonic authority carried with it burdens that could not lightly be shed. As twentieth-century leaders have often reminded their own citizens, a great power cannot, in its own interest, afford to shirk its duties in the international arena.

Pericles and Athens in Context

Thucydides' reader is presented with Pericles' two Assembly speeches, with their complementary (material and conceptual) analyses of Athenian power, in the context of other public speakers' attempts to define the nature of power and the structural relations between states. Some of these speeches, delivered by Athens' enemies, make light of Athenian power, whereas others seem to share the Realist assumptions adumbrated by the Archaeology and elucidated by Pericles. The Corcyrean envoys in book 1 (1.32–36), addressing the Athenian Assembly, take a strong Realist line in explaining their state's policy, past and present, and they closely mirror Pericles' and the Archaeology's focus on the essential distinction between sea and land power. By contrast, the Corinthians, in their address to the Athenians opposing the Corcyrean proposal for an Athenian alliance (1.37–43), concentrate on international justice. That is to say, they are unrealistic in supposing that there is (or could be) some general constraint among nations capable of ensuring the nonviolent resolution of conflicts. The Corinthians fail to persuade the Athenian audience and the alliance is struck. This was, we are told (1.22.5, 1.55.2), among the key incidents that sparked the war.[29]

Corinthian speech makers return to the fray when they attempt, this time successfully, to persuade the Spartans to lead a war effort against Athens (1.68–71). They point to a key "triadic" material factor, alluding to Sparta's error in allowing the Athenians to refortify their city after the Persian Wars and to build the long walls connecting the city with the harbor complex at Piraeus (1.69.1). But at the heart of the Corinthian argument is a claim that the Athenians are by nature fundamentally unlike the Spartans. The former are ambitious, restless, eager to take risks, and endlessly active; the latter are overcautious, slow to act, and tend to ignore danger even when it is nearby. The Corinthian account of Athenian activity meshes well with Pericles' evocation of Athenian exceptionalism in the Funeral Oration and with his emphasis on the unique quality of sea power in the second Assembly speech. But the Corinthians seem to miss the point by attributing the Athenians' hyperactivity solely to their national character, rather than to their rational pursuit of a Realist policy over time.[30] This approach casts the war into the moralistic terms of a conflict between national characters, and blurs the focus on state-actors and their respective sources of power. The strong Realist will simply scoff at the Corinthians' "error"—but what if Corinthians and other Greeks actually acted upon their moral convictions?

The Corinthian position is quickly challenged by the Athenian envoys who happened to be in Sparta on other business and were allowed (in Thucydides' text, anyway) to respond to the Corinthians' charges (1.73–78). Thucydides states that the Athenian envoys were concerned to explain to the assembled Peloponnesians the extent of Athenian *dunamis* (1.72.1) in hopes of preventing a hasty decision for war. The speech itself is a model of strong Realist analysis. The envoys claim that when states aggrandize themselves and seek *arkhê* they are simply acting according to human nature. Nor is the Athenian empire anything new or surprising: it is simply a manifestation of the well-known principle that the weak must submit to the strong—that is, the general principle established in the Archaeology. If Athens is occasionally high-handed in its dealings with its allies, this is to be expected; what people forget is that Athens is more scrupulous in its relations with imperial subjects than it needs to be (in light of the power inequity) and indeed more scrupulous than are other imperial powers toward those they control.

The language of human nature and the inevitability of the rule of the strong over the weak dramatically foreshadows the language used by another set of Athenian envoys much later in Thucydides' history: the anonymous speakers who carry on a policy dialogue with the hapless rulers of the island-polis of Melos in the end of book 5. The Melian Dialogue can be read as a *reductio ad absurdum* of the Realist position developed in book 1: according to the rules of debate established by the Athenians at Melos, no considerations of justice were allowed voice; all arguments were to be grounded in state interest alone. Moreover, the Melians were instructed to explain how their proposed line of action would serve *Athens'* best interests. The Melian Dialogue is indeed a confrontation between a (Melian) view of appropriate behavior based on traditional Greek morality and the (Athenian) sophistic view that justice is nothing more than submission to necessities established by strength (see the essay of G. Crane in this volume). But it may also be read as a confrontation between theory and history. In light of the argument of the Archaeology (the argument assumed by the Athenian speakers), the Melians appear perverse because they fail to act like those self-interested weak communities who submitted to strong Minos. And yet, Thucydides' reader now must consider not only the logic of the Archaeology, but the historical example of Athens' own imperial subjects. Pericles has let on that these no longer regard their original decision to accept Athens as a hegemon as rationally self-interested; as we shall see (below), some are willing to take considerable risks to free themselves from the relationship that the Athenians sought to impose upon the Melians. This helps explain why the Athenians must set such narrow rules on the debate: the weaker state's self-interest is no longer transparently compatible with that of the imperial power. The Melians cannot come up with a Realist reply to the Athenians, but the level of Melian physical resistance to annexation (quite considerable as it turns out: 5.114–116) is the limiting factor to the Athenians' ability to freely extend their own power. Melian resistance to Athens in word and deed points directly to the difficulties Athens can expect to experience when it comes to attempting to annex the much larger and more distant island of Sicily (books 6–7).

To return to the debate at Sparta: the Peloponnesians reject the realistic policies advocated by the Athenian envoys at Sparta and vote for war. A somewhat different Realist argument is subsequently made by King Archidamus in addressing the Spartan citizen Assembly (1.80–85). Like Pericles, Archidamus focuses on the difficulties the Spartans will encounter in conducting the war due to their lack of material resources. Rejecting realistic Archidamus, the Spartans embrace a decision to confront Athens head on. The Peloponnesian and Spartan decisions seems to be precipitated on the one hand by the Corinthian reading of Athenian anti-Peloponnesian activity as an outgrowth of their innate natural character, and on the other by the Spartan ephor Sthenelaides' laconic speech to the effect that the Athenians must be punished as international wrongdoers (1.86). Much more may be going on behind the scenes, of course—we might choose to put less emphasis on Corinthian rhetoric about national character and considerably more weight on the thinly disguised Corinthian threat to secede from the Peloponnesian League and join Athens (1.71.4). In this case we could make the Spartan decision for hot war an outcome of a realistic assessment that their alliance (and thus the basis of their military superiority) could not survive an extended cold war.[31]

On the whole, however, it seems to me that in 1.1–2.64 Thucydides leads his reader to draw a sharp contrast between Athens, a consummately powerful unitary state-actor guided by the clear realistic vision and power-centered foreign policy of Pericles, and the ill-prepared Peloponnesian confederation, which lacked both the material prerequisites of true power and the stable leadership capable of identifying those resources in the first place. According to the premises established by Thucydides Theoretikos, Athens was set to win the war. But then comes section

2.65: a laudatory summary of Periclean policy, the announcement that Athens in fact had lost the war, and the statement of Thucydides' opinion that Athens lost the war because Pericles' inferior successors did not stick to his policy with regard to sea power and empire, and that they gave away Athenian chances by looking to their private good rather than to the common good of the state. Section 2.65 is, in the reading I offer here, the watershed, the point at which the theses developed by Theoretikos are summed up and the long period of Athenian decline from the Periclean acme traced by Histor is first anticipated.

At the beginning of 2.65, having explained that Pericles' second Assembly speech was meant to staunch the Athenians' anger at Pericles and to turn their thoughts away from their present miseries, Thucydides makes a particularly telling comment. With respect to state policy (*dēmosiai*), the Athenians accepted Pericles' words: they did not send a peace embassy to Sparta and they began to prosecute the war with renewed vigor. But with respect to their interests as private individuals (*idiai*) they still felt aggrieved (2.65.1). Taken as individuals, "they" are further broken out by Thucydides into classes: the ordinary people, the *dēmos*, were angry since they had lost what little they had; the powerful elite (*dunatoi*) had lost their landed estates in the country. Thus the potentially dangerous internal divisions signaled in the Funeral Oration have come back to haunt us. Yet as long as Pericles lived, this tendency of the Athenians to fragment into private interest-oriented classes was restrained or at least masked. This is underlined by the resumed narrative: Pericles was reelected general because the Athenians knew he was the best man in the polis for dealing with state interests (2.65.2–4). In this general Athenian opinion our author heartily concurs: Under Pericles and during the peace, he says, Athens had been wisely led and the polis reached its peak. Moreover, "when the war broke out, in this circumstance too, he seemed an accurate prognosticator in respect to power" (*prognous tēn dunamin*: 2.65.2). Pericles did not live long after his second speech: he died two and a half years into the war. Yet after his death, we are told, his forethought (*pronoia*) in respect to the war became clearer than ever, since he had told the Athenians that they would prevail in the war if they avoided risking their sea power, attempting to expand their empire, or otherwise endangering the polis (2.65.6–7). And there it is: the confident statement that assumptions about Athenian power established by Theoretikos *would* have accurately predicted the course of the war if only realistic Periclean leadership had been maintained.

But, continues Histor, post-Periclean leadership was not up to Pericles' standard. The key difference, we are immediately told (2.65.7), is that Pericles' successors failed to pay attention to the interests of the state, and they made policy on the basis of what they supposed would conduce to their private (*idiai*) advantage. As a result, their successes brought advantage only to themselves, while their failures weakened the ability of the state as a whole to conduct the war. And now comes Thucydides' dramatic revelation of the "true" nature of Athenian domestic politics under Pericles: By contrast to his successors, he was a leader, not a mere follower of the will of the *dêmos*. Indeed, under Pericles the polis was "in *logos* a democracy, in *ergon* the rule (*arkhê*) of the foremost man" (2.65.8–9). His successors, more or less equal to one another but each eager to become first man, sought to please the *dêmos*—that is to say, per what has gone before, they looked only toward the selfish interests of one social class within the polis—in so doing they gave up control of the affairs of state (*ta pragmata*). And so (as we, students of Theoretikos, would expect) given these circumstances and given a "great polis with an empire to govern," serious mistakes were made (2.65.10–11). Thucydides takes as his case in point the policy errors made by the Athenians in respect to the Sicilian expedition and its aftermath. He then concludes section 2.65 with a return to the theme of Pericles' excellence: Even after the disaster in Sicily the Athenians held out for another eight years against a staggering array of enemies, including the son of the king of Persia (note the echo of Pericles' second Assembly speech) who provided the necessary capital for a Peloponnesian fleet (2.65.11–12). Thus, in the end it was the Athenians who destroyed themselves by their "private quarrels" (*idiai diaphorai*). "So overwhelmingly great," concludes Theoretikos, "were the resources which Pericles had in mind at the time when he prophesied an easy victory for the polis over the Peloponnesians" (2.65.13).

At the watershed passage of 2.65, Theoretikos and Histor seem to be well matched yokemates: Theoretikos' confident prediction of victory on the basis of Pericles' Realist approach meshes well with Histor's explanation of divisive selfishness among the post-Periclean leaders as the explanation of what went wrong and why Athens eventually lost. But I do not think that Thucydides' text would have its enduring appeal if this rather simplistic explanation for Athenian failure really undergirded the subsequent historical narrative. Nor does 2.65 actually anticipate the rest of Thucydides' narrative as well as it might. It has been pointed out by more than one commentator that Thucydides' description at 2.65 of what went wrong in Sicily (i.e., a failure by the Athenians back home to adequately reinforce their expeditionary forces rather than a misestimation of the nature of the opposition they would face in Sicily [2.65.11–12]) is not borne out by the narrative of books 6 and 7.[32] At 2.65, we might say, Histor concedes too much ground to Theoretikos.

Moreover, upon reflection, the unstinting praise of Pericles at 2.65 seems somewhat misplaced. The Funeral Oration shows that Pericles understood the complexity of the Athenian social and governmental situation. One might therefore suggest that, for a man repeatedly lauded for his insight and his forethought, Pericles was remarkably blind to the conjoined factors of the structure of internal Athenian politics and his own mortality. If "a Pericles" was the indispensable precondition for rational policymaking because Athens was a crypto-monarchy, why did its leader make no provision for his own successor? The reader begins to suspect that either (1) Pericles was inadequately insightful about the importance of his own role, or (2) he was selfishly uninterested in Athens' fate after his death, or (3) the description of Athens as a crypto-monarchy is overdrawn.

Happily for Thucydides' reputation, the value of his work is not actually summed up by section 2.65. Rather than simply aberrant selfishness on the part of a few bad leaders, the long narrative following 2.65 suggests that there were deep structural problems associated with decision-making (and thus policy formation) in the Athenian *dēmokratia*. Pericles' unique leadership qualities may have successfully masked those structural problems for a time. But a Realist analysis that requires Periclean-style leadership as a dependable norm, as a predictable constant of the policy environment, will be worse than useless in the real world in which Pericles is mortal and unable to designate a successor. Rather than simply playing out the results of an unexpected selfishness on the part of a few bad men, Histor's narrative demonstrates the weakness of any theory of power that focuses uniquely on unitary state-actors and ignores internal social forces, the complexities of domestic politics, and "irrational" attachment to a transnational set of norms and values.[33]

A full exposition of how the narrative following 2.65 complicates the strong Realist's vision of Athens as a unitary state-actor, behaving rationally to further its hegemonic interests in a world of similarly motivated state-actors, would much exceed the limits of this chapter. Here I offer just one example of how the Realist calculus established by Theoretikos becomes inextricably tangled in Histor's unfolding analysis of Athenian internal politics and in issues of social diversity, norms, and values.

Mytilenean Debate

In the fourth year of the war (428 B.C.) the great polis of Mytilene on the island of Lesbos declared itself independent of Athenian control, thus confirming Pericles' comment about the attitudes of subject states and fulfilling a prediction about revolts within Athens' empire made by the Corinthian envoys to Athens in the debate over Corcyra.[34] Thucydides points out that the revolt of Mytilene occurred at a difficult time for Athens: the city was suffering from plague and "from the war, which had only just now reached its full strength" (3.3.1). Beset by these problems, most Athenians were reluctant to acknowledge the truth of the report (3.3.1). But eventually the danger of the situation intruded on the Athenian consciousness and a fleet was sent to besiege Mytilene. The Mytileneans were soon shut up behind their city walls. Hard-pressed by an efficient Athenian siege and betrayed in their hopes of reinforcements from Sparta, they ran low on food. The aristocratic leaders of the revolt consequently armed the Mytilenean lower classes in anticipation of a battle. Yet once armed, the *dēmos* of Mytilene failed to support the insurrection. Faced by insurrection and the specter of civil war, the leaders of Mytilene hurriedly surrendered (3.28). The crisis over, the Athenian Assembly set about deciding who had been responsible for the revolt and who should be punished.

The material factors detailed by Theoretikos played key roles in the Athenian suppression of the revolt: Athenian capital reserves and sea power proved fully capable of overwhelming or scaring off all opposition, whether it was a small Peloponnesian fleet in the Aegean or the army of the revolting Mytileneans. But irrational emotions, confused Athenian perceptions regarding the wellsprings of their own power, and stark domestic social divisions were made manifest in the ensuing debate over the fate of Mytilene.

Thucydides' account convinces his reader that, although Athenian *dunamis* had stood the test, suppressing the revolt of Mytilene consumed considerable Athenian time, effort, and cash. The revolt coincided with a Spartan naval expedition to Ionia (3.29–33). Although the Spartan admiral with his small fleet accomplished little in military terms, he threw a bad scare into the Athenians and he brutally murdered a number of captives (3.32.1). The mood of the Assembly was not charitable: In a fit of anger the Athenians voted to treat the population of Mytilene as a single, unitary entity (i.e., to treat Mytilene judicially as a state-actor). They ordered their general on Lesbos to kill all adult male Mytileneans and to sell the rest of the polis' population into slavery (3.36.1–3). A trireme was dispatched to Lesbos carrying the grim instructions. But very quickly, while the ship was still en route, Athenian anger was replaced by a sense of remorse and "it became clear" that "most of the citizens" wanted a chance to reconsider their action (3.36.5).

A second Assembly was hastily called and we are told that several speeches were given on either side of the issue. Thucydides presents two orations that, he says, represent the most starkly opposed positions (3.49.1). The first of the pair is spoken by Cleon, son of Kleainetos, the citizen who had been "victorious" (3.36.6) in advocating the general punishment at the first Assembly. Thucydides describes Cleon as "the most violent of the citizens [of Athens] and by far the best trusted by the *dêmos*" (3.36.6). Cleon's speech, which opposes any amelioration of the sentence against the Mytileneans, is attacked (3.41) by a certain Diodotus, the son of Eukrates, about whom we are told nothing other than that he had also spoken against Cleon at the previous meeting. The two speeches are a matched pair and share several themes; both offer a substantial "meta-rhetoric" (i.e., a rhetorical discussion of the nature of public deliberation), and both purport to explain the proper foundations of state policy. Together they offer Thucydides' reader her first detailed insight into the environment and the tenor of post-Periclean democratic politics and policymaking. Although, in line with the analysis of 2.65, the self-interest of Athenian politicians is a key factor, the Mytilenean Debate as a whole suggests that self-interested behavior of politicians was only one factor in a much more complicated social and political situation.

Cleon begins his speech with an implicit rejection of one of Pericles' points in the Funeral Oration: he had often noticed that a *dêmokratia* is incapable of running an empire (3.37.1). Why is this? Because the Athenians fail to see that their empire really is a tyranny (not just "like a tyranny," per Pericles), and because of their indecisiveness (3.37.2). The root of the problem is overclever public speech makers. Athens has no need for these men; indeed,

> ignorance (*amathia*) mixed with moderate sobriety is more beneficial [to the polis] than cleverness mixed with insubordination. Ordinary men, when compared with the more gifted, actually administer poleis better. For the latter [the

gifted] wish to appear wiser than the laws and to excel at speaking about the public weal, since they are unable to express a clear opinion in regard to more important things. And thus they often ruin poleis. But the former [ordinary men], not trusting overmuch in their own sharp wits, are content to remain less learned than the laws, and are unable to pick apart a speech by a good speaker. These are men who can judge impartially rather than [acting like] rival athletes, and so things go well. It would be best for us [politicians] to do likewise and avoid being so swept away by cleverness and keenness for the competition that we contradict our own true opinions when we advise you, the masses (*to plêthos*). (3.37.3–4)

Thus, if there must be politicians, Cleon suggests that they should act and speak more like ordinary Athenians. But as it is, he goes on to say, instead of politicians acting like ordinary citizens, the ordinary folk all wish that they could be clever speakers themselves. Lacking actual oratorical attainments, they fancy themselves connoisseurs of oratory (3.37.6–7). As a result, debate causes delay, which is to the advantage of wrongdoers. Instead of wasting their time listening to speeches and then endlessly changing their minds about policy, the Athenians would do better to act (that is, vote on policy issues) in the heat of righteous anger and then stick by those decisions (3.38.1).

Cleon implies that there can be no *good* reason for opposing his own policy of general punishment, of treating Mytilene judicially as a unitary state-actor rather than as a diverse society. He sets up a narrow and exclusionary framework to explain the motives of those Athenians who spoke against his proposal: either they hoped to make a public display of their rhetorical powers or they had been bribed to support an inherently bad policy. In either case, Cleon reminds his audience, in these sorts of contests (*agones*) it is the speakers who reap the prizes (i.e., adulation or bribe money) while it is the polis that is exposed to dangers (3.38.2–3, cf. 3.40.3).

But you [citizens] yourselves are the cause of this evil for having set up these contests (*agones*); you have become accustomed to being spectators of orations (*logoi*) while gaining your knowledge of facts (*erga*) from what you hear. You decide what is possible regarding what has to be done in the future by looking to those who speak well. Even regarding events of the past, you don't rate the evidence of what you actually saw above what you have heard in some over-clever bit of verbiage. (3.38.3–4)

This statement, with its emphasis on the priority of experience and *erga*, has clear and obvious affinities to Theoretikos' Realist approach. Cleon claims that his mission is to lead the Athenians away from their foolish habits of speech-spectatorship (3.39.1). But what does he offer instead? Hardly a dispassionate grasp of material realia. Rather Cleon urges the Athenians to recall vividly their emotions at the time the revolt first broke out, and advises them to act according to that mimetically restored emotional state (3.40.7). Like Thucydides of 2.65, Cleon finds much to criticize in democratic Athens and he uses some of the same terminology, but in the end Cleon urges his listeners to take the easy path of relying on their visceral emotions when making decisions. Thucydides' reader, struggling with the complex text, is not offered such an easy road to right judgment.

Cleon's meta-rhetoric results in his claim that political speech is an impediment to action and that strong-felt emotion is a more appropriate wellspring of policy than public debate. Diodotus, on the other hand, stoutly defends reiterated public discussion of especially important affairs. Indeed, he says, anger and overquickness are the two greatest impediments to good policy (*euboulia*: 3.42.1).

> But if someone argues that speeches (*logoi*) are not teachers in regard to affairs (*pragmata*) either he is a fool or he is on the lookout for some private (*idiai*) advantage. He is a fool if he supposes that it is possible to consider the uncertain future by some other means; he is seeking his own advantage if he hopes to propose some shameful thing, and is unable to speak well or convincingly regarding it, yet by slandering well is able to strike fear into both the opposing speakers and the listeners. (3.42.2)

Here, Diodotus reveals the obvious flaw in Cleon's anti-public speech meta-rhetoric: Cleon's attack on clever speech is embedded in a clever speech, and so demonstrates the impossibility of communicating meaning except through the medium of words. Like Cleon, Diodotus attributes to his opponents an illegitimate private interest in personal gain and he claims that those private interests will endanger the state. But having suggested that Cleon is either a fool or is out for personal advantage, Diodotus then attacks the rhetorical practice of claiming that one's enemies place personal gain over the public good (3.42.3–6). It would be much better, Diodotus goes on to say, if the Athenians would abandon their habit of dishonoring those who lose public debates. If they quit punishing losers of oratorical contests, then orators would speak their minds honestly, rather than advocating policies they did not believe in with an eye toward gaining the praise of the many (*to plêthos*). But, he continues, as it is, we do just the opposite, and because speakers have to work under constant suspicion of being bribe-takers, the polis loses the benefit of good advice. Evil and goodwilled speakers alike are forced to lie and the polis is the only entity for whose good it is *impossible* for a citizen to work openly (3.43.1–3).

Although the appeal to the public good initially recalls Pericles, Diodotus' meta-rhetoric is almost as muddled as Cleon's: he accuses Cleon of self-interest, and in the next breath points out how destructive the rhetorical practice of making such accusations is to the political practice of decision-making. Indeed, he claims that slanderous rhetoric is specifically destructive in that *all* speakers, even goodwilled men (like himself) are made into liars. In sum Diodotus willfully embraces the well-known "Cretan Liar" paradox: Since Diodotus is an orator and all orators are liars, the truths he claims to teach through speech are thoroughly compromised and his defense of the value of public discussion becomes paradoxical: What public good can result from reiterated debates among liars? Moreover, his comment on the impossibility of doing good openly for the polis explicitly contra-dicts Pericles' Funeral Oration encomium of Athenian public-spiritedness. If public speakers are liars and unable to do good for the state in any overt way, how are we to take his next statement (meant to refute Cleon's claim that ordinary, ignorant citi-zens are good administrators)? It is necessary, he says, that we speakers look a little bit further ahead than the rest of you citizens who just glance things over in a super-ficial manner (3.43.4). What techniques will the speaker use to gauge the likely course of future events? How will the results of this forethought be translated into good state policy in a democracy, given the necessity of public mendacity? Why should listeners believe that an acknowledged liar is sincere when he claims to seek the public good rather than private advantage? Diodotus does not say, but then, given his argument about the impossibility of doing good openly, he simply can't.

The reader of these convoluted meta-rhetorical arguments may opine that nei-ther Cleon's claims for policymaking by ordinary men nor Diodotus' claims for leadership of the democracy by foresightful public speakers is particularly con-vincing. Our sense of unease is not assuaged by their respective arguments regard-ing the proper basis for policy. Both speakers appear at first to be Realists: they agree that the main determinant in the Assembly's decision must be Athens' impe-rial interests and each claims that his policies will best serve Athenian interests. Each orator acknowledges the fact that Athens' imperial income derives from pros-perous cities, a very salient point given the importance of capital in the "material triad." Cleon comes up with a tortuous argument for linking the utter destruction of Mytilene with imperial prosperity (3.39.7–8) but the economic point clearly favors the case for leniency: dead men don't pay tribute. Yet Diodotus mentions the negative economic consequences of Cleon's policy only in passing (3.46.3). Rather than work through the benefit/loss equation with specific reference to Mytilene, both Cleon and Diodotus emphasize that the issue at hand has more to do with the future than the present and with the empire in general rather than the polis of Mytilene in particular (3.40.7, 3.44.3, 3.48.2). They are theorists of power before being policymakers. They agree that the treatment of Mytilene will be a test case for what happens to insurgents, and thus the issue is not the material and particular results of exterminating a great and prosperous city, but rather how Athe-nian harshness or leniency will be *perceived* by the other subject states. Each, in short, claims theoretical insight and that policy should be made on the basis of his theory of human behavior; neither makes any serious use of actual Greek history.

The issue thus becomes group psychology rather than the "material or historical facts of the matter"—indeed, neither speaker suggests that facts matter very much. And as a result, the two contestants must base their arguments on appeals to human nature, to assumptions about what "men or states are likely to do in a given situa-tion." Human nature (*phusis*) is a major issue for Thucydides. But how, according to these Athenian public speakers, is one to know human nature? Despite his stated confidence in ordinary citizens as decision-makers, Cleon implies that the everyday experiences of the men in his audience are *not* an adequate basis for making judgments about international relations (3.37.2), whereas Diodotus—who believes that only politicians with special insight into affairs can come up with good public policy—claims (45.2–3, 45.6) that the collective behavior of poleis closely mimics the private behavior of individuals. Cleon claims that people only revolt when they have suffered some form of violence (3.39.2). And hence he is able to argue that it would be improper to forgive the Mytileneans on the grounds that "it is human nature to do wrong" (3.40.1). Diodotus disagrees: it *is* in fact the natural tendency of both individuals and poleis to do wrong (3.45.3). Yet for Diodotus it is not only suffering violence that leads people to revolt, but a wide variety of factors: poverty, wealth, hope, chance, and emotion of various sorts (3.45.4–6). Once again, neither offers any historical examples to buttress his opinion.

If they disagree on the wellsprings of human perversity, nor do the two speakers agree on what factors will successfully deter the tendency of people to do wrong (i.e., the tendency of subjects to resist their hegemon). For Cleon, it is a universal truth that people despise kindness but respect harsh treatment (3.39.5). Wrong, according to Diodotus: the harshest treatment is the death sentence, and that has not prevented people from doing wrong in the past. To support this claim, Diodotus develops a quasi-historical argument: It "seems probable" that long ago (3.45.3) sentences for wrongdoing were less strict than they are now, and that they were gradually made harsher in a vain attempt at deterrence. He cites no authority for this opinion, and it is not at all clear that it would fit what most Athenians thought they knew about their own past history.[35] But it did fit his argument nicely, and rhetorical expediency, not historical validity, is obviously the point.

Cleon's position on the value of a deterrent example is part and parcel of his general rule that holding an empire is a matter of raw strength rather than goodwill (*eunoia*: 3.37.2). He therefore refuses to take into cognizance the political implications of social distinctions between Mytileneans: the Athenians must not say that it was "the few" (*hoi oligoi*) who were the cause of the revolt and let the Mytilenean *dêmos* go free; the few and the many were all equally guilty of wronging us (3.39.6). Diodotus, on the other hand, urges that punishment be meted out individually, and to as few individuals as possible (3.46.3). Rather surprisingly, in light of his prior Realist linkage of individual and state behavior, he focuses on internal social distinctions within poleis. He notes that as matters now stand, in every polis of the empire the *dêmos* (meaning the lower classes) is well-disposed (*eunous*) toward Athens. As a result, if in some allied town *hoi oligoi* initiate a revolt, they cannot count on the support of the *dêmos*, and so you (Athenians) have *to plêthos* (the many) as an ally (3.47.2). Obviously if the Athenians treat Mytilene juridically as a state-actor, that is, punish *dêmos* and *oligoi* identically, they will lose this valuable ally. Notably, both speakers use the term "*dêmos*" in its "sociological" or "factional" sense ("lower class" not "citizenry"), and thus they reveal what Pericles' Funeral Oration attempts to conceal—the "fact" (from a Thucydidean perspective) that *dêmokratia* was the self-interested rule of a socially defined faction rather than the rule of all citizens by all citizens. But Diodotus wants to use this sociological fact pragmatically in Athens' interest; Cleon, whom Thucydides has told us is the darling of the Athenian *dêmos,* and whose strongly egalitarian sentiments might be expected to appeal to poorer Athenians, sees no purpose in encouraging a "transnational" lower-class sociopolitical solidarity that would cut across the nationalist sentiments of polis populations.

The two speakers also differ on the question of whether justice has anything to do with the decision. Cleon pretends it does. His conglomerate argument switches back and forth from the language of law and right to that of necessity and advantage: the Athenian empire is a tyranny, he claims, and whether it is just or unjust is irrelevant (3.37.2, 3.40.4); yet by opposing this tyranny the Mytileneans have acted terribly unjustly and so they deserve the general punishment demanded by righteous anger. Diodotus is quick to jump on Cleon's inconsistency. This is not jury trial, but a policy debate, not a case in which justice and national interest go hand in hand, but one in which advantage alone deserves a hearing. Whether guilty or not the Mytileneans are more of a benefit to Athens alive than dead, and that should be the end of it (3.44.1–4, 3.47.5).[36]

So what is the audience—either the original audience of assemblymen or Thucydides' audience of readers—to make of all of this? The contestants in the Mytilenean Debate pose several questions that our reading of sections 1.1–2.64, in light of the assessment of Pericles and his successors at 2.65, has taught us are vitally important: Can a post-Periclean democracy run an empire? What is the relationship between deployment of power in the international arena, domestic politics, social diversity, and Panhellenic norms? What is the proper role of politicians and public debate in the democratic state? What is the relationship between an individual speaker's personal interests and the public interest? How can one determine if there is some discontinuity between public and private interests, and what should be done about it? Can prudent (in this case, restrained) policy and decisive action be reconciled? What is the appropriate basis for a future-oriented policy? How is foresight cultivated?

Arguably, the workability of democracy and the success of Athens as a state-actor depends on having the answers to these questions. Yet neither the speech of Cleon or that of Diodotus gives a convincing answer to any of them. The reader is surely led to prefer Diodotus' position, by the clearly prejudicial introduction to Cleon as "most violent and most influential" of the public speakers, by the brutality of the policy Cleon advocates, and perhaps by the relatively greater degree of rational realism in Diodotus' arguments. Diodotus' approach of exploiting social distinctions within the allied states might be made congruent with the logic developed in the Archaeology by assuming that Diodotus recognized that social divisions made for weakness in communities; Athens might have an easier time ruling weaker subjects. But his rhetoric is a long way from the confident integration of public speech and foreign policy that characterized Thucydides' portrayal of Pericles, and the reader has become very aware that there are deep social divisions in Athens, as well as in the subject states. She is not likely to be sanguine about the chances of the cryptic and necessarily mendacious style of leadership that Diodotus offers his audience and she may not be surprised to find that Diodotus disappears from Thucydides' text after the debate. The unsatisfactory nature of the two speeches is reflected in the final decision: Diodotus prevailed (3.49.1) but, Thucydides points out, the vote was very close.[37]

The complexities entailed by the social diversity and political practices that underpinned the polis as a state-actor are front and center in the Mytilenean Debate. There is no doubt that the division between mass and elite, *dêmos* and *dunatoi*, in Athens and in the poleis Athens must deal with is of key significance both for Athenian internal politics and for Athenian international relations. Yet neither Cleon nor Diodotus offers a coherent vision of the relationship between public speech, national policy action, and the material world or of the relationship between state-actor and polis society. And without a coherent vision, it is hard to imagine that Athens will consistently make the right choices on difficult matters of policy.

Conclusions

It will be a long road from the confused rhetoric of the Mytilenean Debate to the collapse of Athenian power after the Sicilian disaster and then to the final collapse of 404. As in the case of the revolt of Mytilene, Athens' Sicilian policy and the long endgame of the Ionian War (410–404 B.C.—not recorded in Thucydides' text, but well known to him and anticipated in his narrative) will hang on the intertwining of material factors (ships, walls, and money) with the problematic factors of democratic politics, social diversity, and Panhellenic values. But by the end of book 3 of the history, the reader has already begun to grasp a basic lesson: the confident, strong Realist theory developed by Theoretikos in 1.1–2.64 is inadequate to explain what really happens to states in conflict, and a fortiori inadequate to explain the behavior of states that are democratic in fact as well as in name. He who would truly understand why great affairs of the past played out as they did, or how they might play out in the future, must move beyond the tidy theory of unitary, self-interested state-actors. He must, Histor suggests, look to the world of social relations and politics, the world of speech and reception, of human interests that are hopelessly tangled up with competing loyalties to traditional Greek norms, to the polis, to various social groups, to individual selves. Histor does not actually refute the Realist theses presented by Theoretikos. But he shows us that a knowledge of theoretical principles animating states will only be useful when that knowledge is tempered by the close attention to human complexity that comes from learning to think historically.

How might the reading of Thucydides offered here be brought into a fruitful engagement with mid-twentieth-century history? I do not actually suppose that there is a lot to be gained by a direct one-to-one comparison of historical individuals or specific events. Roberts notes that "Gerald Johnson writing in the *New Republic* in 1961 identified Cleon with Joseph McCarthy and the impeachment of

Alcibiades [in 415 B.C.] as the work of a House Committee on un-Athenian activities."[38] Although Johnson's sort of comparison seems to me tendentious, perhaps comparisons drawn rather more broadly may help us to think through both historical periods. I must, however, admit myself to be well out of my professional depth when dealing with the cold war era. The reader must take what follows as provisional and as an invitation to academic colleagues with the appropriate expertise to test and revise the following hypothesis: The citizenry of the United States in the 1950s, I would posit, found itself in a position at least superficially analogous to that of the Athenian citizenry on the threshold of the Peloponnesian War. Like Athens, the United States was then a young, powerful, democratic state with a war-tested political and military leadership—as well as being a diverse society held together in part by a national self-identity strongly influenced by the recent experience of having led a coalition of free states to a dramatic victory against a nightmarish common enemy (Persia, Axis). Like pre–Peloponnesian War Athens, the post–World War II United States enjoyed a remarkably strong "triadic" base: vast national wealth and a secure hegemony, capacity to project its power (in the case of the United States, by air as well as by sea and potentially through the instrumentality of atomic weapons). Moreover, in the relative geographic security of its own hemisphere, the United States approximated the impregnable island nation of Pericles' dreams (1.143.5). Finally, the exacerbation of a rivalry with another hegemonic power under the leadership of a one-time ally (Peloponnesian League/Soviet bloc) encouraged Athenians and Americans to think in starkly bipolar terms. Thus, per the article by Robert Campbell with which this chapter began, there was a strong and perhaps understandable tendency for cold war–era American readers of Thucydides to identify with the powerful vision of Periclean Athens offered by "Thucydides Theoretikos" in 1.1–2.64.

I have argued that there are, however, two identifiable "voices" in Thucydides' text: a historical as well as a theoretical voice.[39] Theoretikos' optimism regarding potential human understanding of state behavior is complicated by Histor's narrative, which shows that a democracy-as-hegemon may be unstable under extreme pressure and might even collapse under the weight of its attempt to square the ideal of democratic rule by public discourse at home with self-interested, "tyrannical" rule abroad. Some modern political analysts (one thinks of George Kennan) may have supposed that the answer to what we might call "the Thucydides dilemma" of squaring democracy and hegemony is implied in Diodotus' vision of a wise but secretive and even publicly mendacious leadership: On this reasoning, a Realist policy elite must find ways to avoid allowing open, public, democratic deliberation from affecting major policy decisions, especially in the sphere of international relations. Such analysts may embrace some version of "democratic elitism" or advance the "end of ideology" arguments that were particularly prevalent in the aftermath of the Korean War.[40] I have argued here that Thucydides, read "in the round," offers little true comfort for such positions. He encourages the reader to take a critical view of democracy and ideology but he also encourages suspicion of elite leadership and raises doubts about the human capacity to transcend ideology or to operate outside a set of "irrational" values and norms.

Thucydides' Athenians ironically (in light of the Athenian celebration of their own freedom of speech) feel most at liberty to talk the language of strong Realism when audiences are oligarchs (e.g., the Athenian spokesmen at Sparta in book 1 and the envoys at Melos in book 5). So too the cold war American policy elite sometimes seemed to envy the freedom from democratic oversight—from political interference, from naive idealism, from extraneous, "unrealistic, irrational" factors like social or international justice—that they perhaps imagined was enjoyed by their policymaking counterparts in totalitarian societies. Accordingly, there have been many efforts to establish buffers between the policymaking environment and public deliberations, to insulate American policymakers from the messy, contradictory ideals of American society and the complexities of American democratic politics.[41] To the extent that the American policy elites have succeeded in achieving insulation and turning away from the complexity of history to the antiseptic world of theory, they have blinded themselves to the meaning of important, if untidy, realities. It is not, on these terms, so surprising that American policymakers so badly misunderstood the potential role of revolutionary China in what they fondly imagined to be a bipolar international regime. Nor that the Korean War was soon followed by the even more disastrous war in Southeast Asia. Nor that the economic power and political coherence of the Soviet empire was so seriously overestimated. In retrospect, and in the light of history, it now seems clear enough that China was likely to react violently to the American move toward the Yalu River and unlikely to remain happily within the Soviet "sphere"; that leaders in Vietnam, Cambodia, and Thailand had local agendas and were not simple "dominoes" in the great game; that the Soviet Union was incapable of maintaining indefinitely a credible pose as a unitary state-actor in the face of deep-rooted social and political contradictions. Perhaps an attentive student of Thucydides Histor, one who did not demand that a complex text tell only one simple story and offer one simple lesson, would have been more sensitive to these matters than were some American cold war analysts and policymakers.

It is tempting, therefore, to imagine that a more careful attention to Thucydides Histor would have served as a corrective by focusing American analysts' attention on the dangerous contradictions and unstable complexities inherent in diverse societies like China and the Soviet Union, especially given that those societies pretended to be democracies, ruled by "the people." But reading Thucydides more carefully is no panacea. A key weakness of American policy in Korea resulted from the excessive focus on Russia and the European sphere and a concomitant failure to take proper account of China as a (potentially at least) independent power with a complex history that would affect both its hegemonic ambitions and its security concerns. Similarly, the Athenian focus on Sparta and the Greek world may have blinded Athenian policymakers to the financial and diplomatic role that Persia might play in an escalating conflict. In book 1, Thucydides establishes a strictly bipolar frame for the war that is to come (esp. 1.1.1). On the basis of my two-voice thesis we might expect that this interpretive straitjacket would engage the critical attention of Histor. Yet it is not clear that it ever did. W.R. Connor notes that the crumbling of the original bipolar framework is evident (inter alia) in the increasingly hectic narrative of book 8.[42] Now, book 8 gives the impression of being a draft and it is here that the text abruptly ends. Would Thucydides have pulled it all together in a revised draft and a completed text? Or had the hypertrophy of historical complexity so overbalanced the initial theoretical order that the entire project of writing history in two voices lost its coherence and simply could not be sustained?

The sort of gross misestimation of other states' strengths and intentions that characterized American policymaking in the cold war era is hardly unique in human history and there is no good reason to suppose that reading Thucydides more carefully will immunize us against error. Few Greeks, for example, seem to have grasped how weak Sparta actually was in the decades after the surrender of Athens in 404.[43] Sparta's catastrophic collapse following the loss of the battle of Leuctra in 371 seems to have come as a real shock to many contemporaries. Among them was the historian Xenophon, who, as self-appointed continuator of Thucydides' history, had ample opportunity to absorb whatever lessons the master had to teach. Thucydides' text remains a "possession," and continues to tantalize those "who wish to understand clearly the events that happened in the past and that (human nature being what it is) will, at some time or other and in much the same ways, be repeated in the future" (1.22). But perhaps his most cogent lesson is the necessity of intellectual humility in the face of the uncertainties, peculiarities, contingencies, and idiosyncrasies that will continue to challenge theorists, historians, and policymakers alike.

Notes

1. Jennifer Tolbert Roberts, *Athens on Trial: The Antidemocratic Tradition in Western Thought* (Princeton: Princeton University Press, 1994), 260–61, cf. 297–98, citing R. Campbell, "How Democracy Died," *Life* 30 (1 January 1951): 88–96.

2. On the "cold war reading" of Thucydides see further W.R. Connor, *Thucydides* (Princeton: Princeton University Press, 1984), 1–11; R. Ned Lebow and Barry S. Strauss, eds., *Hegemonic Rivalry: From Thucydides to the Nuclear Age* (Boulder: Westview Press, 1991), passim.

3. It is important to keep in mind that even if we limit ourselves to Greece, Thucydides is only one possible starting point for the tradition. One might, for example, start with Herodotus, who gives a very different take on the origins of political theorizing and self-conscious history; see for example Pericles Georges, *Barbarian Asia and the Greek Experience: From the Archaic Period to the Age of Xenophon* (Baltimore: Johns Hopkins University Press, 1994), 115–206.

4. See, for example, Laurie M. Johnson, *Thucydides, Hobbes, and the Interpretation of Realism* (Dekalb: Northern Illinois University Press, 1993); Christian Reus-Smit, *The Moral Purpose of the State: Culture, Social Identity, and Institutional Rationality in International Relations* (Princeton: Princeton University Press, 1999). This tradition is analyzed from a

classicist's perspective in a Gregory Crane's *Thucydides and the Ancient Simplicity* (Berkeley: University of California Press, 1998), to whom I am indebted for the Johnson reference.

5. Michael W. Doyle, "Thucydidean Realism," *Review of International Studies* 16 (1990): 223–37; "Thucydides: A Realist?" in Lebow and Strauss, eds., *Hegemonic Rivalry*, 169–88. Doyle rejects the notion that Thucydides is either a "fundamentalist Realist" (one who sees all human behavior as characterized by an interest-based drive for power), or a "structuralist Realist" (one who assumes that all state actors are functionally similar units, among which the most consistently rationally self-interested and power seeking will flourish). The burden of this chapter is to elaborate upon Doyle's thesis of Thucydides as "a Minimalist, but neither a Fundamentalist nor a Structuralist," along the lines suggested by Connor, *Thucydides*—that is, by paying close attention to how literary qualities interact with substantive content to "mislead" and ultimately to educate the reader.

6. Caveat lector: "Theoretikos" and "Histor" are my own labels; they are not terms that Thucydides uses of himself. They are used here as convenient, rather than precise, ways to distinguish between the "strong Realist" and "historical" voices in the text.

7. Cynthia Farrar, *The Origins of Democratic Thinking: The Invention of Politics in Classical Athens* (Cambridge: Cambridge University Press, 1988), argues strongly for the unity of history and theory in Thucydides. Although I disagree with much of what she has to say about Thucydides and democracy, her central insight regarding Thucydides' creation of a historical theory of human action remains powerful and convincing.

8. See M. Pope, "Thucydides and Democracy," *Historia* 37 (1988): 276–96, for a discussion of Thucydides' use of the terms "the Athenians," "the Corinthians," etc. for state-actors.

9. On "real" versus ideological factors in Thucydides: J. Ober, "Thucydides' Criticism of Democratic Knowledge," in R.M. Rosen and J. Farrell, eds., *Nomodeiktes: Greek Studies in Honor of Martin Ostwald* (Baltimore: Johns Hopkins University Press, 1993), 81–98; "Civic Ideology and Counterhegemonic Discourse: Thucydides and the Sicilian Debate," in A.L. Boegehold and A.C. Scafuro, eds., *Athenian Identity and Civic Ideology* (Baltimore: Johns Hopkins University Press, 1994), 102–26.

10. Adam Parry, *"Logos" and "Ergon" in Thucydides* (New York: Arno Press, 1981), remains the fundamental study.

11. On the distinction between the conceptual categories of "society" and "state" and their use in the analysis of the polis: J. Ober, *The Athenian Revolution: Essays on Ancient Greek Democracy and Political Theory* (Princeton: Princeton University Press, 1996), 161–87.

12. The role of capital resources in Thucydides is discussed in detail by Lisa Kallett-Marx, *Money, Expense, and Naval Power in Thucydides' History. 1–5.24* (Berkeley: University of California Press, 1993).

13. Here I am using the term "security" to mean "defense of key urban assets and population from enemy attack." This is, of course, a much more limited sense of the term "security" than that often used by modern international relations theorists.

14. Minos' thalassocracy is something of a deus ex machina in that Thucydides never tells us where Minos got the funds to build his navy. Was he a successful pirate who became a legitimate ruler? Or, as Charles Pazdernik suggests (in an unpublished paper, "Thucydides on Money"), should we imagine on the parallel of Pelops (1.9.2) that Minos inherited wealth from ancestral sources in the (already developed) East?

15. Thucydides' developmental scheme seems limited to the Greek world; it does not explain the rise of Persia—which was long an entirely land-based power. That Persia was a great *arkhê* is explicitly acknowledged in the Archaeology: its strength was a limiting factor in the growth of the power of the Ionians after they had become a sea power: 1.16. It is only later that Darius, with the aid of the Phoenician navy, conquered the Aegean islands: ibid. Cf. below.

16. Athenian imperialism and its problematic association with Persia: Georges, *Barbarian Asia*; D. Rosenbloom, "The Tragedy of Power: Myth, Memory, and Hegemony in the Theater of Aischylos" (Ph.D. diss., Princeton University, 1992); Robert Metcalfe, "Herodotus and Athens" (Ph.D. diss., University of Toronto, 1996).

17. For a detailed history of the Athenian empire, see Russell Meiggs, *The Athenian Empire* (Oxford: Clarendon Press, 1972).

18. N. Loraux, "Thucydide et la sedition dans les mots," *Quaderni di Storia* 12 (1986), 95–134, points out the importance of the theme of the polis as existing in a more or less constant state of civil conflict.

19. I recently heard a thirdhand story of an elderly classicist who told his students, "I have taught Thucydides for 30 years; I have read the text through 30 times and each time I hope that this time Athens will win." This "Foaftale" (the term is used by contemporary folklorists for stories told about "a friend of a friend") nicely captures the tensions that the text encourages.

20. Walls of Athens and Piraeus: esp. 1.89–93; Long walls: 1.107.1 (with Simon Hornblower, *A Commentary on Thucydides. Volume I: Books I-III* [Oxford: Clarendon Press, 1991], ad loc.), 1.107.4, 1.108.3. R.E. Wycherley, *The Stones of Athens* (Princeton: Princeton University Press, 1978), 7–25, is a useful account of the archaeology.

21. Narrative passages on Athenian navy and financial reserves: see Kallet-Marx, *Money*, passim.

22. There is a huge secondary literature on the relationship between the speeches in Thucydides' text and the speeches actually delivered, centered on the ambiguous language of 1.22.1. See, recently, Harvey Yunis, *Taming Democracy: Models of Political Rhetoric in Classical Athens* (Ithaca: Cornell University Press, 1996), 59–86, arguing, with special reference to the speeches of Pericles, that the speeches in the text are plausible fictions, retaining the speaker's intention but otherwise created from whole cloth by Thucydides.

23. The policy of refusing battle was perhaps politically more difficult for Pericles to achieve than Thucydides leads us to suppose: Ober, *Athenian Revolution*, 72–85.

24. Thuc 2.18–23. The deme of Oinoe and its walls: J. Ober, *Fortress Attica* (Leiden: E.J. Brill, 1985), 154–55. On Greek siegecraft and fortifications, see also J. Ober, "National Ideology and Strategic Defense of the Population, from Athens to Star Wars," in Lebow and Strauss, eds., *Hegemonic Rivalry*, 251–67; "Hoplites and Obstacles," in Victor D. Hanson, ed., *Hoplites: The Classical Greek Battle Experience* (New York: Routledge, 1991), 173–96.

25. The following several paragraphs, on the Funeral Oration, are adapted from Ober, "Thucydides' Criticism of Democratic Knowledge."

26. E.g., 2.43.1. Cf. also 2.40.1, 2.42.2.

27. As he might have: cf. 6.39.1, in which Athenagoras of Syracuse defines *dêmokratia* as the rule of the entire citizenry (*dêmos xumpas*), *oligarchia* as rule of a few. Cf. the comments of A.W. Gomme in A.W. Gomme, A. Andrewes, and K.J. Dover, *A Historical Commentary on Thucydides*, 5 vols. (Oxford: Clarendon Press, 1945–81), 2.107–108. R. Sealey, "The Origins of *Demokratia*," *California Studies in Classical Antiquity* 6 (1973): 280–82, seems to me to go too far in reading Pericles' remark as apologizing for democracy. For various readings of this complicated passage, see Hornblower *Commentary*, ad loc.

28. On the radically different nature of Athenian *dunamis* from traditional forms of Greek power, see G. Crane, "The Fear and Pursuit of Risk: Corinth on Athens, Sparta and the Peloponnesians: Thucydides 1.68–71, 120–121," *Transactions of the American Philological Association* 122 (1992): 227–56; "Power, Prestige, and the Corcyrean Affair in Thucydides 1," *Classical Antiquity* 11 (1992): 1–27; and in this volume.

29. Speeches of Corcyreans and Corinthians: Crane, "Power, Prestige"; Ober, "Thucydides' Criticism."

30. Corinthian speeches to Spartans: Hornblower *Commentary*, 108, 114; and especially the insightful analysis of Crane, "Fear and Pursuit."

31. On Corinth as a "third force" in the Peloponnesian War, see M. Sordi, "Scontro di blocci e azione de terze forze scoppio della guerra del peloponneso," in Lebow and Strauss, eds., *Hegemonic Rivalry*, 87–98.

32. Dover in Gomme et al., *Historical Commentary*, 5.43–47; Hornblower, *Commentary*, 348.

33. Forthcoming work by several younger "constructivist" critics of Realist international relations theory—e.g., Alexander Wendt (*Social Theory of International Politics*), Christian Reus-Smit (*The Moral Purpose of the State*), Martha Finnemore (*Defining National Interests in International Society*), and Peter Katzenstein—focuses on the failure of state-actor models to take account of "endogenous dependent variables" like domestic politics and the social context established by transnational values and norms. See also Nicholas Onuf, *World of Our Making* (Columbia: University of South Carolina Press, 1989), and the essays of Evangelista, Kauppi, Lebow, Ober, and Strauss in Lebow and Strauss, eds., *Hegemonic Rivalry*.

34. Secondary bibliography on the debate: Hornblower, *Commentary*, 420–22; add now Yunis, *Taming Democracy*, 87–101. Mytilene was not tribute-paying ally, but contributor of ships, and so in a small privileged category of quasi-free states within the empire.

35. For example, the well-known tradition that the early Athenian law code attributed to Drakon entailed the death penalty for all categories of crime.

36. Note, however, that Diodotus is not actually true to his own ban on arguments based on justice: 3.47.3: if you kill the *dêmos* of Mytilene, you will act unjustly toward your benefactors. In rejecting justice/law arguments Diodotus claims that in order to guard the state the Athenians should not look to harsh laws (*nomoi*) but to realistic administration (*erga*: 3.46.4); the contrast between *nomoi* and *erga* is a play on the familiar *logos/ergon* antithesis.

37. Compare the much more positive reading of Diodotus' speech in, e.g., Yunis, *Taming Democracy*, 87–101; Arlene Saxonhouse, *Athenian Democracy: Modern Mythmakers and Ancient Theories* (Notre Dame: Notre Dame University Press, 1996), 72–86.

38. Roberts, *Athens on Trial*, 298, citing G. Johnson, "God Was Bored," *New Republic* 145 (September 11, 1961):10.

39. I have not answered the obvious question of whether Thucydides the author *deliberately* and self-consciously employed two voices as a didactic device. I have deliberately drawn the contrast quite starkly because I do think that the author intentionally establishes a confrontation between the confident assumptions arising from theoretical speculation (e.g., the sort of thinking encouraged by the contemporary Sophists) and the messy reality of the historical narrative he worked so hard to establish. I suppose that he seriously thought that reading his text would be good for a policymaker and that he regarded policy made on the basis of theories that could be summed up in slogans like "justice is the will of the stronger" to be bad policy. On Thucydides' self-consciousness as a writer and his didactic relationship with the imagined reader, see Connor, *Thucydides*, passim.

40. Dangers of democratic elitism: C. Wright Mills, *The Power Elite* (New York: Oxford University Press, 1956); Peter Bachrach, *The Theory of Democratic Elitism: A Critique* (Boston: Little, Brown, 1967); M.I. Finley, *Democracy Ancient and Modern*, 2d ed. (New Brunswick: Rutgers University Press, 1985). "End of ideology": Daniel Bell, *The End of Ideology: On the Exhaustion of Political Ideas in the Fifties*, rev. ed. with afterword (Cambridge: Harvard University Press, 1988), originally published in 1960. Notably, the American decision to intervene in Korea is one of the "big decisions" that Mills attributes to the cryptic power elite, while Bell (54) assigns it to the individual president on the basis of his constitutional executive authority.

41. Of course this impulse much antedates the cold war era. The U.S. Constitution itself

reflects the Federalists' fear that the "mob" would unduly influence policymaking by elites; their fears were in part stimulated by their reading of Athenian history in the Peloponnesian War era (directly or indirectly based on Thucydides): Roberts, *Athens on Trial*, 179–93.

42. Connor, *Thucydides*, 213–17; cf. Connor, "Polarization in Thucydides," in Lebow and Strauss, eds., *Hegemonic Rivalry*, 53–69.

43. For good analyses of this period, see Charles D. Hamilton, *Sparta's Bitter Victories: Politics and Diplomacy in the Corinthian War* (Ithaca: Cornell University Press, 1979); Paul A. Cartledge, *Agesilaus and the Crisis of Sparta* (Baltimore: Johns Hopkins University Press, 1987); Barry S. Strauss, *Athens After the Peloponnesian War: Class, Faction, and Policy, 403–386 B.C.* (Ithaca: Cornell University Press, 1987).

14

Father of All, Destroyer of All: War in Late Fifth-Century Athenian Discourse and Ideology

Kurt A. Raaflaub

War is both king of all and father of all, and it has revealed some as gods, others as men, it has made some slaves, others free.

—Heraclitus, no. 22 fragm. B53 Diels-Kranz

Surely, the wise man must avoid war.

—Euripides, Trojan Women 400

War is a violent teacher.

—Thucydides 3.82.2

1. Introduction

This chapter[1], clearly still a work in progress, represents an effort to understand how the Athenians reacted to war and talked about war at a time when their experience of war was most intense and difficult: during the last part of the Peloponnesian War (431–404).[2] I use "discourse" here in a very broad way, to encompass all forms of communication of thought and ideas and of public interchange that are still accessible to us through literary, epigraphic, artistic, and monumental sources. The period in question is especially suitable, relatively speaking, for such an attempt because it is one of the best-documented periods in ancient Greek history:

the archaeological record is unparalleled in the Greek world; most of the extant tragedies and comedies were performed during the war; political theory and historiography were coming into their own, and rhetoric was discovered as an art and instrument of power.

I consider it important to analyze the Athenian discourse on war not in isolation but in its larger context, and to understand such context broadly, including not only discourse on peace, but also on power and empire, if not on politics and democracy. In order to enable the reader as best as possible to experience the ideological environment and to imagine and visualize what the Athenian citizens would have seen and heard in their everyday lives in public places and at public events and how this might have affected their reactions to the public discourse on war, I shall use a framing story. It will guide us through the city and its harbor, help us observe what was to be seen, and help us think, at various places, about various forms of discourse used there. This, I hope, will help us understand the overall nature and impact of such discourse and the Athenians' attitudes to war and empire. First, though, a few general remarks.

a. The Transformation of Warfare and Society in Athens

The Greek experience of war after the Persian Wars differed greatly from that of earlier times.[3] During the archaic period, wars were mostly short and fought between neighboring communities; they consisted of brief, violent, and almost ritualistic encounters of hoplite armies: citizens fighting on their land for their land. Destruction of cities was rare, and the concept of subjecting another city for the purpose of ruling over it and extracting tribute or other resources from it was virtually non-existent. Rather, formations of power were based on expansion through absorption of conquered territory, and on hegemonial leadership in a system of alliances.[4]

The introduction of naval warfare on an unprecedented scale during and after the Persian Wars changed the experience of war fundamentally. Campaigns now extended over longer periods, the crews needed intensive training, and these crews mostly consisted of lower-class citizens and mercenaries; this entailed a massive change in the demography of Greek warfare. Interconnected with these military and social changes were equally massive political changes: first, the formation of a large naval empire—the first empire established by a Greek polis—encompassing nearly two hundred communities in the Aegean and Black Sea areas by the 460s and 450s, and second, the realization of full democracy in the late 460s and 450s.[5] The citizens who fought the wars for their city were now fully in charge of decisions about these wars and everything connected with them. Democracy further changed the nature of warfare.[6] Wars, requiring the maintenance of a permanent fleet and the absence of thousands of citizens for weeks, months, and occasionally even years at a time, and causing a heavy death toll, now became a regular feature: this was the price the Athenians paid for their empire and its benefits. These experiences, and their rationalization in political theory, form the background to Plato's statement, "What most men call 'peace' is really only a fiction (*onoma*), and in cold fact all states are by nature (*kata phusin*) fighting an undeclared war against every other state" (Plato, *Laws* 626a, tr. T.J. Saunders).

In connection with all these changes Athens was fundamentally transformed. In the course of less than two generations (ca. 500–450), it developed from an almost entirely agrarian, aristocratically ruled, mostly self-contained polis to a large, demographically varied and differentiated community and the economic center of the Aegean world and far beyond.[7] To give one particularly important example, by the mid-480s, Athens still had a relatively small number (hundreds rather than thousands) of professional craftsmen and traders and a relatively small population of metics and slaves. Great resources in money and manpower needed to be committed to the project of building and operating a large fleet. A vast infrastructure was put in place to build and outfit so many ships, to repair and maintain them, to keep them dry in the winter, to organize the labor required for all these tasks, to support and house the personnel, and to keep track of everything (below, section 2). A whole industry and sizable bureaucracy must have sprung up for this purpose around the shipyards in the Piraeus. Part of this workforce consisted of Athenian citizens who migrated from the rural hinterland to the metropolitan area; another part of immigrants who settled permanently in Attica (called "co-inhabitants": metics); a third part of slaves. As a result, the population of Attica and, among this population, the proportion of non-farming citizens, metics, and slaves increased massively, necessitating, by the middle of the century at the latest, importation of grain on a regular basis.[8] Most of those whose livelihood wholly or largely depended on the fleet probably settled near the navy yards. Markets and shops to supply these people, taverns and other establishments to entertain them, attached themselves to this growing area. Within a short time the Piraeus became a city of its own right, planned, on the basis of amazingly abstract and egalitarian principles, by Hippodamus of Miletus (below, section 2). Simultaneously, the villages forming Athens and still recognizable in the city-districts (demes) of Cleisthenes' tribal system, grew together, and Athens with the Piraeus became a huge metropolitan agglomeration, tied together and secured in the 450s by the Long Walls (below, section 4)—themselves a massive undertaking with considerable economic impact.

At the same time, Athens developed into a democratic power that pursued an aggressively imperialist policy and from at least the late 460s was locked into a competitive and mostly hostile relationship with the older hegemonial power, Sparta. Empire, democracy, and changes in the nature of warfare were interconnected.[9] Such deep and rapid changes could not fail to cause tensions and conflicts. It was the function of some parts of public discourse, especially in the theater, to help the citizens cope with these problems, while other parts of that discourse served the purpose of re-affirming the shared identity and ideology of the Athenian citizens.

b. The Framing Story

At the beginning of Plato's *Republic* we hear that Socrates has walked down from Athens to the Piraeus to attend the festival of the Thracian goddess Bendis whose cult has just been introduced to the community. The prayers and spectacles are over and Socrates is about to walk back to the city, when his friend Polemarchus sees him. He convinces Socrates to stay, watch in the evening the torch-race on horseback in honor of the goddess, participate in the subsequent all-night festival (*pannuchis*), and spend the time before dinner at his father's home in the Piraeus (Plato, *Rep.* 327a–28b). This sets the stage for the discussion of justice that is the subject of the first book of the *Republic*. We shall join Socrates and his friends in the morning on their return trip to the city.

The dramatic date of the first book of the *Republic* is uncertain. The cult of Bendis was introduced in Athens in the late 430s, perhaps at the beginning of the Peloponnesian War (section 9 below). A temple mentioned by Xenophon may date to that time. An inscription with additional regulations, perhaps instituting the *pannuchis*, is dated to 413/12. For the purposes of this essay, we shall assume that the expanded festival took place for the first time in the spring of 411 and that this was the occasion of Socrates' trip to the Piraeus (Xen. *Hell.* 2.4.11).[10] In 411 Socrates, born ca. 470, was almost sixty years old. He was a baby when Cimon secured Athenian supremacy in the Aegean with his double victory over the Persians at the Eurymedon, a boy when Ephialtes in 462 proposed the reforms that fully established democracy, a teenager when Athens fought its first war with Sparta in the 450s and a large Athenian expeditionary force was destroyed in Egypt in 454, in his twenties when Pericles and Thucydides son of Melesias in the mid-440s fought their intense political battle over the building program on the Acropolis and the direction Athenian policies should take, almost thirty when he heard Pericles propose the building of the "Middle Wall" to the Piraeus (Plato, *Gorg.* 455e), and almost forty when Pericles convinced his fellow citizens in 432 that they should accept the Spartan challenge and go to war again (Thuc. 1.139–45). Socrates fought as a hoplite in several campaigns in the early part of the Peloponnesian War (Plato, *Symposium* 219e–221e; Diogenes Laërtius 2.22–23). By our dramatic date of 411 he would have witnessed the debates about the Sicilian campaign and the departure of the great armada in 415, learned about the destruction of that army in 413, and lived through two very difficult years during which the Athenians built an entirely new fleet, were confined to the city and Piraeus by the constant threat posed by the Spartan garrison at Decelea, and tried, not without success, to restore some balance in the war (Thuc. books 6–8).[11]

2. Military Power and Conditioning

We meet Socrates and his friends as they emerge from Polemarchus' house in the Piraeus after a long night of festive reveling and a few hours of sleep. The Piraeus at that time was heavily urbanized (one of the models of the "Hippodamic" style of urban planning, with a rectangular grid of streets and house lots of fairly equal size) and densely populated, a center of manufacture and trade.[12] Visitors would notice especially the three war harbors—in fact, the road to the sanctuary of Bendis began near the Zea Harbor and ended close to the Mounichia Harbor. In the spring of 411, the fleet was stationed on Samos, but work on constructing new ships and repairing or maintaining old ones was going on all the time. The shipyards and the sheds to accommodate a fleet of three to four hundred wooden triremes were enormous; even when not on campaign, the fleet must have given work to many thousands of Athenians.[13]

In seeing all this, Socrates and his friends would remember how crucial this fleet was for the community's fortune in war and what decisive contribution it had made to Athens' rise to power and greatness. They would also remember the hectic activity that broke out whenever a war fleet was preparing to depart:

> Ships' captains shouting for sailors, pay being distributed, figureheads of Athena being gilded, the Piraeus corn market all a-bustle with rations being doled out everywhere; people buying water-bottles and jars and rowlock thongs at one stall, garlic, onions and olives at another; farewell parties being held, with the best anchovies and flute-girls, and maybe later a bloody nose or two; and down at the docks they'd have been planing spars into oars, driving in rivets, putting on thongs, and the masters of the oarsmen meanwhile practising trills and runs on their flutes. (Aristoph. *Acharnians* 544–54, tr. A.H. Sommerstein)

No less vivid probably was the memory of the departure, only four years earlier, of the armada to Sicily. As Thucydides emphasizes, this was "by a long way the most costly and the finest-looking force of Hellenic troops that up to that time had ever come from a single city" (Thuc. 6.31.1, tr. R. Warner).

> The Athenians themselves and any of their allies who were in Athens at the time went down to Piraeus at dawn on the day appointed and manned the ships for putting out to sea. The rest of the people, in fact almost the entire population of Athens, citizens and foreigners, went down to Piraeus with them. Those who were natives of the country all had people to see off on their way, whether friends or relatives or sons, and they came full of hope and full of lamentation at the same time, thinking of the conquests that might be made and thinking, too, of those whom they might never see again. . . . At this moment when they were really on the point of parting from each other with all the risks ahead, the danger of the situation came more home to them than it had at the time when they voted for the expedition. Nevertheless they were heartened with the strength they had and with the sight of the quantities of every kind of armament displayed before their eyes. As for the foreigners and the rest of the crowd, they came merely to see the show. . . . When the ships were manned and everything had been taken aboard, . . . silence was commanded by the sound of the trumpet, and the customary prayers made before putting to sea were offered up, not by each ship separately, but by them all together following the words of a herald. The whole army

had wine poured out into bowls and officers and men made their libations from cups of gold and of silver. The crowds on the shore also . . . joined together in the prayers. Then, when the hymns had been sung and the libations finished, they put out to sea. (Thuc. 6.30.1–31.1, 32.1–2)

The return of the fleet at the end of a campaign, often after heavy losses, would stir mixed emotions, even if victory had been gained. Xenophon describes what happened only a few years later after the disastrous defeat at Aegospotami in 405 (Xenophon, *Hellenica* 2.2.3, tr. R. Warner).[14]

It was at night that the Paralus [the messenger ship] arrived at Athens. As the news of the disaster was told, one man passed it on to another, and a sound of wailing arose and extended first from Piraeus, then along the Long Walls until it reached the city. That night no one slept. They mourned for the lost, but more still for their own fate.

A great deal could be said here, and has been said elsewhere in this volume, about the sociology, psychology, and ideology of warfare, both hoplite and naval. Two points, however, are essential for my present purpose. First, virtually all the citizens who participated in the religious or political activities of their city and voted upon its policies, were or had been engaged in its wars, often many times. They knew the nuts and bolts of warfare, and most of them were familiar with large parts of their empire. They knew, therefore, what they were dealing with. Moreover, they were eminently proud of their own and their ancestors' achievements. This tradition—that most citizens are closely familiar with politics and war—has an analogy in modern Israel, but, because of the remoteness of war or politics (or both) in the experience of the average citizen, not in most other modern nations. It is therefore especially hard for us to imagine the impact this experience of long and intense conditioning for war must have had on the outlook and mentality of the citizens and on their reaction to each new challenge.[15]

Second, Aeschylus' funeral epigram, supposedly composed by the poet himself, is cited in later sources: "This monument in wheatbearing Gela hides an Athenian dead: Aeschylus, son of Euphorion. Of his noble courage the sacred field of Marathon could tell, and the longhaired Mede, who had good cause to know." Already Pausanias noted the remarkable fact that Aeschylus, the most famous poet of his time, did not mention his poetry at all, only his participation in the glorious victory of Marathon (Anon. *Life of Aeschylus* 11, tr. J. Herington; Paus. 1.14.5). In his essay *The Glory of the Athenians*, Plutarch observes, "If you take away the men of action, you will have no men of letters" (Plutarch, *Moralia* 345c–51b, at 345c). Without heroes, no history, no heroicizing paintings or poetry. More importantly, great as such artistic achievements might be, no one would ever think of rating them higher than the actual deeds. Even tragedy, the crown jewel among Athens' cultural achievements—what profit did it bring the city

to compare with the shrewdness of Themistocles which provided the city with a wall, with the diligence of Pericles which adorned the Acropolis, with the liberty which Miltiades bestowed, with the supremacy to which Cimon advanced her? If in this manner the wisdom of Euripides, the eloquence of Sophocles, and the poetic magnificence of Aeschylus rid the city of any of its difficulties or gained for her any brilliant success, it is but right to compare their tragedies with trophies of victory, to let the theatre rival the War Office, and to compare the records of dramatic performances with the memorials of valour. (Plutarch, *Moralia* 348c–d, tr. F.C. Babbitt)

On the Roman side, Sallust and Cicero had the same experience: try as they might, there simply was no way to deny that the prestige and glory accruing to military achievement prevailed over that resulting from rhetorical, juridical, or literary excellence (Sallust, *Catilinarian Conspiracy* 1–3; *Jugurtha* 2–4; Cicero, *Pro Murena* 9.22–13.29). Ever since Homer—and presumably generations of singers before him—poets praised the fame of heroic deeds on the battlefield, and the heroes of Marathon and Plataea were still extolled in song and annual celebration decades after the events.[16] As Herodotus, Thucydides, and hosts of their successors demonstrate, wars provided most of the material for history as well (below, section 7). With few exceptions (as in the cases of Solon the lawgiver or the tyrant slayers, Harmodios and Aristogeiton), military accomplishments were seen as the only way to acquire eternal fame. Aspiring leaders, who grew up in an environment shaped by these traditions and whose education consisted largely of absorbing them by way of song,[17] inevitably were conditioned to think of war and military success as the best way to the top. Such conditioning was bound to influence their thoughts and actions. Unlike in Rome, however, in Greece in the century or two before the Persian Wars there had been relatively few opportunities to realize such ambitions.[18] In Athens after the Persian Wars they were plentiful.

To some extent, therefore, war, occurring frequently, lasting longer and playing a much larger role in life and thought of citizens and community than was previously the case, created its own dynamics and perpetuated itself. Thucydides seems to indicate that this happened in 415 when the citizens decided, against Nicias' serious warnings, to intervene in Sicily:

> It was all the easier to provide for everything as the city had just recovered from the plague and the years of continuous war, and as a number of the young had grown to manhood, and capital had accumulated as a result of the truce [i.e., the Peace of Nicias].
>
> There was a passion for the enterprise which affected everyone alike. The older men thought that they would either conquer the places against which they were sailing or, in any case, with such a large force, could come to no harm; the young had a longing for the sights and experiences of distant places and were confident that they would return safely; the general masses and the average soldier himself saw the prospect of getting pay for the time being and of adding to the empire so as to secure permanent paid employment in future. The result of this excessive enthusiasm of the majority was that the few who actually were opposed to the expedition were afraid of being thought unpatriotic if they voted against it, and therefore kept quiet. (Thuc. 6.26.2; 24.3–4)

Such reactions are understandable only against a background of long conditioning in naval warfare and imperial rule.

3. Economic Aspects

The last of the three motives mentioned by Thucydides leads us into another sphere, that of economy. Thucydides' statement, implying that the decision for the Sicilian campaign was influenced heavily by personal economic considerations, is by far the strongest of its kind. Scholars have long debated whether political or economic motives were primary in Greek wars. Even if, as seems likely, wars were rarely conducted primarily for economic reasons, there is much evidence to show that economic arguments played a role in discussions of war—not least because war itself, and the industries necessary to support a large navy as well as a large hoplite army, gave employment to many thousands of persons.[19] Typically, the peace-making efforts of Aristophanes' comic heroes are opposed, not only by warmongers among citizens and generals, but also by the arms manufacturers who are put out of business, and peace is seen as delightful not least because it opens easy access again to sorely missed delicacies from the lands of former enemies (such as eels from Lake Copais in Boeotia and piglets from Megara: Aristoph. *Acharnians* 729–835, 880–94. Arms manufacturers: *Peace* 447–49, 545–49, 1210–64).

This brings us back to Socrates and his friends who would know, and perhaps be walking through, the "entertainment and red light district" surrounding the harbor in the Piraeus like every other harbor in the world. They would notice the hustle and bustle in the commercial harbor, the "Emporion" on the east side of the Kantharos or Grand Harbor, which was surrounded by large warehouses and markets. With Athens' domination of the seas, the Piraeus increasingly became the center of trade in the Aegean and far beyond; it served as point of entry for a great variety of goods imported to Athens (among which not least was the grain needed to feed the growing population of Attica) and as transfer station to other destinations.[20] A fragment of the comic poet Hermippus gives us a vivid impression:

> From Cyrene silphium-stalks and oxhides, from the Hellespont tunny and saltfish, from Italy salt and ribs of beef . . . Syracuse offers porc and cheese . . . from Egypt sailcloth and raw materials for ropes, from Syria frankincense. Fair Crete sends cypress wood for the gods, Libya plentiful ivory to buy, and Rhodes raisins and figs sweet as dreams; from Euboea come pears and big apples, slaves from Phrygia, mercenaries from Arcadia. Pagasae provides slaves with or without tattoos, and the Paphlagonians dates that come from Zeus and shiny almonds . . . Phoenicia supplies the fruit of the date-palm and fine wheat-flour, Carthage rugs and cushions of many colours. (Hermippus, *Phormophoroi*, fragm. 63 *PCG* 5, cited by Athenaeus 1.27e–28a; tr. M. Ostwald)

The Athenians were perfectly aware that their access to such abundance was a direct result of their city's power. The "Old Oligarch," an anonymous author of an antidemocratic pamphlet, writes, "Because of their control of the sea, the Athenians have mingled with peoples in different areas and discovered various gastronomic luxuries; the specialities of Sicily, Italy, Cyprus, Egypt, Lydia, Pontus, the Peloponnese or any other area have all been brought back to Athens because of their control of the sea." The Thucydidean Pericles confirms this in the Funeral Oration: "The greatness of our city brings it about that all the good things from all over the world flow in to us, so that to us it seems just as natural to enjoy foreign goods as our own local products" (Pseudo-Xenophon, *Constitution of the Athenians* 2.7; see also 2.11–13, tr. J.M. Moore; Thuc. 2.38.2).

Again, much more could be said on this topic, not least on the great number of visitors who came to Athens for legal, political, or business purposes and further fueled the Athenian economy (Ps. Xenophon, *Const. of the Athenians* 1.16–18).[21] But the essential point is clear. Despite losses, constraints, and hardships, in the Athenians' experience war and the empire gained by war were profitable. They provided employment, directly and indirectly. Apart from the many thousands who served on the fleet and worked in the manufacturing and supply businesses that supported the fleet, thousands more earned at least part of their living in military and political offices necessitated by war and empire (Aristot., *Const. of the Athenians* 24.3). War and empire dramatically increased the wealth and standing of many Athenians: perhaps ten thousand lower-class citizens acquired land abroad, mostly in settlements on the territories of "allied" cities, and thereby reached hoplite status, and the gains of the upper classes are highlighted by Aristophanes' sarcastic comments (especially in *Knights*) and by the records of auctions at which the property of those convicted of involvement in the religious scandals of 415 was sold.[22] War and empire thus contributed to improving the quality and enjoyment of life and, as the monumental decoration of the civic and religious centers of the city showed, brought many other benefits to individuals and community alike.[23]

4. Military Theory

Passing through the fortifications of the Piraeus, Socrates' group would enter the area enclosed by the Long Walls that connected the harbors of Piraeus and Phaleron with the city of Athens over a distance of more than five miles. They had been constructed in 458/7 and supplemented by an internal parallel wall (167 meters off the north wall) in the late 440s.[24] The men walking through this space, which in 411 must have been crowded with evacuees from Attica, inevitably were reminded of Pericles' strategy of sacrificing the countryside, thus blunting Spartan superiority on land, and relying entirely on the fleet to bring in the necessary supplies and harm the enemies by raiding their territory from the sea.

This concept of warfare, explained by Thucydides' Pericles in a programmatic speech advocating the war, was based in several respects on theoretical considerations (Thuc. 1.141.2–143).[25] It assumed, on the one hand, the superiority of an island state that was defended by a fleet controlling the seas over a landbound state relying on a hoplite army and, on the other hand, the superiority of a centrally controlled power with concentrated decision-making procedures and vast financial resources over a federally organized power with an entirely agrarian economic base and no monetary resources. I focus here on what we might call the "theory of sea-power."

> Sea-power is of enormous importance. . . . Suppose we were an island, would we not be absolutely secure from attack? As it is we must try to think of ourselves as islanders; we must abandon our land and our houses, and safeguard the sea and the city. . . . What we should lament is not the loss of houses or of land, but the loss of men's lives. Men come first; the rest is the fruit of their labour. (Pericles in Thuc. 1.143.5)

Two years later, when the Athenians were demoralized by plague and Spartan incursions, Thucydides lets Pericles say:

> The whole world before our eyes can be divided into two parts, the land and the sea. . . . Of the whole of one of these parts you are in control. . . . With your navy as it is today there is no power on earth . . . which can stop you from sailing where you wish. This power of yours is something in an altogether different category from all the advantages of houses or of cultivated land. You may think that when you lose them you have suffered a great loss, but in fact . . . you should weigh them in the balance with the real source of your power. (Thuc. 2.62.2–3)

Similarly, the "Old Oligarch" writes,

> There is one weakness in the Athenian position: as rulers of the sea, if they lived on an island, it would be open to them to harm their enemies if they wished while remaining themselves immune from devastation of their land or invasion as long as they controlled the sea. . . . In addition to this, if they lived on an island, they would also be freed from the fear of the city being betrayed by oligarchs, or the gates opened, or the enemy being let in. . . . Again, there would be no chance of anyone's staging a coup against the common people; in the present situation, if anyone planned a coup, he would do so in the hope of bringing in the city's enemies by land; if Athens were an island, this fear also would be removed. Since it happens that the city was not founded on an island, they handle the situation as follows: they deposit all their property on islands, relying on their control of the sea, and they disregard any devastation of Attica, realising that if they allow themselves to be moved by this they will be deprived of other greater benefits. (Pseudo-Xenophon, *Const. of the Athenians* 2.14–16)

I have cited this passage in full because it reveals the theorizing nature of this treatise. According to H. Frisch, the "theory of sea-power," as it appears in these two authors, comprises the following elements:

> To the ruler of the sea there are no distances. He has the initiative and may attack at any time. He may use compulsion solely to blockade and control. He has time as

his ally if only he refrains from any decisive trial of strength by land. He cannot be starved, but may starve others. Through the thalassocracy timber and other materials for ship-building may be controlled, so that no other sea-power may develop without the assent of the ruler of the sea. Sea-power for that matter cannot be improvised. The ruler of the sea alone has funds on a large scale at his command. Finally, . . . it were the best thing if the sea-power were situated in an island.[26]

The "Old Oligarch" is usually dated to the 420s, but an earlier date in the late 440s is not impossible.[27] He is influenced in other passages as well by ideas discussed by some Sophists, and it is quite probable, although no evidence survives, that among their theories on society and politics some of these migrating teachers also analyzed the nature and advantages of sea-power. In Athens, such theories would have found an especially interested audience. Here long before the advent of the Sophists, citizens and leaders must have been thinking about the advantages and disadvantages of various types of warfare and about possible ways of neutralizing the awesome power of Sparta's hoplite army. Some of the programs later authors attribute in this context to Themistocles, Aristides, and Cimon are of dubious authenticity (e.g., Plutarch, *Themistocles* 4; *Cimon* 11; Aristot., *Const. of the Athenians* 24). But already in 472 Aeschylus formulated the lesson, derived from the battle of Salamis, that without a superior fleet the strongest land army might be doomed (*Persians* 728), and the historical facts speak for themselves.[28] The transfer of the naval base from Phaleron to the Piraeus already before the Persian Wars and the fortification and expansion of the latter, the use of an effective standing navy to control an alliance system rapidly transformed into an empire, the concept of the Long Walls—realized after a successful experiment with similar walls in Megara in 459–58 (Thuc. 1.103.4)—and the strategic decisions underlying the Thirty-Years' Peace with Sparta in 446—these all required careful planning and the conceptualization of specific political and military principles.[29]

At any rate, by the last third of the century, such ideas were widespread. Herodotus, for example, argues that Sparta's plans to fortify and defend the isthmus of Corinth necessarily would have failed if the Persians had succeeded in neutralizing Athenian sea-power and thus controlling the seas themselves. The same author lets a Persian general comment on the senselessness of the Greeks' fighting wars against each other without exploiting, in each case, their own strengths and the opponents' weaknesses (Herodotus 7.139; 7.9b).[30] It seems therefore that Pericles' claim, in the last speech given him by Thucydides, that so far he had not sufficiently drawn the Athenians' attention to the enormous significance of their control of the seas, must be understood as a reference within Thucydides' work rather than to the novelty of such ideas at that time. Frisch rightly concludes that the entire debate about the Peloponnesian War must have turned on the advantages and disadvantages of sea power, "and any educated Athenian who was in the least interested in these issues must have had all the arguments at his finger tips" (cf. Thuc. 2.62.1).[31]

That Athenian warfare was based to a remarkable degree on theoretical concepts is visible in another aspect of Pericles' war plan as well: his insistence that the outcome of wars, like anything else, could be planned and predicted with great certainty. This is why Thucydides, who was especially interested in these aspects, lets Pericles discuss the resources, strengths, and weaknesses of both powers in great detail, concluding, "I could give you many other reasons why you should feel confident in ultimate victory, if only you will make up your minds not to add to the empire while the war is in progress, and not to go out of your way to involve yourselves in new perils" (Thuc. 1.144.1).[32] Such confidence in his own intellectual ability and in the predictability of war seems to have been characteristic of Pericles, and Thucydides agrees with him: "When the war broke out . . . [Pericles] appears to have accurately estimated what the power of Athens was. . . . After his death his foresight with regard to the war became even more evident." His successors, so says Thucydides, violated the principles he had set, more concerned with their own power than with the good of the city. And yet, despite egregious mistakes and setbacks—not least among the latter the plague, the only danger Pericles had not anticipated—the city was able to protract the war for many years. "And in the end it was only because they had destroyed themselves by their own internal strife that finally they were forced to surrender. So overwhelmingly great were the resources which Pericles had in mind at the time when he prophesied an easy victory for Athens over the Peloponnesians alone." Whatever the accuracy of Thucydides' assessment, such confidence in perceiving the essential factors, calculating the resources and predicting the course and outcome of events is an essential component of what Christian Meier calls the "consciousness of ability," typical of fifth-century political thought and the closest ancient equivalent to our modern concept of progress (Thuc. 2.65.5–13; plague: 2.61.2–3, 64.1).[33]

Such optimism prevailed in the early phases of the war. By 411, the Sicilian disaster, the worst of the Athenians' violations of Pericles' strategy (Thuc. 2.65.11), had shaken such confidence. But the war was not over yet.

5. Political Rhetoric

While pondering these issues, Socrates and his friends reach the junction of the Long Walls and city walls. The appearance of the latter would remind them of how Themistocles' trickery enabled the Athenians after the Persian Wars to build these walls in a great hurry, against Spartan objections, and thus to secure their independence from potential Spartan interference. This anecdote too, though not entirely fictitious, underscores later concepts of independence and self-reliance that played a major role in the debates about whether or not to go to war with Sparta (Thuc. 1.90; Plut. *Themistocles* 19.1–3).[34] Crossing this wall through the gate on a saddle between the Pnyx and the Hill of the Nymphs,[35] our group turns right and reaches the meeting place of the Assembly on the Pnyx. There they pause, sit down, and look over the city. Soon they are overcome by the magic of the place itself.

At that time, the *bêma* (the speaker's platform) was located at the bottom of the big half-round space carved out of the rock, so that the assembled men would see the city. At the end of the fifth century this arrangement was reversed, supposedly at the instigation of the "Thirty Tyrants"; henceforth the citizens looked up and away from the monuments of the city's imperial greatness.[36] Socrates could not know this, of course. Nor was he aware that here, only five years hence, he would be president of the assembly when it decided, against his objection, to vote on an illegal proposal and, by a single vote rather than separately, impose the capital penalty upon the generals who had failed to rescue the shipwrecked sailors after the battle of Arginusae. Next to the conviction of Socrates, this was the one decision that would be used most often by posterity to support the claim that democracy was an irresponsible and irrational system (Xenophon, *Hellenica* 1.7.12–15).[37]

The Pnyx was the place where the citizens met to deliberate and decide on their city's policies. Here twenty years earlier Pericles had convinced his fellow citizens that it was better to fight a war than to give in to any Spartan ultimatum. Among equals, he said, arbitration was the only honorable way to settle disputes. Yielding to demands would be interpreted as a sign of fear, elicit greater demands, and by necessity result in the "enslavement" of the yielding side (Thuc. 1.140.2–141.1, cf. 2.62.3).[38] Here again, two years later, Pericles had persuaded the Athenians, bitter and disgruntled though they were, to persevere in their quest for glory and power. As Thucydides formulates his words, "If one has a free choice and can live undisturbed, it is sheer folly to go to war. But suppose the choice was forced upon one—submission and immediate slavery or danger with the hope of survival: then I prefer the man who stands up to danger rather than the one who runs away from it." The policy pursued by Athens

> entails suffering, . . . yet you must remember that you are citizens of a great city and that you were brought up in a way of life suited to her greatness. . . . Remember, too, that the reason why Athens has the greatest name in all the world is because she has never given in to adversity, but has spent more life and labour in warfare than any other state, thus winning the greatest power that has ever existed in history, such a power that will be remembered for ever by posterity: . . . that of all Hellenic powers we held the widest sway over the Hellenes, that we stood firm in the greatest wars against their combined forces and against individual states, that we lived in a city . . . which was the greatest in Hellas. (Thuc. 2.61.1–4)

In Aristotle's time, forty regular assembly meetings were scheduled every year—an average of one every nine days (Aristot. *Const. of the Athenians* 43.3–6)![39] In the fifth century, there may have been fewer fixed meetings but in wartimes they may have been even more frequent. Of course, not every assembly had to deal with subjects of the same magnitude. But issues concerning war were constantly on the agenda, for the assembly decided on policies, principles, and details, and the competition among ambitious leaders made sure that a close watch was maintained on every issue and every campaign. Foreign policy and wars provided the bulk of the assembly's agenda and most of the contentious issues. This was the sphere in which politicians could distinguish themselves and here they fought their rhetorical battles. In such an atmosphere proposals for activist and aggressive policies a priori had a better chance: glory and a great reputation for leadership depended on success in action and victory, not on caution, quietism, and peace. All this helps us understand why types like Cleon and Hyperbolus could rise to such prominence, and why Alcibiades prevailed over Nicias in the spring of 415. Paired with the appeal to patriotism, chauvinism, and exaggerated expectations, proposals for war easily prevailed over those for peace—as is documented by several peace opportunities squandered in the 420s and in the final phase of the war.[40]

6. Funeral Ideology

Leaving the Pnyx, Socrates and his friends descend to the saddle below the Areopagus, turn left, soon reach the Dipylon Gate,[41] and step out onto the broad road leading to the precinct of the Hero Akademos, later the site of Plato's Academy. This road was flanked by public burials, including the tombs of distinguished leaders (such as Cleisthenes and Pericles) and those of the citizens who died in Athens' wars. The latter were buried every year in a common tomb crowned by stelae with a laudatory epigram and a list of all their names. The "ancestral custom" (*patrios nomos*) of burying the community's heroes in a solemn public ceremony following a traditional pattern and in a public burial ground (*dêmosion sêma*) was probably instituted relatively soon after the Persian Wars, at a time when, in connection with the emerging empire and democracy, communal warfare took on new meaning and significance.[42] Thucydides describes this *patrios nomos* as follows:

> Two days before the ceremony the bones of the fallen are brought and put in a tent which has been erected, and people make whatever offerings they wish to their own dead. Then there is a funeral procession in which coffins of cypress wood are carried on wagons. There is one coffin for each tribe, which contains the bones of members of that tribe. One empty bier is decorated and carried in the procession: this is for the missing, whose bodies could not be recovered. Everyone who wishes to . . . can join in the procession, and the women who are related to the dead are there to make their laments at the tomb. The bones are laid in the public burial-place. (Thuc. 2.34)

One of the best-preserved examples is the stele of the state burial of 447/6.[43] Here in the left-hand column the epigraph ("In the Chersonese of the Athenians these men died") is followed by the name of the leader ("Epiteles the general") and the names of all the other war dead, ordered by tribe ("of the tribe Erechtheïs: Pythodoros, Aristodikos, Telephos, Pythodoros; of the tribe Aigeïs: Epichares, Mnesiphilos, Phaidimides, Laches, Nikophilos," etc., for a total of twenty-seven). The right-hand column lists twelve men who died in Byzantium. An additional list, engraved in smaller letters at the bottom of both columns, records nineteen men who fell "in the other campaigns." Finally, the heroes' praise is eternalized by an honorary epigram:

These men lost their flowering youth at the Hellespont,
fighting, and they brought glory upon their fatherland,
while the enemies groaned as they carried away the harvest of war,
but for themselves these men set up an eternal memorial of bravery.

This monument and tomb stood among dozens of similar ones in a public cemetery located, not as in Washington in a park-like setting across the river and separated from the city, but in the middle of one of the busiest industrial quarters and best residential areas.

At the beginning of the *dēmosion sēma*, just outside the Dipylon Gate, a wide space was left open for the annual public ceremony connected with the state burial. On the platform erected there a leading citizen, "a man chosen by the city for his intellectual gifts and for his general reputation," delivered the funeral oration in praise of the dead (Thuc. 2.34).[44] Pericles and Demosthenes were among those speakers. If we can generalize from the surviving examples, the "Funeral Oration" soon became a distinct genre of oratory with its own traditional *topoi*. In many ways, it was the embodiment of civic ideology.[45] War naturally was predominant here: the speakers recounted the ancestors' contributions to the glory and greatness of their city, in myth and history (especially in the Persian Wars); they then extolled the present generation's achievement in preserving and enlarging the city's power by achieving yet more victories, and they praised the *aretê* of those to be buried who had sacrificed themselves for their community. Shared pride in communal glory helped overcome grief and loss and prepared the citizens for further ordeals.

In the funeral oration given to Pericles, Thucydides deliberately condenses this traditional eulogy by omitting "ancient history" (which we moderns call "myth") and briefly summarizing the accomplishments of three generations: the forefathers who (in the Persian Wars) preserved the freedom of the country, the fathers who created the empire and handed it down, "not without blood and toil," to the present generation. "And then we ourselves . . . have, in most directions, added to the power of our empire" and made the community virtually self-sufficient for war and peace. Pericles skips these all too familiar issues: "I shall say nothing about the warlike deeds by which we acquired our power or the battles in which we or our fathers gallantly resisted our enemies, Greek or foreign." Rather, he wants to discuss the spirit, way of life, and institutions that made Athens great, that is, he wants to focus on the factors and structures that made martial accomplishment and imperial greatness possible (Thuc. 2.36.1–4). The character portrait of the Athenian citizen sketched in this speech is justly famous—not least in view of the fact that it was about to be shattered under the pressures of plague and war. In our present context, it is remarkable especially for its emphasis on the citizens' total dedication to the common good. In dying, these citizens have proved their *aretê*; they have acted like lovers (*erastai*) of their city, completely subordinating their self-interest to the needs and demands of their polis and thereby overcoming whatever limitations they may have had individually. The common goal, worthy of receiving the highest priority in each citizen's life, is to contribute to the city's greatness, glory, and power (Thuc. 2.43.1; cf. 2.60).[46]

This, in a nutshell, is civic *aretê*, this is the core of the civic ideology promoted by the democratic and imperial city. This message was repeated annually in the ceremonies of the *patrios nomos,* echoed regularly in the rhetoric of the assembly, occasionally brought to the fore in the theater, and powerfully present in public monuments and images.[47] It is not difficult to imagine the cumulative impact of this ideology on the collective psyche of the Athenian citizens.

7. History

In a reflective mood Socrates and his friends turn around, enter the city again, and follow the Panathenaic Way from the Pompeion at the Dipylon Gate, where the Panathenaic procession began, to the Agora.[48] They reach it at its northwest corner and stop to take in the sight. Immediately to their left stood the "Painted Stoa" (*stoa poikilê*), erected by Cimon's family between 475 and 450 and named after the series of large paintings hanging on its inside walls. Executed by outstanding artists, these depicted mythical and historical military successes of the Athenians: the defeat of the Amazons, the Athenians at Troy, the victory over the Spartans at Oinoë, and the battle of Marathon. The Stoa also displayed trophies of notable Athenian victories, among them shields of the Spartans captured at Pylos in 425/4, inscribed with "The Athenians [took those] from the Lacedaemonians from Pylos" (Paus. 1.15).[49]

The victories over the Persians at Marathon and over the Spartans at Pylos were historical events, one featuring prominently in Herodotus' and the other in Thucydides' *Histories.* If the tradition is correct that Herodotus received a prize in Athens for reciting parts of his Persian War *logoi,* Socrates might well have heard him.[50] What matters more for our present purpose is that the writing of history, succeeding in the tradition of heroic epic and stimulated by the monumental experience of the confrontation between the giant Persian empire under its absolute king and the tiny but free Hellenic city-states, and by doubts and worries about contemporary developments, from the beginning focused on war. War defined the greatness of the subject and the claim that the war to be described was the greatest that had ever happened belonged to the typical features of Greco-Roman historiography ever since Herodotus and Thucydides set that standard. "*History*, as the ancients elementally defined it, provides the descriptions of men's 'deeds,' or *praxeis*; . . . it is the *expositio rerum gestarum*, 'the narration of deeds.'"[51] "To be the greatest in war and speaking": thus the Homeric hero defines the qualities determining status in early Greek society. But the primacy clearly belongs to fighting. From the beginning, "men's deeds" were understood in terms of war. Heroic exploits on the battlefield, success and victory in war: these were the surest ways to glory (*kleos*) in song and later in written history. Such glory shone on the heroes' descendants too. Hence the eagerness of Cimon's family to display in "their" Stoa a heroicizing painting of their father's greatest achievement, the victory of the Athenians over the Persians at Marathon—a painting that showed Miltiades fighting in the front rank, even though the people would not permit his name to be written on the painting. Hence too, court orations referred emphatically to the deeds of the speaker's ancestors for the city, whether authentic or not.[52]

8. Monuments and Images

On the west side of the Agora, to the right of Socrates' group, stood the "Stoa of the King" (*stoa basileios*), probably one of the earliest public buildings on the square.[53] Just south of it the Athenians had recently (around 430) erected another, much larger Stoa, dedicated to Zeus Eleutherios (Zeus the Liberator), magnificently decorated and crowned by akroteria in the form of flying Victory Goddesses or Nikai. This building most probably was designed to reaffirm, in response to Sparta's battle cry of "liberty for the Hellenes" (Thuc. 2.8.4–5 and often), Athens' claim to be the true liberator of Hellas (against the Persians) and the "greatest and freest city" (Thuc. 7.69.2; 6.89.6).[54] Farther south on the same side of the Agora stood the Bouleuterion (the meeting place of the democratic Council of Five Hundred) and the Tholos (the meeting place of the *prutaneis*, the executive committee of the Council, one-third of whom were in attendance at the Tholos days and nights); both buildings were intimately connected with deliberations about foreign policy and war.[55] Looking up to their right, Socrates would see the temple of Hephaestus. Its splendid decorations included the labors of Heracles and of Theseus, a battle watched by gods, and the battle between Centaurs and Lapiths, a popular theme that by then probably symbolized the confrontation between Greeks and Persians.[56]

In the Agora stood the monument of the Eponymous Heroes, one of the founding monuments of democracy. The foundations excavated near the Bouleuterion date to the mid-fourth century; those of an earlier version, attested from the 420s and erected probably in the Periclean era, perhaps at the very beginning of the war, have not been found or identified with certainty. Elsewhere (on the Athenian Marathon dedication at Delphi, on the Parthenon frieze, perhaps also on the shield of Athena Parthenos) these heroes, eponymous of the ten Cleisthenic tribes, represented the Athenian citizen army. It is fitting, therefore, that among the public announcements posted on this monument were the lists with the names of the hoplites who were summoned to march out on the next campaign.[57] Among other monuments that crowded the square, Socrates and his group would notice, in the area in front of and near the "Painted Stoa," a number of commemorative inscribed herms.[58] Famous among these were the three herms that were dedicated to celebrate the Athenian victory, under Cimon's leadership, over the Persians at Eion in Thrace in 476/5:

> They too were men of stout heart, who beside the swift current of Strymon
> under Eion's walls fought with the sons of the Mede:
> pitiless famine and fire they brought to beleaguer the city,
> death-dealing Ares they followed, and harried their foes to despair.

This shall stand as the tribute which Athens paid to her leaders,
    Homage to hard-fought victories, earned by their valiant deeds.
Those who come after may read and from this memorial take courage,
    and in their country's cause march no less bravely to war.
From this city Menestheus once marched out with the Atridae,
    leading his army to war on the divine plains of Troy.
None knew better than he, of the bronze-armoured Greeks, Homer tells us,
    how to manoeuvre the line or draw up the battle array.
So the proud title has clung ever since to the children of Athens,
    masters of warlike arts and leaders of valiant men.

As Plutarch, who cites these epigrams, emphasizes, "Although Cimon's name does not appear in any of these inscriptions, his contemporaries regarded this memorial as a supreme mark of honor for him." The anecdote Plutarch tells next is worth our attention as well: when Miltiades asked the assembly for an olive crown as public reward for his victory at Marathon, one citizen supposedly responded, with great applause from the crowd, "When you have fought and conquered the barbarians by yourself, Miltiades, then you can ask to be honoured by yourself" (Plutarch, *Cimon* 7.4–8.1, tr. Scott-Kilvert).[59]

"Leaders in war and bravery," this is the highest praise bestowed upon the victorious Athenians; their deeds and *aretê* are supposed to inflame the desire of future generations to toil with equal dedication for the community's affairs (*xuna pragmata*). "The memorials of all our noble deeds stand dedicated in the Agora," says Aeschines. He mentions the painting of Marathon and the herms of Eion (Aeschines 3.181–86). We know of one other famous monument that perhaps stood in the Agora and was dedicated to those who defeated the enemy and saved their city: the Persian War monument with a series of epigrams praising the heroes of various battles—again collectively, without mentioning the leaders. Statues honoring outstanding individuals were rare in the fifth century—but they proliferated in the fourth century, under changing conditions, and especially in the Hellenistic and Roman periods. Later generations considered it remarkable that not even Miltiades and Themistocles had received one; nor, for that matter, did Cimon, Pericles, or Nicias (Aeschines 3.181; Demosthenes 23.196).[60] It was the demos as collectivity of the citizens who claimed the glory for, and bore the burden of, the city's wars and victories; it was the collective *aretê* of the demos that was intended to serve as model for future generations (see also section 6 above).[61] Not by accident such reluctance to let the individual stand out was paralleled, during the first three-quarters of the fifth century, by the virtual disappearance of monumental tombs from Athenian cemeteries.[62]

These stelae and monuments were not the only reminders of Athens' martial achievements and imperial power. Dozens of inscriptions cluttered the spaces in front of and between the official buildings; they recorded the texts of treaties and assembly decrees, many of them passed by the Athenians alone but binding on the subject cities of the empire.[63]

Continuing their walk across the Agora and up the slope, Socrates and his friends would finally arrive on the Acropolis, greeted by other monuments reminding them of their city's wars and victories: to their right, the elegant temple of Athena Nike, built in the early years of the war, with battle scenes on the friezes and Nike figures on the parapet; straight ahead the giant statue of Athena Promachos, surrounded by private and public votive offerings many of which commemorated military exploits as well, and the whole scene dominated by the Parthenon, which might be called, without exaggeration, the prime monument of Athens' imperial greatness.[64] Here the spectator might well stand in awe and reflect, not only on the buildings and monuments, but also on the rituals that supported and emphasized the city's martial prowess.

9. Religion and Rituals

This is a huge topic. Because war is so fraught with death, it was surrounded by and interwoven with vows, rituals, and sacrifices: indeed, writes Burkert, "war may almost appear like one great sacrificial action."[65]

> There was sacrifice before setting off, then adornment and crowning with wreaths before battle—all as if it were a festival. A slaughtered victim introduced the subsequent deadly action. . . . Afterward, a monument, a tropaion, was set up on the battlefield as a consecrated, enduring witness. This was followed by the solemn burial of the dead. . . . The burial, almost as important as the battle itself, was far more lasting in its consequences, for it left an enduring "monument." It almost seems as though the aim of war is to gather dead warriors. . . . The erected and consecrated monument . . . embodies the duty of the following generation. For war, necessary yet controlled because it is ritual, has this function above all: it must integrate the young into the patriotic community. . . . As a rule, the Greeks' *spondai* [truces or treaties] were for a period of thirty years at most. Each generation has the right and the obligation to have its war.[66]

I shall not deal here with the rituals performed during campaigns once the army left the community.[67] Nor shall I say much about the procedures and ceremonies used in preparing the young men (*ephebes*) for their life as warriors and citizens. Although I consider it virtually certain that the institution of the *ephêbeia* was firmly established in the fifth century and played an important role in the individual citizens' and the community's life, we know close to nothing about the details; practically all the extant evidence comes from the fourth century, and form and rituals may have changed considerably over time.[68] I shall focus briefly on five important aspects: civic rituals connected with actual warfare, military components of cults and festivals, festivals and cults commemorating important military victories, interaction between religion and warfare, and innovations in cult and religion due to warfare.

First, then, civic rituals connected with actual warfare: The Roman festival calendar listed many rituals that were connected with war and, for example, marked the beginning and end of the campaigning season, the separation of the warrior from the community and his reintegration into it. Very little of that is visible in the Athenian calendar, and it is uncertain whether, as was sometimes believed, the Apatouria in the fall really contained a marked element of campaign-ending ritual.[69] It is clear, however, that the army, before leaving town for war, offered sacrifices at the shrine of heroicized maidens called Hyakinthides; their cult was connected with the myth of the sacrifice of the daughters of Erechtheus, Athens' first king, in a war against Eleusis; their death had brought Athens victory, and thus the army's sacrifice, just as the annual communal cattle-sacrifice, celebrated with choruses of maidens, was supposed to guarantee victory in battle (Euripides, *Erechtheus* fr. 65.65–89; Austin 1968, 37–39).[70] Equally clearly, all campaigning of the year ultimately ended with the elaborate and solemn rituals of the *patrios nomos*, the state burial of the war dead described above (section 6).

Second, military components of cults and festivals: Mars was one of Rome's principal deities, ancestor of its early kings, central in cult and ideology. By contrast, the cult of Ares is extremely rare in Greece overall and not attested in the city of Athens before the late first century B.C. Eirene, the Goddess of Peace, received a cult only after 375/4 (Isocrates 15 [*Antidosis*].110).[71] But Nike, the Goddess of Victory, was present everywhere, particularly during the Peloponnesian War: in her sanctuary on the Acropolis, on the roof of the Stoa of Zeus at the Agora (above, section 8), in a great bronze statue set up on the Acropolis after the victory at Pylos, and in golden statues celebrating other Athenian triumphs. Even the treasury reserve, collected from the subject cities' tribute, dedicated to Athena and kept in her sanctuary on the Acropolis, was cast in golden Nikai.[72] The Athenian Nike was one of many impersonations of Athena herself, and Athena, besides being the city's protectress (Polias), stood proudly on the Acropolis, in her full war gear, as the embodiment of martial virtue (above, section 8) in her functions as Promachos ("leader in war") and Parthenos ("maiden").[73] Her nature as a warrior goddess was underscored by the fact that every four years, at the Great Panathenaea, she received a panoply (the hoplite's arms and armor) from each of the cities of the empire (see below).

In his sketch of an ideal city, Plato says, "A man must spend his life playing at certain pastimes—sacrificing, singing, and dancing—so as to win the favor of the gods and be able to repel the enemy in battle." Accordingly, once a month the men of his city would come together for a day of sacrifices, games, and military exercises. Since hunting and athletics were seen as closely related to fighting, even such non-military activities were considered useful preparation for warfare (Plato, *Laws* 7.803c, 8.829b–c).[74] On the other hand, games and competitions at festivals often contained a distinctly military component. In the case of Athens, such occasions included the hoplite race and several equestrian events at the Panathenaic Festival and perhaps part of the Funeral Games connected with the *patrios nomos* as well. War-dances, often called "the Pyrrhic," were considered especially important for military training and performed at various occasions, most significantly at the Panathenaea in honor of Athena. Even without competitions and performances, military aspects were present at other festivals too, for example in some obscure ritual performed at the Thesmophoria by the women of Athens to secure the success of their men at war and at the Great Dionysia when the war orphans who had reached adulthood were equipped with a panoply and dismissed from state custody, and the tribute collected from the empire was displayed in the theater.[75]

Third, festivals and cults commemorating important victories: In his essay *The Glory of the Athenians,* cited before (section 2), Plutarch compares the esteem bestowed on poets with that of generals. Prizes for military victories include whole cities or islands, hundreds of triremes, thousands of talents of silver, glory and trophies. "These are the things the city celebrates in her festivals, for them she sacrifices to the gods," not for the dramatic successes of Aeschylus or Sophocles. "Even now the polis celebrates the victory at Marathon on the sixth of Boëdromion. On the third they won the battle of Plataea. The sixteenth of Munichion they dedicated to Artemis, for on that day the goddess shone with full moon upon the Greeks as they were conquering at Salamis. . . . These are the things which have uplifted Athens to heights of glory and greatness." As the mention of Artemis shows, these festivals were days of thanksgiving for divine help in gaining victory, an aspect that remained especially strong in the commemoration of the battle of Salamis.[76] Herodotus mentions a selection of new cults and rituals that were established in recognition of divine assistance in the Persian Wars: prayers for the Plataeans who had fought with the Athenians at Marathon, a sanctuary for Pan who had revealed his support for the Athenians to the runner Pheidippides on the way to Sparta, and a cult for Boreas and a new title for Poseidon after the storm that badly damaged the Persian fleet at Artemisium (Herodotus 6.105.3, 111.2; 7.189, 192.2).[77]

Fourth, religion and politics: This again is a large and complex topic.[78] I mention only two examples. On the one hand, the Athenians used cults and rituals to control and, to a certain extent, integrate their empire. The "allies" were required to make contributions to some of Athens' major festivals; for example, a decree of 425 stipulated, not for the first time, that all cities were to bring a cow and a panoply to the Great Panathenaea. Similarly, they had to supply a phallus for the Dionysia and first fruit offerings to the Eleusinia.[79] Among others, the cult of "Athena, the guardian goddess of Athens" (*Athêna Athênôn medeousa*) was established in some other cities, most probably upon Athens' instigation.[80] On the other hand, Socrates and his friends would vividly remember one of the most blatant cases of the impact of religion on politics: the mutilation of the herms in 415, shortly before the departure of the Athenian armada to Sicily. While the preparations were going on,

> it was found that in one night nearly all the stone Hermae in the city of Athens had had their faces disfigured by being cut about. These are a national institution, the well-known square-cut figures, of which there are great numbers both in the porches of private houses and in the temples. No one knew who had done this, but large rewards were offered by the state in order to find out who the criminals were. . . . The whole affair, indeed, was taken very seriously, as it was regarded as

an omen for the expedition, and at the same time as evidence of a revolutionary conspiracy to overthrow the democracy. (Thuc. 6.27)

The scandal was compounded by news about the profanation of the mysteries of Demeter in private households. It led to the imprisonment and death of dozens of citizens from leading families, resulted in Alcibiades' defection to Sparta and thus massively influenced not only the Sicilian expedition but the continuation of the war in Greece as well (Thuc. 6.27–29, 53, 60–61; Andocides 1 [*On the Mysteries*]).[81]

Finally, the impact of war on religion: We might distinguish between intensification and innovation. In the *Iliad* (6.269–76), Hector urges his mother:

> Go to Athena's shrine, the queen of plunder,
> go with offerings, gather the older noble women
> and take a robe, the largest, loveliest robe
> that you can find throughout the royal halls,
> a gift that far and away you prize most yourself
> and spread it out across the sleek-haired goddess' knees.
> Then promise to sacrifice twelve heifers in her shrine,
> yearlings never broken, if only she'll pity Troy,
> the Trojan wives and all our helpless children . . .

The Athenian women brought Athena a new robe every year.[82] Although no source describes it for us, we can be sure that they also approached the goddess with special gifts and prayers in critical stages of the war, when religious feelings would naturally be intensified. These very feelings at least partly explain the emergence of signs of religious intolerance in the last third of the fifth century.[83]

Intensification of religious sentiments often resulted in innovation. The Thracian hunter- and warrior-goddess Bendis whose festival in the Piraeus Socrates visits in Plato's *Republic* (above, section 1), was introduced to Athens, it seems, at the very beginning of the Peloponnesian War. The cult of Athena Nike had long existed on the Acropolis; in the 440s the demos assumed control over her priesthood, and in the first part of the war she received a beautiful marble temple (above, section 8). Asclepius, the healing god, was brought from his center in Epidaurus to the Piraeus and Athens in the late 420s.[84] Also during the Peloponnesian War the cult of the Asian Mother (*Mêtêr*) was officially installed at the Agora in the "Old Bouleuterion," which around the same time became the official state archive.[85] The Phrygian god Sabazios, often paralleled with Dionysus, is first attested in Athens around the same time. It is not difficult to think of reasons that prompted the Athenians, in those very years, to introduce into their community so many deities connected with healing, fertility, and religious mystery,[86] but, as the plot of Euripides Bacchae and official resistance to Asclepius illustrate, enthusiasm about these new cults was not universal.[87]

10. Comedy

Stepping over to the south wall of the Acropolis, Socrates and his friends would look down into the Theater of Dionysus (see section 11 below). A few months earlier, probably in early February 411 at the festival of the Lenaea, Aristophanes had performed there his *Lysistrata*.[88] The plot runs as follows. The heroine of the play, Lysistrata, convinces the Athenian women and delegations of women from Sparta, Thebes, and other cities to save Greece by organizing a sex-strike. Deprived of their marital pleasures, she argues, the men, who stubbornly continue to fight a fratricidal war, will soon succumb and conclude peace. In addition, the Athenian women occupy the Acropolis, thereby cutting off the city's financial resources and making it impossible materially as well to continue the war. They repel the attack of an official (*proboulos*) and his policemen and utterly defeat the chorus of old citizens who are represented as warmongers and blindly obedient to the authorities. The women also resist the temptation of joining their increasingly sex-starved and desperate husbands. Capitulating to such unbearable pressure, the men of Sparta and Athens finally agree to a negotiated peace.

This is the third peace play among Aristophanes' extant comedies; two or more are lost (*Georgoi, Olkades*, dated after *Acharnians*). In *Acharnians* (performed in 425) the farmer Dikaiopolis, tired of his city's politics and the assembly's unwillingness to consider peace, concludes a private thirty-year peace with Sparta, defends it stubbornly against hostile attacks by his warmongering countrymen and various profit seekers, and enjoys its benefits, happily and selfishly, while his neighbor, a general and war hero, marches out on a wintery campaign and is brought back, wounded in an—accident.

In *Peace*, performed in 421 shortly before the signing of a peace treaty with Sparta (the "Peace of Nicias"), another farmer, Trygaios, raises a giant beetle and flies up to the gods to ask Zeus why he wants to destroy Greece. The gods, it turns out, fed up with the Greeks' squandering of numerous peace opportunities, have gone off and left the War God in charge of Hellas. He has imprisoned the Goddess of Peace in a cave and intends to pound every Greek city to a pulp in his giant mortar. Fortunately, he lacks a pestle, and the main warring powers, Athens and Sparta, have lost theirs (through the recent deaths of their leading generals). While War is gone to make one himself, Trygaios and the Athenian farmers, most persistent among the men of Greece, succeed in liberating Peace. As a reward, her two attendants are brought to Athens, Festival to be returned to the community, Harvest to be Trygaios' wife. Back in Athens, Trygaios prepares for his wedding and fends off a number of war profiteers who attack him for having ended the war.

All three plays raise important questions. I shall focus on *Lysistrata* but make connections with the other plays when appropriate. Before talking about politics in Athenian comedy, one needs to address two basic questions: that of the political function and seriousness of comedy, and that of the poet's political standpoint. The latter question can be answered more easily. Aristophanes most likely belonged to the elite or was at least closely familiar with members of the elite.[89] While de Ste. Croix attributes to the poet an elite point of view, others have argued, more plausibly, that the perspective presented in Aristophanes' plays, whether or not it corresponds to the poet's own, is likely to be one shared by a majority of the audience.[90] On the seriousness of political comedy opinions are deeply divided. If the parabasis in *Frogs* can be believed, comedy indeed was supposed to be able to give serious advice. But such advice typically was packaged in a thick layer of sexual jokes, political and social exaggeration, and outright fantasy. Advice thus is rarely expressed directly and bluntly; rather, important issues are brought up and the audience is made to think about them indirectly, through comic effects and laughter. Whether or not he expects to have a lasting impact, the comic poet can express truths, even sad and upsetting ones, and he can criticize individuals and groups with impunity—or almost. In this, his function is, at least to some extent, comparable to that of Thersites in book 2 of the *Odyssey*, of a medieval "Hofnarr," and of the masked figures participating in Basel's traditional "Fasnacht" (Aristophanes, *Frogs* 686–737; Thersites: *Iliad* 2.211–77).[91]

What, then, are the issues raised by *Lysistrata*? I shall focus here on four particularly important contrasts: war vs. peace, individual vs. community, male vs. female, and private vs. public.[92]

a. War vs. Peace

In *Acharnians* and *Peace*, despite the differences of perspective, peace is defined and presented primarily in terms of ending war-imposed destruction and restrictions and restoring normal life in farm and community, normal trade with neighbors, normal festivals in the countryside. Even in *Peace*, which is dominated initially by Panhellenic concerns (58–63, 93–94, 105–6, 292–98), this lofty premise is soon abandoned. Even though the "Panhellenes" begin to pull Peace out of her cave, it is the farmers of Attica who succeed in accomplishing the task (508–11). After the goddess' liberation, Trygaios focuses one last time on all of Greece: "Look now, look how happy all the cities of Greece are now! All reconciled, chatting merrily away and laughing!" (538–40, tr. A.H. Sommerstein). But then it is the Attic farmers again (551–55) and finally only Trygaios, celebrating his wedding and enjoying the blessings of peace in his own private world (see 556–81, 918–21, 974–1016). In *Lysistrata*, these economic and bucolic aspects are missing almost entirely. The heroes are women and wives, leaving their homes and concerned about their homes. The men have been absent for months (99–112). The women's purpose in occupying the political realm is to stop the war and save Hellas in order to restore the integrity of their homes and family lives. The war and its deprivations are seen here from the perspective of society at large, and we come close to getting a good look at the misery they cause.

The crucial passage is a confrontation between the women and the official (*proboulos*) as representative of government. The *proboulos* challenges the women's right to concern themselves with public issues, because they "have no share in the war" (588). This is an ambivalent statement: the *proboulos* means, "You are not fighting in the war," which is correct, but the women understand, "The war is none of your business," which Lysistrata refutes: by bearing the children and then sending them out as soldiers, the women carry more than double the burden of the war (588–90). Here the *proboulos* interrupts her: "Be silent! Don't bring up bad memories!" (590). Whether this refers to the recent disaster in Sicily or more generally to the fact that in war many sons do not return home, here we have at least an allusion to the losses caused by the war—an issue otherwise carefully avoided in comedy. Instead, Lysistrata goes on, with warm feelings, to describe the sad lot of the women who have lost their husbands or have not even found one and whose short bloom of life passes in solitude (593–97). We shall return to this passage later.[93]

Politics certainly is not lacking in this picture. Profiteers and warmongers among politicians, generals, and industrialists are attacked and ridiculed left and right (e.g., *Peace* 435ff., 543ff., 1045ff.); the people and their leaders are criticized for starting the war for absurd reasons (*Acharnians* 496ff., *Peace* 603ff.) and for squandering numerous opportunities for peace (*Peace* 657ff.); there are complaints about the unfair distribution of the burden of war (*Acharnians* 598ff.), and so on. What is never challenged in principle is the basis of Athens' power, its empire. In the year after the defeat of the rebellious allied city, Mytilene (Thuc. 3.1–19, 26–28, 35–50 [the "Mytilenian Debate"]), Aristophanes seems to have brought the subject cities on stage in the chorus of his first victorious play, *Babylonians*—as his rival, Eupolis, did soon afterward in his *Poleis*—but the fragments of both plays are too scarce to allow certainty about their plots and tendency.[94] Otherwise, Athens' imperial policies are taken for granted—though heartily parodied in *Knights* and *Birds*, where the Athenian creator of the birds' new city immediately proceeds to create an empire and to challenge the gods for overall power.[95] Nor is democracy attacked in principle. What is ridiculed, exaggerated, and criticized are abuses and peculiar ways of behaving, on the part of both individuals and the demos as a whole.[96]

In sum, then, the war is criticized because it disrupts normal and happy life, because it threatens the integrity of the *oikos* (household), because it serves the interests of few individuals more than those of the community, and because it wreaks havoc among Greek cities. Peace is advocated because it avoids these disastrous consequences and restores normal life in family, economy, community, and the entire Greek world. In the *Lysistrata*, at a time when peace was virtually unobtainable, Aristophanes is most specific about why peace is a moral obligation for both parties (1124–34, 1136–61),[97] and about his ideal vision of a Hellenic peace: it is to restore the dual leadership of Sparta and Athens as it supposedly existed in Cimon's time and failed to work again during the Peace of Nicias. Although the poet sees the past through rosy glasses, he anticipates here an important element of fourth-century political debate.[98]

b. Individuals vs. Community

Comedy is in part blame poetry, directed at collectivities and individuals. Aristophanes is perfectly aware that the demos as a whole shares the responsibility for entering the war, continuing the war, and failing to exploit opportunities for peace (see esp. *Acharnians* 1–133, 215–36, 496–571). Nevertheless, the blame more often lies with influential leaders (politicians like Pericles or Cleon, profiteers, war-hungry generals and officials). And rightly so: to them war meant gain in glory and personal power, if not in wealth; to the ambitious politician war was almost indispensable.[99]

In *Lysistrata*, the poet makes an unusually strong effort to separate the war-crazy leaders and their followers from the ordinary citizens. While the old men, whose livelihood depends on politics, without thinking play the game of the hawkish leaders, the younger men fight the war, suffer, and are presented as willing to make peace—albeit only after this thought has been forced upon them by their wives' sex-strike. Individual interests thus are shown to contradict communal interests. The crucial passage here is the extended metaphor of weaving in the play's pseudo-parabasis:

> When a ball of wool becomes tangled, like the city at war, we use our spindles to put it in order. In the same way, through embassies, can the tangled threads of the war be properly arranged. We must wash the filth from the city as from a ball of wool, beating out the bad and useless parts and picking away the burrs—those who clump in caucuses and knot themselves together to obtain positions of power must be combed out and their leaders plucked away. Then the wool must be carded into the basket of peace and goodwill where all useful people shall be: citizens, metics, friends. And all of our colonial cities, now strewn like fragments of a whole, must be drawn into the common ball and woven together into a mantle for all the people. (*Lysistrata* 565–86, as summarized by Henderson 1980, 199)

c. Men vs. Women

Lysistrata is the first play in which Aristophanes uses gender role reversal as a device to bring out essential concerns. Lysistrata is given several character traits that mark her as the quintessential good citizen—in sharp contrast to those who represent the male citizen body and turn out to be bad citizens, in the sense that they are not concerned with what really is the good of the community. In fact, Lysistrata and her fellow women here assume roles and embody values that are explicitly praised as those of Athenian men in the prime document of Athenian civic ideology, Pericles' Funeral Oration in Thucydides (Thuc. 2.35–46).[100]

In order to assume this role, which requires them to be actively involved and eloquent about it, the comic women must break out of the role assigned them by their male masters in real society: to be silent and subservient.[101] Lysistrata rejects this view, because women make a major contribution to the communal war effort (588–90, cited above; cf. Euripides, *Medea* 248–51; *Erechtheus*, fr. 50.14–15 [Austin 1968, 26]), because men have mismanaged affairs and squandered the polis' patrimony, and because women too have common sense, intelligence, and knowledge. So Lysistrata speaks up (531–54): the women will now be the talkers, the men silent; they will take care of the war; they will go to any length to achieve *aretê*; they are the ones who have "character, charm, daring, wisdom, and an intelligent patriotic valour" (545–49, tr. A.H. Sommerstein).[102] In assuming this role and responsibility, Lysistrata proves *andreiotatê* ("bravest," literally "most manly": 549, 1108). All these are character traits of the male democratic citizen as praised by Pericles.

d. Private vs. Public

The women do not assume this role of good citizen and interfere in the public realm in order to exert power permanently. They have a clearly defined goal: to stop the war and establish peace as a necessary precondition to reintegrating the family and household. Having achieved this goal, they will go home and slip back into their previous roles. This means that the poet sees the war and its disastrous impact on the community as the result of a detrimental separation in the polis of the private sphere of the *oikos* from the public sphere of politics, as disintegration of something that needs to remain integrated if the polis is to remain sound and healthy. In other words, the principles and values of the political sphere have taken on a life of their own, have lost their link with the values of the domestic sphere and thus turned against the latter by pursuing policies that are harmful to the *oikos* and the community as a whole. This only makes sense if we remember that Greek thought from Hesiod to Aristotle assumed a very close correspondence between *oikos* and polis: their structures and economies, and the qualities required to run them, were always considered similar.[103]

Now it has long been recognized that democracy made it necessary precisely to separate the political sphere from the private and domestic one. Christian Meier in particular has emphasized how the democratic citizens developed a "political identity" that tended and was expected to take precedence over their private identity.[104] According to Thucydides and other thinkers of the time, democracy bred and required a specific type of citizen who thought and behaved in specific ways and pursued specific policies, all of which were different from those in an oligarchic polis. The democratic citizen was supposed to put the interests of the polis above his private interests, to be a "lover of his city."[105] Here in the *Lysistrata*, we see the other side of the coin, the reaction to this ideology: the recovery and long-term health of the community requires precisely the reintegration of the citizen's social and political identity—even at the expense of the public and martial glory extolled by Pericles. The wholeness and health of home and family is indispensable for the survival of the community! It became possible to recognize this, it seems, only under the impact of a long and bitter war and of extreme forms of the pursuit of communal power.[106]

There is yet another dimension of ideological reversal. The conflict between

Athens and Sparta had increased a deep and multidimensional split in the Greek world: between Attic/Ionian and Doric Greeks, between democratic and oligarchic societies. Lysistrata and the Athenian women wipe such distinctions off the table. Their concerns are shared by women across Greece; they are elementary for any Greek community. The war is thus recognized as what it really is: fratricidal, self-destructive madness. Only a few years earlier, Herodotus had published his *Histories*, letting both Spartans and Athenians express principles that were as valid half a century after the events as they were in 479 and that utterly contradicted the warring parties' present behavior (see esp. Herodotus 8.141–44). Such inversion of official ideology, I think, was an important part of the political function of comedy; it helped raise political awareness.

11. Tragedy

I shall focus on Euripides' *Trojan Women*, a play performed in late March 415, a few weeks before the departure of the Athenian armada to Sicily.[107] Again, a few preliminary remarks are necessary.

By that time, tragedies at the Great Dionysia were performed in the theater of Dionysus on the south slope of the Acropolis. Although the theater existing then was much more modest than the extant one that was built in the fourth century, plays were attended by many thousands of citizens, probably considerably more than the assembly. Given the proverbial equation of citizens and polis, the theater audience indeed was equivalent to the polis—whether or not women were allowed into the theater.[108] The "congruity between the political and theatrical arenas meant that the responses of Athenian citizens as jurors and assemblymen were inevitably influenced by the fact of their having been members of theatrical audiences, and vice versa."[109]

Dramatic performances took place in the context of a public religious festival that was organized under direction of state officials, supervised by the assembly, and financed by "liturgies" (a method of indirect taxation by which each performance was sponsored by a wealthy member of the community). The competition among the poets who were "assigned a Chorus" was adjudicated, according to highly refined procedures, by a panel of judges who were selected by lot; the victors were honored by prizes and a public inscription and entitled to set up a monument. The festival with its performances lasted several days and was opened by unique public ceremonies. These included libations, not by the priest of Dionysus, but by the most influential civic officials, the ten generals (*stratêgoi*); the announcement of distinctions bestowed upon meritorious citizens and other benefactors of the community; the display of the tribute collected from the subject cities; and the presentation of the panoply to adult war orphans who had been raised at public expense.[110] The dramatic performances thus took place in a civic and political context with strong military and imperial connotations.[111]

We have discussed the political function of comedy. Tragedy was political too, but less overtly and directly. It used, with very few exceptions, not contemporary situations but myth to dramatize, conceptualize, and analyze important communal problems, including political issues. The surviving plays reveal intense concern with the impact of the democratic system on Athenian society, with Athens' imperial role and, especially in the last third of the century, with aspects of war. Even more than the comic poet, the tragedian does not take sides or give advice on specific issues. Rather, he induces the audience to think about important aspects of public issues that tend to be overlooked or suppressed in the heat of political debate; he encourages the citizens to look at the other side of things. In all this, the tragedian assumes a traditional role: that of the poet as advisor and educator of his society.[112]

The *Trojan Women* was not Euripides' first war play. Around the outbreak of the Peloponnesian War he staged *Children of Heracles*, a play that raises some of the very issues that dominate Pericles' first speech in Thucydides, especially the need to refute any ultimatum by a foreign power in order to preserve the liberty and dignity of one's own state (Thuc. 1.140–41; see above section 5).[113] The play is thus easily misunderstood as a patriotic affirmation of Pericles' intransigent policy toward Sparta, but such impressions are dispelled at the end of the play. The *Suppliant Women* of ca. 424 reveals a much more ambivalent attitude toward war and democracy; the play's protagonists question the validity of the concept of a just war and criticize the martial fervor of young and hotheaded leaders who advocate war without concern for the good of the community.[114] The final section of the play, in which a desperate war widow throws herself onto her husband's pyre, brings to mind all the misery war causes for family, society, and community. Earlier in the war Euripides also wrote two plays on Trojan themes. In *Andromache* he describes the tragic fate of Hector's wife, who becomes the slave and concubine of Achilles' son and the victim of his wife's resentment. Vehemently anti-Spartan in tendency, this play puts on stage one aspect of the misery of war victims. *Hecuba*, performed in 425, focuses entirely on these miseries. Hecuba, widow of King Priam of Troy, has lost all but one of her sons in the war. She is deprived of her daughter, Polyxena, whom the Greeks sacrifice on Achilles' tomb to appease the hero's spirit, and of her last son, when the friend to whom she had sent him to be saved from the war kills the boy for material gain. Overcome by grief, Hecuba turns into a fury of revenge, lures the treacherous friend into a trap, blinds him, and kills his children.

Euripides' war plays operate on several levels. They question the justification of wars and the motives of the political leaders promoting wars. They question the glory and benefits accruing to the victors and demonstrate that they too lose out by paying a price that is too high. The plays dramatize the plight of the victims of war and focus on the dehumanizing impact of war on those fighting it. All these elements are combined most forcefully in the *Trojan Women*. As Justina Gregory writes,

The action of *Trojan Women* unfolds a series of disasters, each more gratuitous than the last. The play opens when the city of Troy has fallen; the men have been killed, the women reduced to slavery. The Greeks then proceed to additional outrages in the sacrifice of Polyxena, the murder of Astyanax [Hector's son], and the firing of Troy. There is no relief, no variation from the relentless litany of horrors. Although the audience learns from the prologue that the Greeks will soon be punished for their sacrilegious crimes, that knowledge is not shared by the Trojans, who despair of any justice emanating from the gods. In terms of the action the play is one of the darkest Euripides ever wrote.[115]

Like all tragedies, this too is a complex play. It can be read on many different levels, focusing on various aspects.[116] I shall limit myself here to a few general observations and again to the poet's efforts to reverse official ideology.

a. Causation and Responsibility

The women of Troy (Hecuba, Cassandra, Andromache), each approaching the problem from a different perspective, blame individuals: mostly Helen, but also Agamemnon. Helen, in a speech obviously inspired by sophistic rhetoric,[117] blames everybody but herself: the gods (especially Aphrodite), Hecuba, Priam, Paris, Menelaus, and her third husband. In the prologue Poseidon holds Athena responsible. Yet in a remarkable prayer to Zeus (884–88) Hecuba expresses her belief that what humans attribute to Zeus, that is, to divine interference, is caused either by nature's necessity (*anankê phuseôs*) or by the mortal's mind (*nous brotôn*: 886). So the gods are out. What about individuals? They are guilty, yes, but not alone. Athena turned from friend to foe of the Greeks because Ajax violated her sanctuary at Troy *and* because "the Achaeans did and said nothing to him about it" (71). In Cassandra's view, the Trojans' fate is preferable to that of the Greeks because they "for one woman's sake . . . threw thousands of lives away" (366–69). Thus "the Greeks," that is, the community as a whole, are as responsible as the leaders and other individuals. We are reminded here of contemporary statements attesting to an intense debate on the complex relationship between individual and communal responsibility for political decisions. A vivid example is provided by the contrast between the collective enthusiasm for the Sicilian expedition and the Athenians' rage against individual supporters of the plan after it had failed:

> When the news reached Athens, for a long time people would not believe it. . . . And when they did recognize the facts, they turned against the public speakers who had been in favour of the expedition, as though they themselves had not voted for it, and also became angry with the prophets and soothsayers and all who at the time had . . . encouraged them to believe that they would conquer Sicily. (Thuc. 8.1)[118]

b. Play and Contemporary Events

Scholars have interpreted the play as an indictment of the Peloponnesian War or of the rape of the island of Melos by the Athenians (made famous by Thucydides' "Melian Dialogue" on the nature of power politics) or as a warning against the impending Sicilian expedition. It is true that Euripides must have been working on his play and perhaps finished it before Melos fell, its men were killed, and its women and children sold into slavery. But Euripides did not need to wait for the news from Melos to know or imagine what would happen to the community. Mytilene was nearly executed in the same way in 427 Scione was in 421, and there was no lack of other examples.[119] It is also true that the first assemblies debating a possible intervention in Sicily probably took place in the very month the play was performed. But the issue had been alive for ten years, the request for help had arrived months before, and Athenian ambassadors had been in Sicily to check things out since then. It did not take much to imagine a large Athenian force inflicting death, slavery, and destruction on distant cities under a flimsy pretense of just war. Such, at any rate, had been the attack on Melos.[120] Finally, it is true as well that Euripides was not the first poet to choose the losers' perspective. Aeschylus had done the same in his *Persians* more than fifty years earlier, there too to invoke empathy and illustrate the complexity of human vicissitudes, combined with a serious warning of potentially detrimental policies. Should we conclude, therefore, that "Euripides' concentration on the victims rather than the victors of war seems likely to reflect the ethical complexity characteristic of the genre rather than a specific political reference"?[121] I do not think so: the play contains, perhaps not a specific, but a general reference to contemporary events that the audience cannot possibly have overlooked. It is here, I think, that the inversion of official ideology begins.

c. Inversion of Ideology

In Aristophanes' *Knights*, the Athenian demos is saluted, in the figure of rejuvenated old man Demos, as king (*basileus*) and monarch of the Hellenes, whose rule (*arkhē*) is feared by all people just like that of a tyrant (Aristoph. *Knights* 1111–14, 1330, 1333). In his last speech in Thucydides Pericles tells the Athenians to remember the eternal glory their city had acquired through great achievements in war and their rule over many other Hellenic cities (Thuc. 2.64.3, cited above in section 5). In the Funeral Oration Pericles boasts of the "mighty marks and monuments of empire" the Athenians have created, to be admired by contemporaries and posterity. "We do not need the praises of a Homer, or of anyone else whose words may delight us for the moment, but whose estimation of facts will fall short of what is really true" (Thuc. 2.41.4). Indeed, what is really true? The Trojan women, defeated and enslaved but not broken, have another perspective on eternal glory. Hecuba says: "And yet had not the very hand of God gripped and crushed this city deep in the ground, we should have disappeared in darkness, and not given a theme for music and the songs of men to come" (1242–45; tr. R. Lattimore). As Cassandra puts it, the Greeks came to Troy "and died day after day, though none sought to wrench their land from them nor their own towering cities." Those who died were not buried at home by their families—which were in disarray anyway, due to their men's long absence. The Trojans, by contrast, "have that glory which is loveliest: they died for their own country!" Those who survived came home to their families each night; those who fell were buried in home soil by loving hands; they proved their valor and became heroes. "Though surely the wise man will forever shrink from war, yet if war come, the hero's death will lay a wreath not without lustre on the city" (374–402).

We are reminded of Homer's Hector, of Tyrtaeus' Spartans: the hero who dies in defense of his city brings true glory upon his name, family, and community (*Iliad* 15.494–99; Tyrtaeus, fr. 12 West). The Homeric formula, "to ward off the day of slavery" (*Iliad* 6.463),[122] is used consciously in monuments memorializing Hellas' greatest days of glory. The Athenians and all the other Hellenic soldiers who died in the Persian Wars received heroic burial and honors precisely because, as, among others, the epigram on the Persian War Monument in Athens proclaimed, "they checked, as footsoldiers [and on swift-faring ships]/all of Hellas from [witnessing the day] of slavery."[123] In an unusually serious statement Herodotus defends the view—which apparently was rather unpopular in his days in many parts of Hellas—that "Greece was saved by the Athenians. It was the Athenians who held the balance: whichever side they joined was sure to prevail. It was the Athenians who, having chosen that Greece should live and preserve her freedom, roused to battle the other Greek states which had not yet submitted. It was the Athenians who—after the god—drove back the Persian king." These accomplishments formed the foundation of Athens' claim to leadership and rule in Hellas, trumpeted at every occasion (Herodotus 7.139.5, tr. A. de Sélincourt).[124] At Marathon, Salamis, and Plataea the Athenians had been defending their own land and that of their friends and brothers;[125] on Melos and Sicily they took on the part of the barbarian invader. The contrast was sharp and unflattering: the inversion of ideology in the *Trojan Women* is unmistakable.

On the other hand, the conquerors cannot avoid being tainted by sacrilege and cruelty. The Athenians and Ionians had destroyed the sanctuaries in Sardis; Xerxes burned the Athenian sanctuaries; revenge for this and other sacrileges and devastations was one of the stated purposes of the Delian League (Thuc. 1.96.2; 6.76.3). When the Athenians were about to lose the Peloponnesian War, Xenophon writes, "they could see no future for themselves except to suffer what they had made others suffer, people of small states whom they had injured not in retaliation for anything they had done but out of the arrogance of power" (Xen. *Hell.* 2.2.10, tr. R. Warner). Achaeans, says Hecuba in *Trojan Women*, after they have thrown little Astyanax, Hector's son, from the walls, "all your strength is in your spears, not in your minds!" Fear and emotion have expelled reason. The little son of a hero must be destroyed because he is feared as a symbol of future restoration. What will the poet inscribe on the boy's funeral monument? "Here lies a little child the Argives killed, because they were afraid of him! A horrible inscription [or monument] for Hellas!" (1158–91). Again we think of Melos: here lies a little island the Athenians destroyed because they were afraid of it as a symbol of independence!

In this play the victors, momentarily basking in their power and the glory of victory, will meet disaster on their way home—all of them, not just Ajax! The Trojans say of their own city: "I see the work of gods who pile tower-high the pride of those who are nothing and dash present grandeur down!" (612–13). What was noble becomes slave (614–15). The great city (*megalopolis*) becomes a non-city (*apolis*), and there is no Troy anymore (1291–92). This too can be applied to the victors. Even Athens, the greatest city of all, might one day be crushed into a pile of smoking rubble. The poets and historians, aware of the rise and fall of empires and cities and remembering Athens' meteoric rise to world power in only one generation after the Persian Wars, worried acutely about this possibility: Sophocles' *Antigone* and *Oedipus* as well as Herodotus' *Histories* attest to that; Thucydides, writing and revising his work after 404, lets Pericles consider it. Eleven years after the performance of *Trojan Women* Athens escaped this fate by a hair (Xenophon, *Hellenica* 2.2.19–20).[126]

12. Conclusion and Questions

I have argued that in the late fifth century the Athenians were surrounded by remind-ers of their city's tradition of martial success and imperial greatness. Their own experience, and that of their fathers and grandfathers, of participating in their community's wars and of their mighty fleet roaming the seas of their empire; the goods on the markets in the Piraeus and the city; the constant presence of foreigners who came to Athens for economic, political, or legal reasons; the public monuments on Agora, Acropolis, and in many other places; and the images and inscriptions on these monuments, the funeral monuments, and inscriptions in the public cemetery outside the Dipylon Gate; the religious festivals, ceremonies, and rituals they at-tended; the political meetings of council and assembly in which they participated; and all the other political activities (such as committees and panels of judges) in which they were involved—these all combined to impress on them the same mes-sage: their city had achieved unique power and greatness through constant involve-ment in war, through a unique series of victories and successes, and through unique dedication and sacrifice on the part of the citizens. Hence they were called to con-tinue this tradition and to live up to the example and standards set by their ancestors. These constant reminders of their community's civic ideology conditioned the Athe-nian citizens from youth on to accept war as inevitable and even desirable.

No wonder that the public discourse on war was one-sided, that the need to maintain and enlarge the empire and for this goal constantly to go to war was virtually taken for granted. At least in politics serious questions about this ideol-ogy were rarely raised—or so it seems. Thucydides' tendency to pair antithetical speeches and his frequent allusion, in summarizing the debates in the assembly, to a variety of speakers and opinions suggest, however, that the ultimately victorious proposal did not necessarily pass uncontested. The evidence gleaned from the corpus of the fourth-century orators confirms this assumption. Conversely, ap-peals to patriotism, outbursts of emotion and chauvinism, and massive peer pres-

sure must often have forced those opposed to the dominant view to keep silent—
what Thucydides reports about the debate before the Sicilian expedition was hardly
unique (Thuc. 6.24.3–4).[127] Hence what the same citizens experienced, at least
occasionally, in a different sphere, that of the theater, was perhaps less surprising
and antithetical than it appears to us.

Clearly, both Aristophanic comedy and Euripidean tragedy did ask unsettling
questions about, or present outright inversions of, various aspects of the official
ideology, and they showed on stage the consequences, for their own society and for
the victims, of their city's policies. Given the small number of preserved plays (less
than 50 out of about 1,500 tragedies and comedies performed in the fifth century)
and the random character of fragments and information extant from the lost plays,
we cannot know how regular and frequent such efforts were and whether Aristophanes'
and Euripides' concerns were fully shared by other poets. Even if they were—which,
at any rate, would have been the case much more during the Peloponnesian War than
before—what was the effect of such efforts on the audience and on politics? Legends circulated in antiquity about Euripides' popularity that even saved some of the
Athenian prisoners of war in Sicily from death, but despite such legends Athens
survived in 404, not because a Phocian man cited Euripides, but because the Spartans remembered Athens' merits in the Persian Wars, or perhaps because they found
it preferable to use a weakened Athens as a counterweight to Theban aspirations.[128]
In *Frogs*, performed at the very end of the war, Dionysus brings Aeschylus back
from the underworld to give sound advice and save the city (Aristoph. *Frogs* 1418–
21, 1433–36, 1500–3). The Athenians might have done well to listen to their poets
long before 405. Why did they not hear their message? Why did words uttered in the
theater have little visible impact on debates on the Pnyx?

Various answers seem possible. Powerful and specific factors kept propelling
the Athenian community on its track of war in the service of power politics: glory
and pride, shared among all citizens, worries about unpredictable repercussions of
a drastic change in policies, undeniable profits and advantages, and the collective
need of the democratic citizen body to prove themselves and reaffirm the legitimacy of their shared power.[129] In addition, theirs was an age of almost limitless
optimism, of a sense of achievement and possibilities without boundaries, perhaps
comparable with much of the nineteenth-century European experience—before
such optimism was crushed in the trenches of Verdun. Moreover, to sound a rather
cynical note, it is perhaps a general rule in human history that reason and moral
precepts do not usually prevail over self-interest, hunger for power, and martial
frenzy. More importantly, war seems to correspond to a deep human need that, as
many events in today's world attest sadly and pervasively, is not easily controlled
by rational arguments. As Walter Burkert writes,

> History, as far back as we can trace it, is the history of conquests and wars. Ever
> since Thucydides, historians have tried to understand the necessity of these events
> and, if possible, make them predictable. But it is precisely the irrational, compul-

sive character of this behavior mechanism that confronts us more clearly today than ever before. War is ritual, a self-portrayal and self-affirmation of male society. Male society finds stability in confronting death, in defying it through a display of readiness to die, and in the ecstasy of survival.[130]

Be that as it may, perhaps the function of the theater was precisely one of balancing the picture without changing it. The ancients recognized as one of its purposes a ritual of "cleansing" (*katharsis*), in order to re-establish emotional balance (Aristotle, *Poetics* 1449b, 24–28).[131] We might feel reminded of one of the functions that churchgoing and other organized forms of religion seem to play for many today: to make the worshipers feel good, to silence their worries and bad conscience—and to allow them to go on as before. Virtually all politicians claim to be religious, but their actions more often than not blatantly contradict the precepts of their religion.

Still, should we really compare the poets with the proverbial prophets who are not honored in their own country—honored in the sense of being taken seriously by corresponding action? Like the historians who shared their insights, they may have felt sometimes like that noble Persian who foresaw the Persian defeat at Plataea and told his Greek table companion during a banquet on the eve of the battle: "No one believes warnings, however true. Many of us . . . know our danger. . . . Verily it is the sorest of all human ills to abound in knowledge and yet have no power over action" (Herodotus 9.16.4–5). Yet neither tragedians nor comedians should be thought of as far-sighted or idealistic "left-wing" intellectuals whose views were largely at odds with those of the majority of their fellow citizens. This is the picture modern scholars occasionally draw of Euripides.[132] But neither he nor Aristophanes could have been so successful and won so many competitions for the right to perform their plays—not to speak of their first prizes—unless the views they presented on stage, however these corresponded to their personal beliefs, were broadly acceptable to their audiences. Bucolic vistas of peace and escapist utopias could easily be attributed to Athenian men, especially if these were trying to avoid not war but the law courts (as in *Birds*). War profiteers, overzealous magistrates, corrupt politicians, and warmongering generals could be ridiculed and beaten up to hearty applause. More serious criticism and problems preferably were presented through the mouths and myths, not of Athens or Athenian men, but of Thebans, Argives, Trojans—and Athenian women. Such distancing made it possible for the theater to fulfill one of its essential functions—to deal with important communal concerns—without offending patriotic feelings and stirring up angry reactions.[133]

Notes

1. Shorter versions of this chapter were presented in Australia at the University of New England in Armidale NSW; the University of Melbourne; and at Macquarie University in Sydney; at Loyola College, Baltimore; at the University of California, Santa Barbara; at Loyola Marymount University in Los Angeles; and, as the 29th Gail A. Burnett Lecture in Classics, at San Diego State University (published as "Leaders in War and Bravery: The

Ideology of War in Late Fifth-Century Athens," San Diego: Department of Classics and Humanities, San Diego State University, 1998). I thank colleagues at these universities for their hospitality and stimulating discussions, and the participants in the Wilson Center conference for helpful comments.

2. All dates are B.C. Abbreviated source collections: *IG: Inscriptiones Graecae* (vols. 1.1–2, ed. David M. Lewis and Lilian Jeffery, 3rd ed. [Berlin: De Gruyter, 1981/94]); ML: David M. Lewis and Russell Meiggs, eds., *A Selection of Greek Historical Inscriptions* (rev. ed., Oxford: Clarendon Press, 1988); Fornara: Charles W. Fornara, ed., *Archaic Times to the End of the Peloponnesian War*, Translated Documents of Greece and Rome, 2nd ed., vol. 1 (Cambridge: Cambridge University Press, 1983); PCG 3.2 and 5: Rudolf Kassel and Colin Austin, eds., *Poetae Comici Graeci*, vols. 3.2 and 5 (Berlin: De Gruyter, 1984, 1986); Edmonds: John M. Edmonds, *The Fragments of Attic Comedy*, vol. 1 (Leiden: Brill, 1957); Diels-Kranz: Hermann Diels and Walther Kranz, eds., *Die Fragmente der Vorsokratiker*, 11th ed. (Zurich: Weidmann, 1964); Freeman: Kathleen Freeman, *Ancilla to the Presocratic Philosophers* (Cambridge: Harvard University Press, 1948).

3. Raaflaub 1994, 114–18, 126–30.

4. Hoplite warfare: Connor 1988; Hanson 1989, 1991. Destruction of cities: Ducrey 1968, 112; Karavites 1982, 33–35, 130. Subjection of cities: Raaflaub 1985, 82–96. Formation of power: Raaflaub 1990.

5. Meiggs 1971; Powell 1988; Rhodes 1992; Bleicken 1994; Raaflaub 1997.

6. See the contribution of Victor Hanson to this volume and, generally, Finley 1986; Meier 1990b.

7. Frost 1976; Garnsey 1988, pt. 3; Davies 1992; Raaflaub 1998.

8. Garnsey 1985. On the demographic shifts, see the bibliography cited in n. 7.

9. Raaflaub 1994; Hanson, this volume.

10. *IG* I³ 383. 143; *IG* I³ 136; see Bingen 1959; Burkert 1985, 179; Simms 1988; Garland 1992, 111–14; Parker 1996, 170–75. The festival of Bendis was celebrated in the early summer (on Thargelion 19; Mikalson 1975, 158). In 411, not much later, the Four Hundred entered office (Aristot. *Constitution of the Athenians* 32.1), and the events leading up to the oligarchic coup had been going on for a while (Kagan 1987, ch. 5–6). Thus the city was in political turmoil. In my essay I shall ignore these events which, of course, were deeply connected with the Athenians' outlook on the war (Thuc. 8.53–54).

11. Cf. Kagan 1981, pt. 2; 1987, ch. 1–4.

12. Hippodamus of Miletus: McCredie 1971; Garland 1987; Hoepfner and Schwandner 1994, 22–50.

13. Amit 1965; Jordan 1975; Garland 1987, 95–100; Gabrielsen 1994.

14. Losses: Strauss 1986, 179–82.

15. Sociology, psychology, and ideology of warfare: see Hanson, Cartledge, this volume. Furthermore, e.g., on Aristophanes' attitude to war, Ehrenberg 1962, ch. 11; on the hoplites: Anderson 1970; Ridley 1979 (who rightly emphasizes that the hoplites continued to play an important role even during the Peloponnesian War); Vidal-Naquet 1986, 85–105; Hanson 1989, 1991, and 1996; on the sailors and navy, apart from the titles cited above: Morrison and Coates 1986; Casson 1991; Strauss 1996. We should not forget the cavalry: Bugh 1988; Spence 1993. Pritchett 1971–1991 is a vast resource on many aspects of warfare. See in general also Vernant 1968; Finley 1986; Meier 1990b; Rich and Shipley 1993. "They knew what they were dealing with": see, e.g., Kagan 1981, 165, refuting Thuc. 6.1.1.

16. Boedeker 1995, 220–25; 1998a; see section 9 below.

17. Marrou 1956, 36–45.

18. Raaflaub 1986, 29–34.

19. Employment: Garland 1987, 98. Economic motives: summary and bibliography in Raaflaub 1994, 132–35.

20. Emporion: Garland 1987, 83–95; see also von Reden 1995. Markets on and around

the Athenian Agora: Wycherley 1978, ch. 3, and the testimonia in Wycherley 1957, ch. 4. Import of grain: Garnsey 1985.

21. Ostwald 1992, 310–12.

22. ML 79; Fornara 147. On the scandals of 415, see section 9 below.

23. Meiggs 1972, ch. 14; Finley 1982, 41–61.

24. Wycherley 1978, 15–16; Garland 1987, 23–26; Conwell 1992.

25. On Thucydides the theoretician, see Ober this volume. On Pericles' war strategy: Kagan 1974, ch. 1; Ober 1985; Spence 1990. On the economic impact: Foxhall 1993.

26. Frisch 1942, 78; cf. Starr 1979a.

27. Bowersock 1967, 33–38.

28. See Momigliano 1944, 1.

29. Frisch 1941, ch. 3.

30. Cf. Shimron 1989, 89–92.

31. Frisch 1942, 85; contra: Momigliano 1944, 3.

32. See Kallet-Marx 1993a, ch. 3; Ober this volume.

33. Meier 1990a, ch. 8.

34. On the walls see Travlos 1971, 158–74; Wycherley 1978, 10–15. Later debates: Euripides, *Children of Heracles* 1–380; Thuc. 1.140.2–141.1; see Raaflaub 1985, 234–37.

35. XV on Travlos' plan, 1971, 169.

36. Travlos 1971, 466–73; cf. Forsén and Stanton 1996. Plutarch, *Themistocles* 19.6 focuses on the speaker first looking out to the sea and then being forced to look inland, "for they believed that Athens' naval empire had proved to be the mother of democracy and that an oligarchy was more easily accepted by men who tilled the soil." This surely is wrong in thinking of the speaker rather than the citizens (see also Frost 1980, 178).

37. See Ostwald 1986, 434–45. Posterity: Roberts 1994, ch. 3–4.

38. On political deliberation and decision-making, see generally Ober 1989; Yunis 1991.

39. Hansen 1987.

40. See, e.g., the debates in 425, after Cleon's victory at Pylos: Thuc 4.15–22, 27–28, 41; more generally, see 3.37–38, 42–43; cf. Raaflaub 1994, 136–38.

41. IV on Travlos' plan, 1971, 169.

42. Clairmont 1983, 15. On *patrios nomos* and *dēmosion sēma*, see Stupperich 1977; Clairmont 1983; Loraux 1986.

43. *IG* I³ 1162; ML 48; Clairmont 1983, no. 32b.

44. Clairmont 1983, 36–37.

45. Lysias, *or.* 2; Demosthenes *or.* 60; Hyperides, *or.* 6; see also Gorgias, no. 82 fragm. B6 Diels-Kranz, Freeman 130; Plato, *Menexenus*; Isocrates, 4 (*Panegyric*) and 12 (*Panathenaic*). See Strasburger 1958; Kierdorf 1966; Loraux 1986.

46. Plague: Thuc. 2.47–55, 59, 65.1–4; cf. generally 3.82–84. For the Athenian character portrait, see also 1.70; Raaflaub 1994.

47. Assembly: Raaflaub 1985, 218–23. Theater: e.g., Aeschylus, *Persians*; Euripides, *Heraclidae* and *Suppliants*; Zuntz 1955. Monuments: below, section 8.

48. Travlos 1971, 422–28.

49. Cf. Hölscher 1973, 50–84; Camp 1986, 66–72; Castriota 1992, 76–89.

50. Ostwald 1992, 328–29, but see Podlecki 1977, 247.

51. Fornara 1983, 1–2, referring to Aristot. *Rhetoric* 1.1360a 35, and Quintilian, *Training in Oratory* 2.4.2. Tradition of epic: Fornara 1983, 29–32; see also Herington 1991. Greek-Persian confrontation: Meier 1987. Critical attitude: Boedeker 1998a. See also Moles 1993, esp. 100.

52. Thomas 1989, esp. ch. 2, 5. Miltiades, Aeschines 3 (*Against Ctesiphon*).186; cf. the Eion epigrams, below section 8.

53. Camp 1986, 53–57, 100–5. See generally on this section also Hölscher 1991, 368–75; 1996; 1998.

54. See Raaflaub 1985, 233–51. On the Stoa of Zeus, see Travlos 1971, 527–33; Camp 1986, 105–7.

55. See, e.g., Pseudo-Xenophon, *Const. of the Athenians* 3.2: the Council "has multifarious business to deal with concerning war, revenue, legislation, the day-to-day affairs of the city and matters affecting their allies, and has to receive the tribute and look after the dockyards and shrines." Further, Aristotle, *Const. of the Athenians* 46–49, and Rhodes 1972. For the sake of illustration, I cite a passage from Demosthenes' oration *On the Crown*, describing the Athenians' reaction to the attack of Philip II of Macedon in 339: "Evening had already fallen when a messenger arrived bringing to the prytaneis the news that Elatea had been taken. They were sitting at supper, but they instantly rose from table, cleared the booths in the Agora of their occupants, and unfolded the hurdles, while others summoned the generals and ordered the attendance of the trumpeter. The commotion spread through the whole city. At daybreak on the morrow the prytaneis summoned the Council to the Bouleuterion, and the citizens flocked to the Assembly. Before the Council could introduce the business and prepare the agenda, the whole body of citizens had taken their places on the hill. The Council arrived, the prytaneis formally reported the intelligence they had received, and the courier was introduced. As soon as he had told his tale, the herald put the question, Who wishes to speak?" (18.169–70). On the Bouleuterion: Travlos 1971, 191–95; Camp 1986, 52–53; Shear 1995. Tholos: Travlos, 553–61; Camp, 94–97; Seiler 1986, 29–35.

56. Travlos 1971, 261–73; Camp 1986, 82–87. This temple was long believed to be the Theseion. In fact, the shrine erected for the Athenian national hero by Cimon, who brought his bones back from the island of Skyros to Athens in 476 (Plutarch, *Cimon* 8.5–6; *Theseus* 36), was located to the southeast of the excavated portion of the Agora (Pausanias 1.17; Plut. *Thes.* 36.2; cf. Travlos, 233–41 under "Gymnasium of Ptolemy"; Wycherley 1978, 64; see Shapiro 1989, 143 with n. 11 for the possibility that there were two sanctuaries: the Theseion proper and another shrine, perhaps in the Agora, that housed the hero's bones). As was fitting for Theseus, his shrine was decorated with depictions of the battle against the Amazons and that between Centaurs and Lapiths (Wycherley 1957, 113–19 for the testimonial; Barron 1972; Shapiro 1989, 143–49; Castriota 1992, 33–63, on the paintings and the function of the cult; Castriota 1997, 209–13).

57. Shear 1970; Kron 1976, 13–31, 202–41; Camp 1986, 97–100.

58. Stone pillars with male genitals and a portrait of the god Hermes. See Camp 1986, 74–77; the testimonia in Wycherley 1957, 103–8.

59. For the principle of not letting the leaders inscribe their own names upon such monuments, "in order that the inscription might not seem to be in honour of the generals, but of the people," see also Aeschines 3 (*Ag. Ctesiphon*), 178–88.

60. For the honors awarded to Cleon after his surprise victory at Pylos, see Gauthier 1985, 95–96. Marathon monument: ML no. 26; Fornara no. 51; *IG* I³ 503 (with new fragments); see West 1970; Barron 1990. Later proliferation: Gauthier 1985, 120–28; Wycherley 1957, ch. 5.

61. Such emphasis on the collective achievement of the citizen body, attested for the immediate aftermath of the Persian Wars, might be interpreted as the first indication of emerging "democratic sentiments." Private dedications honoring outstanding individuals were not barred; see, e.g., ML no. 18, Fornara no. 49, on the memorial of Callimachus, the polemarch at Marathon, and Lonis 1989, 286–93, on the difference between public and private dedications.

62. Stupperich 1977, 71–135; Morris 1992, ch. 5.

63. For the most important examples, see ML and Fornara.

64. Acropolis: Travlos 1971, 52–71; Wycherley 1978, ch. 4. Athena Nike: Travlos, 148–57; Mark 1993; Jameson 1994. Votive offerings: Travlos 1971, 69 fig. 88; cf. Raubitschek 1949, esp. 455–67. Parthenon: Herington 1955; Travlos, 444–57; Meiggs 1972, ch. 15; Castriota 1992, ch. 4; Jenkins 1994; Osborne 1994, but see Giovannini 1997.

65. Burkert 1985, 267. On Athenian religion in the fifth century: Burkert 1992; Parker 1996, ch. 8–10; Jameson 1997, and a brief survey in Raaflaub 1998, 36–40.

66. Burkert 1983, 48.

67. Pritchett 1971–1991, 3; Lonis 1979; Jameson 1991.

68. Vidal-Naquet 1986, 106–28; 1992, 215–51; recent bibl. in Raaflaub 1994, 141, with n. 90.

69. Pritchett 1971–1991 vol. 3, 157, n. 6, with bibl.; on Rome, see now Rüpke 1990.

70. Burkert 1983, 64–66; Lonis 1979, 206–9.

71. Pritchett 1971–1991 vol. 3, 161. Ares: Burkert 1985, 169–70; Pritchett, 157–61. The temple of Ares erected on the Athenian Agora in the late first century B.C. actually is a mid-fifth-century building, transported there perhaps from the Attic town of Acharnai (Travlos 1971, 104–11; Camp 1986, 184–85; contra: Pritchett, 159–61).

72. Thompson 1944. Pylos: Pausanias 4.26.6. See Lonis 1979, ch. 12.

73. See Ridgway 1992.

74. Pritchett 1971–1991 vol. 3, 154–56; Lonis 1979, ch. 2.

75. Great Dionysia: below, section 11. Thesmophoria: Deubner 1932, 59–60. Pyrrhic: Lonsdale 1993, ch. 5; Ceccarelli forthcoming. Panathenaea: Kyle 1992. *Patrios nomos*: Stupperich 1977, 54–56; Clairmont 1983, ch. 3.

76. Pritchett 1971–1991 vol. 3, 168–83; Plutarch, *Moralia* 349c–50a. On the shrine of Artemis Eukleia that was built from the spoils of Marathon, see Francis 1990, 98.

77. See further Boedeker 1998b.

78. See, e.g., Nilsson 1951; Starr 1979b; Shapiro 1994; Jameson 1997.

79. Panathenaea: ML 69 (Fornara 136), lines 55–58; cf. 46 (Fornara 98), lines 41–43. Dionysia: Burkert 1992, 261. Eleusinia: Simms 1975.

80. Meiggs 1972, ch. 16; Smarczyk 1990; Parker 1994.

81. Cf. MacDowell 1962; Gomme, Andrewes, and Dover 1970, 264–88; Kagan 1981, ch. 8; Furley 1996; Munn 2000.

82. Barber 1992. *Iliad* 6.269–76, tr. R. Fagles.

83. Dover 1988; Mikalson 1983, ch. 12.

84. Bendis: section 1 with n. 10 above. Priesthood of Nike: *IG* I³ 35; ML 44; Fornara 93; cf. Garland 1992, 102–3; temple: n. 64 above. Asclepius: Ostwald 1992, 313–14; Garland 1992, ch. 6; Parker 1996, 175–85.

85. Burkert 1985, 177–79; Travlos 1971, 352; Camp 1986, 91–94; Parker 1996, 188–94. It is fairly generally accepted that the cult was introduced in Athens at the end of the sixth century (Nilsson 1967, 725–27). Since the terra-cotta figurines of Meter found on the Acropolis are votives that suggest private cult (I thank Lynn E. Roller for this information; her book on the Anatolian origins of the Meter cult and its spread to the Greek and Roman worlds is forthcoming), the assumption of a late sixth-century public cult rests entirely on the identification as an early Meter temple of the foundation of a small, rectangular, elongated building just north of the Old Bouleuterion. Shear, arguing against the thesis of Miller (1995) that the so-called Old Bouleuterion (constructed in the first half of the fifth century) itself really was not a Bouleuterion but a Meter temple facing out to the Agora, now demonstrates that the small building north of it could not possibly have been a Meter temple either (1995, 171–78; cf. Miller 1995, 137, n. 6). Unless Miller's thesis is accepted, no evidence thus seems to exist for a public cult of the goddess in Athens before the late fifth century. See, however, Robertson 1996; Roller 1996.

86. See also Dodds 1951, ch. 6, esp. 192–95. Sabazios: Burkert 1985, 179; Parker 1996, 194.

87. Ostwald 1992, 314.

88. Henderson 1987, xv–xxv, esp. xv–xvi; on the Lenaea: Pickard-Cambridge 1968, 25–42.

89. Dow 1969; Lind 1985; Cartledge 1990, xv–xvi.

90. De Ste. Croix 1972, 355–76; contra: Heath 1987; Cartledge 1990, 46–47.

91. Fasnacht: Gelzer 1992. For discussion of various aspects of this issue, see Gomme 1938; de Ste. Croix 1972, 355–76; Heath 1987a (with Alan H. Sommerstein's review, *Journal of Hellenic Studies* 109 [1989] 222–23); Henderson 1980, 1993, 1998; Reckford 1987, ch. 5; Redfield 1990; Cartledge 1990; Sommerstein et al. 1993; Konstan 1995; Dobrov 1997.

92. See in general Vaio 1973; Henderson 1980; Newiger 1980; Taaffe 1993, ch. 2; Konstan 1995, ch. 3.

93. See Seel 1960, 79–89; Henderson 1980, 195–200.

94. *Babylonians*: Edmonds 1, 589–96; *PCG* 3.2, 62–77; Schmid and Stählin 1946, 183–84; Norwood 1930; Forrest 1975; Cartledge 1990, 44–45. *Poleis*: Edmonds 1, 386–96; *PCG* 5, 424–41; Schmid and Stählin, 118; Storey 1990, 18–20.

95. In 414, the allusion to the Sicilian expedition could hardly be overlooked; see Konstan 1995, ch. 2. See also Forrest 1975. More generally on the question of debates about and criticism of Athens' imperial policies in the fifth century, see Meiggs 1972, ch. 21; Raaflaub 1985, 215–18.

96. On Aristophanes and democracy, see Ehrenberg 1962, ch. 13; Landfester 1975; Stark 1975; also Cartledge 1990, ch. 5; Henderson 1993. Generally on the debate about democracy in late fifth-century Athens: Raaflaub 1989.

97. See also Herodotus 8.141–44.

98. See Ryder 1965.

99. See, e.g., the remarks about the role of leaders in *Peace* 685–87; cf. Raaflaub 1994, 136–38.

100. Aristophanes and women: Rossellini 1979; Loraux 1991; Taaffe 1993.

101. See Lysistrata herself, 506–30, and Pericles' advice to the widows: "Your glory is great if you do not fail to live up to your own nature, and if there is the least possible talk of you among men either for praise or for blame" (Thuc. 2.45.2, tr. P.J. Rhodes); see Seel 1960, ch. 7. For a recent interpretation of Pericles' statement, see Kallet-Marx 1993b (with bibl.).

102. Cf. Henderson 1987, 138 *ad* 549.

103. See, e.g., Herodotus 5.29; Xenophon, *Memorabilia* 3.4.6–12; Spahn 1980.

104. Meier 1990a, ch. 6; Humphreys 1978, 201. Contra: Strauss 1993.

105. *Erastês, philopolis, philodêmos:* cf. Connor 1971, ch. 3; Monoson 1994; Raaflaub 1994, 104–14.

106. On the tension between *oikos* and polis in tragedy, see Saïd 1998, 285–95.

107. Gomme, Andrewes, and Dover 1970, 276.

108. Theater of Dionysus: Travlos 1971, 537–40. Citizens are the polis: Thuc. 7.77.7, *andres polis;* cf. Alcaeus 426 Campbell; Herodotus 8.61. Theater audience and polis: Kolb 1979, 530. Women: Csapo and Slater 1995, 286–87.

109. Ober and Strauss 1990, 238; cf. Connor 1996.

110. Pickard-Cambridge 1968, pt. 2; Goldhill 1990; Csapo and Slater 1995, 103–21; Kallet 1998.

111. Why were war and empire so intimately connected with this particular festival? The fact that it took place at the beginning of the sailing season, so that allies and foreigners could attend, seems a superficial explanation. The question requires further scrutiny.

112. As expressed in Aristophanes' *Frogs*, e.g., 1008–10; Konstan 1995, 63. See generally, Ober and Strauss 1990; Gregory 1991; Meier 1993; Croally 1994; Goff 1995; Pelling 1997; Saïd 1998. Specifically on tragedy and empire: Rosenbloom 1995; on tragedy and democracy: Raaflaub 1989, 49–54.

113. Zuntz 1955; Raaflaub 1985, 245–46.

114. Burian 1985; Raaflaub 1989, 50–52; Michelini 1994.

115. Gregory 1991, 156–57.

116. For recent interpretations, see Scodel 1980; Gregory 1991, 155–81; Croally 1994, and the bibl. cited there.

117. Perhaps even by Gorgias' famous *Encomium of Helen* (no. 82, fragm. B11, Diels-Kranz; Freeman 131–33).

118. Compare the enthusiasm of Thuc. 6.24 (cited above in section 2). Individual and communal responsibility: see, e.g., the Mytilenian Debate, 3.37–38, 42–43.

119. Van Erp Taalman Kip 1987. Melian Dialogue: Thuc. 5.85–113. The end of Melos: 5.116. Other cities: 3.50 (Mytilene), 5.32 (Scione); see Ducrey 1968.

120. Cf. Menelaos' statement in *Trojan Women* 864–68. For a discussion of the Athenian attack on Melos, see Gomme, Andrewes, and Dover 4, 156–58; Seaman 1997. Date of play and debate about Sicilian expedition: Thuc. 6.8.1; Kagan 1981, 166; Gregory 1991, 179, n. 2. Earlier debates: Kagan, ch. 7.

121. Gregory 1991, 179, n. 2. *Persians*: Meier 1993, 63–78; Raaflaub 1988, 284–86.

122. Cf. Raaflaub 1985, 29–30.

123. ML no. 26; Fornara no. 51; cf. Raaflaub 1985, 72–79.

124. Raaflaub 1985, 215–23.

125. Illustrated most impressively by the battle cry of the Greeks in Aesch. *Persians* 402–5: "Forward, you sons of Hellas! Set your country free! Set free your sons, your wives, tombs of your ancestors, and temples of your gods. All is at stake: now fight!" (tr. P. Vellacott).

126. See Kagan 1987, ch. 15. Pericles: Thuc. 2.64.3. Sophocles and Herodotus: Knox 1979, ch. 8; Raaflaub 1987; 1988, 296–99.

127. Cited above in section 2. Fourth century: Ober 1989; see also Ober and Strauss 1990.

128. See above n. 126; Plutarch, *Nicias* 29; *Lysander* 15.

129. Raaflaub 1994, 130–46.

130. Burkert 1983, 47; cf. 48, cited above in section 9.

131. Cf. Heath 1987b, 124–25.

132. For discussion, see Stevens 1956; Kovacs 1987, ch. 1, esp. 9–21.

133. See Zeitlin 1990.

Bibliography

Amit, M. 1965. *Athens and the Sea: A Study in Athenian Sea-Power*. Collection Latomus 74. Brussels: Latomus.

Anderson, John K. 1970 *Military Theory and Practice in the Age of Xenophon*. Berkeley: University of California Press.

Austin, Cole. 1968. *Nova Fragmenta Euripidea*. Berlin: De Gruyter.

Barber, E.J.W. 1992. "The Peplos of Athena." In *Goddess and Polis: The Panathenaic Festival in Ancient Athens*, ed. Jennifer Neils, 103–17. Hanover, NH: Hood Museum of Art, Dartmouth College; and Princeton: Princeton University Press.

Barron, J.P. 1972. "New Light on Old Walls: The Murals of the Theseion." *J. of Hellenic St.* 92: 20–45.

———. 1990. "All for Salamis." In *"Owls to Athens": Essays on Classical Subjects Presented to Sir Kenneth Dover*, ed. Elizabeth M. Craik, 133–41. Oxford: Oxford University Press.

Bingen, Jean. 1959. "Le décret SEG X 64 (Le Pirée, 413/2?)." *Revue belge de philologie et d'histoire* 37: 31–44.

Bleicken, Jochen. 1994. *Die athenische Demokratie*. 2nd ed. Paderborn: Schöningh.

Boedeker, Deborah. 1995. "Simonides on Plataea: Narrative Elegy, Mythodic History." *Zeitschrift für Papyrologie und Epigraphik* 107: 217–29.

————. 1998a. "Presenting the Past in Fifth-Century Athens." In *Democracy, Empire, and the Arts in Fifth-Century Athens*, ed. Deborah Boedeker and Kurt Raaflaub, 185–203. Cambridge: Harvard University Press.

————. 1998b. "The New Simonides and Heroization at Plataia." In *Archaic Greece: New Approaches and New Evidence*, ed. Nick Fisher and Hans van Wees, 231–49. London: Duckworth; and Swansea: Classical Press of Wales.

Boedeker, Deborah, and John Peradotto, eds. 1987. *Herodotus and the Invention of History. Arethusa* 20.

Boedeker, Deborah, and Kurt Raaflaub, eds. 1998. *Democracy, Empire, and the Arts in Fifth-Century Athens*. Cambridge: Harvard University Press.

Bowersock, Glen W. 1967. "Pseudo-Xenophon." *Harvard Studies in Classical Philology* 71: 33–55.

Bugh, Glenn R. 1988. *The Horsemen of Athens*. Princeton: Princeton University Press.

Burian, Peter. 1985. "*Logos* and *Pathos*: The Politics of the *Suppliant Women*." In *Directions in Euripidean Criticism: A Collection of Essays*, ed. Peter Burian, 129–55. Durham: University of North Carolina Press.

Burkert, Walter. 1983. *Homo necans: The Anthropology of Ancient Greek Sacrificial Ritual and Myth*. Berkeley: University of California Press.

————. 1985. *Greek Religion, Archaic and Classical*. Cambridge: Harvard University Press.

————. 1992. "Athenian Cults and Festivals." In *Cambridge Ancient History*, 2nd ed., vol. 5, 245–67. Cambridge: Cambridge University Press.

Camp, John M. 1986. *The Athenian Agora: Excavations in the Heart of Classical Athens*. London: Thames and Hudson.

Cartledge, Paul. 1990. *Aristophanes and His Theatre of the Absurd*. Bristol: Bristol Classical Press.

Casson, Lionel. 1991. *The Ancient Mariners: Seafarers and Sea Fighters of the Mediterranean in Ancient Times*. Princeton: Princeton University Press.

Castriota, David. 1992. *Myth, Ethos, and Actuality: Official Art in Fifth-Century b.c. Athens*. Madison: University of Wisconsin Press.

————. 1997. "Democracy and Art in Late-Sixth- and Fifth-Century b.c. Athens." In *Democracy 2500?: Questions and Challenges*. Archaeological Institute of America, Colloquia and Conference Papers 2, ed. Ian Morris and Kurt A. Raaflaub, 197–216. Dubuque, IA: Kendall/Hunt, 1997.

Ceccarelli, Paola. 1998. *La pirrica nell'antichità greco romana. Studi sulla danza armata*. Pisa: Istituti editoriali e poligrafici internazionali.

Clairmont, Christoph W. 1983. *Patrios Nomos: Public Burial in Athens During the Fifth and Fourth Centuries b.c.* British Archaeol. Reports, Intern. Ser. 161. 2 vols. Oxford: British Archaeological Reports.

Connor, W.R. 1971. *The New Politicians of Fifth-Century Athens*. Princeton: Princeton University Press.

————. 1988. "Early Greek Land Warfare as Symbolic Expression." *Past & Present* 119: 3–29.

————. 1996. "Festivals and Democracy." In *Colloque international: Démocratie athénienne et culture*, ed. Michael B. Sakellariou, 79–89. Athens: Ekdotike Athenon.

Conwell, D.H. 1992. "The Athenian Long Walls: Chronology, Topography and Remains." Ph.D. diss., University of Pennsylvania.

Coulson, W.D.E., O. Palagia, T.L. Shear, H.A. Shapiro, and F.J. Frost, eds. 1994. *The Archaeology of Athens and Attica Under the Democracy*. Oxford: Oxbow.

Croally, N.T. 1994. *Euripidean Polemic: The Trojan Women and the Function of Tragedy*. Cambridge: Cambridge University Press.

Csapo, Eric, and William J. Slater. 1995. *The Context of Ancient Drama.* Ann Arbor: University of Michigan Press.

Davies, John K. 1992. "Greece After the Persian Wars," and "Society and Economy." *Cambridge Ancient History*, 2nd ed, vol. 5: 15–33, 287–305. Cambridge: Cambridge University Press.

de Ste. Croix, Geoffrey E.M. 1972. *The Origins of the Peloponnesian War.* Ithaca: Cornell University Press.

Deubner, Ludwig. 1932 *Attische Feste.* Berlin: Akademie-Verlag.

Dobrov, Gregory W., ed. 1997. *The City as Comedy: Society and Representation in Athenian Drama.* Chapel Hill: University of North Carolina Press.

Dodds, E.R. 1951. *The Greeks and the Irrational.* Berkeley: University of California Press.

Dover, Kenneth J. 1988. "The Freedom of the Intellectual in Greek Society" (1976). In *The Greeks and Their Legacy: Collected Papers*, vol. 2, ed. Kenneth J. Dover, 135–58. Oxford: Blackwell.

Dow, Sterling. 1969. "Some Athenians in Aristophanes." *American Journal of Archaeology* 73: 234–35.

Ducrey, Pierre. 1968. *Le traitement des prisonniers de guerre dans la Gréce antique des origines à la conquête romaine.* Paris: De Boccard.

Ehrenberg, Victor. 1962 *Aristophanes and the People of Athens: A Sociology of Old Attic Comedy.* 3rd ed. New York: Schocken Books.

Erp Taalman Kip, A. Maria van. 1987. "Euripides and Melos." *Mnemosyne* 40: 414–19.

Finley, Moses I. 1982. *Economy and Society in Ancient Greece.* New York: Viking.

———. 1986. "War and Empire." In *Ancient History: Evidence and Models*, Moses I. Finley, ed., 67–87. New York: Viking.

Fornara, Charles W. 1983. *The Nature of History in Ancient Greece and Rome.* Berkeley: University of California Press.

Forrest, W.G. 1975. "Aristophanes and the Athenian Empire." In *The Ancient Historian and His Materials (Essays Presented to C.E. Stevens)*, ed. B. Levick, 17–29. Farnborough, Hants.: Gregg.

Forsén, Björn, and Greg Stanton, eds. 1996. *The Pnyx in the History of Athens.* Helsinki: Foundation of the Finnish Institute at Athens.

Foxhall, Lin. 1993. "Farming and Fighting in Ancient Greece." In *War and Society in the Greek World*, ed. John Rich and Graham Shipley, 134–45. London: Routledge.

Francis, E.D. 1990. *Image and Idea in Fifth-Century Greece: Art and Literature After the Persian Wars*, ed. Michael Vickers. London: Routledge.

Frisch, Hartvig. 1942. *The Constitution of the Athenians: A Philological-Historical Analysis of Pseudo-Xenophon's Treatise De re publica Atheniensium.* Copenhagen: Gyldendal.

Frost, Frank J. 1976. "Tribal Politics and the Civic State." *American Journal of Ancient History* 1: 66–75.

———. 1980. *Plutarch's Themistocles: A Historical Commentary.* Princeton: Princeton University Press.

Furley, William D. 1996. *Andokides and the Herms: A Study of Crisis in Fifth-Century Athenian Religion*, Bulletin Suppl. 65. London: Institute of Classical Studies.

Gabrielsen, Vincent. 1994. *Financing the Athenian Fleet: Public Taxation and Social Relations.* Baltimore: Johns Hopkins University Press.

Garland, Robert. 1987. *The Piraeus from the Fifth to the First Century B.C.* Ithaca: Cornell University Press.

———. 1992. *Introducing New Gods: The Politics of Athenian Religion.* Ithaca: Cornell University Press.

Garnsey, Peter. 1985. "Grain for Athens." In *Crux: Essays Presented to G.E.M. de Ste.*

Croix on His 75th Birthday, ed. Paul Cartledge and F.D. Harvey, 62–75. Exeter: Imprint Academic.

————. 1988. *Famine and Food Supply in the Graeco-Roman World: Responses to Risk and Crisis.* Cambridge: Cambridge University Press.

Gauthier, Phillipe. 1985. *Les cités grecques et leurs bienfaiteurs. Bulletin de Correspondance Hellénique*, Suppl. vol. 12. Athens: Ecole française d'Athénes.

Gelzer, Thomas. 1992. "Die Alte Komödie in Athen und die Basler Fasnacht." In *Klassische Antike und Neue Wege der Kulturwissenschaften: Symposium Karl Meuli*, ed. Fritz Graf, 29–61. Basel: Verlag der schweizerischen Gesellschaft für Volkskunde.

Giovannini, Adalberto. 1997. "La participation des alliés au financement du Parthénon: aparchè ou tribut?" *Historia* 46: 145–57.

Goff, Barbara, ed. 1995. *History, Tragedy, Theory: Dialogues on Athenian Drama*. Austin: University of Texas Press.

Goldhill, Simon. 1990. "The Great Dionysia and Civic Ideology." In *Nothing to Do with Dionysos? Athenian Drama in Its Social Context*, ed. John J. Winkler and Froma I. Zeitlin, 97–129. Princeton: Princeton University Press.

Gomme, Arnold W. [1938] [1962] 1987. "Aristophanes and Politics." *Classical Review* 52: 97–109. Reprinted in Arnold W. Gomme, *More Essays in Greek History and Literature*. Oxford: Blackwell, 70–91; reprint, New York: Garland Publishing.

Gomme, Arnold W., Antony Andrewes, and Kenneth J. Dover. 1970. *A Historical Commentary on Thucydides*. Vol. 4. Oxford: Clarendon Press.

Gregory, Justina. 1991. *Euripides and the Instruction of the Athenians*. Ann Arbor: University of Michigan Press.

Hansen, Mogens H. 1987. *The Athenian Assembly in the Age of Demosthenes*. Oxford: Blackwell.

Hansen, Mogens H., and Kurt A. Raaflaub, eds. 1995. *Studies in the Ancient Greek Polis*, Historia Einzelschrift 95. Stuttgart: Steiner.

Hanson, Victor D. 1989. *The Western Way of War: Infantry Battle in Classical Greece*. New York: Oxford University Press.

————, ed. 1991. *Hoplites: The Classical Greek Battle Experience*. London: Routledge.

————. 1996. "Hoplites into Democrats: The Ideology of Athenian Infantry." In *DEMOKRATIA: A Conversation on Democracies, Ancient and Modern*, ed. Josiah Ober and Charles Hedrick, 289–312. Princeton: Princeton University Press.

Heath, Malcolm. 1987a. *Political Comedy in Aristophanes*, Hypomnemata 87. Göttingen: Vandenhoeck and Ruprecht.

————. 1987b. *The Poetics of Greek Tragedy*. Stanford: Stanford University Press.

Henderson, Jeffrey. 1980. "Lysistrata: The Play and Its Themes." In *Aristophanes: Essays in Interpretation. Yale Classical Studies* 26, ed. Jeffrey Henderson, 153–218.

————. 1987. *Aristophanes' Lysistrata, Edited with Introduction and Commentary*. Oxford: Clarendon Press.

————. 1993. "Comic Hero Versus Political Elite." In *Tragedy, Comedy and the Polis*, ed. Alan H. Sommerstein, Stephen Halliwell, Jeffrey Henderson, and Bernhard Zimmermann, 307–20. Bari: Levante Editori.

————. 1998. "Attic Old Comedy, Frank Speech, and Democracy." In *Democracy, Empire, and the Arts in Fifth-Century Athens*, ed. Deborah Boedeker and Kurt Raaflaub, 255–73. Cambridge: Harvard University Press.

Herington, C.J. 1955. *Athena Parthenos and Athena Polias: A Study in the Religion of Periclean Athens*. Manchester: Manchester University Press.

————. 1991. "The Poem of Herodotus." *Arion* 3rd ser. 1.3: 5–16.

Hoepfner, Wolfram, and Ernst-Ludwig Schwandner. 1994. *Haus und Stadt im klassischen Griechenland*. 2nd ed. Munich: Deutscher Kunstverlag.

Hölscher, Tonio. 1973. *Griechische Historienbilder des 5. und 4. Jahrhunderts v. Chr.* Würzburg: Triltsch.

―――. 1998. "Images and Political Identity: The Case of Athens." In *Democracy, Empire, and the Arts in Fifth-Century Athens*, ed. Deborah Boedeker and Kurt Raaflaub, 153– 83. Cambridge: Harvard University Press.

―――. 1991. "The City of Athens: Space, Symbol, Structure." In *Athens and Rome, Florence and Venice: City-States in Classical Antiquity and Medieval Italy*, ed. Anthony Molho, Kurt Raaflaub, and Julia Emlen, 355–80. Stuttgart: Steiner; and Ann Arbor: University of Michigan Press.

―――. 1996. "Politik und Öffentlichkeit im demokratischen Athen: Räume, Denkmäler, Mythen." In *Colloque international: Démocratie athénienne et culture*, ed. Michael B. Sakellariou, 171–87. Athens: Ekdotike Athenon.

Humphreys, Sally C. 1978. *Anthropology and the Greeks*. London: Routledge and Kegan Paul.

Jameson, Michael H. 1991. "Sacrifice Before Battle." In *Hoplites: The Classical Greek Battle Experience*, ed. Victor D. Hanson, 197–227. London: Routledge.

―――. 1994. "The Ritual of the Athena Nike Parapet." In *Ritual, Finances, Politics: Athenian Democratic Accounts Presented to David Lewis*, ed. Robin Osborne and Simon Hornblower, 307–24. Oxford: Clarendon Press.

―――. 1997. "Religion in the Athenian Democracy." In *Democracy 2500?: Questions and Challenges*. Archaeological Institute of America, Colloquia and Conference Papers 2, ed. Ian Morris and Kurt A. Raaflaub, 171–95. Dubuque, IA: Kendall/Hunt.

Jenkins, Ian. 1994. *The Parthenon Frieze*. Austin: University of Texas Press.

Jordan, Borimir. 1975. *The Athenian Navy in the Classical Period*. Berkeley: University of California Press.

Kagan, Donald. 1974. *The Archidamian War*. Ithaca: Cornell University Press.

―――. 1981. *The Peace of Nicias and the Sicilian Expedition*. Ithaca: Cornell University Press.

―――. 1987. *The Fall of the Athenian Empire*. Ithaca: Cornell University Press.

Kallet (-Marx), Lisa. 1998. "Accounting for Culture in Fifth-Century Athens." In *Democracy, Empire, and the Arts in Fifth-Century Athens*, ed. Deborah Boedeker and Kurt Raaflaub, 43–58. Cambridge: Harvard University Press.

Kallet-Marx, Lisa. 1993a. *Money, Expense, and Naval Power in Thucydides' History 1– 5.24*. Berkeley: University of California Press.

―――. 1993b. "Thucydides 2.45.2 and the Status of War Widows in Periclean Athens." In *Nomodeiktes: Greek Studies in Honor of Martin Ostwald*, ed. Ralph M. Rosen and Joseph Farrell, 133–43. Ann Arbor: University of Michigan Press.

Karavites, Peter. 1982. *Capitulations and Greek Interstate Relations: The Reflection of Humanistic Ideals in Political Events*, Hypomnemata 71. Göttingen: Vandenhoeck and Ruprecht.

Kierdorf, Wilhelm. 1966. *Erlebnis und Darstellung der Persekriege. Studien zu Simonides, Pindar, Aischylos und den attischen Rednern*, Hypomnemata 16. Göttingen: Vandenhoeck and Ruprecht.

Knox, Bernard M.W. 1979. *Word and Action: Essays on the Ancient Theater*. Baltimore: Johns Hopkins University Press.

Kolb, Frank. 1979. "Polis und Theater." In *Das griechische Drama*, ed. G.A. Seeck, 504– 45. Darmstadt: Wissenschaftliche Buchgesellschaft.

Konstan, David. 1995. *Greek Comedy and Ideology*. New York: Oxford University Press.

Kovacs, David. 1987. *The Heroic Muse*. Baltimore: Johns Hopkins University Press.

Kron, Uta. 1976. *Die zehn attischen Phylenheroen: Geschichte, Mythos, Kult und Darstellungen*. Berlin: Gebrüder Mann.

Kyle, Donald G. 1992. "The Panathenaic Games: Sacred and Civic Athletics." In *Goddess and Polis: The Panathenaic Festival in Ancient Athens*, ed. Jennifer Neils, 77–101. Hanover: Hood Museum of Art, Dartmouth College; and Princeton: Princeton University Press.

Landfester, M. 1975. "Aristophanes und die politische Krise Athens." In *Krisen in der Antike: Bewusstsein und Bewältigung*, ed. Géza Alföldy et al., 27–45. Düsseldorf: Schwann.

Lane, Eugene N. ed. 1996.. *Cybele, Attis and Related Cults: Essays in Memory of M.J. Vermaseren*. Leiden: Brill.

Lind, Hermann. 1985. "Neues aus Kydathen: Beobachtungen zum Hintergrund der 'Daitales' und der 'Ritter' des Aristophanes." *Museum Helveticum* 42: 249–61.

Lonis, Raoul. 1979. *Guerre et religion en Grèce à l'époque classique. Recherches sur les rites, les dieux, l'idéologie de la victoire*. Paris: Belles Lettres.

Lonsdale, Steven H. 1993. *Dance and Ritual Play in Greek Religion*. Baltimore: Johns Hopkins University Press.

Loraux, Nicole. 1986. *The Invention of Athens. The Funeral Oration in the Classical City*. Cambridge: Harvard University Press.

———. 1991. "Aristophane, les femmes d'Athènes et le théâtre." In *Aristophane*, ed. Olivier Reverdin and Bernard Grange, Entretiens sur l'antiquité classique 38. Vandoeuvres-Geneva: Fondation Hardt.

McCredie, James R. 1971. "Hippodamos of Miletus." In *Studies Presented to George M.A. Hanfmann*, ed. David G. Mitten et al., 95–100. Mainz: von Zabern.

MacDowell, Douglas M. 1962. *Andokides, On the Mysteries*. Oxford: Clarendon Press.

Mark, Ira S. 1993. *The Sanctuary of Athena Nike in Athens: Architectural Stages and Chronology*. Hesperia Suppl. 26. Princeton: American School of Classical Studies at Athens.

Marrou, Henri I. 1956. *A History of Education in Antiquity*, tr. George Lamb. London: Seed and Ward.

Meier, Christian. 1987. "Historical Answers to Historical Questions: The Origins of History in Ancient Greece." In *Herodotus and the Invention of History. Arethusa* 20, ed. Deborah Boedeker and John Peradotto, 41–57.

———. 1990a. *The Greek Discovery of Politics*, tr. David McLintock. Cambridge: Harvard University Press.

———. 1990b. "Die Rolle des Krieges im klassischen Athen." *Historische Zeitschrift* 251: 555–605.

———. 1993. *The Political Art of Greek Tragedy*, tr. Andrew Webber. Baltimore: Johns Hopkins University Press.

Meiggs, Russell. 1972. *The Athenian Empire*. Oxford: Clarendon Press.

Merkelbach, Reinhold, and C.H. Youtie. 1968. "Ein Michigan-Papyrus über Theramenes." *Zeitschrift für Papyrologie und Epigraphik* 2: 161–69.

Michelini, Ann N. 1994. "Political Themes in Euripides' *Suppliants*." *American Journal of Philology* 115: 219–52.

Mikalson, Jon D. 1975. *The Sacred and Civil Calendar of the Athenian Year*. Princeton: Princeton University Press.

———. 1983. *Athenian Popular Religion*. Chapel Hill: University of North Carolina Press.

Miller, Stephen G. 1995. "Old Metroon and Old Bouleuterion in the Classical Agora of Athens." In *Studies in the Ancient Greek Polis*, Historia Einzelschrift 95, ed. Mogens H. Hansen and Kurt A. Raaflaub, 133–56. Stuttgart: Steiner.

Moles, J.L. 1993. "Truth and Untruth in Herodotus and Thucydides." In *Lies and Fiction in the Ancient World*, ed. Christopher Gill and T.P. Wiseman, 88–121. Austin: University of Texas Press.

Momigliano, Arnaldo. 1944. "Sea-Power in Greek Thought." *Classical Review* 58: 1–7.

Monoson, S. Sara. 1994. "Citizen as Erastes: Erotic Imagery and the Idea of Reciprocity in the Periclean Funeral Oration." *Political Theory* 22: 253–76.

Moore, J.M. 1975. *Aristotle and Xenophon on Democracy and Oligarchy.* Berkeley: University of California Press.

Morris, Ian. 1992. *Death-Ritual and Social Structure in Classical Antiquity.* Cambridge: Cambridge University Press.

Morris, Ian, and Kurt A. Raaflaub, eds. 1997. *Democracy 2500?: Questions and Challenges.* Archaeological Institute of America, Colloquia and Conference Papers 2. Dubuque, IA: Kendall/Hunt.

Morrison, J.S., and J.F. Coates. 1986. *The Athenian Trireme: The History and Reconstruction of an Ancient Greek Warship.* Cambridge: Cambridge University Press.

Munn, Mark. 2000. *The School of History: Athens in the Age of Socrates.* Berkeley: University of California Press.

Neils, Jennifer, ed. 1992. *Goddess and Polis: The Panathenaic Festival in Ancient Athens.* Hanover: Hood Museum of Art, Dartmouth College; and Princeton: Princeton University Press.

Newinger, Hans-Joachim. 1980. "War and Peace in the Comedy of Aristophanes." In *Aristophanes: Essays in Interpretation. Yale Class. St.* 26, ed. Jeffrey Henderson, 219–37.

Nilsson, Martin P. [1951] 1972. *Cults, Myths, Oracles, and Politics in Ancient Greece.* Lund: Gleerup. Reprint New York: Cooper Square Publishers.

———. 1967. *Geschichte der griechischen Religion,* 3rd ed., vol. 1. Munich: Beck.

Norwood, G. 1930. "The Babylonians of Aristophanes." *Classical Philology* 25: 1–10.

Ober, Josiah. [1985] 1996. "Thucydides, Pericles, and the Strategy of Defense." In *The Craft of the Ancient Historian: Essays in Honor of Chester G. Starr,* ed. John W. Eadie and Josiah Ober, 171–88. Lanham, MD: University Press of America. Reprinted in *The Athenian Revolution: Essays on Ancient Greek Democracy and Political Theory,* ed. Josiah Ober, 72–85. Princeton: Princeton University Press.

———. 1989. *Mass and Elite in Democratic Athens: Rhetoric, Ideology, and the Power of the People.* Princeton: Princeton University Press.

Ober, Josiah, and Charles Hedrick, eds. 1996. *DEMOKRATIA: A Conversation on Democracies, Ancient and Modern.* Princeton: Princeton University Press.

Ober, Josiah, and Barry S. Strauss. 1990. "Drama, Political Theory, and the Discourse of Athenian Democracy." In *Nothing to Do with Dionysos? Athenian Drama in Its Social Context,* ed. John J. Winkler and Froma I. Zeitlin, 237–70. Princeton: Princeton University Press.

Osborne, Robin. 1994. "Democracy and Imperialism in the Panathenaic Procession: The Parthenon Frieze in Its Context." In *The Archaeology of Athens and Attica Under the Democracy,* ed. W.D.E. Coulson, O. Palagia, T.L. Shear, H.A. Shapiro, and F.J. Frost, 143–50. Oxford: Oxbow.

Osborne, Robin, and Simon Hornblower, eds. 1994. *Ritual, Finances, Politics: Athenian Democratic Accounts Presented to David Lewis.* Oxford: Clarendon Press.

Ostwald, Martin. 1986. *From Popular Sovereignty to the Sovereignty of Law: Law, Society, and Politics in Fifth-Century Athens.* Berkeley: University of California Press.

Parker, Robert. 1994. "Athenian Religion Abroad." In *Ritual, Finances, Politics: Athenian Democratic Accounts Presented to David Lewis,* ed. Robin Osborne and Simon Hornblower, 339–46. Oxford: Clarendon Press.

———. 1996. *Athenian Religion: A History.* Oxford: Clarendon Press.

Pelling, Christopher ed. 1997. *Greek Tragedy and the Historian.* Oxford: Clarendon Press.

Pickard-Cambridge, Arthur. [1968] 1988. *The Dramatic Festivals of Athens,* 2nd ed., rev. John Gould and D.M. Lewis. Oxford: Clarendon Press. Reprint with supplement and corrections, Oxford: Clarendon Press.

Podlecki, Anthony J. 1977. "Herodotus in Athens?" In *Greece and the Eastern Mediterranean in Ancient History and Prehistory: Studies Presented to Fritz Schachermeyr*, ed. Konrad H. Kinzl, 246–65. Berlin: De Gruyter.

Powell, Anton. 1988. *Athens and Sparta: Constructing Greek Political and Social History from 478 B.C.* London: Routledge.

Pritchett, W. Kendrick. 1971–1991 *The Greek State at War.* 5 vols. Berkeley: University of California Press.

Raaflaub, Kurt A. 1985. *Die Entdeckung der Freiheit*, Vestigia 37. Munich: Beck.

———. 1986. "The Conflict of the Orders in Archaic Rome: A Comprehensive and Comparative Approach." In *Social Conflicts in Archaic Rome: New Perspectives on the Conflict of the Orders*, ed. Kurt A. Raaflaub, 1–51. Berkeley: University of California Press.

———. 1987. "Herodotus, Political Thought, and the Meaning of History." In *Herodotus and the Invention of History. Arethusa* 20, ed. Deborah Boedeker and John Peradotto, 221–48.

———. 1988. "Politisches Denken im Zeitalter Athens." In *Pipers Handbuch der politischen Ideen*, vol. 1: *Frühe Hochkulturen und europäische Antike*, ed. Iring Fetscher and Herfried Münkler, 273–368. Munich: Piper.

———. [1989] 1990. "Contemporary Perceptions of Democracy in Fifth-Century Athens." *Classica et Mediaevalia* 40: 33–70. Also in *Aspects of Athenian Democracy*, ed. W. Robert Connor et al., 37–70. Copenhagen: Museum Tusculanum Press.

———. 1990. "Expansion und Machtbildung in frühen Polis-Systemen." In *Staat und Staatlichkeit in der frühen römischen Republik*, ed. Walter Eder, 511–45. Stuttgart: Steiner.

———. 1994. "Democracy, Power, and Imperialism in Fifth-Century Athens." In *Athenian Political Thought and the Reconstruction of American Democracy*, ed. J. Peter Euben, John R. Wallach, and Josiah Ober, 103–46. Ithaca: Cornell University Press.

———. 1997. "Power in the Hands of the People: Foundations of Athenian Democracy," and "The Thetes and Democracy." In *Democracy 2500?: Questions and Challenges*, Archaeological Institute of America, Colloquia and Conference Papers 2, ed. Ian Morris and Kurt A. Raaflaub, 31–66, 87–103. Dubuque, IA: Kendall/Hunt.

———. 1998. "The Transformation of Athens in the Fifth Century B.C." In *Democracy, Empire, and the Arts in Fifth-Century Athens*, ed. Deborah Boedeker and Kurt Raaflaub, 15–41. Cambridge: Harvard University Press.

Raubitschek, Antony E. 1949. *Dedications from the Athenian Akropolis*. Cambridge: Harvard University Press.

Reckford, Kenneth J. 1987. *Aristophanes' Old and New Comedy*. Chapel Hill: University of North Carolina Press.

Reden, Sitta von. 1995. "The Piraeus—A World Apart." *Greece & Rome* 42: 24–37.

Redfield, James. 1990. "Drama and Community: Aristophanes and Some of His Rivals." In *Nothing to Do with Dionysos? Athenian Drama in Its Social Context*, ed. John J. Winkler and Froma I. Zeitlin, 314–35. Princeton: Princeton University Press.

Rhodes. Peter J. 1972. *The Athenian Boule*. Oxford: Clarendon Press.

———. 1992. "The Delian League to 449 B.C.," and "The Athenian Revolution." In *Cambridge Ancient History*, 2nd ed., vol. 5: 34–61. Cambridge: Cambridge University Press.

Rich, John, and Graham Shipley, eds. 1993. *War and Society in the Greek World*. London: Routledge.

Ridgway, Brunilde Sismondo. 1992. "Images of Athena on the Akropolis." In *Goddess and Polis: The Panathenaic Festival in Ancient Athens*, ed. Jennifer Neils, 199–242. Hanover: Hood Museum of Art, Dartmouth College; and Princeton: Princeton University Press.

Ridley, Ronald T. 1979. "The Hoplite as Citizen: Athenian Military Institutions in Their Social Context." *L'antiquité classique* 48: 508–48.

Roberts, Jennifer T. 1994. *Athens on Trial: The Antidemocratic Tradition in Western Thought*. Princeton: Princeton University Press.

Robertson, Noel. 1996. "The Ancient Mother of the Gods: A Missing Chapter in the History of Greek Religion." In *Cybele, Attis and Related Cults: Essays in Memory of M.J. Vermaseren*, ed. Eugene N. Lane, 239–304. Leiden: Brill.

Roller, Lynn. 1996. "Reflections of the Mother of the Gods in Attic Tragedy." In *Cybele, Attis and Related Cults: Essays in Memory of M.J. Vermaseren*, ed. Eugene N. Lane, 305–21. Leiden: Brill.

Rosenbloom, David. 1995. "Myth, History, and Hegemony in Aischylos." In *History, Tragedy, Theory: Dialogues on Athenian Drama*, ed. Barbara Goff, 91–130. Austin: University of Texas Press.

Rossellini, Michéle. 1979. "Lysistrata: une mise en scène de la féminité." In *Aristophane, les femmes et la cité*, Les cahiers de Fontenay 17: 11–32.

Rüpke, J. 1990. *Domi militiae: Die religiöse Konstruktion des Krieges in Rom*. Suttgart: Steiner.

Ryder, T.T.B. 1965. *Koine Eirene: General Peace and Local Independence in Ancient Greece*. Oxford: Clarendon Press.

Saïd, Suzanne. 1998. "Tragedy and Politics." In *Democracy, Empire, and the Arts in Fifth-Century Athens*, ed. Deborah Boedeker and Kurt Raaflaub, 275–95. Cambridge: Harvard University Press.

Sakellariou, Michael B., ed. 1996. *Colloque international: Démocratie athénienne et culture*. Athens: Ekdotike Athenon.

Schmid W., and O. Stählin. 1946. *Geschichte der griechischen Literatur*. Pt. 1 vol. 4. Munich: Beck.

Scodel, Ruth. 1980. *The Trojan Trilogy of Euripides*, Hypomnemata 60. Göttingen: Vandenhoeck and Ruprecht.

Seaman, Michael G. 1997. "The Athenian Expedition to Melos in 416." *Historia* 46: 385–418.

Seel, Otto. 1960. *Aristophanes oder Versuch über Komödie*. Stuttgart: Klett.

Seiler, Florian. 1986. *Die griechische Tholos*. Mainz: von Zabern.

Shapiro, H. Alan. 1989. *Art and Cult Under the Tyrants in Athens*. Mainz: von Zabern.

———. 1994. "Religion and Politics in Democratic Athens." In *The Archaeology of Athens and Attica Under the Democracy*, ed. W.D.E. Coulson, O. Palagia, T.L. Shear, H.A. Shapiro, and F.J. Frost, 123–30. Oxford: Oxbow.

Shear, T. Leslie Jr. 1970. "The Monument of the Eponymous Heroes in the Athenian Agora." *Hesperia* 39: 145–222.

———. 1995. "Bouleuterion, Metroon, and the Archives at Athens." In *Studies in the Ancient Greek Polis*, Historia Einzelschrift 95, ed. Mogens H. Hansen and Kurt A. Raaflaub, 157–90. Stuttgart: Steiner.

Shimron, Binyamin. 1989. *Politics and Belief in Herodotus*, Historia Einzelschrift 58. Stuttgart: Steiner.

Simms, Robert M. 1975. "The Eleusinia in the Sixth to Fourth Centuries B.C." *Greek, Roman, and Byzantine Studies* 16: 269–79.

Simms, Ronda M. 1988. "The Cult of the Thracian Goddess Bendis in Athens and Attica." *Ancient World* 18: 59–76.

Smarczyk, Bernhard. 1990. *Untersuchungen zur Religionspolitik und politischen Propaganda Athens im Delisch-Attischen Seebund*. Munich: tuduv Verlagsgesellschaft.

Sommerstein, Alan H., Stephen Halliwell, Jeffrey Henderson, and Bernhard Zimmermann, eds. 1993. *Tragedy, Comedy and the Polis*. Bari: Levante Editori.

Spahn, Peter. 1980. "Oikos und Polis. Beobachtungen zum Prozess der Polisbildung bei Hesiod, Solon und Aischylos." *Historische Zeitschrift* 231: 529–64.

Spence, I.G. 1990. "Perikles and the Defence of Attika during the Peloponnesian War." *Journal of Hellenic Studies* 110: 91–109.

————. 1993. *The Cavalry of Classical Greece.* Oxford: Clarendon Press.

Stark, Isolde. 1975. "Das Verhältnis des Aristophanes zur Demokratie der athenischen Polis." *Klio* 57: 329–64.

Starr, Chester G. 1979a. "Thucydides on Sea Power." *Mnemosyne* 31: 343–50.

————. 1979b. "Religion and Patriotism in Fifth-Century Athens." In *Panathenaia: Studies in Athenian Life and Thought in the Classical Age*, ed. T.E. Gregory and A.J. Podlecki, 11–22. Lawrence: University of Kansas Press.

Stevens, P.T. 1956. "Euripides and the Athenians." *Journal of Hellenic Studies* 76: 87–94.

Storye, Ian C. 1990. "Dating and Re-dating Eupolis." *Phoenix* 44: 1–30.

Strasburger, Hermann. 1958. "Thukydides und die politische Selbstdarstellung der Athener." *Hermes* 86: 17–40.

Strauss, Barry S. 1986. *Athens After the Peloponnesian War: Class, Faction and Policy, 403–386 B.C.* Ithaca: Cornell University Press.

————. 1993. *Fathers and Sons in Athens: Ideology and Society in the Era of the Peloponnesian War.* Princeton: Princeton University Press.

————. 1996. "The Athenian Trireme: School of Democracy." In *DEMOKRATIA: A Conversation on Democracies, Ancient and Modern*, ed. Josiah Ober and Charles Hedrick, 313–25. Princeton: Princeton University Press.

Stupperich, Reinhard. 1977. *Staatsbegräbnis und Privatgrabmal im klassischen Athen*, Diss. Münster.

Taaffe, Lauren K. 1993. *Aristophanes and Women.* London: Routledge.

Thomas, Rosalind. 1989. *Oral Tradition and Written Record in Classical Athens.* Cambridge: Cambridge University Press.

Thompson, Dorothy B. 1944. "The Golden Nikai Reconsidered." *Hesperia* 13: 173–209.

Travlos, John. 1971. *Pictorial Dictionary of Ancient Athens.* New York: Praeger.

Vaio, John. 1973. "The Manipulation of Theme and Action in Aristophanes' Lysistrata." *Greek, Roman and Byzantine Studies* 14: 369–80.

Vernant, Jean-Pierre, ed. 1968. *Problèmes de la guerre en Grèce ancienne.* The Hague: Mouton.

Vidal-Naquet, Pierre. 1986. *The Black Hunter: Forms of Thought and Forms of Society in the Greek World*, tr. A. Szegedy-Maszak. Baltimore: Johns Hopkins University Press.

————. 1992. "Retour au Chasseur Noir." In *La Grèce ancienne*, vol. 3: *Rites de passage et transgressions*, Jean-Pierre Vernant and P.Vidal-Naquet, 215–51. Paris: Editions du Seuil.

West, William C. 1970. "Saviors of Greece." *Greek, Roman and Byzantine Studies* 11: 271–82.

Winkler, John J., and Froma I. Zeitlin, eds. 1990. *Nothing to Do with Dionysos? Athenian Drama in Its Social Context.* Princeton: Princeton University Press.

Wycherley, R.E. 1957. *The Athenian Agora*, vol. 3: *Literary and Epigraphical Testimonia.* Princeton: American School of Classical Studies at Athens.

————. 1978. *The Stones of Athens.* Princeton: Princeton University Press.

Yunis, Harvey. 1991. "How Do the People Decide? Thucydides on Periclean Rhetoric and Civic Instruction." *American Journal of Philology* 112: 179–200.

Zeitlin, Froma I. 1990. "Thebes: Theater of Self and Society in Athenian Drama.," In *Nothing to Do with Dionysos? Athenian Drama in Its Social Context*, ed. John J. Winkler and Froma I. Zeitlin, eds., 130–67. Princeton: Princeton University Press.

Zuntz, Günther. 1955. *The Political Plays of Euripides.* Manchester: Manchester University Press.

15

Characters and Characteristics of Korean War Novels

Dong-Wook Shin

Prose fiction accounts of the Korean War (June 25, 1950–July 13, 1953), like their composers, are finally individual. And yet they all attempt to describe, to somehow make sense of, the same historical event. It is from this event and the perceptions of it that the present study will depart. If there can be said to be a common denominator for the disparate fictional accounts of the war, it would be their common awareness of it as a national disaster. Some authors chose to foreground the international context, the clash of the Soviet and U.S. ideologies and armies played out on Korean soil. Korea and its citizens were mere pawns, which at once made the war ineluctable and, at least partially, absolved individual participants for their often savage acts. Others took a more microscopic view and treated the war through its effects, physical and emotional, on individuals or isolated rural communities. Finally, and perhaps predictably, the majority of works opted for a balance between these two poles. None, however, could avoid affording it the serious treatment it deserved as a tragedy that scarred an entire nation.

An entire generation of writers had already come of age prior to liberation from Japan and subsequent Korean independence in 1945. Many of these authors were called upon to serve in 1950. Some were fortunate enough to be drafted for their literary skills, others were expected to fight, but all experienced directly the personal and national dislocation brought on by the conflict. But they were hardened by the trials of colonial government, world war, and the social upheaval following liberation; their take on this latest calamity reflects a certain stoic reserve. Yŏm Sang-sŏp (1897–1963), one of Korea's most celebrated novelists, provides a representative example of his generation's take on the war. Yŏm was a South Korean navy lieutenant commander while he worked on his novel *A Sudden Shower* (1952–53).

The story unfolds over a three-month period and deals with the distorted lives and experiences of four main characters detained in Seoul following the capital's seizure by North Korean forces. The protagonists, not so ironically, are all affiliated with the Korea-U.S. Trading Company: they are company president Kim Hak-su; Kang Sun-jae, his secretary; Sin Yŏng-sik, a section chief; and Im Il-sŏk, a laborer. Under the newly ascendant communists, their fortunes have been reversed; the irony increases as Kim secretly prepares to abscond with what were his own company's funds. He is discovered and reported, however, by Im and then dragged off by the North Korean military police. A parallel thread in the story also serves to demonstrate the dislocation from previously well-ordered lives. Sin breaks his promise to his fiancée, who has already fled south, and falls in love with Kang Sun-jae.

The story does point out a few systematic injustices accompanying the war: the inhuman conditions of soldiers on both sides, confiscation of private property, mandatory conscription, and so on. But as to the historical significance of the war, and of the international power struggle that went with it, the story remains mute. In fact, *A Sudden Shower* adequately sums up the way Seoul residents thought of the war, at least in Yŏm's rather impassive view. Shortly after the North Koreans capture the city the narrator relates, "They weren't the footsteps of people with some urgent business to attend to, they were those of window shoppers who just came out to have a look around."[1] Such stoicism does not apply only to the denizens of the capital; the narrator and the protagonists are equally unconvinced by the ostensibly ideological motives for the savagery of their northern brethren. Yŏm's characters, far from being swayed by strident propaganda, pejoratively refer to the repetitive and hateful "vibrating voices" and to feeling "nauseated at their public lectures."[2] Thus, even while the Korean War violently shifts the otherwise well-planned direction of each person's life, the story confirms the somewhat passive wisdom of a generation that had already faced and overcome many calamities and was quietly confident that this too would pass.

The stories of Hwang Sun-wŏn (1915–2000), a widely anthologized and translated author, share certain aspects with that of Yŏm, while at the same time showing a deeper, more personal attachment to and concern for characters and their predicaments. Hwang wrote several acclaimed war stories, including *Cranes* (1953) and *Trees on a Cliff* (1960). In *Cranes*, he tells the story of two close friends, Sŏng-sam and Tŏk-jae, who grew up together in a small village knowing nothing of ideology. When the North Korean army occupies the village, Tŏk-jae is appointed vice chairman of the Village Farmers' Union. But after the village is reclaimed by the South Korean forces, he is arrested for his treasonous activity by Sŏng-sam, who is now working as a member of the security police. Following the arrest and during the escort, however, Sŏng-sam recalls how they used to catch cranes in those very fields and releases his friend, telling him to go hunt for cranes.

It is obvious that the writer is determined to portray the possibility of human reconciliation as more important than the political significance of ideological conflict in the village. Finally, at least for these two friends, it is complete trust in their communal existence in the village devoid of anything resembling political inclination that resolves their conflict. And, although not cynical enough to use a brief bout of rain as metaphor, Hwang does share with Yŏm both a certain skepticism concerning the staying power of imported ideology on Korean soil and a measured confidence in the nation's endurance and ability to survive yet another tragedy.

Sŏnu Hwi (1922–1986) is perhaps the last author who can be said to have cut his teeth under the Japanese and to have experienced the war as an adult. He treats the subject of the Korean War in several stories, including *Flowers of Fire* (1957), *Vengeance* (1958), and *The Tale of the Ssarigol Pass* (1962). *Flowers of Fire* is the story of protagonist Ko Hyŏn's firm determination and resolute action set against the backdrop of the Soviet army's crossing the northern border into Korea and the heightened ideological conflict on the peninsula. We need to understand the different philosophical positions of his father and grandfather in order to fully appreciate Ko Hyŏn's actions of disaffiliation from both. His father, perhaps too idealistically, fought against the Japanese before Liberation and, as a consequence, was forced to flee the country. His grandfather, on the other hand, represents the opposite position; he believes that compromise and conformity, regardless of a regime's legitimacy, are the best policies for ensuring personal prosperity, and always encourages his grandson to behave accordingly. However, Ko is neither idealist nor opportunist. Able to see clearly the shortcomings of both generations before him, he opts against blindly taking a side and hiding from the conflict altogether. Rather, he takes what is perhaps the most dangerous of all positions; he stands in the middle, opposed to the crimes perpetrated by both sides against the nation and humanity in the name of ideology. The work is regarded as one of the finest products of the so-called Literature of Engagement, and in seriously taking up, rather than simply dismissing, the question of ideology, it provides a good bridge to the next generation of writers.

Most Korean writers who began their careers in the late 1950s and early 1960s share a common denominator: their direct confrontation with the Korean War. Of course, there is nothing remarkable in that alone. It has already been discussed above that many writers from the previous generation were just as concerned about their war experiences. What is remarkable then about these new writers is that quite a number of their works consistently confront the problems of ideological commitment that their generation faced. Not as callous or jaded as the generation that preceded them, these authors truly struggled with the question of involvement and, even more importantly, with which side. O Sang-wŏn, Sŏ Ki-wŏn, Ch'oe In-hun, and Ch'oe Il-lam belong to this new group of politically conscious writers.

O Sang-wŏn (1930–1985) deals with the strength of an intellectual South Korean platoon leader's commitment throughout a series of terrible ordeals in *A Respite* (1955). While his platoon carries out its long and arduous retreat to the south through the harsh winter, casualties claim his men one by one until he alone remains. His situation grows worse as he is captured by enemy forces and interrogated. His life is repeatedly threatened, yet he refuses to bend. Immediately before his execution he reaffirms his conviction in individual freedom as he says, "I am pleased, pleased that till death I was never machinery nor instrument . . . glad that I lived my life as a human being and now die as the same."[3] Shot in the back, he falls to die in the pure, white snow. The frequent depiction of both the harsh cruelty of winter and the untainted purity of snow in this novel seems to be ironically dualistic in meaning; it adds to the unmitigated cruelty of the war that threatens to weaken the protagonist's determination even as it signifies the impregnable nature of the latter.

Sŏ Ki-wŏn (1930–), another writer active in the 1950s, produced some of the most memorable and realistically detailed war prose ever written with stories such as *A Blank Map* (1956), *An Embrace in This Deep Night* (1960), and *An Eve* (1961). We would, however, do him a great injustice if we were only to appreciate his fine craftsmanship, for it is his penetrating analysis of the absurdities of the war and of the psychology of young soldiers that establishes him as one of the representative writers of the 1950s.

In *An Eve* he creates in Sŏng-ho a fitting spokesman for a generation. Despite his utter disillusionment with the war and subsequent mental agony, Sŏng-ho never loses his cool, detached objectivity. He does not engage in simplistic arguments against war; for him things are much less simple and much more personal. At one point we are even told that "Sŏng-ho might be willing to give up his life and even enjoy his death if the war were not against our own people. Any kind of war except this would be fine with him, even if he had to be on the side of some colonial invaders."[4] But despite his words, he also perceives that the war is the inevitable outcome of the complex unfolding of international political conditions. This objective understanding of the war sustains his detached and somewhat stoical stance on the war throughout the story. In seeming contradition, he continues to fight only because he wants the war to end with a minimum of damage to the nation. His complex views on the situation, though far from perfect, are portrayed as realistic and are contrasted to the naiveté and even cowardice of his civilian friends. Overall, the story is considered to be the forerunner of its kind in portraying the war in an objective manner and with all its contradictory details.

Ch'oe In-hun (1936–) is another author known for his complex depiction of the war and its many after effects. By far the most popular and widely acclaimed novel dealing with the Korean War is his *The Square* (1960). This work instantly established Ch'oe as one of the most critically acclaimed writers in modern Korean literary history. In fact, it is generally agreed upon among critics that no other literary achievement has ever surpassed *The Square* in terms of its uninhibited yet carefully thought out expression of the complex ideological aspects of the hostile relationship bewteen North and South Korea and the subsequent bloody war.

Yi Myŏng-jun begins the story an innocent and harmless university student of philosophy, but is dragged into the conflict when he is arrested and tortured by the police for his possible connection with his communist father, who has gone to North Korea. His individuality is further violated, and the absurdity of the situation magnified, when his interrogator states: "It is your duty as a Korean citizen to take responsibility for his crime. It is also your duty in a way as his son."[5] Myŏng-jun suddenly finds himself stripped of his right to individual freedom and then brought out into the symbolic "square," a politically stigmatized space in which the worth of a human being is judged according to his ideological inclinations.

Overcome partially by fear and despair and partially by his naive longing for an ideal life, he defects to the north. There he finds only further disillusionment. His father, already remarried and leading the central section of the party propaganda apparatus, more closely resembles a common, disgusting bourgeois than a committed revolutionary. Furthermore, he is utterly disappointed to find neither revolutionary fervor nor creative criticism for the party in the north.

After losing his job as a reporter to politics and losing his lover to the war, Myŏng-jun makes a heroic choice. At the close of the war, standing on the 38th parallel, the line dividing one nation into two hostile regimes, he must choose to which he will pledge his allegiance. But, having seen the moral and ideological bankruptcy of both, he chooses neither and boards a ship bound for India. This "nationalism," which transcends the physical nation, is a first in modern Korean literature, and takes Ko Hyŏn's patriotic stance against both regimes even one step further toward a universal human compassion.

Ch'oe Il-lam (1932–) displays this humanity and love beyond national borders in his *Fellow Travellers* (1959). This is the moving story of an African-American soldier lost during the Korean War. Private Dops is separated from his unit during heavy enemy artillery fire and wanders the mountains for several days. Overcome by hunger and cold, but most of all by the desperate need for companionship, he comes down from the mountains and finds a baby who has been deserted in an old house. Dops gradually builds an attachment to the baby and decides to take it with him. They get snowed in, however, before they can make it out of enemy territory. Dops, finding himself slowly subdued by hunger and cold, tries everything he can to save the baby. But this story has no happy ending in the traditional sense; they both eventually die in the forest. Dops, however, does not despair as he whispers, "Oh, my little baby. We might have died alone, never having known each other. . . . It's horrible even to think about it. We must thank God for having each other."[6] Ch'oe does an admirable job of uncovering the possibility of human love even in the midst of the deterministic force of the war.

As seen above, the subject matter of Korean War novels varies over time. In this section, then, our survey continues to move with these changes into the 1970s and 1980s. During these decades we see young writers attempting to further expand the thematic boundaries of the Korean War. Still, there is as much continuity as there are abrupt shifts in direction, with their predominant concerns being for the possibility of healing and reunifying the nation. These are the natural outcome of their continuation of and gradual expanding upon the objective, sober approaches to the war initiated by the previous generations of writers.

Predictably, the subject of war trauma continues to fascinate many writers just as it had in past decades. *Abe's Family* (1979) by Chŏn Sang-guk (1940–) represents an attempt to examine the possibility of reconciliation with the painful past of an entire country by delving into the traumatic records of a single family. Chin-ho's is one of so many Korean emigrant families living in America, forced by postwar poverty and despair to leave Korea. When Chin-ho joins the U.S. army and learns he is to be stationed in Korea, he is at first pleased with the prospect of returning, meeting his old friends, and boasting of his new life as a U.S. citizen. But his experiences in the United States have not been entirely satisfactory. His mother, who used to be so brave and determined to keep the family together, now lives in a state of strange lethargy. So Chin-ho, hoping to discover why, brings her diary with him when he leaves for Korea, and slowly her horrible experiences during the war begin to unfold.

At the outbreak of the war, she gets married and becomes pregnant shortly thereafter. When UN forces enter the village, she and several other women are raped by a group of soldiers. She gives birth prematurely due to the trauma, and her child, Abe, turns out to be mentally retarded. Her husband, who was forced to join the People's Army, never returns home. She is soon remarried, however, to Kim Sang-man, Chin-ho's father, but when they leave for America many years later they are forced to leave Abe behind.

After Chin-ho reads his mother's diary, he realizes that she is deeply tormented by the fact that she could not bring her son Abe to America. It seems appropriate that his awareness of the family identity, based on the act of understanding and forgiveness, is channeled into finding his retarded brother. Needless to say, Abe serves as a ghastly reminder of the war in human form. But he also symbolizes the past itself and the Korea they left behind. Acknowledging Abe's existence and embracing it as a part of the family are indeed the final, courageous acts of making peace with the past in order to ensure peace in the future.

Although Yi Ch'ŏng-jun (1939–), a representative writer of the so-called April 19th Generation, was not old enough to fight in the war, he is nevertheless considered to be one of the few writers of his generation to possess an innate feeling for the ethnic tragedy. His *A Fool and A Jackass* (1966) is the investigation of war trauma through a first-person narrative and an interesting use of metafiction. The narrator's older brother, a doctor, accidentally loses a female patient during surgery and, tormented by an acute sense of guilt, gives up his medical career. He then makes an odd professional turn, shutting himself up in his room and writing his first novel. It is not long before the narrator discovers that his brother is writing about his painful experiences during the war.

As the war breaks out, he enlists in the army and fights under Sergeant O, his ruthless squad leader. When Private Kim, one of his fellow soldiers, is wounded, O suggests to him that they get rid of Kim so that they can flee more efficiently. Although he has always found O's inhumanity repulsive, this time he thinks that his squad would be better off without the wounded soldier. When O actually carries out the murder, however, he is deeply tormented, blaming himself for having failed to stand up against O's brutality. The feeling of guilt has been gnawing at him ever since.

The story concludes with a debate between the two brothers concerning the ending of the older brother's novel. The narrator, who had been secretly following the development of the story with the belief that it would bring him to the final secret of his brother's tormented conscience, hastily finishes the story himself with his brother's killing Private Kim. His brother, however, opts for a different ending in which he kills Sergeant O. In the end, nothing is solved or proven save for the fact that the war, while it raged and after it had ended, both in and out of fiction, demands lives.

The Dream of a Lasting Regret (1971), written by Yi Mun'gu (1940–), seems to arrive at the same conclusion. This work is composed of a series of wartime memoirs written by destitute day laborers who were hired to relocate a large public cemetery. No other writer before him had penetrated so completely the dismal conditions of this lowest rung of the working class. However, it is not only economic hardships that drive them to their present misery. They are violent, avaricious, hateful, and devoid of the mutual understanding that people in such misery might otherwise share. There is something horribly wretched in them, and it gradually turns out that each character had undergone some horrible wartime trauma that still greatly affected his existence. The character of Ku Pon-ch'il, for example, is at first portrayed as the archetype of an evil figure. We soon discover, however, that his cruelty and anger are deeply rooted in his personal contact with the vengeful war. He killed first for ideology and later for revenge, but, ironically, neither he nor his enemies had any real knowledge of the political causes in which they were involved. Rather, war itself has become their ideology and is almost naturalistically conditioned to bring out the bestial instincts in each of its victims.

In his story *The Rainy Spell* (1979), Yun Hŏng-gil (1942–) finds a way not only to criticize the divisive effects of ideology, but also to overcome them. In this story Yun describes the tragedy of a family's ideological confrontation through the eyes of a child. Conflict exists between the child's two grandmothers who are living under the same roof. Each old woman has an unmarried son fighting on opposite sides in the war.

The confrontational mood within the house heightens when the child's maternal grandmother receives the tragic news that her son has been killed at the front. Later the other woman hears that her son, a partisan guerrilla fighter, has also been killed. But this superstitious woman refuses to believe the news, since her shaman fortune-teller had already divined her son's safe return down to details such as the exact date and time. And so she proceeds to prepare a feast for her son's homecoming. When her son fails to appear on the appointed day, however, she faints in utter despair. Then a strange thing happens: a large snake slithers in over the fence and lies resting in a tree. The child's maternal grandmother pleads for its peaceful departure by speaking to it in a ritualistic manner. To her the snake represents the symbolic reincarnation of the child's paternal uncle, who lingers in this world due to unresolved anger and regrets. It is this final act of reconciliation that brings their divided house back together and symbolically soothes the ideological animosity between the two split bodies of Korea.

One might wonder to what extent such a device succeeds in conveying to readers an enduring sense of reconciliation when we well know that this represents a mere symbolic gesture based on shamanic folk customs, and in no way actually ameliorates the military situation on the peninsula. Is it truly sufficient to appease the enmity engendered by civil war? In order to answer this we need to determine whether the possibility of reconciliation in the age of division lies in Korean ethnic solidarity or in something more complex, practical, and external like politico-economic measures. It is obvious that Yun chooses the former. In his eyes, Koreans are almost innately equipped with the desire to live undivided, a desire that lies beyond the reach of ideology and cannot be tarnished even by the unsympathetic movement of history.

Though certain writers, Cho Chŏng-nae for example, do focus on such issues as ideology and class struggle, the trend continues toward overcoming, even discrediting, ideology for the sake of the nation. Yi Mun-yŏl (1948–) has added his take on the problem in his novel *The Age of Heroes* (1984). In this work Yi confronts the problem of the insurmountable gap between abstract idealism and the real world by following a communist idealist to his final realization of the failure of his idealistic cause. Yi Tong-yŏng is the eldest son of a highly esteemed Confucian family in the Yŏngnam region. In the beginning he is portrayed rather positively as a man of strong socialist convictions. He leans toward the radical left while studying in Japan before the Liberation and later, when Korea gains her independence, he becomes a member of the South Korean Workers' Party. However, the actual political activities and machinations of the party soon leave him disillusioned and cut off from his original idealism.

The writer contrasts the story's turbulent and irresponsible masculine actions with the unerring feminine endurance of Tong-yŏng's remaining family. Tong-yŏng's involvement in leftist politics leaves his family entirely defenseless in the face of the physical dangers of war. His wife Chŏng-ŏn and his aged mother become the leaders of the family, unflinchingly confronting the harsh realities of war and dealing with them one by one in order to save the family. Their actual confrontation with the war, which stands in stark contrast to Tong-yŏng and other male party members' abstract ideological concerns and political strategies, is clearly the most heroic portrait in the story.

Tong-yŏng insists the war is "for the people," but when his mother rhetorically asks, "What people are you talking about? I see no such people in the south and don't think there will be any more in the north. [Are they not] something that exists only in your words? And yet for that ghost of yours you dare to sacrifice innocent lives?"[7] We know she is not fooled. It is finally through the simple and honest eyes of Tong-yŏng's mother that the war is exposed without any embellishment of ideological legitimacy or historical inevitability.

Our access to North Korean works is still somewhat limited. Unsurprisingly, however, we find that the subject of the Korean War is important there as well. In works that specifically deal with this subject, such as *The Taedong River* (1961) by Han Sŏr-ya (1900–1962?) and *The Burning Island* (1959) by Hwang Kŏn (??–??), we see a strong, if understandable, anti-American bias. More troubling, however, is the near complete absence of doubt as to righteousness of cause and efficacy of ideology. Even when they do celebrate "the people," phrases such as "spirit and ideology" are never far behind.[8] So often it is not a universal faith in humanity, or even the Korean people, being expressed, but a particularistic and exclusive belief in those of like political leaning. Han Sŏr-ya does laudably describe the heroic resistance and clever ingenuity of North Korean citizens, but he portrays them perhaps too heroically and unidimensionally, as throughout the story they never once doubt their ideological commitment and fight steadfastly in its defense.

The Burning Island, set during the UN forces' landing at Inchon in September of 1950, portrays the brave resistance and eventual fall of a North Korean navy battery stationed at Wŏlmido. Although an American artillery attack makes resistance all but impossible, the battery commander, Yi Tae-hun, refuses to leave his trench and fights until the very end. Navy correspondent An Chŏng-hŭi also refuses to leave her post in order to fulfill her duties until death. Even after the American navy has landed, An stays on the battleground to send information to headquarters. In the last scene of the story, the division commander reads An's final telegram and comments, "with blazing upright souls like these on our side, those bastards will never be able to defeat us."[9] Such characters and characterizations are meant to verify the invincible fighting spirit and unwavering loyalty of the North Koreans, but in their extremity they often suffer a loss of credibility. In any case, these works serve as an informative contrast to those of South Korea.

Though we explored South Korean writers generationally and attempted to flesh out where and how they differed, we also looked for common themes and continuities. Prevalent among these was a sense of frustration and disillusionment with the false promise of ideology. Nearly all writers had to come to grips with the fact that abstract ideals can never stand in isolation from, nor be considered more important than, the people they are to serve. The history of the literary treatment of the war helps us to understand the different strategies of the Korean people for coping with their traumatic experiences, and for exploring paths toward ethnic reconciliation. We traced the development of these themes up to the present day but certainly, as the peninsula and the nation remain divided, the final chapter in this saga has yet to be written.

Notes

1. Yŏm Sang-sŏp, *Yŏm Sang-sŏp chŏnjip* [*The Complete Works of Yŏm Sang-sŏp*], vol. 7 (Seoul: Minŭmsa, 1987), p. 38.

2. Ibid., p. 192.

3. O Sang-wŏn, *Complete Edition of Modern Korean Literature*, vol. 7 (Seoul: Sin'gu, 1966), p. 192.

4. Sŏ Ki-wŏn, *Complete Edition of Modern Korean Literature*, vol. 7 (Seoul: Sin'gu, 1966), p. 234.

5. Ch'oe In-hun, *Complete Works of Ch'oe In-hun*, vol. 1 (Seoul: Chisŏng sanŏpsa, 1976), p. 68.

6. Ch'oe Il-lam, *New Korean Short Stories*, vol. 2 (Seoul: Ŭmun'gak, 1979), p. 408.

7. Yi Mun-yŏl, *Yŏngung sidae* [*The Age of Heroes*] (Seoul: Minŏmsa, 1984), p. 121.

8. Han Sŏr-ya, *Taedong'gang* [*The Taedong River*] (Pyongyang: Chosŏn Writers Publishing Union, 1961), p. 87.

9. Hwang Kŏn, "Pul t'anŭn sŏm" [The Burning Island], in *Mok ch'ukki* [The Stock Farm] (Pyongyang: Chosŏn Writers Publishing Union, 1959), p. 206.

Chronology

Barry S. Strauss and David R. McCann

Ancient Greece (all dates are B.C.)

508   Cleisthenic revolution, democracy founded in Athens

490, 480–479 Persian Wars: Persian empire attempts to extend its rule to mainland Greece.

490   Battle of Marathon: Athenian and Plataean infantrymen defeat relatively small invading Persian army.

480   Under leadership of King Xerxes, massive Persian land-sea force invades Greece. Defeats Spartans at Thermopylae and fights a draw with Greek navy at Artemisium. Battle of Salamis: Greek fleet, spearheaded by Athenian contingent, defeats Persian navy off Athenian coast after Persian army burns city of Athens to the ground. Xerxes withdraws from Greece.

479   Battle of Plataea: Spartan-led Greek infantry defeats Persian land army, remainder of Persian expedition withdraws from Greece. Battle of Mycale: Greek land-sea victory over Persians on Anatolian coast.

477   Delian League founded, Athens becomes leader of naval confederacy. Athens is leading sea power of Greece, Sparta leading land power.

461–456   Long Walls built, connecting Athens and Piraeus, establishing a land-sea fortress.

460–445   So-called First Peloponnesian War: Sparta and its allies defeat Athenian attempts to establish hegemony over central Greece but are forced to accept existence of Athens' naval alliance. Thirty Years' Peace of 446/445 commits both sides to arbitrate future conflicts.

431–404	Peloponnesian War: in spite of peace treaty, war breaks out again between Athenians and Peloponnesians.
ca. 430	Pericles delivers Funeral Oration in winter at end of first year of war.
430–427	Siege of Plataea
428–427	Rebellion of Mytilene; its suppression in 427 marks dramatic date of Thucydides' Mytilenian Debate.
427	Civil war in Corcyra
421	Peace of Nicias: establishes uneasy peace and alliance between Athenians and Peloponnesians.
418	Battle of Mantinea: Spartans defeat coalition army and reassert their control of Peloponnesian alliance.
416	Melos conquered by Athens, dramatic date for Thucydides' Melian Dialogue.
415–413	Sicilian Expedition: Athenians attack Syracuse with large expedition and ultimately suffer disastrous defeat.
414–413	Sparta declares war again on Athens.
415	Euripides' *Trojan Women* performed.
413	Large-scale revolt by Athenian allies. Peloponnesians establish permanent fort at Decelea in northern Attica, from which they harass Athenian territory. In following years, ca. 20,000 Athenian slaves desert to the enemy.
412	Persian-Spartan treaty brings Persia into anti-Athenian coalition, lending ships and funding to Peloponnesians.
411	Oligarchy temporarily in power in Athens; Aristophanes' *Lysistrata* performed.
404	Defeated by Sparta on sea and besieged on land, Athens surrenders, gives up fleet, walls, and naval empire.
404–403	The Thirty in power in Athens, replacing democracy with oligarchy.
403	Restoration of Athenian democracy after successful guerrilla war by democrats.
403–371	Spartan hegemony: Sparta the leading state in Greece, but others continually challenge its power.
399	Trial of Socrates
395–386	Corinthian War: Coalition of Corinth, Thebes, and Athens attacks Spartans; Persians at first support the allies and then throw support to Sparta after Athens attempts to regain sea power.

371   Battle of Leuctra: Thebes defeats Sparta, ends Spartan hegemony, and in following years frees helots (Spartan serfs).

Modern Korea

1864   Beginning of the reign of King Kojong and the regency of the Taewŏn'gun.

1866   Destruction of the American merchant ship General Sherman.

1871   American troops invade briefly in delayed response to the destruction of the *General Sherman.*

1876   Japanese naval vessels threaten; Kanghwa Treaty gives Japan special new interests in Korea.

1882   Modern military training begins in Korea. Treaty of Chemulp'o (Inchon) with United States. Taewŏn'gun abducted by Chinese and taken to China.

1884   Progressive coup; Chinese intervention. Progressives flee to Japan.

1894   Beginning of Tonghak rebellion. Request for Chinese assistance and provision of Chinese troops leads to Japanese intrusion and beginning of Sino-Japanese War.

1895   End of Sino-Japanese War. Queen Min assassinated by Japanese-led force attacking the Korean royal palace.

1896   King Kojong flees to Russian legation.

1904   Last-ditch effort to reach a compromise between Russian and Japanese interests in Korea suggests 39th parallel as dividing line. Beginning of Russo-Japanese War.

1905   Treaty of Portsmouth concludes Russo-Japanese War but fails to include provisions for Russian reparations payments to Japan. Japanese interests in Korea encouraged by way of compensation.

1907   Kojong abdicates; press and public security laws passed; Korean army disbanded. Beginning of anti-Japanese Righteous Armies campaigns.

1910   Treaty of Annexation turns over control of Korea to Japan.

1918   American President Woodrow Wilson's "Fourteen Points." The goal of national self-determination inspires Korean delegation to journey to the Paris treaty conference to present Korea's case, but the delegation is not heard.

1919   Death of Kojong; state funeral. March 1 Independence Movement, nationwide demonstrations for Korean independence and against continued Japanese colonial rule.

1945	End of World War II and defeat of Japan. Division of Korea at 38th parallel for accepting surrender of Japanese forces in the North by the Russian and in the South by American forces. Establishment of United States Military Government in Korea. Moscow Agreement on trusteeship for Korea leads to widespread anti-trusteeship demonstrations. Assassination of political leader Song Chin-u.
1946	Strikes, demonstrations; labor uprising in Taegu. South Korean Workers' Party established.
1947	Supreme People's Assembly established in North; Central People's Committee; South Korean interim government. Assassination of political leaders Yŏ Un-hyŏng and Chang Tŏk-su.
1948	Kim Ku, Kim Kyu-sik, and other political leaders travel from Seoul to Pyongyang for planning meeting; UN-sponsored general elections take place in South; Syngman Rhee elected president by the National Assembly. Protests on Cheju Island against the elections lead to insurrection; South Korean military troops and paramilitary forces dispatched to Cheju. Yŏsu-Sunch'ŏn military insurrection against orders to go to Cheju Island. Democratic People's Republic established in North. South Korean Workers' Party outlawed. UN recognition of Republic of Korea. Soviet troops withdrawn from North Korea.
1949	U.S. troops withdrawn from South Korea. Suppression of Cheju uprising ends with 30,000–60,000 civilians killed. Kim Ku assassinated.
1950	Dean Acheson's speech at the Tokyo Press Club states that Korea lies outside the perimeter of American defense interests. June 25, North Korean forces cross the 38th parallel and attack the South. Seoul falls; retreat to Pusan. UN forces under General Douglas MacArthur enter the war; U.S. blockade of Taiwan Strait. Inchon landing September 15; recapture of Seoul. UN troops cross the 38th parallel; Pyongyang falls. Chinese People's Liberation Army attacks UN Forces in the North; UN forces retreat. Fall of Seoul. MacArthur dismissed. Truce negotiations begin.
1952	Truce negotiations continue. Air campaign levels North Korea. Rhee re-elected by popular vote. Dwight Eisenhower promises to "go to Korea."
1953	Death of Joseph Stalin. Conclusion of armistice.
1953–2000	Since 1953 there have continued to be periodic aggressive acts, interspersed with efforts to negotiate various agreements between the Republic of Korea and the Democratic People's Republic of Korea. Negotiations started in mid-2000 continue and seem most promising.

Contributors

Paul Cartledge is reader in Greek history in the University of Cambridge, and fellow and director of studies in classics, Clare College. He is author, co-author, editor, and co-editor of a dozen books, including *Agesilaos and the Crisis of Sparta* (1987), *Xenophon: Hiero the Tyrant and Other Treatises* (1997), *Hellenistic Constructs: Essays in Culture, History and Historiography* (1997), *Democritus and Atomistic Politics* (1999), and *The Cambridge Illustrated History of Ancient Greece* (1998). He is currently writing, for the Cambridge University Press "Key Themes in Ancient History" series that he co-edits, *Elite and Mass: Political Thought in Greece from Homer to Plutarch*.

Gregory Crane is professor of classics at Tufts University. He is the author of two books on Thucydides as well as a book on Homer and numerous articles on other aspects of Greek literature. His current projects range from research into digital libraries to a study of Athenian imperialism in the fifth century.

Bruce Cumings teaches East Asian political economy and international history at the University of Chicago. He is the author of several books on modern Korean history, including the two-volume *Origins of the Korean War* and, most recently, *Korea's Place in the Sun. A Modern History*.

Victor D. Hanson is professor of classics at California State University–Fresno and the author of some fifty articles and reviews and seven books on the military and agrarian history of Ancient Greece. He recently published *The Soul of Battle: From Ancient Times to the Present Day, How Three Great Liberators Vanquished Tyranny* (1999), and *Wars of the Ancient Greeks* (1999). He lives with his wife and three children on his family farm south of Selma, California, where he grew up.

Robert Kagan is senior associate at the Carnegie Endowment for International Peace and director of the U.S. Leadership Project. He writes a monthly column on foreign policy for the *Washington Post* and is a contributing editor at the *Weekly Standard* and the *New Republic*. He is the author of *A Twilight Struggle: American Power and Nicaragua, 1977–1990* (1996), and his articles have appeared in the *New York Times*, the *Washington Post*, the *Wall Street Journal, Foreign Affairs, Foreign Policy*, the *National Interest*, the *New Republic, Commentary, Policy Review*, the *New Criterion*, and the *Weekly Standard*. Mr. Kagan served in the State Department from 1984 to 1988. He is currently at work on a history of American foreign policy.

David R. McCann is the author or translator of a number of books on Korean literature and literary and political culture, including *Form and Freedom in Korean Poetry, The Middle Hour: Selected Poems of Kim Chi Ha, Selected Poems of Sŏ Chŏngju*, (with Sung-il Choi), *Prison Writings by Kim Dae Jung* and *Early Korean Literature: Selections and Introductions* (2000). He recently edited *Korea Briefing: Toward Reunification* (1997). He taught Korean and Japanese literature at Cornell University, and is presently Korea Foundation Professor of Korean Literature at Harvard University.

Josiah Ober is the David Magie Professor of Ancient History in the classics department of Princeton University. He is the author of books and articles in Greek history and political theory, including *Fortress Attica* (1985), *Mass and Elite in Democratic Athens* (1989), *The Athenian Revolution* (1996), and *Political Dissent in Democratic Athens* (1998).

Kongdan (Katy) Oh is a research staff member at the Institute for Defense Analyses, a not-for-profit private think tank for the Office of Secretary of Defense; nonresident senior fellow at the Brookings Institution; lecturer at George Mason University's graduate program on international commerce and policy; and co-principal of Oh & Hassig Pacific Rim Consulting, a small consulting firm specializing in East Asian policy research.

Kurt A. Raaflaub is professor of classics and history at Brown University and was joint director of the Center for Hellenic Studies in Washington D.C. His main interests are the social, political, and intellectual history of archaic and classical Greece and the Roman republic. He has written *Dignitatis contentio* (1974), a book on political strategies in Caesar's civil war, and *The Discovery of Freedom in Ancient Greece* (1985, 2d ed. in preparation); is co-author of *Aspects of Athenian Democracy* (1990) and *Ancient History: Recent Work and New Directions* (1997); and has edited or co-edited collected volumes on *Social Struggles in Archaic Rome* (1986), *Between Republic and Empire: Interpretations of Augustus and His Principate* (1990), *City-States in Classical Antiquity and Medieval Italy*

(1993), *Democracy 2500? Questions and Challenges* (1997), and *Democracy, Empire, and the Arts in Fifth-Century Athens* (1998). His current projects focus on early Greek political thought and on the historical interpretation of "Homeric society."

Jennifer T. Roberts is professor of classics and history at the City College of New York and the City University Graduate Center. She has served as the president of the New York Classical Club and is currently vice president for outreach of the American Philological Association. The recipient of grants from the National Endowment for the Humanities and the American Council of Learned Societies, she is the author of two books on Athens, *Accountability in Athenian Government* (1982) and *Athens on Trial: The Antidemocratic Tradition in Western Thought* (1994). She has also co-authored and edited several textbooks, including, most recently, with Sarah Pomeroy, Stanley Burstein, and Walter Donlan, *Ancient Greece: A Political, Social, and Cultural History* (1999) and, with Walter Blanco, the *Norton Critical Edition of Thucydides' Peloponnesian War* (1998).

Ellen Schrecker is professor of history at Yeshiva University. She is the author of several books and articles about the McCarthy era, including *No Ivory Tower: McCarthyism and the Universities* (1986); *The Age of McCarthysim: A Brief History with Documents* (1994); and *Many Are the Crimes: McCarthyism in America* (1998). She is also the editor of *Academe*, the magazine of the American Association of University Professors.

Dong-Wook Shin is professor of Korean Literature, Kumamoto Gakuen University, Japan. He received a B.A. in Korean literature from Seoul National University, and M.A. and Ph.D. degrees in Korean literature from Korea University. He has taught at Korea University and Yonsei University, and was a visiting research professor at the University of Illinois, and Harvard-Yenching Fellow, Harvard University. He is the author of many books on Korean literature, including *Korean Poetry, an Historical Study* (1981), *Studies on Korean Narrative Literature* (1981), *Voices and Poetic Imagination* (1991), and *Korean Novels of the 1930s* (1994).

Ronald Steel is professor of international relations at the University of Southern California. A frequent contributor to numerous publications on international and public affairs, he is a columnist for the *New Republic*, and is on the advisory boards of *World Policy Journal* and *New Perspectives Quarterly*. Most recently the author of *Temptations of a Superpower*, his earlier books include *Walter Lippmann and the American Century*, *Imperialists and Other Heroes*, *Pax Americana*, and *The End of Alliance: America and the Future of Europe*. A former Guggenheim fellow, he is also the recipient of several book awards, including the Bancroft Prize in American History, the National Book Award, and the National Critics Circle Award.

Barry S. Strauss is professor of history and classics, and director of the Peace Studies Program at Cornell University. He is the author of some thirty articles and reviews and six books, among them such works in ancient history as *Fathers and Sons in Athens* (1993) and *Athens After the Peloponnesian War* (1987). He is a frequent contributor to *MHQ: The Quarterly Journal of Military History*, where he writes on ancient, medieval, and early modern history in Europe and East Asia. He is currently completing a book on citizen-soldiers in contemporary America.

Dae-Sook Suh is professor of political science at the University of Hawaii, and former director of the Center for Korean Studies. He is the author of a number of books, including *The Korean Communist Movement, 1918–1948* (1967), *Documents of Korean Communism* (1970), *Korean Communism, 1945–1980: A Reference Guide to the Political System* (1981), *Koreans in the Soviet Union* (1987), *Kim Il Sung: The North Korean Leader* (1988), and other works. He was a senior fellow at the East-West Center, a Fulbright scholar, and guest scholar at the Woodrow Wilson International Center for Scholars at the Smithsonian Institution.

Kathryn Weathersby, a historian of Soviet foreign relations, is currently an independent scholar based in Washington, D.C. The author of numerous articles on the Soviet role in the Korean War based on research in Moscow archives, she is currently completing a book on Soviet policy toward Korea from 1945 to 1953.

Stephen J. Whitfield holds the Max Richter Chair in American Civilization at Brandeis University. He has also served as a visiting professor at the Hebrew University of Jerusalem and at the Sorbonne. His books include *The Culture of the Cold War* (second edition, 1996) and *In Search of American Jewish Culture* (1999).

Index